continued fractions and
orthogonal functions

LECTURE NOTES IN PURE AND APPLIED MATHEMATICS

1. *N. Jacobson,* Exceptional Lie Algebras
2. *L.-Å. Lindahl and F. Poulsen,* Thin Sets in Harmonic Analysis
3. *I. Satake,* Classification Theory of Semi-Simple Algebraic Groups
4. *F. Hirzebruch, W. D. Newmann, and S. S. Koh,* Differentiable Manifolds and Quadratic Forms
5. *I. Chavel,* Riemannian Symmetric Spaces of Rank One
6. *R. B. Burckel,* Characterization of C(X) Among Its Subalgebras
7. *B. R. McDonald, A. R. Magid, and K. C. Smith,* Ring Theory: Proceedings of the Oklahoma Conference
8. *Y.-T. Siu,* Techniques of Extension on Analytic Objects
9. *S. R. Caradus, W. E. Pfaffenberger, and B. Yood,* Calkin Algebras and Algebras of Operators on Banach Spaces
10. *E. O. Roxin, P.-T. Liu, and R. L. Sternberg,* Differential Games and Control Theory
11. *M. Orzech and C. Small,* The Brauer Group of Commutative Rings
12. *S. Thomier,* Topology and Its Applications
13. *J. M. Lopez and K. A. Ross,* Sidon Sets
14. *W. W. Comfort and S. Negrepontis,* Continuous Pseudometrics
15. *K. McKennon and J. M. Robertson,* Locally Convex Spaces
16. *M. Carmeli and S. Malin,* Representations of the Rotation and Lorentz Groups: An Introduction
17. *G. B. Seligman,* Rational Methods in Lie Algebras
18. *D. G. de Figueiredo,* Functional Analysis: Proceedings of the Brazilian Mathematical Society Symposium
19. *L. Cesari, R. Kannan, and J. D. Schuur,* Nonlinear Functional Analysis and Differential Equations: Proceedings of the Michigan State University Conference
20. *J. J. Schäffer,* Geometry of Spheres in Normed Spaces
21. *K. Yano and M. Kon,* Anti-Invariant Submanifolds
22. *W. V. Vasconcelos,* The Rings of Dimension Two
23. *R. E. Chandler,* Hausdorff Compactifications
24. *S. P. Franklin and B. V. S. Thomas,* Topology: Proceedings of the Memphis State University Conference
25. *S. K. Jain,* Ring Theory: Proceedings of the Ohio University Conference
26. *B. R. McDonald and R. A. Morris,* Ring Theory II: Proceedings of the Second Oklahoma Conference
27. *R. B. Mura and A. Rhemtulla,* Orderable Groups
28. *J. R. Graef,* Stability of Dynamical Systems: Theory and Applications
29. *H.-C. Wang,* Homogeneous Branch Algebras
30. *E. O. Roxin, P.-T. Liu, and R. L. Sternberg,* Differential Games and Control Theory II
31. *R. D. Porter,* Introduction to Fibre Bundles
32. *M. Altman,* Contractors and Contractor Directions Theory and Applications
33. *J. S. Golan,* Decomposition and Dimension in Module Categories
34. *G. Fairweather,* Finite Element Galerkin Methods for Differential Equations
35. *J. D. Sally,* Numbers of Generators of Ideals in Local Rings
36. *S. S. Miller,* Complex Analysis: Proceedings of the S.U.N.Y. Brockport Conference
37. *R. Gordon,* Representation Theory of Algebras: Proceedings of the Philadelphia Conference
38. *M. Goto and F. D. Grosshans,* Semisimple Lie Algebras
39. *A. I. Arruda, N. C. A. da Costa, and R. Chuaqui,* Mathematical Logic: Proceedings of the First Brazilian Conference
40. *F. Van Oystaeyen,* Ring Theory: Proceedings of the 1977 Antwerp Conference
41. *F. Van Oystaeyen and A. Verschoren,* Reflectors and Localization: Application to Sheaf Theory
42. *M. Satyanarayana,* Positively Ordered Semigroups
43. *D. L Russell,* Mathematics of Finite-Dimensional Control Systems
44. *P.-T. Liu and E. Roxin,* Differential Games and Control Theory III: Proceedings of the Third Kingston Conference, Part A
45. *A. Geramita and J. Seberry,* Orthogonal Designs: Quadratic Forms and Hadamard Matrices
46. *J. Cigler, V. Losert, and P. Michor,* Banach Modules and Functors on Categories of Banach Spaces

Additional Volumes in Preparation

continued fractions and orthogonal functions
theory and applications

edited by

S. Clement Cooper
University of Colorado
Boulder, Colorado

W. J. Thron
Washington State University
Pullman, Washington

Marcel Dekker, Inc.   **New York•Basel•Hong Kong**

Library of Congress Cataloging-in-Publication Data

Continued fractions and orthogonal functions: theory and applications / edited by S. Clement Cooper, W. J. Thron.
     p. cm. -- (Lecture notes in pure and applied mathematics; v. 154)
     "Proceedings of a research seminar-workshop held in Loen, Norway, June 21–July 4, 1992"--Pref.
     Includes bibliographical references and index.
     ISBN 0-8247-9071-5 (acid-free)
     1. Continued fractions--Congresses.   2. Functions, Orthogonal--Congresses.
I. Cooper, S. Clement. II. Thron, Wolfgang J. III. Series.
QA295.C65 1994
515'.243--dc20   93-33234
                                                                        CIP

The publisher offers discounts on this book when ordered in bulk quantities. For more information, write to Special Sales/Professional Marketing at the address below.

This book is printed on acid-free paper.

MARCEL DEKKER, INC.
270 Madison Avenue, New York, New York 10016

Current printing (last digit):
10 9 8 7 6 5 4 3 2 1

PRINTED IN THE UNITED STATES OF AMERICA

Preface

This volume contains the proceedings of a research Seminar-Workshop held in Loen, Norway. For this, the fourth such workshop, we returned to the setting of the first one which was held in 1981. The second was held in Pitlochry and Aviemore, Scotland, in 1985, while the third was in Redstone, Colorado, in 1988. The motivation for the Seminar-Workshops is to bring together a small group (15 or fewer) of mathematicians who normally are separated geographically, but whose research interests are close, to work together without any distractions. For such an undertaking, a peaceful, rural setting with good food and nice walks is desirable. The fact that we returned to Loen is a testimony to the superb location and accommodations. In this setting, we exchanged recent results, bounced emerging ideas off one another, worked together, and sketched out future joint research projects. Four of the seven participants from the first workshop returned to this one. One original participant, Arne Magnus, was particularly missed and we dedicate this volume to his memory. Arne passed away October 19, 1991. An obituary appeared in *Communications in the Analytic Theory of Continued Fractions*, **1**, 1992.

The contents of this volume fall primarily into two main categories:

1. strong moment problems, orthogonal polynomials and Laurent polynomials and applications, and

2. sequences of linear fractional transformations, convergence results including truncation error bounds, and special functions.

There is a survey paper in the first category. Other topics in the first category include discrete distributions and limit functions arising from indeterminate moment problems, Szegö polynomials and their application to frequency analysis, a quadrature formula arising from q-starlike functions and a general formulation for recurrence relations for orthogonal functions. In the second category, general sequences of linear fractional transformations as well as continued fractions are considered. Convergence theorems were central to most studies and truncation error bounds are found in some. Two studies focussed on convergence regions. A substantial contribution involving continued fraction representations for functions related to the gamma function is also included.

This most recent Seminar-Workshop was organized by Haakon Waadeland, William B. Jones, Lisa Lorentzen and Olav Njåstad. Grateful acknowledgement is made for the financial support from a number of sources. Different individuals received support from their respective universities and, in some cases, from the United States Educational Foundation in Norway (Fulbright Grant), the Norwegian Research Council (NAVF), and the U.S. National Science Foundation. In addition to supporting its own participants, the University of Trondheim supported the conference as a whole. We also wish to thank the director and staff of the Hotel Alexandra in Loen for providing excellent working facilities and a cordial atmosphere for the Workshop. Finally, we would like to thank Maria Allegra of Marcel Dekker, Inc., for accepting this volume for publication in the *Lecture Notes in Pure and Applied Mathematics Series*.

The loss of Arne Magnus between the last two workshops is a reminder of how precious our time is to work together. The editors thought it would be of interest to

give a brief historical account of the collaboration of the participants in these workshops. We have emphasized the connections of members of this group with one another and, in particular, with Arne.

It all began in 1940 when Wolf Thron started working with Walter Leighton on a thesis concerned with convergence of continued fractions at the Rice Institute. In early 1946, when Thron was still in the U.S. Army, Leighton, who had become chairman of the Washington University Mathematics Department in St. Louis, Missouri, approached Thron about joining the department. Thus, Wolf was on the faculty of Washington University when Arne Magnus arrived in St. Louis as a teaching assistant in 1947. They became friends. Magnus did his thesis on a problem in several complex variables under Z. Nehari. In 1954 Thron moved to Boulder. The department was growing and so Wolf asked Arne, who was then at the University of Nebraska, whether he would be interested in coming to the University of Colorado. With everybody's approval, Arne came to Boulder in the fall of 1956.

Wolf had two graduate students (Jerry Lange and Bill Jones) working with him on topics in continued fraction theory. Arne joined the seminar that probably was started in the fall of 1958, soon after Wolf returned from a year in Munich where he had worked with O. Perron. Arne became interested in continued fractions (and Padé tables) and wrote a number of papers on P-fractions and their relationship to the Padé table. Jerry Lange finished his Ph.D. in 1960 and went on to the University of Missouri.

Through Sigmund Selberg, who was in Boulder in 1957-1958, Arne received an invitation to teach at the Norges Tekniske Høgskole (NTH) in Trondheim. So the Magnus family spent the year 1961-1962 in Trondheim, where Arne became friendly with Haakon Waadeland (they had met years earlier in Oslo). Haakon had written his Dr. Philos. thesis on univalent functions, but had also done some work on continued fractions in earlier years.

In 1963, Arne became chairman at Boulder and invited Haakon (and his family) to spend the academic year 1963-1964 at the University of Colorado. Haakon joined the Boulder seminar and began to investigate whether properties such as holomorphy and boundedness of a function $f(z)$ would imply results concerning convergence and limit periodicity for its continued fraction expansions.

During 1962-1963, while Wolf was in India, Arne advised Bill. Since Bill Jones technically received his Ph.D. from Vanderbilt he was eligible for and received a tenure track position in the applied mathematics department when he finished his Ph.D. in 1963. Somewhat later Wolf's main interest had shifted to general topology, but Bill kept Wolf from giving up continued fractions altogether. By now they have been joint authors of one book and 34 articles. After a while Bill had some Ph.D. students of his own in continued fractions. The first were Bob Snell (1968), David Field (1971) and Walter Reid (1978).

Arne moved on to Fort Collins (Colorado State University) in 1966. He had become interested in Padé tables through his work on P-fractions. He assigned two of his early Ph.D. students, Norman Franzen (1966) and Jan Wynn (1972), thesis topics in this area.

In the middle fifties Wolf had started a study of the convergence behavior of sequences $\{T_n(w)\}$ of linear fractional transformations of which continued fractions are a special

case. Of particular interest is the limit periodic case. Define $t_n(w) = T_{n-1}^{-1}(T_n(w))$ and $t(w) = \lim_{n \to \infty} t_n(w)$. dePree (1962) studied the case where $t(w) = w$ for his thesis with Wolf. Arne also got interested in this problem and had Mike Mandell (1969) and John Gill (1971) study cases where $t(w)$ is not the identity transformation. John Gill has continued to work actively in this and related fields (convergence acceleration and modification, among others).

Independently of John, Haakon also was investigating convergence acceleration, but from a different perspective. He used nearness properties in both convergence acceleration and analytic continuation.

Thron's first long visit to Trondheim came in 1978-1979. During this year Haakon and he jointly conducted a seminar on continued fractions. One of the students was Lisa Jacobsen. Haakon and Wolf worked together on convergence acceleration and modification of continued fractions (overlapping the work of Gill). They also began to discuss the possibility of having a workshop on continued fractions.

Haakon's efforts to organize and finance a three-and-a-half-week long workshop came to fruition when seven mathematicians met at the Hotel Alexandra in Loen in the summer of 1981. The participants were: Arne, Bill, Elaine Puléo (Arne's M.S. student), Haakon, Lisa, Walter Reid and Wolf.

Olav Njåstad was Haakon's colleague from 1966 onward and met Arne during his 1972-1973 sabbatical. Olav's contact with Bill and Wolf began when he spent a sabbatical in Boulder in 1973-1974. He was at that time mainly interested in general topology, in particular, compactification theory which overlapped Thron's work on general extension theory. Olav returned to Boulder for the academic year 1980-1981. At this time he was interested in shifting his research activities into real and complex analysis and Bill and Wolf were looking for collaborators in the investigations into strong moment problems that they had just started. Olav became interested and joined the work on strong moment problems and related topics.

Arne and Wolf were in Trondheim once more for the academic year 1982-1983. Haakon had encouraged Lisa to try for a Dr. Philos. degree. With a thesis mainly devoted to modified continued fractions and how these can be applied in convergence theory, Lisa completed her degree in 1983. Wolf, Arne and Hans Wallin constituted the examination committee. Lisa has continued an active research program on her own as well as frequently in collaboration with Haakon and occasionally with Arne, Wolf, Bill and others. Haakon and Lisa (now Lorentzen) coauthored the book *Continued Fractions with Applications* (1992).

The introduction of general T-fractions (Wolf, Bill and Perron) and M-fractions (Murphy and McCabe) suggested a study of two-point Padé tables. Arne and John McCabe independently initiated investigations of such tables. Later, Arne's student Sandy Clement worked with both Arne and John on the nonnormal two-point table.

John McCabe's first Ph.D. student, Sri Ranga (1985), investigated the strong moment problem and this emphasis continues. John began to collaborate regularly with Arne and the two developed a close friendship. John and Arne's investigations, following the work on two-point Padé tables, dealt with special functions. This tied in with research being done by Nancy Wyshinski and Bill resulting in a joint publication.

Bill Jones had several doctoral students this past decade: Chris Baltus (1984), Sandy (Clement) Cooper (1988) and Nancy Wyshinski (1991). Chris worked with Bill on best convergence regions, Sandy worked jointly with Bill and Arne on continued fraction solutions to Riccati differential equations and Nancy worked with Bill (as well as Arne, Wolf and John McCabe) on special functions.

The concept of "separate convergence" goes back to Śleszyński in 1888. Arne translated his articles and reported on them in the Boulder seminar in 1987-1988. Members of the seminar agreed on the name separate convergence and explored its occurrence in other authors' writings. Wolf, Lisa and Haakon made some new contributions to the subject. Actually, Haakon had proved a special result about separate convergence in 1966 without using that terminology. The concept also underlies some of the recent work of Arne, Bill, Sandy, Nancy Wyshinski and Cathy Bonan-Hamada.

Arne had several master's students in the 1980's, one of whom was Cathy Bonan. Cathy wrote a master's thesis in 1988, then spent a year at the University of Southern Colorado with John Gill before returning to graduate school (this time in Boulder). Cathy Bonan-Hamada has worked on orthogonal Laurent polynomials as well as on indeterminate strong moment problems with Bill, Wolf and Arne.

Following graduation in 1988, Sandy Cooper accepted a position at Washington State University. Her first master's student, Cathleen Craviotto, worked with her on δ-fractions (which had been introduced by Jerry Lange). Cathleen's Ph.D. work with Bill involves best truncation error bounds. Sandy's first two Ph.D. students, Lyle Cochran (1993) and Phil Gustafson (anticipated 1994) are focussing on topics in the area of strong moment problems and orthogonal Laurent polynomials on the real line.

In addition to working with strong moment problems and continued fraction theory, Haakon has retained his interest in univalent functions. Frode Rønning worked with Haakon and S. Ruscheweyh on univalent function theory, finishing his Ph.D. in 1990. Recently continued fractions have crept into Frode's work. Lately, Haakon has also investigated value regions with probabilistic information.

Johan Karlsson was a student of Hans Wallin's at Umeå. Continued fractions and Padé approximants are among their fields of interest. Both participated in a conference in Köja (Sweden) in 1982 in which Arne, Lisa, Haakon, Olav and Wolf also took part. Karlsson and Wallin's work on continued fractions is closely related to (although independent of) Haakon's studies of boundary versions of known convergence results and his experiments with finite convergence sets.

Much of the effort expended by the group members over this past decade has centered on moment problems, orthogonal polynomials and orthogonal Laurent polynomials (both on the real line and the unit circle). This shift motivated the editors to adopt the title *Continued Fractions and Orthogonal Functions: Theory and Applications* instead of *Analytic Theory of Continued Fractions, IV* as might have been anticipated.

The participants in the Seminar-Workshops have changed over the years, which is a reflection of the natural dynamics of an active, vital research group. This group has a long history of collaboration, a generous attitude towards the sharing of ideas and a very free, open style of working together. This attitude has not only contributed to the success of the group, but has been the catalyst of some important deep research. We

look forward to many more years of working together and to finding new collaborators along the way.

S. Clement Cooper
W. J. Thron

Contents

Contributors

CATHERINE M. BONAN-HAMADA, University of Colorado, Boulder, Colorado, USA

A. BULTHEEL, K. U. Leuven, Leuven, BELGIUM

LYLE COCHRAN, Washington State University, Pullman, Washington, USA

S. CLEMENT COOPER, Washington State University, Pullman, Washington, USA

C. M. CRAVIOTTO, University of Colorado, Boulder, Colorado, USA

JOHN GILL, University of Southern Colorado, Pueblo, Colorado, USA

P. GONZALEZ-VERA, Universidad de La Laguna, La Laguna, Tenerife, Canary Islands, SPAIN

E. HENDRIKSEN, University of Amsterdam, Amsterdam, THE NETHERLANDS

WILLIAM B. JONES, University of Colorado, Boulder, Colorado, USA

JOHAN KARLSSON, Chalmers Institute of Technology and University of Göteborg, Göteborg, SWEDEN

L. J. LANGE, University of Missouri, Columbia, Missouri, USA

LISA LORENTZEN, University of Trondheim (NTH), Trondheim, NORWAY

ARNE MAGNUS*, Colorado State University, Fort Collins, Colorado, USA

OLAV NJÅSTAD, University of Trondheim (NTH), Trondheim, NORWAY

FRODE RØNNING, Trondheim College of Education, Charlottenlund, NORWAY

W. J. THRON, University of Colorado, Boulder, Colorado, USA

HAAKON WAADELAND, University of Trondheim (AVH), Dragvoll, NORWAY

HANS WALLIN, University of Umeå, Umeå, SWEDEN

---

* deceased

Participants

CATHERINE M. BONAN-HAMADA, Department of Mathematics, Box 395, University of Colorado, Boulder, Colorado 80309-0395, USA

LYLE COCHRAN, Department of Pure and Applied Mathematics, Washington State University, Pullman, Washington 99164-3113, USA

S. CLEMENT COOPER, Department of Pure and Applied Mathematics, Washington State University, Pullman, Washington 99164-3113, USA

C. M. CRAVIOTTO, Department of Mathematics, Box 395, University of Colorado, Boulder, Colorado 80309-0395, USA

WILLIAM B. JONES, Department of Mathematics, Box 395, University of Colorado, Boulder, Colorado 80309-0395, USA

JOHAN KARLSSON, Department of Mathematics, Chalmers Institute of Technology and University of Göteborg, S-412 96 Göteborg, SWEDEN

L. J. LANGE, Department of Mathematics, University of Missouri, Columbia, Missouri 65211, USA

LISA LORENTZEN, Division of Mathematical Sciences, University of Trondheim (NTH), N-7034 Trondheim, NORWAY

MARGARET H. MAGNUS, 708 Dartmouth Trail, Fort Collins, Colorado 80525, USA

JOHN H. MCCABE, Department of Mathematical and Computational Sciences, Mathematical Institute, University of St. Andrews, North Haugh, St. Andrews KY16 9SS, Fife, Scotland, UK

OLAV NJÅSTAD, Department of Mathematical Sciences, University of Trondheim (NTH), N-7034, Trondheim, NORWAY

FRODE RØNNING, Trondheim College of Education, Rotvoll allé, N-7050 Charlottenlund, NORWAY

W. J. THRON, Department of Mathematics, Box 395, University of Colorado, Boulder, Colorado 80309-0395, USA

HAAKON WAADELAND, Department of Mathematics and Statistics, University of Trondheim (AVH), N-7055 Dragvoll, NORWAY

NANCY J. WYSHINSKI, Department of Mathematics, Trinity College, Hartford, CT 06106, USA

1

Discrete Distribution Functions for Log-Normal Moments

CATHERINE M. BONAN–HAMADA   Department of Mathematics, University of Colorado, Boulder, CO 80309–0395, U.S.A.

WILLIAM B. JONES   Department of Mathematics, University of Colorado, Boulder, CO 80309–0395, U.S.A.

W. J. THRON   Department of Mathematics, University of Colorado, Boulder, CO 80309–0395, U.S.A.

ARNE MAGNUS[†]

Dedicated to the memory of Arne Magnus

1   INTRODUCTION

For each pair (a, b) such that $-\infty \le a < b \le \infty$, let $\Phi(a, b)$ denote the family of distribution functions (i.e., real-valued, bounded, non-decreasing functions) with infinitely many points of increase on $a < t < b$. The *classical log-normal distribution function* $\varphi(t) \in \Phi(0, \infty)$ is defined by

$$\varphi'(t) := \frac{q^{\frac{1}{2}}}{2\kappa\sqrt{\pi}} \, e^{-(\frac{\ln t}{2\kappa})^2}, \quad 0 < t < \infty, \quad 0 < q = e^{-2\kappa^2} < 1. \qquad (1.1)$$

$\varphi(t)$ is said to *generate* the *sequence of log-normal moments* $\{c_k\}_{k=-\infty}^{\infty}$ defined by

$$c_k := \int_0^{\infty} (-t)^k \varphi'(t) dt = (-1)^k q^{-\frac{k^2}{2}-k}, \quad k = 0, \pm 1, \pm 2, \dots . \qquad (1.2)$$

Other distribution functions generating the same moment sequence (1.2) are known (see for example [2, 3]).

For each $r \in \mathbf{Z}$, there is a closely related distribution function $\varphi_r(t) \in \Phi(0, \infty)$ defined by

$$\varphi'_r(t) = t^r \varphi'(t), \quad r = 0, \pm 1, \pm 2, \dots . \qquad (1.3)$$

Research supported in part by the National Science Foundation under Grant DMS–9103141 and INT–9113400.

[†]Late Professor of Mathematics, Colorado State University, Fort Collins, CO 80523, U.S.A.

The moments $c_k^{(r)}$ associated with $\varphi_r(t)$ are given by

$$c_k^{(r)} := \int_0^\infty (-t)^k \varphi_r'(t) dt = \int_0^\infty (-1)^r (-t)^{k+r} \varphi'(t) dt = (-1)^r c_{k+r}, \quad k = 0, \pm 1, \pm 2, \ldots .$$
(1.4)

The sequence $\{c_k^{(r)}\}_{k=-\infty}^\infty$ is called the *r-translate* of the sequence $\{(-1)^r c_k\}_{k=-\infty}^\infty$. Thus if r is even, $\{c_k^{(r)}\}_{k=-\infty}^\infty$ is the r-translate of the log-normal sequence of moments (1.2). Hereafter it is understood that statements involving r are valid for all $r \in \mathbb{Z}$ unless otherwise noted.

Some recent applications of log-normal distributions in atmospheric sciences, biology, ecology, geology, economics and business can be found in [6]. In 1894 Stieltjes [15, pp. 507–508] considered the log-normal distribution function as an example for which the Stieltjes moment problem is indeterminate. Wigert [17] studied polynomials orthogonal with respect to the weight function (1.1). Laurent-polynomials (L-polynomials) orthogonal with respect to (1.1) were introduced in 1985 by Pastro [13]. In 1990, Cooper, Jones and Thron [4] studied positive T-fractions associated with the log-normal distribution (1.1). They gave another proof of the indeterminacy of the strong Stieltjes moment problem by showing that the even and odd order approximants of the positive T-fraction associated with (1.1) converge to different limits. The authors of [4] derived closed form expressions for the orthogonal L-polynomials associated with (1.1) by using the fact that these orthogonal L-polynomials are essentially the denominators of the aforementioned positive T-fraction.

In [5] it is shown that subsequences of the orthogonal L-polynomials of even and odd order converge to separate limits. The notation employed in the present paper conforms to a large extent with that used in [7]. We summarize the results obtained in [7] which are used subsequently.

With each distribution function $\varphi_r(t)$ defined by (1.3) and its associated moment sequence $\{c_k^{(r)}\}_{k=-\infty}^\infty$ given by (1.4) we consider the pair $(L_{r,0}, L_{r,\infty})$ of formal power series (fps)

$$L_{r,0}(z) := \sum_{k=1}^\infty -c_{-k}^{(r)} z^k \quad \text{and} \quad L_{r,\infty}(z) := \sum_{k=0}^\infty c_k^{(r)} z^{-k}.$$
(1.5)

Corresponding to the pair $(L_{r,0}, L_{r,\infty})$ is a PC-fraction (Perron–Carathéodory continued fraction)

$$\frac{\alpha_{r,1}}{1} + \frac{1}{\beta_{r,2} z} + \frac{\alpha_{r,3} z}{\beta_{r,3}} + \frac{1}{\beta_{r,4} z} + \frac{\alpha_{r,5} z}{\beta_{r,5}} + \frac{1}{\beta_{r,6} z} + \cdots$$
(1.6a)

whose coefficients $\alpha_{r,k}$ and $\beta_{r,k}$ are given by

$$\alpha_{r,1} := c_0^{(r)} = (-1)^r q^{-\frac{r^2}{2} - r}, \quad \alpha_{r,2n+1} := 1 - q^{-n}, \quad n = 1, 2, 3, \ldots ,$$
(1.6b)

$$\beta_{r,2n} := q^{rn + \frac{n}{2}}, \quad \beta_{r,2n+1} := q^{-rn - \frac{3n}{2}}, \quad n = 1, 2, 3, \ldots ,$$
(1.6c)

[7, (3.3) in Theorem 3.2]. The nth numerator $A_{r,n}(z)$ and denominator $B_{r,n}(z)$ of (1.6) are polynomials in z defined by the second order linear difference equations

$$A_{r,0}(z) := 0, \quad B_{r,0}(z) := 1, \quad A_{r,1}(z) := \alpha_{r,1}, \quad B_{r,1}(z) := 1,$$
(1.7a)

$$A_{r,2n}(z) = \beta_{r,2n} z A_{r,2n-1}(z) + A_{r,2n-2}(z), \qquad n = 1,2,3,\dots, \qquad (1.7b)$$

$$B_{r,2n}(z) = \beta_{r,2n} z B_{r,2n-1}(z) + B_{r,2n-2}(z), \qquad n = 1,2,3,\dots, \qquad (1.7c)$$

$$A_{r,2n+1}(z) = B_{r,2n+1} A_{r,2n}(z) + \alpha_{r,2n+1} z A_{r,2n-1}(z), \qquad n = 1,2,3,\dots, \qquad (1.7d)$$

$$B_{r,2n+1}(z) = \beta_{r,2n+1} B_{r,2n}(z) + \alpha_{r,2n+1} z B_{r,2n-1}(z), \qquad n = 1,2,3,\dots. \qquad (1.7e)$$

The PC-fraction (1.6) is uniquely determined by the following correspondence properties:

$$L_{r,0}(z) - \frac{A_{r,2n}(z)}{B_{r,2n}(z)} = O(z^{n+1}), \quad L_{r,\infty}(z) - \frac{A_{r,2n}(z)}{B_{r,2n}(z)} = O\left(\left(\frac{1}{z}\right)^n\right), \quad n = 0,1,\dots,$$

$$(1.8a)$$

$$L_{r,0}(z) - \frac{A_{r,2n+1}(z)}{B_{r,2n+1}(z)} = O(z^n), \quad L_{r,\infty}(z) - \frac{A_{r,2n+1}(z)}{B_{r,2n+1}(z)} = O\left(\left(\frac{1}{z}\right)^{n+1}\right), \quad n = 0,1,\dots,$$

$$(1.8b)$$

(see for example [9]). Here the symbol $O(z^n)$ denotes a fps starting with z to a power greater than or equal to n.

For each fixed $r \in \mathbb{Z}$ and $n = 0,1,2,\dots$, we define L-polynomials by

$$U_{r,2n}(z) := \gamma_{r,2n} \frac{A_{r,4n}(-z)}{z^n}, \quad U_{r,2n+1}(z) := \gamma_{r,2n+1} \frac{A_{r,4n+2}(-z)}{z^{n+1}}, \qquad (1.9a)$$

$$V_{r,2n}(z) := \gamma_{r,2n} \frac{B_{r,4n}(-z)}{z^n}, \quad V_{r,2n+1}(z) := \gamma_{r,2n+1} \frac{B_{r,4n+2}(-z)}{z^{n+1}}. \qquad (1.9b)$$

$$X_{r,2n}(z) := \delta_{r,2n} \frac{A_{r,4n+1}(-z)}{z^n}, \quad X_{r,2n+1}(z) := \delta_{r,2n+1} \frac{A_{r,4n+3}(-z)}{z^n}, \qquad (1.9c)$$

$$Y_{r,2n}(z) := \delta_{r,2n} \frac{B_{r,4n+1}(-z)}{z^n}, \quad Y_{r,2n+1}(z) := \delta_{r,2n+1} \frac{B_{r,4n+3}(-z)}{z^n}, \qquad (1.9d)$$

where the normalizing constants $\gamma_{r,m}$ and $\delta_{r,m}$ are given by

$$\gamma_{r,2n} := (-1)^n q^{-rn+n^2-\frac{n}{2}} \prod_{j=1}^{\infty}(1 - q^j),$$

$$(1.9e)$$

$$\gamma_{r,2n+1} := (-1)^{n+1} q^{-r(n+1)+n^2+\frac{(n-1)}{2}} \prod_{n=1}^{\infty}(1 - q^j),$$

and

$$\delta_{r,2n} := (-1)^n q^{rn+n^2+\frac{3n}{2}} \prod_{j=1}^{\infty}(1 - q^j),$$

$$(1.9f)$$

$$\delta_{r,2n+1} := (-1)^n q^{r(n+1)+n^2+\frac{5n+3}{2}} \prod_{j=1}^{\infty}(1 - q^j).$$

In [7, Theorems 3.5 and 4.2] it is shown that the following limits exist and are equal to the indicated values, the convergence in each case being uniform on compact subsets of the punctured plane $\mathbb{C} - \{0\}$:

$$V_r^{(0)}(z) := \lim_{n\to\infty} V_{r,2n}(z) = \sum_{j=-\infty}^{\infty} q^{j^2}(-q^{r+\frac{1}{2}}z)^j = \lim_{n\to\infty} Y_{r,2n+1}(z) =: Y_r^{(1)}(z), \quad (1.10a)$$

$$V_r^{(1)}(z) := \lim_{n \to \infty} V_{r,2n+1}(z) = \sum_{j=-\infty}^{\infty} q^{j^2} (-q^{r+\frac{3}{2}} z)^j = \lim_{n \to \infty} Y_{r,2n}(z) =: Y_r^{(0)}(z), \quad (1.10b)$$

$$U_r^{(0)}(z) := \lim_{n \to \infty} U_{r,2n}(z) = c_0^{(r)} \sum_{j=-\infty}^{\infty} q^{j^2} \left(\sum_{k=0}^{\infty} (-1)^k q^{\frac{k^2}{2} + (2j-\frac{1}{2})k} \right) (-q^{r+\frac{1}{2}} z)^j \quad (1.10c)$$

$$= \lim_{n \to \infty} X_{r,2n+1}(z) =: X_r^{(1)}(z),$$

$$U_r^{(1)}(z) := \lim_{n \to \infty} U_{r,2n+1}(z)$$

$$= c_0^{(r)} \sum j = -\infty^\infty q^{j^2} \left(\sum_{k=0}^{\infty} (-1)^k q^{\frac{k^2}{2} + (2j-\frac{1}{2})k} \right) (-q^{r+\frac{3}{2}} z)^j \quad (1.10d)$$

$$= \lim_{n \to \infty} X_{r,2n}(z) =: X_r^{(0)}(z).$$

Also established in [7, Theorem 4.2] is the identity

$$U_r^{(1)}(-z)V_r^{(0)}(-z) - U_r^{(0)}(-z)V_r^{(1)}(-z) = (-1)^r q^{-\frac{r^2}{2} - r} \left(\prod_{j=1}^{\infty} (1 - q^j) \right)^3. \quad (1.11)$$

In the present paper we construct for each $r \in \mathbb{Z}$ two discrete distribution functions $\psi_r^{(0)}(t), \psi_r^{(1)}(t) \in \Phi(0, \infty)$ which generate the sequence $\{c_k^{(r)}\}_{-\infty}^{\infty}$ defined by (1.4). Every convex linear combination $(\psi_r(\alpha, t) := (1-\alpha)\psi_r^{(0)}(t) + \alpha\psi_r^{(1)}(t), 0 \le \alpha \le 1)$ of these distribution functions is a distribution function which generates the same sequence of moments. Each distribution function $\psi_r(\alpha, t)$ is a step function whose jumps are given explicitly (Theorem 3.2). To determine the locations and sizes of the jumps we derive in Section 2 some properties of the zeros of $V_{r,2n+\sigma}(z)$ and $V_r^{(\sigma)}(z)$ for $\sigma = 0, 1$. In particular, it is shown that the zeros of $V_r^{(\sigma)}(z)$ are given explicitly by $\{q^{2p-r-\frac{3}{2}+\sigma}\}_{p=-\infty}^{\infty}$ and that each of these zeros is the limit of a sequence of zeros of the L-polynomials $V_{r,2n+\sigma}(z)$. Properties of the zeros of $U_{r,2n+\sigma}(z)$, $V_{r,2n+\sigma}(z)$ and $V_r^{(\sigma)}(z)$ established in Section 2 are used in Section 3 to construct the step functions $\psi_r^{(0)}(t)$ and $\psi_r^{(1)}(t)$ which generate $\{c_k^{(r)}\}_{k=-\infty}^{\infty}$.

2   ZEROS OF $U_{r,2n+\sigma}(z)$, $V_{r,2n+\sigma}(z)$ AND $V_r^{(\sigma)}(z)$, $\sigma = 0, 1$.

We begin by obtaining the zeros of $V_r^{(0)}(z)$ and $V_r^{(1)}(z)$.

THEOREM 2.1.
   (A) *The zeros* $x_{r,p}^{(0)}$ *of* $V_r^{(0)}(z)$ *are given by*

$$x_{r,p}^{(0)} := q^{2p-r-\frac{3}{2}} > 0, \qquad p = 0, \pm 1, \pm 2, \dots \quad (2.1a)$$

and each zero $x_{r,p}^{(0)}$ *is simple.*

(B) *The zeros $x_{r,p}^{(1)}$ of $V_r^{(1)}(z)$ are given by*

$$x_{r,p}^{(1)} := q^{2p-r-\frac{1}{2}} > 0, \qquad p = 0, \pm 1, \pm 2, \dots \tag{2.1b}$$

and each zero $x_{r,p}^{(1)}$ is simple.

(C) *For each $\sigma = 0, 1$, the only limit points of $\{x_{r,p}^{(\sigma)}\}_{p=-\infty}^{\infty}$ are 0 and ∞. More-over,*

$$\lim_{p \to \infty} x_{r,p}^{(\sigma)} = 0 \quad \text{and} \quad \lim_{p \to -\infty} x_{r,p}^{(\sigma)} = +\infty. \tag{2.2}$$

(D) *The zeros $x_{r,p}^{(0)}$, $p = 0, \pm 1, \pm 2, \dots$ of $V_r^{(0)}(z)$ interlace the zeros $x_{r,p}^{(1)}$, $p = 0, \pm 1, \pm 2, \dots$ of $V_r^{(1)}(z)$; that is,*

$$0 < \cdots < x_{r,2}^{(0)} < x_{r,1}^{(1)} < x_{r,1}^{(0)} < x_{r,0}^{(1)} = q^{-r-\frac{1}{2}} < x_{r,0}^{(0)} < x_{r,-1}^{(1)} < x_{r,-1}^{(0)} < \cdots. \tag{2.3}$$

Proof (A): The Jacobi triple product formula [1, p. 348 (1.14)] is given by

$$V(x) := \sum_{j=-\infty}^{\infty} (-1)^j q^{j^2} x^j = \prod_{p=0}^{\infty}(1 - q^{2p+1}x) \prod_{p=0}^{\infty}(1 - q^{2p+1}x^{-1}) \prod_{p=0}^{\infty}(1 - q^{2p}). \tag{2.4}$$

From (2.4) we see that the zeros of $V(x)$ are q^{2p-1}, $p = 0, \pm 1, \pm 2, \dots$, and that each zero is simple. Substituting $w = q^{r+\frac{1}{2}}z$ into (1.10a) and using (2.4) yields

$$V_r^{(0)}(z) = V_r^{(0)}(q^{-r-\frac{1}{2}}w) = V(w) \tag{2.5}$$

It follows that the zeros of $V_r^{(0)}(z)$ are $q^{-r-\frac{1}{2}}q^{2p-1} = q^{2p-r-\frac{3}{2}} =: x_{r,p}^{(0)}$, $p = 0, \pm 1, \pm 2, \dots$, and that each zero $x_{r,p}^{(0)}$ is simple.

(B): From (1.10a,b) one sees that $V_r^{(1)}(z) = V_r^{(0)}(qz)$ and hence the zeros of $V_r^{(1)}(z)$ are $q^{-1}q^{2p-r-\frac{3}{2}} = q^{2p-r-\frac{5}{2}}$, $p = 0, \pm 1, \pm 2, \dots$, and each zero is simple. For the sake of ordering we set $x_{r,p}^{(1)} := q^{2(p+1)-r-\frac{5}{2}} = q^{2p-r-\frac{1}{2}}$, $p = 0, \pm 1, \pm 2, \dots$.

(C), (D): Since $0 < q < 1$, (2.2) and (2.3) follow from (A) and (B). $\square$

THEOREM 2.2. *The n zeros of $V_{r,n}(z)$ are simple, real and positive.*

Proof: From [7, Theorem 2.1] we know that the sequence $\{Q_{r,n}(z)\}_{n=0}^{\infty}$ defined by $Q_{r,n}(z) := V_{r,n}(z)/\gamma_{r,n}$, $n = 0, 1, 2, \dots$, is an orthogonal L-polynomial sequence with respect to $\varphi_r(t)$. Since the normalizing constants $\gamma_{r,n}$ are nonzero, $\{V_{r,n}(z)\}_{n=0}^{\infty}$ is also an orthogonal L-polynomial sequence with respect to $\varphi_r(t)$. The assertion of the theorem then follows from [8, Theorem 3.1 and the remark after Corollary 3.2] and [7, Theorem 3.1]. $\square$

LEMMA 2.3. *The Laurent polynomials $V_{r,n}(z)$, $n = 0, 1, 2, \dots$, satisfy the recurrence*

relations

$$V_{r,0}(z) = \prod_{j=1}^{\infty}(1-q^j), \qquad V_{r,1}(z) = \left(1 - \frac{1}{q^{r+\frac{1}{2}}z}\right)\prod_{j=1}^{\infty}(1-q^j), \qquad (2.6a)$$

$$V_{r,2n}(z) = q^n(1-q^{r+\frac{1}{2}}z)V_{r,2n-1}(z) + (1-q^{2n-1})V_{r,2n-2}(z), \; n=1,2,\ldots, \tag{2.6b}$$

$$V_{r,2n+1}(z) = q^n\left(1 - \frac{1}{q^{r+\frac{1}{2}}z}\right)V_{r,2n}(z) + (1-q^{2n})V_{r,2n-1}(z), \; n=1,2,\ldots. \tag{2.6c}$$

Proof: To verify (2.6a) substitute (1.9e) and (1.7a) into (1.9b). For $n = 1, 2, 3, \ldots$, we have from (1.7c, e)

$$B_{r,4n+2}(z) = \beta_{r,4n+2}zB_{r,4n+1}(z) + B_{r,4n}(z) \tag{2.7a}$$

$$B_{r,4n+1}(z) = \beta_{r,4n+1}B_{r,4n}(z) + \alpha_{r,4n+1}zB_{r,4n-1}(z) \tag{2.7b}$$

$$B_{r,4n}(z) = \beta_{r,4n}zB_{r,4n-1}(z) + B_{r,4n-2}(z). \tag{2.7c}$$

Solving (2.7c) for $B_{r,4n-1}(z)$, substituting the resulting expression and (2.7b) into (2.7a) and simplifying gives

$$B_{r,4n+2}(z) = \left(\beta_{r,4n+2}\beta_{r,4n+1}z + \frac{\beta_{r,4n+2}\alpha_{r,4n+1}z}{\beta_{r,4n}} + 1\right)B_{r,4n}(z)$$
$$- \frac{\beta_{r,4n+2}\alpha_{r,4n+1}z}{\beta_{r,4n}}B_{r,4n-2}(z). \tag{2.7d}$$

Replacing z by $-z$ and using (1.9b) we get

$$\frac{z^{n+1}}{\gamma_{r,2n+1}}V_{r,2n+1}(z) = \left(-\beta_{r,4n+2}\beta_{r,4n+1}z - \frac{\beta_{r,4n+2}\alpha_{r,4n+1}z}{\beta_{r,4n}} + 1\right)\frac{z^nV_{r,2n}(z)}{\gamma_{r,2n}}$$
$$+ \frac{\beta_{r,4n+2}\alpha_{r,4n+1}z}{\beta_{r,4n}}\frac{z^nV_{r,2n-1}(z)}{\gamma_{r,2n-1}} \tag{2.7e}$$

and so

$$V_{r,2n+1}(z) = \left(-\beta_{r,4n+2}\beta_{r,4n+1}z - \frac{\beta_{r,4n+2}\alpha_{r,4n+1}z}{\beta_{r,4n}} + 1\right)\frac{\gamma_{r,2n+1}}{\gamma_{r,2n}z}V_{r,2n}(z)$$
$$+ \frac{\beta_{r,4n+2}\alpha_{r,4n+1}\gamma_{r,2n+1}}{\beta_{r,4n}\gamma_{r,2n-1}}V_{r,2n-1}(z). \tag{2.7f}$$

Substituting (1.9e) and (1.6b,c) into (2.7f) and simplifying gives

$$V_{r,2n+1}(z) = (-q^{r-2n+\frac{1}{2}}z - q^{r+\frac{1}{2}}(1-q^{-2n})z + 1)\frac{(-1)q^{-r+n-\frac{1}{2}}}{z}V_{r,2n}(z)$$
$$- q^{2n}(1-q^{-2n})V_{r,2n-1}(z)$$

$$= \left(q^{-n} + q^n(1-q^{-2n}) - \frac{q^{-r+n-\frac{1}{2}}}{z}\right)V_{r,2n}(z) - (q^{2n}-1)V_{r,2n-1}(z)$$

$$= q^n\left(1 - \frac{1}{q^{r+\frac{1}{2}}z}\right)V_{r,2n}(z) + (1-q^{2n})V_{r,2n-1}(z). \tag{2.7g}$$

One proves (2.6c) in a similar manner. $\square$

LEMMA 2.4. *The L-polynomials $V_{r,n}(z)$ satisfy*

$$
V_{r,2n}(z)V'_{r,2n-1}(z) - V'_{r,2n}(z)V_{r,2n-1}(z) = q^{n+r+\frac{1}{2}}V^2_{r,2n-1}(z)
$$
$$
+ \frac{q^{n-1}(1-q^{2n-1})}{q^{r+\frac{1}{2}}z^2}\, V^2_{r,2n-2}(z)
$$
$$
+ (1-q^{2n-1})(1-q^{2n-2})[V_{r,2n-2}(z)V'_{r,2n-3}(z) - V'_{r,2n-2}(z)V_{r,2n-3}(z)],
$$
$$
n = 2,3,\dots \ . \tag{2.8}
$$

Proof: For $n = 2,3,\dots$, we have from the difference equation (2.6)

$$
V'_{r,2n-1}(z) = q^{n-1}\left(1 - \frac{1}{q^{r+\frac{1}{2}}z}\right)V'_{r,2n-2}(z) + \frac{q^{n-1}}{q^{r+\frac{1}{2}}z^2}V_{r,2n-2}(z)
$$
$$
+ (1-q^{2n-2})V'_{r,2n-3}(z) \tag{2.9a}
$$

and

$$
V'_{r,2n}(z) = q^n(1-q^{r+\frac{1}{2}}z)V'_{r,2n-1}(z) - q^{n+r+\frac{1}{2}}V_{r,2n-1}(z) + (1-q^{2n-1})V'_{r,2n-2}(z). \tag{2.9b}
$$

Using equations (2.6) and (2.9) one arrives at

$$
V_{r,2n}(z)V'_{r,2n-1}(z) - V'_{r,2n}(z)V_{r,2n-1}(z) = [q^n(1-q^{r+\frac{1}{2}}z)V_{r,2n-1}(z)
$$
$$
+ (1-q^{2n-1})V_{r,2n-2}(z)]V'_{r,2n-1}(z) - [q^n(1-q^{r+\frac{1}{2}}z)V'_{r,2n-1}(z)
$$
$$
- q^{n+r+\frac{1}{2}}V_{r,2n-1}(z) + (1-q^{2n-1})V'_{r,2n-2}(z)]V_{r,2n-1}(z)
$$
$$
= (1-q^{2n-1})V_{r,2n-2}(z)V'_{r,2n-1}(z) + q^{n+r+\frac{1}{2}}V^2_{r,2n-1}(z)
$$
$$
- (1-q^{2n-1})V'_{r,2n-2}(z)V_{r,2n-1}(z)
$$
$$
= (1-q^{2n-1})V_{r,2n-2}(z)\left[q^{n-1}\left(1 - \frac{1}{q^{r+\frac{1}{2}}z}\right)V'_{r,2n-2}(z) + \frac{q^{n-1}}{q^{r+\frac{1}{2}}z^2}\,V_{r,2n-2}(z)\right.
$$
$$
+ \ \left.(1-q^{2n-2})V'_{r,2n-3}(z)\right] + q^{n+r+\frac{1}{2}}V^2_{r,2n-1}(z)
$$
$$
- (1-q^{2n-1})V'_{r,2n-2}(z)V_{r,2n-1}(z)
$$
$$
= (1-q^{2n-1})V'_{r,2n-2}(z)\left[q^{n-1}\left(1 - \frac{1}{q^{r+\frac{1}{2}}z}\right)V_{r,2n-2}(z) - V_{r,2n-1}(z)\right]
$$
$$
+ (1-q^{2n-1})\frac{q^{n-1}}{q^{r+\frac{1}{2}}z^2}\,V^2_{r,2n-2}(z)
$$
$$
+ (1-q^{2n-1})(1-q^{2n-2})V_{r,2n-2}(z)V'_{r,2n-3}(z) + q^{n+r+\frac{1}{2}}V^2_{r,2n-1}(z)
$$

$$= -(1 - q^{2n-1})V'_{r,2n-2}(z)(1 - q^{2n-2})V_{r,2n-3}(z)$$
$$+ (1 - q^{2n-1})(1 - q^{2n-2})V_{r,2n-2}(z)V'_{r,2n-3}(z)$$
$$+ q^{n+r+\frac{1}{2}}V^2_{r,2n-1}(z) + (1 - q^{2n-1})\frac{q^{n-1}}{q^{r+\frac{1}{2}}z^2} V^2_{r,2n-2}(z)$$
$$= (1 - q^{2n-1})(1 - q^{2n-2})(V_{r,2n-2}(z)V'_{r,2n-3}(z) - V'_{r,2n-2}(z)V_{r,2n-3}(z))$$
$$+ q^{n+r+\frac{1}{2}}V^2_{r,2n-1}(z) + \frac{(1 - q^{2n-1})q^{n-1}}{q^{r+\frac{1}{2}}z^2} V^2_{r,2n-2}(z). \quad \square$$

THEOREM 2.5. *For $n = 1, 2, \ldots$, the zeros of $V_{r,n}(z)$ separate the zeros of $V_{r,n+1}(z)$.*

Proof: One easily checks that if we define $V_{r,-1}(z) := 0$, then (2.6c) holds for $n = 0, 1, 2, \ldots$, and (2.8) is true for $n = 1, 2, 3, \ldots$. We first show by induction that

$$V_{r,2n}(x)V'_{r,2n-1}(x) - V'_{r,2n}(x)V_{r,2n-1}(x) > 0 \text{ for } x \in (0,\infty), \ n = 1, 2, \ldots \ . \quad (2.10)$$

Setting $n = 1$ in (2.8) and using the fact that $0 < q < 1$ we get

$$V_{r,2}(x)V'_{r,1}(x) - V'_{r,2}(x)V_{r,1}(x) = q^{r+\frac{3}{2}}V^2_{r,1}(x) + \frac{(1-q)}{q^{r+\frac{1}{2}}x^2} V^2_{r,0}(x) > 0.$$

If we assume that for some $n \geq 2$, $V_{r,2n-2}(x)V'_{r,2n-3}(x) - V'_{r,2n-2}(x)V_{r,2n-3}(x) > 0$, then from (2.8) we see that (2.10) is true. Let $x_{r,2n,p}$ and $x_{r,2n,p+1}$ denote two consecutive zeros of $V_{r,2n}(x)$. It follows from (2.10) that $-V'_{r,2n}(x_{r,2n,\ell})V_{r,2n-1}(x_{r,2n,\ell}) > 0$, $\ell = p, p+1$ and hence $V'_{r,2n}(x_{r,2n,\ell})V_{r,2n-1}(x_{r,2n,\ell}) < 0$, $\ell = p, p+1$. By Rolle's theorem there exists $c_{r,2n}$, $x_{r,2n,p+1} < c_{r,2n} < x_{r,2n,p}$ such that $V'_{r,2n}(c_{r,2n}) = 0$. Since $V'_{r,2n}(x)$ changes sign between the zeros of $V_{r,2n}(x)$, $V_{r,2n-1}(x)$ must change sign also. Hence there exists at least one zero of $V_{r,2n-1}(x)$ between consecutive zeros of $V_{r,2n}(x)$. But since $V_{r,2n-1}(x)$ has $2n - 1$ distinct zeros and $V_{r,2n}(x)$ has $2n$ distinct zeros, there is exactly one zero of $V_{r,2n-1}(x)$ between consecutive zeros of $V_{r,2n}(x)$. By a similar argument one shows that the zeros of $V_{r,2n}(x)$ interlace those of $V_{r,2n+1}(x)$. $\quad \square$

THEOREM 2.6. *For $n = 2, 3, \ldots$, the zeros of $U_{r,n}(z)$ separate the zeros of $V_{r,n}(z)$.*

Proof: Proceeding as in Lemma 2.3 one shows that the L-polynomials $U_{r,n}(z)$, $n = 0, 1, \ldots$, satisfy the recurrence relations

$$U_{r,0}(z) = 0, \qquad U_{r,1}(z) = (-1)^r q^{-\frac{r^2}{2}-r} \prod_{j=1}^{\infty}(1 - q^j), \quad (2.11a)$$

$$U_{r,2n}(z) = q^n(1 - q^{r+\frac{1}{2}}z)U_{r,2n-1}(z) + (1 - q^{2n-1})U_{r,2n-2}(z), \ n = 1, 2, \ldots, \quad (2.11b)$$

$$U_{r,2n+1}(z) = q^n \left(1 - \frac{1}{q^{r+\frac{1}{2}}z}\right) U_{r,2n}(z) + (1 - q^{2n})U_{r,2n-1}(z), \ n = 1, 2, \ldots \ . \quad (2.11c)$$

It follows from (2.6) and (2.11) that $U_{r,n}(z)$ and $V_{r,n}(z)$ are the nth numerator and denominator respectively of the continued fraction

$$\overset{\infty}{\underset{n=1}{K}} \left(\frac{a_{r,n}}{b_{r,n}} \right) \tag{2.12a}$$

where

$$a_{r,1} := (-1)^r q^{-\frac{r^2}{2}-r} \prod_{j=1}^{\infty}(1-q^j), \qquad b_{r,1} := -\left[1 + \frac{1}{q^{r+\frac{1}{2}}z}\right]\prod_{j=1}^{\infty}(1-q^j) \tag{2.12b}$$

$$a_{r,n} := 1 - q^{n-1}, \ n = 2,3,\ldots, \qquad b_{r,2n} := q^n(1 - q^{r+\frac{1}{2}}z) \text{ and}$$

$$b_{r,2n+1} := q^n\left(1 - \frac{1}{q^{r+\frac{1}{2}}z}\right), \ n = 1,2,\ldots. \tag{2.12c}$$

From the determinant formula for continued fractions [10] we have

$$U_{r,n}(z)V_{r,n-1}(z) - V_{r,n}(z)U_{r,n-1}(z) = (-1)^{n-1}\prod_{k=1}^{n} a_{r,k}$$

$$= (-1)^{n+r-1}q^{-\frac{r^2}{2}-r}\prod_{j=1}^{\infty}(1-q^j)\prod_{k=2}^{n}(1-q^{k-1}). \tag{2.13}$$

We fix n and suppose that $n + r - 1$ is even. Then from (2.13) we see that $U_{r,n}(z)V_{r,n-1}(z) - V_{r,n}(z)U_{r,n-1}(z) > 0$ for all $z \neq 0$. Let $x_{r,n,k}$ and $x_{r,n,k+1}$ be two consecutive zeros of $V_{r,n}(z)$. Then $U_{r,n}(x_{r,n,\ell})V_{r,n-1}(x_{r,n,\ell}) > 0$ for $\ell = k, k+1$. Since the zeros of $V_{r,n}(n)$ separate those of $V_{r,n+1}(z)$ there exists exactly one zero $x_{r,n-1,k}$ of $V_{r,n-1}(z)$ such that $x_{r,n,k+1} < x_{r,n-1,k} < x_{r,n,k}$. At $x_{r,n,k}$ and $x_{r,n,k+1}$, $U_{r,n}(z)$ and $V_{r,n-1}(z)$ have the same sign. $V_{r,n-1}(x)$ changes sign at $x_{r,n-1,k}$ and thus $U_{r,n}(x)$ must change sign an odd number of times between $x_{r,n,k+1}$ and $x_{r,n,k}$. That is, there exists at least one zero of $U_{r,n}(z)$ between $x_{r,n,k+1}$ and $x_{r,n,k}$. Now $V_{r,n}(z)$ has n zeros while it can be seen from the recurrence relations that for $\sigma = 0, 1$, $z^{n-1+\sigma}U_{r,2n+\sigma}(z)$ is a polynomial of degree $2n + \sigma - 1$. It follows that since $U_{r,n}(z)$ has $n - 1$ zeros, there exists exactly one zero of $U_{r,n}(z)$ between $x_{r,n,k+1}$ and $x_{r,n,k}$. A similar argument holds if $n + r - 1$ is odd. $\square$

THEOREM 2.7.

(A) $V_{r,2n+1}(q^{-r-\frac{1}{2}}) = 0$ for all $n = 0, 1, 2, \ldots$.

(B) For each $n = 1, 2, \ldots$, let $x_{r,2n,p}$, $-n + 1 \leq p \leq n$, denote the zeros (all are simple) of $V_{r,2n}(z)$ arranged so that $x_{r,2n,p+1} < x_{r,2n,p}$ for $p = -n+1, \ldots, n-1$; then

$$0 < x_{r,2n,n} < x_{r,2n,n-1} < \cdots < x_{r,2n,1} < q^{-r-\frac{1}{2}} < x_{r,2n,0} < \cdots < x_{r,2n,-n+1}. \tag{2.14}$$

(C) For each $n = 0, 1, 2, \ldots$, let $x_{r,2n+1,p}$, $-n \leq p \leq n$, denote the zeros (all are simple) of $V_{r,2n+1}(z)$ arranged so that $x_{r,2n+1,p+1} < x_{r,2n+1,p}$ for $p = -n, -n+1, \ldots, n-1$; then

$$0 < x_{r,2n+1,n} < x_{r,2n+1,n-1} < \cdots < x_{r,2n+1,1} < x_{r,2n+1,0}$$

$$= q^{-r-\frac{1}{2}} < x_{r,2n+1,-1} < \cdots < x_{r,2n+1,-n} \tag{2.15}$$

(D) *For $n \geq 1$,*

$$x_{r,2n+1,p} < x_{r,2n,p} < x_{r,2n+1,p-1}, \quad p = -n+1, \ldots, n. \tag{2.16}$$

Proof (A): From (2.6a) we see that $V_{r,1}(z) = \left(1 - \dfrac{1}{q^{r+\frac{1}{2}} z}\right) \prod\limits_{j=1}^{\infty}(1 - q^j)$ and so

$V_{r,1}(q^{-r-\frac{1}{2}}) = 0$. Suppose that $V_{r,2n-1}(q^{-r-\frac{1}{2}}) = 0$. Then using (2.6c) we have
$V_{r,2n+1}(q^{-r-\frac{1}{2}}) = 0$. The result follows by induction.

(B), (C) and (D) are immediate consequences of (A) and Theorem 2.5. □

THEOREM 2.8.

(A) *If $1 \leq p \leq n-1$, then $x_{r,2n-2,p} < x_{r,2n,p} < q^{-r-\frac{1}{2}}$.*
(B) *If $-n+2 \leq p \leq 0$, then $q^{-r-\frac{1}{2}} < x_{r,2n,p} < x_{r,2n-2,p}$.*
(C) *If $1 \leq p \leq n-1$, then $x_{r,2n-1,p} < x_{r,2n+1,p} < q^{-r-\frac{1}{2}}$.*
(D) *If $-n+1 \leq p \leq -1$, then $q^{-r-\frac{1}{2}} < x_{r,2n+1,p} < x_{r,2n-1,p}$.*

Proof: In [7, Theorem 3.4] it is shown that

$$B_{r,2n}(z) = \sum_{j=0}^{n} q^{j^2+(r-n+\frac{1}{2})j} \begin{bmatrix} n \\ j \end{bmatrix} z^j \tag{2.17a}$$

where

$$\begin{bmatrix} n \\ j \end{bmatrix} := \prod_{k=1}^{n}(1-q^k) / \prod_{k=1}^{j}(1-q^k) \prod_{k=1}^{n-j}(1-q^k). \tag{2.17b}$$

Substituting (2.17) and the value of $\gamma_{r,2n}$ given in (1.9e) into (2.9b) (the definition of $V_{r,2n}(z)$) we have

$$\begin{aligned}
V_{r,2n}(z) &= \gamma_{r,2n} \frac{B_{r,4n}(-z)}{z^n} = \frac{\gamma_{r,2n}}{z^n} \sum_{j=0}^{2n} q^{j^2+(r-2n+\frac{1}{2})j} \begin{bmatrix} 2n \\ j \end{bmatrix} z^j \\
&= \frac{(-1)^n q^{-rn+n^2-\frac{n}{2}}}{z^n} \prod_{j=1}^{\infty}(1-q^j) \sum_{j=-n}^{n} q^{(n-j)^2+(r-2n+\frac{1}{2})(n-j)} \begin{bmatrix} 2n \\ n-j \end{bmatrix} (-z)^{n-j} \\
&= \prod_{j=1}^{\infty}(1-q^j) \sum_{j=-n}^{n} q^{-rn+n^2-\frac{n}{2}+(n-j)^2+(r-2n+\frac{1}{2})(n-j)} \begin{bmatrix} 2n \\ n-j \end{bmatrix} (-1)^{-j}(z)^{-j}.
\end{aligned}$$

Thus

$$V_{r,2n}(z) = \prod_{j=1}^{\infty}(1-q^j) \sum_{j=-n}^{n} q^{j^2} \begin{bmatrix} 2n \\ n-j \end{bmatrix} (-q^{r+\frac{1}{2}} z)^{-j}. \tag{2.18a}$$

Similarly one can show that

$$V_{r,2n+1}(z) = \prod_{j=1}^{\infty}(1-q^j) \sum_{j=-n-1}^{n} q^{j^2} \begin{bmatrix} 2n+1 \\ n-j \end{bmatrix} (-q^{r+\frac{3}{2}} z)^{-j}. \tag{2.18b}$$

By examining (2.18a) one sees that $V_{r,2n}(x)$ has the sign of $(-1)^n$ for $0 < x < x_{r,2n,n}$ and consequently $V_{r,2n}(x)$ has the same sign as $(-1)^{p-1}$ for $x_{r,2n,p} < x < x_{r,2n,p-1}$. Similarly $V_{r,2n+1}(x)$ has the sign of $(-1)^{n+1}$ for $0 < x < x_{r,2n+1,n}$ and the same sign as $(-1)^p$ for $x_{r,2n+1,p} < x < x_{r,2n+1,p-1}$.

(A): Let x be fixed with $0 < x < q^{-r-\frac{1}{2}}$ so that $0 < 1 - q^{r+\frac{1}{2}}x < 1$. Evaluating the difference equation (2.6b) at the zero $x_{r,2n-1,p}$ of $V_{r,2n-1}(x)$ we have

$$V_{r,2n}(x_{r,2n-1,p}) = (1 - q^{2n-1})V_{r,2n-2}(x_{r,2n-1,p}).$$

Consequently the graphs of $V_{r,2n}(x)$ and $(1 - q^{2n-1})V_{r,2n-2}(x)$ intersect at the zeros of $V_{r,2n-1}(x)$. Since $(1 - q^{2n-1}) > 0$ and $x_{r,2n,p-1} < x_{r,2n-1,p} < x_{r,2n,p}$, the sign of $V_{r,2n}(x_{r,2n-1,p})$ and the sign of $V_{r,2n-2}(x_{r,2n-1,p})$ are the same as the sign of $(-1)^p$. For $x_{r,2n-1,p} < x < x_{r,2n-1,p-1}$ we see using (2.6b) that

$$q^n(1 - q^{r+\frac{1}{2}}x)V_{r,2n-1}(x) = V_{r,2n}(x) - (1 - q^{2n-1})V_{r,2n-2}(x).$$

Since $q^n(1 - q^{r+\frac{1}{2}}x) > 0$ and $V_{r,2n-1}(x)$ has the same sign as $(-1)^p$, the graph of $(1 - q^{2n-1})V_{r,2n-2}(x)$ is "below" that of $V_{r,2n}(x)$ when p is even and "above" when p is odd. In either case the zero $x_{r,2n-2,p}$ of $V_{r,2n-2}(x)$ in $(x_{r,2n-1,p}, x_{r,2n-1,p-1})$ lies to the left of the zero $x_{r,2n,p}$ of $V_{r,2n}(x)$.

(B): Let x be fixed with $q^{-r-\frac{1}{2}} < x < \infty$. Then $1 - q^{r+\frac{1}{2}}x$ is negative and so in the argument for part (A) the words "below" and "above" are interchanged giving $x_{r,2n,p} < x_{r,2n-2,p}$.

(C) and (D) follow as above by observing the sign of $1 - \dfrac{1}{q^{r+\frac{1}{2}}x}$ in (2.6c). $\square$

COROLLARY 2.9.

(A) *For each* $p \in \mathbb{Z}$, $\{x_{r,2n,p}\}_{n=|p|+1}^{\infty}$ *converges and*

$$\lim_{n \to \infty} x_{r,2n,p} =: k_{r,p}^{(0)} > 0. \tag{2.19a}$$

(B) *For each* $p \in \mathbb{Z}$, $\{x_{r,2n+1,p}\}_{n=|p|+1}^{\infty}$ *converges and*

$$\lim_{n \to \infty} x_{r,2n+1,p} =: k_{r,p}^{(1)} > 0. \tag{2.19b}$$

(C) $k_{r,0}^{(1)} = q^{-r-\frac{1}{2}}$.

Proof (A): For fixed $p \geq 1$ we see from Theorem 2.8 that $\{x_{r,2n,p}\}_{n=|p|+1}^{\infty}$ is a monotone increasing sequence of positive numbers bounded above by $q^{-r-\frac{1}{2}}$. For fixed $p \leq 0$, $\{x_{r,2n,p}\}_{n=|p|+1}^{\infty}$ is a monotone decreasing sequence of positive numbers bounded below by $q^{-r-\frac{1}{2}}$.

(B): A proof is similar to that given for (A).

(C): Since $x_{r,2n+1,0} = q^{-r-\frac{1}{2}}$ for $n = 1, 2, \dots$, (C) is a trivial consequence of (B). $\square$

For completeness we state a version of Hurwitz' Theorem found in [16], used to prove some of the following results.

THEOREM 2.10. (Hurwitz) *Let $f_n(z)$ be holomorphic and single-valued for each $n > 0$ in an open region R, and let $\{f_n(z)\}$ converge uniformly to $f(z) \not\equiv c$, c a constant, in every compact subset of R. Let $a \in R$. Then $f(a) = 0$ if and only if there exists a sequence $\{z_n\}$ with $\lim z_n = a$ and there exists n_0 such that $f_n(z_n) = 0$ for each $n > n_0$.*

LEMMA 2.11. *For $\sigma = 0, 1$, let $V_r^{(\sigma)}(z)$ be defined as in (1.10) and let $k_{r,p}^{(\sigma)}$, $p = 0, \pm 1, \pm 2, \ldots$, be defined as in (2.19). Then*

$$V_r^{(\sigma)}(k_{r,p}^{(\sigma)}) = 0, \qquad p = 0, \pm 1, \pm 2, \ldots . \tag{2.20}$$

Proof: the Lemma follows immediately from Theorem 2.10. □

LEMMA 2.12. *For $\sigma = 0, 1$, let $k_{r,p}^{(\sigma)}$, $p = 0, \pm 1, \pm 2, \ldots$ be defined as in (2.19). Then*

$$0 < k_{r,p}^{(\sigma)} < k_{r,p-1}^{(\sigma)}, \qquad p = 0, \pm 1, \pm 2, \ldots . \tag{2.21}$$

Proof: Fix $p \in \mathbb{Z}$. Since from (2.16) we have $x_{r,2n+\sigma,p} < x_{r,2n+\sigma,p-1}$ for all $n = |p| + 1, |p| + 2, \ldots$, it follows that $\lim_{n \to \infty} x_{r,2n+\sigma,p} = k_{r,p}^{(\sigma)} \leq k_{r,p-1}^{(\sigma)} = \lim_{n \to \infty} x_{r,2n+\sigma,p}$. The zeros of $U_{r,2n+\sigma}(z)$ separate the zeros of $V_{r,2n+\sigma}(z)$ and so there exists a sequence $\{y_{r,2n+\sigma,p}\}_{n=|p|+1}^{\infty}$ of zeros of $U_{r,2n+\sigma}(z)$ with $x_{r,2n+\sigma,p} < y_{r,2n+\sigma,p} < x_{r,2n+\sigma,p-1}$. Suppose $k_{r,p}^{(\sigma)} = k_{r,p-1}^{(\sigma)}$. Then $\lim_{n \to \infty} x_{r,2n+\sigma,p} = \lim_{n \to \infty} y_{r,2n+\sigma,p} = \lim_{n \to \infty} x_{r,2n+\sigma,p-1} = k_{r,p}^{(\sigma)}$. From (2.20), $V_r^{(\sigma)}(k_{r,p}^{(\sigma)}) = 0$ and using Theorem 2.10 one can show that $U_r^{(\sigma)}(k_{r,p}^{(\sigma)}) = 0$ also. But this contradicts (1.11) with $-z = k_{r,p}^{(\sigma)}$. Hence (2.21) holds. □

THEOREM 2.13. *For $\sigma = 0, 1$ and $p = 0, \pm 1, \pm 2, \ldots$ let $x_{r,p}^{(\sigma)}$ be defined as in (2.1) and let $k_{r,p}^{(\sigma)}$ be defined as in (2.19). Then*

$$k_{r,p}^{(\sigma)} := \lim_{n \to \infty} x_{r,2n+\sigma,p} = x_{r,p}^{(\sigma)} := q^{2p-r-\frac{3}{2}+\sigma}, \qquad p = 0, \pm 1, \pm 2, \ldots . \tag{2.22}$$

Proof: From Lemma 2.11 we know that $\{k_{r,p}^{(\sigma)} : p = 0, \pm, 1 \pm 2, \ldots\} \subseteq \{x_{r,p}^{(\sigma)} : p = 0, \pm 1, \pm 2, \ldots\}$. The only limit points of $\{x_{r,p}^{(\sigma)}\}_{p=-\infty}^{\infty}$ are 0 and ∞ and so the only possible limit points of $\{k_{r,p}^{(\sigma)}\}_{p=-\infty}^{\infty}$ are 0 and ∞. In fact, since $0 < k_{r,p}^{(\sigma)} < k_{r,p-1}^{(\sigma)}$ for all $p \in \mathbb{Z}$, we have $\lim_{p \to \infty} k_{r,p}^{(\sigma)} = 0$ and $\lim_{p \to -\infty} k_{r,p}^{(\sigma)} = \infty$. We now show $\{x_{r,p}^{(\sigma)} : p = 0, \pm 1, \pm 2, \ldots\} \subseteq \{k_{r,p}^{(\sigma)} : p = 0, \pm 1, \pm 2, \ldots\}$. Suppose there exists an integer $m_r^{(\sigma)}$ such that $x_{r,m_r^{(\sigma)}}^{(\sigma)} \notin \{k_{r,p}^{(\sigma)} : p = 0, \pm 1, \pm 2, \ldots\}$. Then there exists an integer $p(m_r^{(\sigma)})$

with $k_{r,p(m_r^{(\sigma)})+1}^{(\sigma)} < x_{r,m_r^{(\sigma)}}^{(\sigma)} < k_{r,p(m_r^{(\sigma)})}^{(\sigma)}$. For simplicity we let $m := m_r^{(\sigma)}$ and $p_m := p(m_r^{(\sigma)})$. Fix $0 < \varepsilon < \min \frac{1}{10} \{k_{r,p_m}^{(\sigma)} - x_{r,m}^{(\sigma)}, x_{r,m}^{(\sigma)} - k_{r,p_m+1}^{(\sigma)}\}$. Since $\lim_{n\to\infty} x_{r,2n+\sigma,p} = k_{r,p}^{(\sigma)}$ for each $p \in \mathbb{Z}$, there exist integers $N_{r,m,1}^{(\sigma)}$ and $N_{r,m,2}^{(\sigma)}$ so that $|x_{r,2n+\sigma,p_m+1} - k_{r,p_m+1}^{(\sigma)}| < \varepsilon$ for all $n \geq N_{r,m,1}^{(\sigma)}$ and $|x_{r,2n+\sigma,p_m} - k_{r,p_m}^{(\sigma)}| < \varepsilon$ for all $n \geq N_{r,m,2}^{(\sigma)}$. Now $V_{r,m}^{(\sigma)}(x_{r,m}^{(\sigma)}) = 0$ and so by Hurwitz' Theorem there exists an integer $N_{r,m,3}^{(\sigma)}$ so that for $n \geq N_{r,m,3}^{(\sigma)}$, $V_{r,2n+\sigma}(z)$ has a zero in the neighborhood $N_\varepsilon(x_{r,m}^{(\sigma)})$. Fix $V_{r,m}^{(\sigma)} = \max\{N_{r,m,1}^{(\sigma)}, N_{r,m,2}^{(\sigma)}, N_{r,m,3}^{(\sigma)}\}$. Then $|x_{r,2N_{r,m}^{(\sigma)}+\sigma,p_m+1} - k_{r,p_m+1}^{(\sigma)}| < \varepsilon$, $|x_{r,2N_{r,m}^{(\sigma)}+\sigma,p_m} - k_{r,p_m}^{(\sigma)}| < \varepsilon$ and there exists a zero $x_{r,2N_{r,m}^{(\sigma)}+\sigma,\ell}$ of $V_{r,2N_{r,m}^{(\sigma)}+\sigma}(z)$ in $N_\varepsilon(x_{r,m}^{(\sigma)})$. That is, $x_{r,2N_{r,m}^{(\sigma)}+\sigma,p_m+1} < x_{r,2N_{r,m}^{(\sigma)}+\sigma,\ell} < x_{r,2N_{r,m}^{(\sigma)}+\sigma,p_m}$ which contradicts (2.14) and (2.15) since $x_{r,2N_{r,m}^{(\sigma)}+\sigma,p_m+1}$ and $x_{r,2N_{r,m}^{(\sigma)}+\sigma,p_m}$ are consecutive zeros of $V_{r,2N_{r,m}^{(\sigma)}+\sigma}(z)$. Hence we must have $x_{r,m_r^{(\sigma)}}^{(\sigma)} = k_{r,M_r^{(\sigma)}}^{(\sigma)}$ for some $M_r^{(\sigma)} \in \mathbb{Z}$. Thus $\{x_{r,p}^{(\sigma)} : p = 0, \pm 1, \pm 2, \dots\} \subseteq \{k_{r,p}^{(\sigma)} : p = 0, \pm 1, \pm 2, \dots\}$ and consequently $\{x_{r,p}^{(\sigma)} : p = 0, \pm 1, \pm 2, \dots\} = \{k_{r,p}^{(\sigma)} : p = 0, \pm 1, \pm 2, \dots\}$. Since $x_{r,p}^{(1)} = q^{2p-r-\frac{1}{2}}$ and $0 < q < 1$ we have

$$0 < \cdots < x_{r,3}^{(1)} < x_{r,2}^{(1)} < x_{r,1}^{(1)} < x_{r,0}^{(1)} = q^{-r-\frac{1}{2}} < x_{r,-1}^{(1)} < x_{r,-2}^{(1)} < \cdots.$$

From (2.21) and Corollary 2.9(C) we have

$$0 < \cdots < k_{r,3}^{(1)} < k_{r,2}^{(1)} < k_{r,1}^{(1)} < k_{r,0}^{(1)} = q^{-r-\frac{1}{2}} < k_{r,-1}^{(1)} < k_{r,-2}^{(1)} < \cdots.$$

It follows by the "pigeon hole principle" that $k_{r,p}^{(1)} = x_{r,p}^{(1)}$ for all $p \in \mathbb{Z}$. Similarly $x_{r,p}^{(0)} = q^{2p-r-\frac{3}{2}}$ and so

$$0 < \cdots < x_{r,3}^{(0)} < x_{r,2}^{(0)} < x_{r,1}^{(0)} < q^{-r-\frac{1}{2}} < x_{r,0}^{(0)} < x_{r,-1}^{(0)} < x_{r,-2}^{(0)} < \cdots.$$

In order to conclude that $k_{r,p}^{(0)} = x_{r,p}^{(0)}$ for all $p \in \mathbb{Z}$, it is enough to show that $k_{r,1}^{(0)} < q^{-r-\frac{1}{2}} < k_{r,0}^{(0)}$ since then we have

$$0 < \cdots < k_{r,3}^{(0)} < k_{r,2}^{(0)} < k_{r,1}^{(0)} < q^{-r-\frac{1}{2}} < k_{r,0}^{(0)} < k_{r,-1}^{(0)} < k_{r,-2}^{(0)} < \cdots.$$

We know from (2.14) that $x_{r,2n,1} < q^{-r-\frac{1}{2}} < x_{r,2n,0}$ for all n. Consequently $\lim_{n\to\infty} x_{r,2n,1} = k_{r,1}^{(0)} \leq q^{-r-\frac{1}{2}} \leq k_{r,0}^{(0)} = \lim_{n\to\infty} x_{r,2n,0}$. We cannot have $k_{r,1}^{(0)} = q^{-r-\frac{1}{2}}$ or $k_{r,0}^{(0)} = q^{-r-\frac{1}{2}}$ since $k_{r,1}^{(0)}$ and $k_{r,0}^{(0)}$ are zeros of $V_r^{(0)}(z)$ and the zeros of $V_r^{(0)}(z)$ are exactly $\{q^{2p-r-\frac{3}{2}} : p = 0, \pm 1, \pm 2, \dots\}$. Hence $k_{r,1}^{(0)} < q^{-r-\frac{1}{2}} < k_{r,0}^{(0)}$. It follows that $k_{r,p}^{(0)} = x_{r,p}^{(0)}$ for all $p \in \mathbb{Z}$. $\square$

3   CONSTRUCTION OF TWO NEW DISTRIBUTION FUNCTIONS FOR LOG-NORMAL MOMENTS

In this section we construct explicit step functions $\psi_r^{(0)}(t)$ and $\psi_r^{(1)}(t)$ which generate the moments $c_k^{(r)}$, $k = 0, \pm 1, \pm 2, \ldots$ defined in (1.4).

THEOREM 3.1. (Partial Fraction Decomposition) *For $\sigma = 0, 1$:*

(A)

$$\frac{U_{r,2n+\sigma}(z)}{V_{r,2n+\sigma}(z)} = \sum_{p=-n+1-\sigma}^{n} \frac{z R_{r,2n+\sigma,p}}{z - x_{r,2n+\sigma,p}} \tag{3.1a}$$

where

$$R_{r,2n+\sigma,p} := \frac{U_{r,2n+\sigma}(x_{r,2n+\sigma,p})}{x_{r,2n+\sigma,p} V'_{r,2n+\sigma,p}(x_{r,2n+\sigma,p})}, \quad p = -n+1-\sigma, \ldots, n. \tag{3.1b}$$

(B)

$$\sum_{p=-n+1-\sigma}^{n} R_{r,2n+\sigma,p} = (-1)^r q^{-\frac{r^2}{2}-r}. \tag{3.2}$$

(C)

$$(-1)^r R_{r,2n+\sigma,p} > 0 \quad \text{for all} \quad n = 0, 1, 2, \ldots, \quad k = 0, \pm 1, \pm 2, \ldots \tag{3.3}$$

(D) $\lim_{n \to \infty} R_{r,2n+\sigma,p}$ *exists and*

$$\lim_{n \to \infty} R_{r,2n+\sigma,p} =: R_{r,p}^{(\sigma)} = \frac{U_r^{(\sigma)}(x_{r,p}^{(\sigma)})}{x_{r,p}^{(\sigma)} V_r^{(\sigma)'}(x_{r,p}^{(\sigma)})}, \quad \sigma = 0, 1. \tag{3.4}$$

(E)

$$(-1)^r R_{r,p}^{(\sigma)} > 0, \quad p = 0, \pm 1, \pm 2, \ldots, \quad \sigma = 0, 1. \tag{3.5}$$

(F)

$$0 < \sum_{p=-\infty}^{\infty} (-1)^r R_{r,p}^{(\sigma)} \leq q^{-\frac{r^2}{2}-r}, \quad \sigma = 0, 1. \tag{3.6}$$

Remark. Later we will show that $\leq$ in (3.6) can be replaced by $=$.

Proof (A): From the recurrence relations (2.6) and (2.11) it is readily seen that for $\sigma = 0, 1$, $z^{n+\sigma} V_{r,2n+\sigma}(z)$ is a polynomial of degree $2n + \sigma$ and $z^{n+\sigma-1} U_{r,2n+\sigma}(z)$ is a polynomial of degree $2n + \sigma - 1$. Since the zeros of $U_{r,2n+\sigma}(z)$ separate the zeros $x_{r,2n+\sigma,p}$ of $V_{r,2n+\sigma}(z)$ there exists a partial fraction decomposition of the form

$$\frac{z^{n+\sigma-1} U_{r,2n+\sigma}(z)}{z^{n+\sigma} V_{r,2n+\sigma}(z)} = \frac{U_{r,2n+\sigma}(z)}{z V_{r,2n+\sigma}(z)} = \sum_{p=-n+1-\sigma}^{n} \frac{R_{r,2n+\sigma,p}}{z - x_{r,2n+\sigma,p}}. \tag{3.7a}$$

From (3.7a) we see that

$$\frac{U_{r,2n+\sigma}(z)}{z} = \frac{R_{r,2n+\sigma,-n+1-\sigma}V_{r,2n+\sigma}(z)}{z - x_{r,2n+\sigma,-n+1-\sigma}} + \cdots + \frac{R_{r,2n+\sigma,n}V_{r,2n+\sigma}(z)}{z - x_{r,2n+\sigma,n}}. \qquad (3.7b)$$

Consequently for fixed $k \in \mathbb{Z}$, $-n+1-\sigma \le k \le n$,

$$\frac{U_{r,2n+\sigma}(x_{r,2n+\sigma,k})}{x_{r,2n+\sigma,k}} = \lim_{z \to x_{r,2n+\sigma,k}} \frac{R_{r,2n+\sigma,k}V_{r,2n+\sigma}(z)}{z - x_{r,2n+\sigma,k}} = R_{r,2n+\sigma,k}V'_{r,2n+\sigma}(x_{r,2n+\sigma,k}).$$

$$(3.7c)$$

(B): One can show using the recurrence relations (2.6) and induction that the highest power of z occurring in $V_{r,2n+\sigma}(z)$ is n and the coefficient of z^n is $q^{n^2+n\sigma}(-q^{r+\frac{1}{2}})^n \prod_{j=1}^{\infty}(1 - q^j)$. Similarly, by using equations (2.11) it can be shown that the highest power of z occurring in $U_{r,2n+\sigma}(z)$ is n and the coefficient of z^n is $(-1)^r q^{-\frac{r^2}{2}-r} q^{n^2+n\sigma}(-q^{r+\frac{1}{2}})^n \prod_{j=1}^{\infty}(1 - q^j)$. Consequently

$$(-1)^r q^{-\frac{r^2}{2}-r} = \lim_{z \to \infty} \frac{U_{r,2n+\sigma}(z)}{V_{r,2n+\sigma}(z)} = \lim_{z \to \infty} \sum_{p=-n+1-\sigma}^{n} \frac{R_{r,2n+\sigma,p}}{1 - \frac{x_{r,2n+\sigma,p}}{z}}$$

$$= \sum_{p=-n+1-\sigma}^{n} R_{r,2n+\sigma,p}.$$

(C): Since the zeros of $U_{r,2n+\sigma}(z)$ separate the zeros of $V_{r,2n+\sigma}(z)$, we see that $U_{r,2n+\sigma}(x_{r,2n+\sigma,p})$ and $V'_{r,2n+\sigma}(x_{r,2n+\sigma,p})$ change sign when p changes by one unit. Hence $R_{r,2n+\sigma,p}$ has the same sign for each $p = -n+1-\sigma, \ldots, n$. The result now follows from (B) and (3.1b).

(D): Equation (3.4) is a direct consequence of (3.1b) and the facts that $\{V_{r,2n+\sigma}(z)\}_{n=0}^{\infty}$ and $\{U_{r,2n+\sigma}(z)\}_{n=0}^{\infty}$ converge uniformly with respect to z on compact subsets of $\mathbb{C} - \{0\}$ to $V_r^{(\sigma)}(z)$ and $U_r^{(\sigma)}(z)$ respectively and $\{x_{r,2n+\sigma,p}\}_{n=1}^{\infty}$ converges to $x_{r,p}^{(\sigma)}$.

(E): Fix $p \in \mathbb{Z}$. Since $(-1)^r R_{r,2n+\sigma,p} > 0$, it follows that $(-1)^r R_{r,p}^{(\sigma)} \ge 0$. If $R_{r,p}^{(\sigma)} = 0$, then (3.4) implies that $U_r^{(\sigma)}(x_{r,p}^{(\sigma)}) = 0$. This contradicts equation (1.11) and the fact that $V_r^{(\sigma)}(x_{r,p}^{(\sigma)}) = 0$. Hence $(-1)^r R_{r,p}^{(\sigma)} > 0$.

(F): Fix $k \in \mathbb{N}$. Then for all $n \ge k+1-\sigma$ we have using (3.2) that

$$\sum_{p=-k}^{k}(-1)^r R_{r,2n+\sigma,p} \le \sum_{p=-n+1-\sigma}^{n}(-1)^r R_{r,2n+\sigma,p} = q^{-\frac{r^2}{2}-r}.$$

Thus

$$\sum_{p=-k}^{k}(-1)^r R_{r,p}^{(\sigma)} = \sum_{p=-k}^{k}(-1)^r \lim_{n \to \infty} R_{r,2n+\sigma,p} = \lim_{n \to \infty} \sum_{p=-k}^{k}(-1)^r R_{r,2n+\sigma,p} \le q^{-\frac{r^2}{2}-r}.$$

But k is arbitrary and so $\sum_{p=-k}^{k} (-1)^r R_{r,p}^{(\sigma)} \leq q^{-\frac{r^2}{2}-r}$ for all k and consequently

$$\sum_{p=-\infty}^{\infty} (-1)^r R_{r,p}^{(\sigma)} = \lim_{k\to\infty} \sum_{p=-k}^{k} (-1)^r R_{r,p}^{(\sigma)} \leq q^{-\frac{r^2}{2}-r}. \quad \square$$

For $\sigma = 0,1$ and $n = 1,2,\ldots$, let $\psi_{r,2n+\sigma}(t)$ denote the real-valued step function on $[0,\infty)$ with positive jump $(-1)^r R_{r,2n+\sigma,p}$ at $x_{r,2n+\sigma,p}$, $p = -n+1-\sigma,\ldots,n$, defined by

$$\psi_{r,2n+\sigma}(t) := \begin{cases} 0 & \text{for} \quad 0 \leq t < x_{r,2n+\sigma,n} \\ \sum_{p=n-\ell}^{n} (-1)^r R_{r,2n+\sigma,p} & \text{for} \quad x_{r,2n+\sigma,n-\ell} \leq t < x_{r,2n+\sigma,n-\ell-1} \\ & \qquad \ell = 0,1,\ldots,2n+\sigma-2 \\ \sum_{p=-n+1-\sigma}^{n} (-1)^r R_{r,2n+\sigma,p} & \text{for} \quad x_{r,2n+\sigma,-n+1-\sigma} \leq t < \infty. \end{cases}$$

$$(3.8)$$

Using Helly's Selection Principle [3, Theorem 2.2] the authors were able to prove Theorem 3.2. A more direct proof was suggested to us by Olav Njåstad. The latter proof is used to establish the following theorem.

THEOREM 3.2. *For each* $\sigma = 0,1$ *and* $0 < t < \infty$, *the sequence* $\{\psi_{r,2n+\sigma}(t)\}_{n=1}^{\infty}$ *converges to the step function* $\psi_r^{(\sigma)}(t)$ *defined by*

$$\psi_r^{(\sigma)}(t) := \begin{cases} 0 & \text{for} \quad t = 0 \\ \sum_{p=\ell+1}^{\infty} (-1)^r R_{r,p}^{(\sigma)} & \text{for} \quad x_{r,\ell+1}^{(\sigma)} \leq t < x_{r,\ell}^{(\sigma)} \\ & \qquad \ell = 0,\pm 1,\pm 2,\ldots. \end{cases} \quad (3.9)$$

Proof: We first show that for $\sigma = 0,1$,

$$\sum_{p=-n+1-\sigma}^{n} \frac{(-1)^r}{x_{r,2n+\sigma,p}} R_{r,2n+\sigma,p} = \int_0^\infty \frac{1}{t}\, d\psi_{r,2n+\sigma}(t) = (-1)^{r-1} c_{-1}^{(r)} = q^{\frac{1-r^2}{2}}. \quad (3.10)$$

From (1.9), (1.8) and (1.5) we see that

$$\frac{U_{r,2n+\sigma}(z)}{V_{r,2n+\sigma}(z)} = \frac{A_{r,2(2n+\sigma)}(-z)}{B_{r,2(2n+\sigma)}(-z)} = \sum_{m=1}^{2n+\sigma} -c_{-m}^{(r)}(-z)^m + z^{2n+\sigma+1} h_{r,2n+\sigma}(z) \quad (3.11)$$

where $h_{r,2n+\sigma}(z)$ is holomorphic at $z = 0$. It follows from (3.1a), and (3.8) that

$$
\frac{U_{r,2n+\sigma}(z)}{V_{r,2n+\sigma}(z)} = \sum_{p=-n+1-\sigma}^{n} \frac{zR_{r,2n+\sigma,p}}{z - x_{r,2n+\sigma,p}} = (-1)^r \int_0^\infty \frac{z}{z-t}\, d\psi_{r,2n+\sigma}(t)
$$

$$
= (-1)^r \int_0^\infty \left(-\frac{z}{t} - \frac{z^2}{t^2} - \frac{z^3}{t^3} - \cdots - \frac{z^{2n+\sigma}}{t^{2n+\sigma}} + \frac{z^{2n+\sigma+1}}{t^{2n+\sigma}(z-t)} \right) d\psi_{r,2n+\sigma}(t)
$$

$$
= (-1)^{r-1} \sum_{m=1}^{2n+\sigma} z^m \int_0^\infty \frac{1}{t^m}\, d\psi_{r,2n+\sigma}(t)
$$

$$
+ (-1)^r z^{2n+\sigma+1} \int_0^\infty \frac{1}{t^{2n+\sigma}} \frac{d\psi_{r,2n+\sigma}(t)}{(z-t)}. \tag{3.12}
$$

All integrals in (3.12) are proper Stieltjes integrals since $\psi_{r,2n+\sigma}(t) = 0$ for $0 \le t < x_{r,2n+\sigma,n}$ and $\psi_{r,2n+\sigma}(t) = q^{-\frac{r^2}{2}-r}$ for $x_{r,2n+\sigma,-n+1-\sigma} \le t < \infty$. Subtracting (3.12) from (3.11) yields

$$
0 = \left(c_{-1}^{(r)} + (-1)^r \int_0^\infty \frac{1}{t}\, d\psi_{r,2n+\sigma}(t) \right) z
$$

$$
+ \sum_{m=2}^{2n+\sigma} \left((-1)^{m+1} c_{-m}^{(r)} + (-1)^r \int_0^\infty \frac{1}{t^m}\, d\psi_{r,2n+\sigma}(t) \right) z^m \tag{3.13}
$$

$$
+ \left(h_{r,2n+\sigma}(z) + (-1)^{r-1} \int_0^\infty \frac{d\psi_{r,2n+\sigma}(t)}{t^{2n+\sigma}(z-t)} \right) z^{2n+1+\sigma}.
$$

Dividing both sides of (3.13) by z one obtains

$$
-c_{-1}^{(r)} = (-1)^r \int_0^\infty \frac{1}{t}\, d\psi_{r,2n+\sigma}(t)
$$

$$
+ \sum_{m=2}^{2n+\sigma} \left((-1)^{m+1} c_{-m}^{(r)} + (-1)^r \int_0^\infty \frac{1}{t^m}\, d\psi_{r,2n+\sigma}(t) \right) z^{m-1} \tag{3.14}
$$

$$
+ \left(h_{r,2n+\sigma}(z) + (-1)^{r-1} \int_0^\infty \frac{d\psi_{r,2n+\sigma}(t)}{t^{2n+\sigma}(z-t)} \right) z^{2n+\sigma}.
$$

Equation (3.10) follows by letting $z \to 0$ in (3.14) and using the definition of $c_{-1}^{(r)}$ found in (1.4). We now show that for fixed t, $0 \le t < \infty$ we have $\lim_{n\to\infty} \psi_{r,2n+\sigma}(t) = \psi_r^{(\sigma)}(t)$. Fix $t > x_{r,0}^{(\sigma)}$, $t \ne x_{r,\ell}^{(\sigma)}$, $\ell = -1,-2,\ldots$. Since $x_{r,p}^{(\sigma)} \to +\infty$ as $p \to -\infty$, there exists a positive integer $s_{r,t}^{(\sigma)}$ so that

$$
x_{r,-s_{r,t}^{(\sigma)}+1}^{(\sigma)} < t < x_{r,-s_{r,t}^{(\sigma)}}^{(\sigma)}. \tag{3.15}
$$

To simplify notation we let $s := s_{r,t}^{(\sigma)}$ and keep in mind that although s is fixed it depends on r, σ, and t. For each $k = 0, -1, -2, -3, \ldots$, the sequence $\{x_{r,2n+\sigma,k}\}_{n=|k|+1}^\infty$

decreases monotonically to $x_{r,k}^{(\sigma)}$. Consequently for each $k = 0, -1, -2, \ldots, -s+1$ there exists a positive integer $N_{r,t,k}^{(\sigma)}$ so that

$$x_{r,2n+\sigma,k} < t \quad \text{for all} \quad n \geq N_{r,t,k}^{(\sigma)}, \quad k = 0, -1, \ldots, -s+1. \tag{3.16}$$

Set $N_{r,t}^{(\sigma)} = \max\{s - \sigma, N_{r,t,k}^{(\sigma)}, k = 0, -1, \ldots, -s+1\}$. Then we have

$$x_{r,-s+1}^{(\sigma)} < x_{r,2n+\sigma,-s+1} < t < x_{r,-s}^{(\sigma)} < x_{r,2n+\sigma,-s} \quad \text{for all } n \geq N_{r,t}^{(\sigma)}. \tag{3.17}$$

It follows from (3.8) that

$$\psi_{r,2n+\sigma}(t) = \sum_{p=-s+1}^{n} (-1)^r R_{r,2n+\sigma,p} \quad \text{for all} \quad n \geq N_{r,t}^{(\sigma)}. \tag{3.18}$$

Now for $n \geq N_{r,t}^{(\sigma)}$ we have by (3.10)

$$\sum_{p=-s+1}^{n} \frac{(-1)^r R_{r,2n+\sigma,p}}{x_{r,2n+\sigma,p}} \leq \sum_{p=-n+1-\sigma}^{n} \frac{(-1)^r R_{r,2n+\sigma,p}}{x_{r,2n+\sigma,p}} = q^{\frac{1-r^2}{2}}. \tag{3.19}$$

From Theorem 2.8(A), (C), equation (2.19) and Theorem 2.13 we can conclude that for $p = 1, 2, 3, \ldots$,

$$x_{r,2n+\sigma,p} \leq x_{r,p}^{(\sigma)} = q^{2p-r-\frac{3}{2}+\sigma} \quad \text{for all} \quad n. \tag{3.20}$$

It follows from (3.19) and (3.20) that for $n \geq N_{r,t}^{(\sigma)}$ and $p = 1, 2, \ldots, n$

$$(-1)^r R_{r,2n+\sigma,p} \leq \left(\sum_{q=-s+1}^{n} \frac{(-1)^r R_{r,2n+\sigma,q}}{x_{r,2n+\sigma,q}} \right) x_{r,2n+\sigma,p} \leq q^{\frac{1-r^2}{2}} x_{r,2n+\sigma,p} \tag{3.21}$$

$$\leq q^{\frac{1-r^2}{2}} q^{2p-r-\frac{3}{2}+\sigma} = q^{-\frac{r^2}{2}-r-1+\sigma+2p}.$$

For each $n = 1, 2, \ldots$, we define

$$R_{r,2n+\sigma,p} := 0 \quad \text{for} \quad p = n+1, n+2, n+3, \ldots. \tag{3.22}$$

Since $0 < (-1)^r R_{r,2n+\sigma,p} \leq q^{-\frac{r^2}{2}-r-1+\sigma+2p}$ and

$$\sum_{p=1}^{\infty} q^{-\frac{r^2}{2}-r-1+\sigma+2p} = q^{-\frac{r^2}{2}-r-1+\sigma} \sum_{p=1}^{\infty} q^{2p} < \infty$$

we obtain by (3.22) and the Lebesgue Dominated Convergence Theorem [14]

$$\lim_{n \to \infty} \sum_{p=1}^{n} (-1)^r R_{r,2n+\sigma,p} = \lim_{n \to \infty} \sum_{p=1}^{\infty} (-1)^r R_{r,2n+\sigma,p} = \sum_{p=1}^{\infty} \lim_{n \to \infty} (-1)^r R_{r,2n+\sigma,p}$$

$$= \sum_{p=1}^{\infty} (-1)^r R_{r,p}^{(\sigma)}. \tag{3.23}$$

From (3.18), (3.22) and (3.23) one gets

$$\lim_{n \to \infty} \psi_{r,2n+\sigma}(t) = \lim_{n \to \infty} \sum_{p=-s+1}^{n} (-1)^r R_{r,2n+\sigma,p}$$

$$= \lim_{n \to \infty} \sum_{p=-s+1}^{0} (-1)^r R_{r,2n+\sigma,p} + \lim_{n \to \infty} \sum_{p=1}^{\infty} (-1)^r R_{r,2n+\sigma,p}$$

$$= \sum_{p=-s+1}^{0} (-1)^r R_{r,p}^{(\sigma)} + \sum_{p=1}^{\infty} (-1)^r R_{r,p}^{(\sigma)} = \sum_{p=-s+1}^{\infty} (-1)^r R_{r,p}^{(\sigma)} = \psi_r^{(\sigma)}(t).$$

Similar arguments show that $\lim_{n \to \infty} \psi_{r,2n+\sigma}(t) = \psi_r^{(\sigma)}(t)$ for $t < x_{r,0}^{(\sigma)}$ and $t = x_{r,\ell}^{(\sigma)}$, $\ell = 0, \pm 1, \pm 2, \ldots$. □

THEOREM 3.3.

(A) *For $\sigma = 0, 1$ we have*

$$\frac{U_r^{(\sigma)}(z)}{V_r^{(\sigma)}(z)} = \lim_{n \to \infty} \frac{U_{r,2n+\sigma}(z)}{V_{r,2n+\sigma}(z)} = \int_0^\infty \frac{(-1)^r z \, d\psi_r^{(\sigma)}(t)}{z - t} = \sum_{p=-\infty}^{\infty} \frac{z R_{r,p}^{(\sigma)}}{z - x_{r,p}^{(\sigma)}} \qquad (3.24)$$

where the series on the right converges uniformly on compact subsets of $\mathbb{C} - \{x_{r,p}^{(\sigma)}\}_{p=-\infty}^{\infty} - \{0\}$.

(B) *For $\sigma = 0, 1$,*

$$c_k^{(r)} = \int_0^\infty (-t)^k \, d\psi_r^{(\sigma)}(t), \qquad k = 0, \pm 1, \pm 2, \ldots. \qquad (3.25)$$

Proof (A):

$$\frac{U_r^{(\sigma)}(z)}{V_r^{(\sigma)}(z)} = \lim_{n \to \infty} \frac{U_{r,2n+\sigma}(z)}{V_{r,2n+\sigma}(z)}$$

since $\{U_{r,2n+\sigma}(z)\}_{n=1}^{\infty}$ and $\{V_{r,2n+\sigma}(z)\}$ converge uniformly on compact subsets of $\mathbb{C} - \{0\}$ to $U_r^{(\sigma)}(z)$ and $V_r^{(\sigma)}(z)$ respectively. It follows from (3.1a) and (3.8) that

$$\frac{U_{r,2n+\sigma}(z)}{V_{r,2n+\sigma}(z)} = \sum_{p=-n+1-\sigma}^{n} \frac{z R_{r,2n+\sigma,p}}{z - x_{r,2n+\sigma,p}} = \int_0^\infty \frac{(-1)^r z}{z - t} \, d\psi_{r,2n+\sigma}(t).$$

Fix $z \in \mathbb{C} - [0, \infty)$. Clearly $g(t) := \dfrac{z}{z - t}$ is a continuous complex-valued function of the real variable t on $(0, \infty)$ such that $\lim_{t \to +\infty} g(t) = 0$. Thus by an extension of Helly's second theorem (see for example [12, p. 240, Theorem 6])

$$\lim_{n \to \infty} \int_0^\infty \frac{(-1)^r z}{z - t} \, d\psi_{r,2n+\sigma}(t) = \int_0^\infty \lim_{n \to \infty} \frac{(-1)^r z}{z - t} \, d\psi_{r,2n+\sigma}(t) = \int_0^\infty \frac{(-1)^r z}{z - t} \, d\psi_r^{(\sigma)}(t).$$

Hence $\lim\limits_{n\to\infty}\dfrac{U_{r,2n+\sigma}(z)}{V_{r,2n+\sigma}(z)} = \displaystyle\int_0^\infty \dfrac{(-1)^r z}{z-t}\, d\psi_r^{(\sigma)}(t)$. By the definition of $\psi_r^{(\sigma)}(t)$ in (3.9) we have

$$\int_0^\infty \frac{(-1)^r z}{z-t}\, d\psi_r^{(\sigma)}(t) = \sum_{p=-\infty}^{\infty} \frac{zR_{r,p}^{(\sigma)}}{z - x_{r,p}^{(\sigma)}}.$$

The series converges uniformly on compact subsets of $\mathbb{C} - \{x_{r,p}^{(\sigma)}\}_{p=-\infty}^{\infty} - \{0\}$ since $|z - x_{r,p}^{(\sigma)}|$ is bounded away from zero and since from (3.6)

$$\sum_{p=-\infty}^{\infty} |R_{r,p}^{(\sigma)}| \le q^{-\frac{r^2}{2}-r}$$

(B): From equation (3.24) and Theorem 4.1 of [11] we see that

$$c_{k+r} = \int_0^\infty (-t)^k (-1)^r d\psi_r^{(\sigma)}(t).$$

Thus

$$c_k^{(r)} := (-1)^r c_{k+r} = \int_0^\infty (-t)^k d\psi_r^{(\sigma)}(t). \quad \square$$

In the following theorem we show that we actually have equality in (3.6).

THEOREM 3.4. *For $\sigma = 0,1$ and $p = 0,\pm1,\pm2,\ldots$, let $R_{r,p}^{(\sigma)}$ be defined as in (3.4). Then*

$$\sum_{p=-\infty}^{\infty} (-1)^r R_{r,p}^{(\sigma)} = q^{-\frac{r^2}{2}-r}. \tag{3.26}$$

Proof: From (3.9) and (3.25) we have

$$\sum_{p=-\infty}^{\infty} \frac{(-1)^r R_{r,p}^{(\sigma)}}{1 - \frac{x_{r,p}^{(\sigma)}}{z}} = \sum_{p=-\infty}^{\infty} \frac{(-1)^r z R_{r,p}^{(\sigma)}}{z - x_{r,p}^{(\sigma)}} = \int_0^\infty \frac{z}{z-t}\, d\psi_r^{(\sigma)}(t)$$

$$= \int_0^\infty \left(1 + \frac{t}{z-t}\right) d\psi_r^{(\sigma)}(t)$$

$$= \int_0^\infty d\psi_r^{(\sigma)}(t) + \frac{1}{z}\int_0^\infty \frac{t}{1 - \frac{t}{z}}\, d\psi_r^{(\sigma)}(t)$$

$$= c_0^{(r)} + \frac{1}{z}\int_0^\infty \frac{t}{1 - \frac{t}{z}}\, d\psi_r^{(\sigma)}(t)$$

where the series converges uniformly on compact subsets of $\mathbb{C} - \{x_{r,p}^{(\sigma)}\}_{p=-\infty}^{\infty} - \{0\}$. The result follows by letting $z \to \infty$ with $|\arg z| > 0$ and using (1.4) and (1.2) to find $c_0^{(r)} = (-1)^r c_r = q^{-\frac{r^2}{2}-r}$. $\quad\square$

REFERENCES

1. R. Askey, Ramanujan's Extensions of the Gamma and Beta Functions, *Amer. Math. Monthly* **87** (May 1980), 335–426.

2. R. Askey, Beta integrals and q-extensions, in *Proceedings, Ramanujan Centennial International Conference*, Anuamalainagar, December 15–18, 1987.

3. T. S. Chihara, *Introduction to Orthogonal Polynomials*, Gordon, New York, 1978.

4. S. Cooper, W. B. Jones, and W. J. Thron, Orthogonal Laurent polynomials and continued fractions associated with log-normal distributions, *J. Comput. Appl. Math.* **32** (1990), 39–46.

5. S. Cooper, W. B. Jones, and W. J. Thron, Asymptotics of orthogonal Laurent polynomials for log-normal distributions, *Constr. Approx.* **8** (1992), 59–67.

6. E. L. Crow and K. Shimizu (Eds.), *Log-normal Distributions, Theory and Applications*, Dekker, New York, 1988.

7. W. B. Jones, A. Magnus, and W. J. Thron, PC-fractions and orthogonal Laurent polynomials for log-normal distributions, *J. Math. Anal. Appl.* **170** (1992), 225–244.

8. W. B. Jones, O. Njåstad, and W. J. Thron, Orthogonal Laurent polynomials and the strong Hamburger moment problem, *J. Math. Anal. Appl.* **98** (1984), 528–554.

9. W. B. Jones, O. Njåstad, and W. J. Thron, Continued fractions associated with the trigonometric and other strong moment problems, *Constr. Approx.* **2** (1986), 197–211.

10. W. B. Jones and W. J. Thron, *Continued Fractions: Analytic theory and applications*, in Encyclopedia of Mathematics and Its Applications, Vol. 11, Addison–Wesley, Reading, Mass., 1980. Distributed now by the Cambridge University Press.

11. W. B. Jones, W. J. Thron and H. Waadeland, A strong Stieltjes moment problem, *Trans. Amer. Math. Soc.* **261** (1980), 503–528.

12. I. P. Natanson, *Theory of Functions of a Real Variable*. I, Ungar, New York, 1955.

13. P. I. Pastro, Orthogonal polynomials and some q-beta integrals of Ramanujan, *J. Math. Anal. Appl.* **112** (1985), 517–540.

14. H. L. Royden, *Real Analysis*, Macmillen, New York, 1988.

15. T. J. Stieltjes, Recherches sur les fractions continues, *Ann. Fac. Sci. Toulouse* **8** (1894), J, 1–122; **9** (1894), A, 1–47; *Oeuvres*, Vol. 2, pp. 402–566. Also published in *Mémoires présentés par divers savants à l'Académie de sciences de l'Institut National de France*, Vol. 33, pp. 1–196.

16. W. J. Thron, *Introduction to the Theory of Functions of a Complex Variable*, Wiley, New York, 1953.

17. S. Wigert, Sur les polynomes orthogonaux et l'approximation des fractions continues, *Ark. Mat. Astonom. Fys.* **17**, 1923.

2

Recurrence Relations for Orthogonal Functions

A. Bultheel, Department of Computer Science, K.U. Leuven, B-3001 Leuven, Belgium.

P. Gonzalez-Vera, Facultad de Mathemáticas, Universidad de La Laguna, La Laguna, Tenerife, Canary Islands, Spain.

E. Hendriksen, University of Amsterdam, Department of Mathematics, Plantage Muidergracht 24, 1018 TV Amsterdam, The Netherlands.

O. Njåstad, Department of Mathematical Sciences, University of Trondheim-NTH, N-7034 Trondheim, Norway.

Dedicated to the memory of Arne Magnus

1 INTRODUCTION

In this paper we shall use the following notations: $D = \{z \in C : |z| < 1\}$, $E = \{z \in C : |z| > 1\}$, $T = \{z \in C : |z| = 1\}$, $H_+ = \{z \in C : Im\, z > 0\}$, $H_- = \{z \in C : Im\, z < 0\}$, $R = \{z \in C : Im\, z = 0\}$.

The Cayley transform τ shall here be defined by the formulas

$$Z = \tau(z) = i\frac{1+z}{1-z}, z = \tau^{-1}(Z) = \frac{Z-i}{Z+i}, \quad z \in D, \ Z \in H_+ \tag{1.1}$$

The transform $z \to Z$ maps D onto H_+, E onto H_- and T onto R.

The substar conjugate f_* of a function f is defined as

$$f_*(z) = \overline{f(1/\bar{z})}. \tag{1.2}$$

When f is a rational function, this may be written as

$$f_*(z) = \overline{f}(1/z), \tag{1.3}$$

where the bar denotes complex conjugates of the coefficients. Similarly the subtilde conjugate $f_\sim$ is defined by

$$f_\sim(z) = \overline{f(\bar{z})}. \tag{1.4}$$

When f is a rational function, this may be written as

$$f_\sim(z) = \overline{f}(z). \tag{1.5}$$

The setting of this paper will be certain spaces of rational functions with a functional M defined on them. The functional M gives rise to an inner product $<,>$ (not necessarily definite) through the formula

$$<f, g> = M(f(z) \cdot g_*(z)) \tag{1.6}$$

or

$$<f, g> = M(f(z) \cdot g_\sim(z)). \tag{1.7}$$

When $<f, f> \neq 0$ for all $f \neq 0$ in the space under consideration, we shall call M and $<,>$ quasi-definite, and when $<f, f> >> 0$ for these f we call M and $<,>$ positive definite. When $<,>$ is defined by (1.6) we shall talk about the circle-disk situation, and when $<,>$ is defined by (1.7) we shall talk about the line-plane situation.

When μ is a finite (positive) Borel measure with infinite support on $\boldsymbol{T}$ or $\boldsymbol{R}$ (with suitable restrictions), positive definite functionals M (an thereby positive definite inner products) are defined through

$$M(f) = \int_{-\pi}^{\pi} f(e^{i\theta}) d\mu(\theta) \tag{1.8}$$

in the circle-disk situation,

$$M(f) = \int_{-\infty}^{\infty} f(t) d\mu(t) \tag{1.9}$$

in the line-plane situation.

The spaces of rational functions that will be treated arise in connection with Nevanlinna-Pick interpolation theory. The aim of this paper is to study certain properties of orthogonal

bases for these spaces, more specifically recurrence relations connecting elements of these bases.

The theory of orthogonal sequences in these spaces (for the case that all interpolation points are in D) was initiated by Djrbashian in 1969 (see [14]), and independently by Bultheel, Dewilde and Dym (see [3,5,13]). For a general introduction to recent work on this theory, see [6]. Other papers discussing these and related spaces are [9,10,26,30]. For earlier work on recurrence relations in these spaces, see [6,7,8,10,16,17,28,29]. See also [19].

When all interpolation points coalesce at one point, the situation is essentially a polynomial situation, and when the interpolation points consist of two points cyclically repeated, the situation is to a large extent a Laurent polynomial situation. In particular when the one point is the origin in the circle-disk situation, orthogonal polynomials on T are obtained, see [15,22,23,35], while when the one point is the point at infinity in the line-plane situation, orthogonal polynomials on R are obtained (see [1,2,11]). When the two points are the origin and the point at infinity in the line-plane situation, cyclically repeated, orthogonal Laurent polynomials on R are obtained (see [18,20,21,25,33]).

2   INTERPOLATION IN THE UNIT DISK

Let $\{\alpha_n : n = 1, 2, ...\}$ be an arbitrary sequence of (not necessarily distinct) points (interpolation points) in $D \cup E$. We shall always assume that $\alpha_j \neq 1/\overline{\alpha_k}$ for $j, k = 1, 2, ...$ Note that $1/\overline{\alpha_j} \in E$ if and only if $\alpha_j \in D$, hence $\alpha_j \neq 1/\overline{\alpha_k}$ is always satisfied if $\alpha_n \in D$ for all n or $\alpha_n \in E$ for all n.

We define the Blaschke factor ζ_n as the function

$$\zeta_n(z) = \frac{\overline{\alpha_n}(\alpha_n - z)}{|\alpha_n|(1 - \overline{\alpha_n}z)}, \quad n = 1, 2, ... \tag{2.1}$$

(By convention we set $\frac{\overline{\alpha_n}}{|\alpha_n|} = -1$ if $\alpha_n = 0$.)

The Blaschke products B_n are defined by

$$B_0(z) = 1, \quad B_n(z) = \prod_{k=1}^{n} \zeta_k(z), \quad n = 1, 2, ... \tag{2.2}$$

We define the spaces $\mathcal{L}_n$, $\mathcal{L}$ by

$$\mathcal{L}_n = Span\{B_k : k = 0, 1, ..., n\}, \quad \mathcal{L} = \cup_{n=0}^{\infty} \mathcal{L}_n. \tag{2.3}$$

The functions in $\mathcal{L}_n$ are exactly the functions that may be written in the form

$$L(z) = \frac{p_n(z)}{\pi_n(z)}, \tag{2.4}$$

where

$$\pi_n(z) = \prod_{k=1}^{n} (1 - \overline{\alpha_k} z), \quad n = 1, 2, ... \tag{2.5}$$

and $p_n \in \Pi_n$ (the space of polynomials of degree at most n). This follows by partial fraction decomposition. In particular the situation reduces to the polynomial case $\mathcal{L}_n = \Pi_n$ when $\alpha_n = 0$ for all n.

We set $\mathcal{L}_{n*} = \{\mathcal{L} :\in \mathcal{L}_n\}$, $\mathcal{L}_* = \cup_{n=0}^{\infty} \mathcal{L}_{n*}$.

For $f \in \mathcal{L}_n$ we define its superstar conjugate f^* by

$$f^*(z) = B_n(z) f_*(z). \tag{2.6}$$

Note that this transformation depends on n. It must be clear from the context what n is. Also note that $f^* \in \mathcal{L}_n$ when $f \in \mathcal{L}_n$.

Let M be a quasi-definite functional on $\mathcal{L} + \mathcal{L}_*$, and let $<,>$ be defined by (1.6). Let the sequence $\{\Phi_n : n = 0, 1, 2, ...\}$ be obtained by orthogonalization of the sequence $\{B_n\}$ with respect to $<,>$. We shall assume that all Φ_n are monic, which means that the leading coefficient $b_n = b_n^{(n)}$ in the decomposition

$$\Phi_n(z) = \sum_{k=0}^{n} b_k^{(n)} B_k(z) \tag{2.7}$$

equals one, i.e. $b_n^{(n)} = 1$.

The following orthogonality properties are valid:

$$< f, \Phi_n >= 0 \quad \text{for} \quad f \in \mathcal{L}_{n-1} \tag{2.8}$$

$$< g, \Phi_n^* >= 0 \quad \text{for} \quad g \in \zeta_n \mathcal{L}_{n-1}. \tag{2.9}$$

Recall that we may write

$$\Phi_n(z) = \frac{P_n(z)}{\pi_n(z)}, \quad P_n \in \Pi_n. \tag{2.10}$$

By substituting α_n for z in the expression for $\Phi_n^*(z)$ obtained from (2.4), we easily verify that

$$\Phi_n^*(\alpha_n) = 1. \tag{2.11}$$

We shall call the index n and the function Φ_n **degenerate** if $\Phi_n^*(\alpha_{n-1}) = 0$, **non-degenerate** otherwise. We shall call the index n and the function Φ_n **exceptional** if $\Phi_n(\alpha_{n-1}) = 0$, **non-exceptional** otherwise. Note that in the polynomial case ($\alpha_n = 0$ for all n), $\Phi_n^*(\alpha_{n-1}) = \Phi_n^*(\alpha_n) = \Phi_n^*(0) \neq 0$, so that the degenerate case can not occur.

The following Christoffel-Darboux formula can be shown to be valid (see [6], where only the case $\alpha_n \in D$ for all n is explicitly treated) when the inner product $<,>$ is positive definite (here $\kappa_m = < \Phi_m, \Phi_m >^{-\frac{1}{2}}$):

$$\kappa_n^2 [\Phi_n^*(z)\overline{\Phi_n^*(w)} - \Phi_n(z)\overline{\Phi_n(w)}] = [1 - \zeta_n(z)\overline{\zeta_n(w)}] \sum_{k=0}^{n-1} \kappa_k^2 \Phi_k(z)\overline{\Phi_k(w)}. \tag{2.12}$$

Proposition 2.1 *Assume that the functional M is positive definite, and that $\alpha_n \in D$ for all n or $\alpha_n \in E$ for all n. Then all Φ_n are non-degenerate.*

Proof:

By setting $w = z$ in (2.12) we obtain

$$[1 - |\zeta_n(z)|^2]^{-1}[|\Phi_n^*(z)|^2 - |\Phi_n(z)|^2] > 0. \tag{2.13}$$

From this it easily follows that $\Phi_n^*(z) \neq 0$ for $z \in D$ if $\alpha_n \in D$, $\Phi_n^*(z) \neq 0$ for $z \in E$ if $\alpha_n \in E$. Thus $\Phi_n^*(\alpha_{n-1}) \neq 0$ for all n if $\alpha_n \in D$ for all n or $\alpha_n \in E$ for all n. $\quad\square$

Cf. the discussion above of the special case $\alpha_n = 0$ for all n, i.e. the polynomial situation.

3   RECURRENCE IN THE DISK SITUATION

We state a version of the fundamental result on Szegő type recurrence for the orthogonal rational functions. We shall write τ_n for $\frac{\alpha_n}{|\alpha_n|}$.

Theorem 3.1 *The sequence $\{\Phi_n\}$ satisfies the following recurrence relation:*

$$\Phi_n(z) = \varepsilon_n \frac{z - \alpha_{n-1}}{1 - \overline{\alpha_n}z} \Phi_{n-1}(z) + \delta_n \frac{1 - \overline{\alpha_{n-1}}z}{1 - \overline{\alpha_n}z} \Phi_{n-1}^*(z), \quad n = 1, 2, \ldots \tag{3.1}$$

$$\Phi_n^*(z) = -\tau_n \overline{\delta_n} \frac{z - \alpha_{n-1}}{1 - \overline{\alpha_n}z} \Phi_{n-1}(z) - \tau_n \overline{\varepsilon_n} \frac{1 - \overline{\alpha_{n-1}}z}{1 - \overline{\alpha_n}z} \Phi_{n-1}^*(z), \quad n = 1, 2, \ldots \tag{3.2}$$

$$\alpha_0 = 0, \quad \Phi_0 = 1, \quad \Phi_0^* = 1. \tag{3.3}$$

The coefficients δ_n, ε_n are given by

$$\delta_n = \frac{(1 - \alpha_{n-1}\overline{\alpha_n})\Phi_n(\alpha_{n-1})}{(1 - |\alpha_{n-1}|^2)} \tag{3.4}$$

$$\varepsilon_n = -\tau_n \frac{(1 - \overline{\alpha_{n-1}}\alpha_n)\overline{\Phi_n^*(\alpha_{n-1})}}{(1 - |\alpha_{n-1}|^2)}. \tag{3.5}$$

Proof:

The results follows from [7, Theorem 4.1].   □

Corrollary 3.2 *A function Φ_n (or an index n) can not be both degenerate and exceptional at the same time.*

Proof:

Follows from (3.1)-(3.5) since $\Phi_n(z) \not\equiv 0$.   □

We note that in the polynomial case we have $\Phi_n^*(0) = 1$ for all n, hence $\delta_n = \Phi_n(0)$, $\varepsilon_n = 1$ (recall that $\frac{\overline{\alpha_n}}{(\alpha_n)} = -1$ in this case), and (3.1)-(3.2) reduces to the classical Szegő formulas

$$\Phi_n(z) = z\Phi_{n-1}(z) + \Phi_n(0)\Phi_{n-1}^*(z) \tag{3.6}$$

$$\Phi_n^*(z) = \overline{\Phi_n(0)}z\Phi_{n-1}(z) + \Phi_{n-1}^*(z). \tag{3.7}$$

Since in this case $\varepsilon_n = 1$, Φ_n can never be degenerate.

Proposition 3.3 *If the functional M is positive definite, and all $\alpha_n \in D$ or all $\alpha_n \in E$, then $|\delta_n| < |\varepsilon_n|$ for all n.*

Proof:

We recall that in this situation $\varepsilon_n \neq 0$ (see Proposition 2). From (3.4)-(3.5) we get

$$\frac{\delta_n}{\varepsilon_n} = --\overline{\tau_n}\frac{\Phi_n(\alpha_{n-1})}{\Phi_n^*(\alpha_{n-1})}\frac{1-\alpha_{n-1}\overline{\alpha_n}}{1-\overline{\alpha_{n-1}}\alpha_n}. \tag{3.8}$$

The proof of Proposition 2.1 shows that $|\Phi_n^*(\alpha_m)| > |\Phi_n(\alpha_m)|$ for all m, n when all $\alpha_m \in D$ and when all $\alpha_m \in E$. It follows from (3.8) that $|\frac{\delta_n}{\varepsilon_n}| < 1$, hence $|\delta_n| < |\varepsilon_n|$. $\qquad\square$

Theorem 3.4 *Assume that Φ_n is non-degenerate. Then the following recurrence relation is satisfied:*

$$\Phi_n(z) = -\overline{\tau_n}\frac{\delta_n}{\overline{\varepsilon_n}}\Phi_n^*(z) + \frac{1}{\overline{\varepsilon_n}}(|\varepsilon_n|^2 - |\delta_n|^2)\frac{z-\alpha_{n-1}}{1-\overline{\alpha_n}z}\Phi_{n-1}(z), \quad n = 1, 2, ... \tag{3.9}$$

In particular, this is the case when the functional M is positive definite and all $\alpha_n \in D$ or all $\alpha_n \in E$.

Proof:

This formula is obtained by substitution for $\Phi_{n-1}^*(z)$ from (3.2) into (3.1), ε_n being different from zero. $\qquad\square$

We note that in the polynomial case (3.9) reduces to the well-known formula (see e.g. [22])

$$\Phi_n(z) = \delta_n\Phi_n^*(z) + (1 - |\delta_n|^2)z\Phi_{n-1}(z). \tag{3.10}$$

We shall call the sequence $\{\Phi_n\}$ **non-degenerate** if all Φ_n are non-degenerate. We observe that if $\{\Phi_n\}$ is non-degenerate then (3.2) and (3.9) together define a three-term recurrence relation in the sequence $\{\Phi_0, \Phi_1^*, \Phi_1, \Phi_2^*, \Phi_2, ..., \Phi_n^*, \Phi_n, ...\}$, and thus the elements of this sequence are denominators of a continued fraction. (For basic information on continued fractions, see [24].) We state this property formally in a theorem.

Theorem 3.5 *Let $\{\Phi_n\}$ be a non-degenerate sequence, and $|\delta_n| \neq |\varepsilon_n|$ for all n. Define*

$$Q_{2m}(z) = \Phi_m^*(z), \quad m = 1, 2, ... \tag{3.11}$$

$$Q_{2m+1}(z) = \Phi_m(z), \quad m = 0, 1, 2, ... \tag{3.12}$$

Then the terms of the sequence $\{Q_n\}$ form the denominators of a continued fraction $K_{n=1}^{\infty} \dfrac{a_n(z)}{b_n(z)}$, where the elements $a_n(z)$, $b_n(z)$ are given by

$$a_{2m}(z) = -\tau_m \bar{\varepsilon}_m \frac{1 - \overline{\alpha_{m-1}} z}{1 - \overline{\alpha_m} z}, \quad a_{2m+1}(z) = \frac{1}{\bar{\varepsilon}_m}(|\varepsilon_m|^2 - |\delta_m|^2)\frac{z - \alpha_{m-1}}{1 - \overline{\alpha_m} z}, \quad m = 1, 2, ... \tag{3.13}$$

$$b_{2m}(z) = -\tau_m \bar{\delta}_m \frac{z - \alpha_{m-1}}{1 - \overline{\alpha_m} z}, \quad b_{2m+1}(z) = -\bar{\tau}_m \frac{\delta_m}{\bar{\varepsilon}_m}, \quad m = 1, 2, ... \tag{3.14}$$

The recurrence relation satisfied by the denominators $\{Q_n\}$ are thus

$$Q_{2m}(z) = -\tau_m \bar{\delta}_m \frac{z - \alpha_{m-1}}{1 - \overline{\alpha_m} z} Q_{2m-1}(z) - \tau_m \bar{\varepsilon}_m \frac{1 - \overline{\alpha_{m-1}} z}{1 - \overline{\alpha_m} z} Q_{2m-2}(z), , \quad m = 1, 2, ... \tag{3.15}$$

$$Q_{2m+1}(z) = -\bar{\tau}_m \frac{\delta_m}{\bar{\varepsilon}_m} Q_{2m}(z) + \frac{1}{\bar{\varepsilon}_m}(|\varepsilon_m|^2 - |\delta_m|^2)\frac{z - \alpha_{m-1}}{1 - \overline{\alpha_m} z} Q_{2m-1}(z), \quad m = 1, 2, ... \tag{3.16}$$

$$\alpha_0 = 0, \quad Q_0 = 1, \quad Q_1 = 1. \tag{3.17}$$

In particular the above results hold when the functional M is positive definite and all $\alpha_n \in D$ or all $\alpha_n \in E$.

In the polynomial case (all $\alpha_n = 0$) the continued fractions above reduce to the well-known PC-fractions (Perron-Carathéodory fractions), see [22,23]. The PC-fractions are closely related to the Carathéodory Coefficient Problem. The continued fractions described in Theorem 3.5 shall be called NP-fractions (Nevanlinna-Pick fractions). These continued fractions are related to the Nevanlinna-Pick Interpolation Problem (cf. [1,27,34]) in a way that directly generalizes the relationship between PC-fractions and the Carathéodory Coefficient Problem. For connections between the Nevanlinna-Pick Problem and general system theory, see e.g. [12], and references found there.

We shall call the sequence $\{\Phi_n\}$ **non-exceptional** if all Φ_n are non-exceptional.

Theorem 3.6 *Assume that Φ_{n-1} is non-exceptional. Then the following recurrence relation is valid:*

$$\Phi_n(z) = \frac{1}{1 - \overline{\alpha_n} z}[\varepsilon_n(z - \alpha_{n-1}) - \tau_{n-1}\overline{\varepsilon_{n-1}}\frac{\delta_n}{\delta_{n-1}}(1 - \overline{\alpha_{n-1}}z)]\Phi_{n-1}(z) \tag{3.18}$$

$$+ \tau_{n-1}\frac{\delta_n}{\delta_{n-1}}\frac{(z - \alpha_{n-2})}{(1 - \overline{\alpha_n}z)}[|\varepsilon_{n-1}|^2 - |\delta_{n-1}|^2]\Phi_{n-2}(z) \quad n = 2, 3, \dots$$

$$\Phi_n^*(z) = \frac{(-1)}{1 - \overline{\alpha_n}z}\tau_n[\overline{\varepsilon_n}(1 - \overline{\alpha_{n-1}}z) - \overline{\tau_{n-1}}\frac{\varepsilon_{n-1}\overline{\delta_n}}{\delta_{n-1}}(z - \alpha_{n-1})]\Phi_{n-1}^*(z) \tag{3.19}$$

$$+ \tau_n\frac{\overline{\delta_n}}{\delta_{n-1}}[|\varepsilon_{n-1}|^2 - |\delta_{n-1}|^2]\frac{(z - \alpha_{n-1})(1 - \overline{\alpha_{n-2}}z)}{(1 - \overline{\alpha_n}z)(1 - \overline{\alpha_{n-1}}z)}\Phi_{n-2}^*(z) \quad n = 2, 3, \dots$$

$$\Phi_0 = 1, \quad \Phi_0^* = 1, \quad \Phi_1(z) = \frac{\delta_1 + \varepsilon_1 z}{1 - \overline{\alpha_1} z_1}, \quad \Phi_1^*(z) = \frac{\overline{\varepsilon_1} + \overline{\delta_1}z}{1 - \overline{\alpha_1}z}. \tag{3.20}$$

Proof:

Immediately from (3.1)-(3.2) with n replaced by $n - 1$ we obtain

$$\Phi_{n-2}^*(z) = \frac{1}{\delta_{n-1}} \cdot \frac{1 - \overline{\alpha_{n-1}}z}{1 - \overline{\alpha_{n-2}}z}\Phi_{n-1}(z) - \frac{\varepsilon_{n-1}}{\delta_{n-1}}\frac{z - \alpha_{n-2}}{1 - \overline{\alpha_{n-2}}z}\Phi_{n-2}(z) \tag{3.21}$$

$$\Phi_{n-2}(z) = \frac{-\tau_{n-1}}{\overline{\delta_{n-1}}} \cdot \frac{1 - \overline{\alpha_{n-1}}z}{z - \alpha_{n-2}}\Phi_{n-1}^*(z) - \frac{\overline{\varepsilon_{n-1}}}{\overline{\delta_{n-1}}} \cdot \frac{1 - \overline{\alpha_{n-2}}z}{z - \alpha_{n-2}}\Phi_{n-2}^*(z). \tag{3.22}$$

By substituting from (3.21) for Φ_{n-1}^* and Φ_{n-2}^* in (3.2) with n replaced by $n - 1$, we obtain (3.18). Similarly by substituting from (3.22) for Φ_{n-1} and Φ_{n-2} in (3.1) with n replaced by $n - 1$, we obtain (3.19). $\qquad\square$

Theorem 3.7 *Let $\{\Phi_n\}$ be a non-exceptional sequence and let $|\delta_n| \neq |\varepsilon_n|$ for all n. Then the terms of the sequence $\{\Phi_n\}$ form the denominators of a continued fraction $K_{n=1}^{\infty}\frac{a_n(z)}{b_n(z)}$, where the elements $a_n(z)$, $b_n(z)$ are given by*

$$a_n(z) = \tau_{n-1}\frac{\delta_n}{\delta_{n-1}}\frac{z - \alpha_{n-2}}{1 - \overline{\alpha_n}z}(|\varepsilon_{n-1}|^2 - |\delta_{n-1}|^2) \tag{3.23}$$

$$b_n(z) = \frac{1}{1 - \overline{\alpha_n}z}[\varepsilon_n(z - \alpha_{n-1}) - \tau_{n-1}\overline{\varepsilon_{n-1}}\frac{\delta_n}{\delta_{n-1}}(1 - \overline{\alpha_{n-1}}z)], \quad n = 2, 3, \dots, \quad (3.24)$$

and the terms of the sequence $\{\Phi_n^*\}$ form the denominators of a continued fraction $K_{n=1}^{\infty}\frac{c_n(z)}{d_n(z)}$, where the elements $c_n(z)$, $d_n(z)$ are given by

$$c_n(z) = \tau_n \frac{\overline{\delta_n}}{\overline{\delta_{n-1}}}[|\varepsilon_{n-1}|^2 - |\delta_{n-1}|^2]\frac{(z - \alpha_{n-1})(1 - \overline{\alpha_{n-2}}z)}{(1 - \overline{\alpha_n}z)(1 - \overline{\alpha_{n-1}}z)}, \quad (3.25)$$

$$d_n(z) = \frac{-\tau_n}{1 - \overline{\alpha_n}z}[\overline{\varepsilon_n}(1 - \overline{\alpha_{n-1}}z) - \tau_{n-1}\frac{\varepsilon_{n-1}\overline{\delta_n}}{\delta_{n-1}}(z - \alpha_{n-1})], \quad n = 2, 3, \dots . \quad (3.26)$$

The recurrence relations have the form (3.18)-(3.20).

In particular the above results hold when $\{\Phi_n\}$ is non-exceptional and the functional M is positive definite.

Proof:

Immediate from Theorem 3.6 and Proposition 3.3.   □

In the polynomial case ($\alpha_n = 0$ for all n), the formulas (3.18)-(3.19) reduce to the formulas

$$\Phi_n(z) = (z + \frac{\delta_n}{\delta_{n-1}})\Phi_{n-1}(z) - (1 - |\delta_{n-1}|^2)\frac{\delta_n}{\delta_{n-1}}z\Phi_{n-2}(z) \quad (3.27)$$

$$\Phi_n^*(z) = (1 + \frac{\overline{\delta_n}}{\overline{\delta_{n-1}}}z)\Phi_{n-1}^*(z) - \frac{\overline{\delta_n}}{\overline{\delta_{n-1}}}(1 - |\delta_{n-1}1|^2)z\Phi_{n-2}^*(z). \quad (3.28)$$

These are M-fractions and general T-fractions, respectively. (See [24].) Continued fractions of the form (3.18) or (3.19) are instances of MP-fractions (Multipoint Padé continued fractions), see Section 5 (cf. [17,31]).

For a short treatment of continued fractions obtained from the basic recurrence formulas (3.1)-(3.2), see also [6], with reference to [4].

4 INTERPOLATION ON THE UNIT CIRCLE

Let $\{\alpha_n : n = 1, 2, ...\}$ be an arbitrary sequence of (not necessarily distinct) points on T. Note that $1/\overline{\alpha_n} = \alpha_n$. We introduce functions ω_n by

$$\omega_0 = 1, \quad \omega_n(z) = \prod_{k=1}^{n}(z - \alpha_k), \quad n = 1, 2, ..., \tag{4.1}$$

and define the spaces $\mathcal{L}_n$ and $\mathcal{L}$ by

$$\mathcal{L}_n = Span\{\frac{1}{\omega_0}, \frac{1}{\omega_1}, ..., \frac{1}{\omega_n}\}, \quad \mathcal{L} = \cup_{n=0}^{\infty}\mathcal{L}_n. \tag{4.2}$$

We may also write

$$\mathcal{L}_n = Span\{\frac{z^{m_0}}{\omega_0}, \frac{z^{m_1}}{\omega_1(z)}, ..., \frac{z^{m_n}}{\omega_n(z)}\}, \quad 0 \le m_k \le k \tag{4.3}$$

and

$$\mathcal{L}_n = Span\{\sigma_0, \sigma_1, ..., \sigma_n\}, \tag{4.4}$$

where

$$\sigma_0 = 1, \quad \sigma_n = \frac{i^n(1 + z)^n \prod_{k=1}^{n}(1 + \alpha_k)}{\omega_n(z)}, \quad n = 1, 2, ... \tag{4.5}$$

(We have for convenience assumed that $z = -1$ is not among the points α_n. If $\alpha_n = -1$ for some n, a slightly different definition can be used.)

We observe that $\sigma_{n*} = \sigma_n$, and $\mathcal{L}_{n*} = \{L_* : L \in \mathcal{L}_n\} = \mathcal{L}_n$.

The functions in $\mathcal{L}_n$ are exactly the functions that may be written in the form

$$L(z) = \frac{p_n(z)}{\omega_n(z)}, \quad p_n \in \Pi_n. \tag{4.6}$$

(This follows by partial fraction decomposition).

We shall have occasion to work also with the spaces $\mathcal{L}_n \cdot \mathcal{L}_n$ and $\mathcal{L} \cdot \mathcal{L}$. Note that $\mathcal{L}_n \cdot \mathcal{L}_n \subset \mathcal{L}_n$, $\mathcal{L} \cdot \mathcal{L} \subset \mathcal{L}$, since $1 \in \mathcal{L}_n$ for all n.

Let M be a linear functional on $\mathcal{L} \cdot \mathcal{L}$, and assume that M has real values for all functions in $\mathcal{L}$ that are real on $\mathbf{T}$. As before we define the inner product $<,>$ by (1.6), and assume that this inner product is quasidefinite.

Let $\{\Phi_n\}$ be the monic orthogonal system obtained by the Gram-Schmidt process from the sequence $\{\sigma_n\}$ (or equivalently from the sequence $\{\frac{1}{\omega_n}\}$). That Φ_n is monic means that the coefficient of σ_n in the expansion of Φ_n is one. Thus

$$\Phi_n(z) = \sum_{m=0}^{n} b_m^{(n)} \sigma_m(z), \quad b_n^{(n)} = 1, \quad b_m^{(n)} \in R. \tag{4.7}$$

We may write

$$\Phi_n(z) = \frac{P_n(z)}{\omega_n(z)}, \quad P_n \in \Pi_n. \tag{4.8}$$

We easily verify that

$$\Phi_{n*}(z) = \Phi_n(z), \quad n = 0, 1, 2, \dots. \tag{4.9}$$

Note that $P_n(\alpha_n) \neq 0$ for all n. We shall call the function Φ_n and the index n **singular** if $P_n(\alpha_{n-1}) = 0$. (Note that the properties of being degenerate and of being exceptional coincide in the circle situation, because of (4.9), and we use the word singular for these coinciding phenomena in this case.) We shall call Φ_n and n **regular** if $P_n(\alpha_{n-1}) \neq 0$.

5   RECURRENCE IN THE CIRCLE SITUATION

We shall in this section state without proof a general result on three-term recursion in the circle situation, and briefly discuss two special cases with one interpolation point repeated and with two interpolation points cyclically repeated.

Theorem 5.1 *Assume that the system $\{\Phi_n\}$ is regular. Then a recurrence relation of the following form holds:*

$$\Phi_n(z) = \left(\frac{A_n}{z - \alpha_n} + B_n \frac{z - \alpha_{n-2}}{z - \alpha_n}\right)\Phi_{n-1}(z) + C_n \frac{z - \alpha_{n-2}}{z - \alpha_n}\Phi_{n-2}(z), \quad n = 2, 3, \dots \tag{5.1}$$

with the initial condition

$$\alpha_0 = 0. \tag{5.2}$$

The constants A_n, B_n, C_n *satisfy the inequalities*

$$A_n + B_n(\alpha_{n-1} - \alpha_{n-2}) \neq 0, \quad n = 2, 3, \ldots \tag{5.3}$$

$$C_n \neq 0, \quad n = 2, 3, \ldots . \tag{5.4}$$

Proof:

See [10]. Cf. also [16], which treats the cyclic situation in the line case (discussed in Section 6). □

It follows that the orthogonal functions $\{\Phi_n\}$ are denominators of a continued fraction $K_{n=1}^{\infty} \frac{a_n(z)}{b_n(z)}$, with elements

$$a_n(z) = \frac{C_n(z - \alpha_{n-2})}{z - \alpha_n} \tag{5.5}$$

$$b_n(z) = \left(\frac{A_n}{z - \alpha_n} + B_n \frac{z - \alpha_{n-2}}{z - \alpha_n}\right). \tag{5.6}$$

Continued fractions of this form have been called MP-fractions (Multipoint Padé continued fractions). They bear the same relationship to multipoint Padé approximation as general T-fractions and M-fractions to two-point Padé approximation.

See [17] for further discussion in the cyclic case. Note that if $\alpha_n = 1$ for all n, $(n = 1, 2, \ldots)$ then Φ_n is of the form

$$\Phi_n(z) = \sum_{m=0}^{n} \frac{\beta_m}{(z - 1)^m}. \tag{5.7}$$

Corrollary 5.2 *Let* $\alpha_n = 1$ *for* $n = 1, 2, \ldots$. *Then the sequence* $\{\Phi_n\}$ *satisfies a recurrence relation of the following form:*

$$\Phi_n(z) = \left(\frac{A_n}{z - 1} + B_n\right)\Phi_{n-1}(z) + C_n\Phi_{n-2}(z), \quad n = 2, 3, \ldots \tag{5.8}$$

$$A_n \neq 0 \quad for \quad n = 2, 3, ... \tag{5.9}$$

$$C_n \neq 0 \quad for \quad n = 2, 3, ... \tag{5.10}$$

Proof:

Follows directly from Theorem 5.1, when we recall that Φ_n is always regular in this case. (The case $n = 2$ must be handled separately.) □

For technical reasons we introduced the assumption $\alpha_n \neq -1$ for all n. By changing the factor $(z+1)^n$ to $(z-\eta)^n$ and $(1+\alpha_k)$ to $(\alpha_k - \overline{\eta})$ for some $\eta \in T, \eta \neq \alpha_m$ for all m, in the expression (4.5) we see that a recurrence relation of the form (5.1) is valid also in the case that $\alpha_n = -1$ for some n.

Corrolary 5.3 *Let* $\alpha_{2m} = 1, \quad m = 1, 2, ..., \quad \alpha_{2m+1} = -1, \quad m = 0, 1, 2, ...$ *Assume that the sequence* $\{\Phi_n\}$ *is regular. Then* $\{\Phi_n\}$ *satisfies a recurrence relation of the following form:*

$$\Phi_{2m}(z) = (\frac{A_{2m}}{z-1} + B_{2m})\Phi_{2m-1}(z) + C_{2m}\Phi_{2m-2}(z) \quad m = 1, 2, ... \tag{5.11}$$

$$\Phi_{2m+1}(z) = (\frac{A_{2m+1}}{z+1} + B_{2m+1})\Phi_{2m}(z) + C_{2m+1}\Phi_{2m-1}(z) \quad m = 1, 2.. \tag{5.12}$$

$$A_n \neq 0 \quad for \quad n = 2, 3, ... \tag{5.13}$$

$$C_n \neq 0 \quad for \quad n = 2, 3, ... \tag{5.14}$$

Proof:

Follows directly from Theorem 5.1. □

6   ORTHOGONALITY AND RECURRENCE IN TH PLANE SITUATION

Let $\{A_n\}$ be an arbitrary sequence of (not necessarily distinct) points in $\boldsymbol{H}_+ \cup \boldsymbol{H}_-$, and assume that $A_j \neq \overline{A_k}$ for all j, k. Set

$$\alpha_n = \frac{A_n - i}{A_n + i}, \quad z = \frac{Z - i}{Z + i}. \tag{6.1}$$

cf. (1.1).

Then $\alpha_n \in D \cup E$, and $\alpha_j \neq 1/\overline{\alpha_k}$ for $j \neq k$. To the functions $\zeta_n(z)$ correspond the functions $\Gamma_n(Z)$ defined by

$$\Gamma_n(Z) = \frac{Z - A_n}{Z - \overline{A}_n}. \tag{6.2}$$

Note that $\Gamma_n(Z)$ differs by a constant factor from $\zeta_n(z)$. Define Δ_n by

$$\Delta_0 = 1, \quad \Delta_n(Z) = \prod_{k=1}^n \Gamma_k(Z), \quad n = 1, 2, \dots, \tag{6.3}$$

and set

$$\mathcal{M}_n = Span\{\Delta_0, \Delta_1, \dots, \Delta_n\}, \quad \mathcal{M} = \cup_{n=0}^\infty \mathcal{M}_n. \tag{6.4}$$

We set $\mathcal{M}_{n\sim} = \{M_\sim : M \in \mathcal{M}_n\}, \mathcal{M}_\sim = \cup_{n=0}^\infty \mathcal{M}_{m\sim}$. Let M be a linear functional on $\mathcal{M} + \mathcal{M}_\sim$, and let $<, >$ be defined by (1.7). The functions F_n defined by

$$F_n(Z) = \Phi_n(z) = \Phi_n(\frac{Z - i}{Z + i}) \tag{6.5}$$

are orthogonal functions corresponding to the spaces $\{\mathcal{M}_n\}$. (Here $\{\Phi_n\}$ are orthogonal with respect to the inner product (1.6) corresponding to the bases $\{\mathcal{L}_n\}$.)

We define

$$f_n^\sim(Z) = \Delta_n(Z) f_{n\sim}(Z) \tag{6.6}$$

when $f_n \in \mathcal{M}_n$, and note that $f_n^\sim \in \mathcal{M}_n$.

We further observe that

$$F_n^\sim(Z) = \lambda_n \Phi_n^*(z), \quad z = \frac{Z - i}{Z + i}, |\lambda_n| = 1. \tag{6.7}$$

We call F_n **degenerate** if $F_n^\sim(A_{n-1}) = 0$, **exceptional** if $F_n(A_{n-1}) = 0$.

Note that if $Z = \tau(z), \quad A = \tau(\alpha)$ (cf.(1.1)), then

$$z - \alpha = \frac{2i(Z - A)}{(Z + i)(A + i)} \tag{6.8}$$

$$1 - \overline{\alpha}z = \frac{2i(Z - \overline{A})}{(i - \overline{A})(Z + i)}. \tag{6.9}$$

Theorem 6.1 *The sequence $\{F_n\}$ satisfies a recurrence relation of the following form:*

$$F_n(Z) = E_n \frac{Z - A_{n-1}}{Z - A_n} F_{n-1}(Z) + D_n \frac{Z - \overline{A_{n-1}}}{Z - \overline{A_n}} F_{n-1}^{\sim}(Z), \quad n = 1, 2, ... \qquad (6.10)$$

$$F_n^{\sim}(Z) = \overline{D_n} \frac{Z - A_{n-1}}{Z - \overline{A_n}} F_{n-1}(Z) + \overline{E_n} \frac{Z - \overline{A_{n-1}}}{Z - \overline{A_n}} F_{n-1}^{\sim}(Z), \quad n = 1, 2, ... \qquad (6.11)$$

$$A_0 = i, \quad F_0 = 1, \quad F_0^{\sim} = 1. \qquad (6.12)$$

Proof:

Follows from Theorem 3.1 by the substitutions (6.7), (6.8), (6.9).   □

Theorem 6.2 *Assume that F_n is non-degenerate. Then $\{F_n\}$ satisfies a recurrence relation of the following form:*

$$F_n(Z) = \frac{D_n}{\overline{E_n}} F_n^{\sim}(Z) + \frac{1}{\overline{E_n}}(|E_n|^2 - |D_n|^2)\frac{Z - A_n}{Z - \overline{A_n}} F_{n-1}(Z) \quad n = 1, 2, ... \qquad (6.13)$$

In particular this is the case if the functional M is positive definite and all $A_n \in \boldsymbol{H}_+$ or all $A_n \in \boldsymbol{H}_-$.

Proof:

Follows as above from Theorem 3.4.   □

It follows by combining (6.11) and (6.13) that if the sequence $\{F_n\}$ is non-degenerate and $|D_n| \neq |E_n|$ for all n, then the elements of the sequence $\{F_0, F_1^{\sim}, F_1, F_2^{\sim}, F_2, ..., F_n^{\sim}, F_n, ...\}$ are the denominators of a continued fraction, just as in the disk case. We shall call also these continued fractions NP-fractions.

Theorem 6.3 *Assume that $\{F_n\}$ is non-exceptional. Then $\{F_n\}$ and $\{F_n^{\sim}\}$ satisfy recurrence relations of the following form:*

$$F_n(Z) = [\frac{E_n(Z - A_{n-1})}{Z - \overline{A_n}} + \frac{\overline{E_{n-1}}D_n(Z - \overline{A_{n-1}})}{D_{n-1}(Z - \overline{A_n})}]F_{n-1}(Z) \qquad (6.14)$$

$$- \frac{D_n(Z - A_{n-2})}{D_{n-1}(Z - \overline{A_n})}[|E_{n-1}|^2 - |D_{n-1}|^2]F_{n-2}(Z), \quad n = 2, 3, \ldots$$

$$F_n^{\sim}(Z) = [\frac{\overline{E_n}(Z - \overline{A_{n-1}})}{Z - \overline{A_n}} + \frac{E_{n-1}\overline{D_n}(Z - A_{n-1})}{D_{n-1}(Z - \overline{A_n})}F_{n-1}^{\sim}(Z)] \qquad (6.15)$$

$$- \frac{\overline{D_n}}{D_{n-1}}[|E_{n-1}|^2 - |D_{n-1}|^2]\frac{(Z - A_{n-1})(Z - \overline{A_{n-2}})}{(Z - \overline{A_n})(Z - \overline{A_{n-1}})}F_{n-2}^{\sim}(Z),$$

$$n = 2, 3, \ldots$$

$$F_0 = 1, \quad F_0^{\sim} = 1, \quad F_1(Z) = \frac{D_1 + E_1 Z}{Z - \overline{A_1}}, F_1^{\sim} = \frac{\overline{E_1} + \overline{D_1}Z}{Z - \overline{A_1}}. \qquad (6.16)$$

Proof:

Follows as above from Theorem 3.6. ☐

It follows that if the sequence $\{F_n\}$ is non-exceptional and $|D_n| \neq |E_n|$ for all n, then the elements of the sequences $\{F_n\}$ and $\{F_n^{\sim}\}$ are the denominators of two continued fractions. These are again instances of MP-fractions. (See Section 4,5.)

In the special case $A_n = i$ for all n (corresponding to the polynomial case in the disk situation) these formulas reduce to

$$F_n(Z) = [E_n\frac{Z - i}{Z + i} + \frac{\overline{E_{n-1}}D_n}{D_{n-1}}]F_{n-1}(Z) - \frac{D_n}{D_{n-1}}[|E_{n-1}|^2 - |D_{n-1}|^2]\frac{Z - i}{Z + i}F_{n-2}(Z) \quad (6.17)$$

$$F_n^{\sim}(Z) = [\overline{E_n} + \frac{E_{n-1}\overline{D_n}}{D_{n-1}}\frac{(Z - i)}{(Z + i)}]F_{n-1}^{\sim}(Z) - \frac{\overline{D_n}}{D_{n-1}}[|E_{n-1}|^2 - |D_{n-1}|^2]\frac{Z - i}{Z + i}F_{n-2}^{\sim}(Z). \quad (6.18)$$

7   ORTHOGONALITY AND RECURRENCE IN TH LINE SITUATION

Let $\{A_n\}$ be an arbitrary sequence of (not necessarily distinct) points on $\boldsymbol{R}$. Define the points α_n as in Section 6:

$$\alpha_n = \frac{A_n - i}{A_n + i}. \tag{7.1}$$

Then $\alpha_n \in \boldsymbol{T}$. To the functions $\omega_n(z)$ defined in (4.1) correspond the functions $\Omega_n(Z)$ defined by

$$\Omega_0 = 1, \quad \Omega_n = \prod_{k=1}^{n}(Z - A_k), \quad n = 1, 2, \dots . \tag{7.2}$$

(By this we mean that these functions are basic functions for the spaces obtained by transforming the spaces $\mathcal{L}_n, \mathcal{L}$.)

We set

$$\mathcal{M}_n = Span\{\frac{1}{\Omega_0}, \ \frac{1}{\Omega_n}, \dots, \frac{1}{\Omega_n}\}, \ \mathcal{M} = \cup_{n=0}^{\infty}\mathcal{M}_n. \tag{7.3}$$

Let $\mathcal{L}_n$, $\mathcal{L}$ be the spaces defined by (4.2).

Let M be a functional on $\mathcal{M} \cdot \mathcal{M}$ with $<, >$ defined by (1.7). Assume that M has real values for all functions in $\mathcal{M}$ that are real on R. The functions F_n defined by

$$F_n(Z) = \Phi_n(z) = \Phi_n(\frac{Z - i}{Z + i}) \tag{7.4}$$

are orthogonal functions corresponding to the spaces $\{\mathcal{M}_n\}$. (Here $\{\Phi_n\}$ are orthogonal with respect to the inner product (1.6), corresponding to the bases $\{\mathcal{L}_n\}$.)

We may write F_n in the form

$$F_n(Z) = \frac{P_n(Z)}{\Omega_n(Z)}, \quad P_n \in \Pi_n. \tag{7.5}$$

We shall call F_n **singular** if $P_n(A_{n-1}) = 0$. (Here as in the circle situation, the properties of being degenerate and of being exceptional coincide, and we use the word singular for these coinciding phenomena in this case.) We call F_n **regular** if $P_n(A_{n-1}) \neq 0$.

Theorem 7.1 *Assume that the system $\{F_n\}$ is regular. Then a recurrence relation of the following form holds:*

$$F_n(Z) = [\frac{a_n}{Z - A_n} + \frac{b_n(Z - A_{n-2})}{Z - A_n}]F_{n-1}(Z) + c_n\frac{Z - A_{n-2}}{Z - A_n}F_{n-2}(Z), \quad n = 2, 3, ..., \quad (7.6)$$

with initial condition

$$A_0 = i. \quad (7.7)$$

The constants a_n, b_n, c_n satisfy the inequalities

$$a_n + b_n(A_{n-1} - A_n) \neq 0, \quad n = 2, 3, ... \quad (7.8)$$

$$c_n \neq 0, \quad n = 2, 3, ... \quad (7.9)$$

Proof:

Follows from Theorem 5.1 by the substitutions (1.1) and (7.4). (Note that $\dfrac{u_n + v_n Z}{Z - A_n}$ can always be written in the form $\dfrac{a_n + b_n(Z - A_{n-2})}{Z - A_n}$). □

Let $A_n = \infty$ for all n. By (1.1) this corresponds to $\alpha_n = 1$ for all n. In this case $\Phi_n(z)$ are of the form

$$\Phi_n(z) = \sum_{k=0}^{n} \frac{\beta_k}{(z - 1)^k}, \quad (7.10)$$

see (5.7). By substituting for z from (1.1) we get

$$F_n(Z) = \sum_{k=0}^{n} \gamma_k Z^k, \quad (7.11)$$

which is a polynomial. Thus $A_n = \infty$ for all n represents the polynomial situation.

Corrollary 7.2 *Let $A_n = \infty$ for all n. Then the sequence $\{F_n\}$ satisfies a recurrence relation of the form*

$$F_n(Z) = (a_n Z + b_n)F_{n-1}(Z) + c_n F_{n-2}(Z), \quad n = 2, 3, 4... \quad (7.12)$$

$$a_n \neq 0, \quad n = 2, 3, ... \quad (7.13)$$

$$c_n \neq 0, \quad n = 2, 3, ... \quad (7.14)$$

Proof:

Follows from Corollary 5.2 by the substitutions (1.1) and (7.4). □

Let $A_{2m} = \infty$, $m = 1, 2, ...$, $A_{2m+1} = 0$, $m = 0, 1, 2, ...$. Then the orthogonal functions $\Phi_n(z)$ are of the form

$$\Phi_{2m}(z) = \sum_{k=0}^{m} \frac{\beta_k}{(z-1)^k} + \sum_{k=0}^{m} \frac{\gamma_k}{(z+1)^k}, \tag{7.15}$$

$$\Phi_{2m+1}(z) = \sum_{k=0}^{m} \frac{\beta_k}{(z-1)^k} + \sum_{k=0}^{m+1} \frac{\gamma_k}{(z+1)^k}. \tag{7.16}$$

By substituting in (7.11)-(7.12) from (1.1) and (7.4) we get

$$F_{2m}(Z) = \sum_{k=-m}^{m} \eta_k Z^k \tag{7.17}$$

$$F_{2m+1}(Z) = \sum_{k=-(m+1)}^{m} \eta_k Z^k. \tag{7.18}$$

These functions are orthogonal Laurent polynomials.

Corrollary 7.3 *Let $A_{2m} = \infty$ for $m = 1, 2, ...$, $A_{2m+1} = 0$ for $m = 0, 1, 2, ...$. Assume that the sequence $\{F_n\}$ is regular. Then $\{F_n\}$ satisfies a recurrence relation of the following form:*

$$F_{2m}(Z) = (a_{2m}Z + b_{2m})F_{2m-1}(Z) + c_{2m}F_{2m-2}(Z), \quad m = 1, 2, 3, ... \tag{7.19}$$

$$F_{2m+1}(Z) = (\frac{a_{2m+1}}{Z} + b_{2m+1})F_{2m}(Z) + c_{2m+1}F_{2m-1}(Z), \quad m = 0, 1, 2, ... \tag{7.20}$$

$$a_n \neq 0 \quad \text{for} \quad n = 2, 3, ... \tag{7.21}$$

$$c_n \neq 0 \quad \text{for} \quad n = 2, 3, ... \tag{7.22}$$

Proof:

Follows from Corollary 5.3 by the substitutions (1.1) and (7.4). □

The above recurrence relations are essentially the well-known recurrence relations for regular orthogonal Laurent polynomials. (See [21,33]. Cf. also [32].)

REFERENCES

1. N.I. Akhiezer, *The classical moment problem and some related questions in analysis,* Hafner, New York, 1965.

2. C. Brezinski, *Padé-Type Approximants and Orthogonal Polynomials,* Birkhäuser, Basel/Boston/Stuttgart, 1980.

3. A. Bultheel, Orthogonal matrix functions related to the multivariable Nevanlinna-Pick problem, *Tijdschr. Belgisch Wisk. Genootenshap,* Ser. B. 32 (2) (1980) 149-170.

4. A. Butheel. *Laurent series and their Padé approximations,* volume 27 of Operator Theory: Advances and Applications. Birkhäuser, Basel, 1987.

5. A. Bultheel and P. Dewilde, Orthogonal functions related to the Nevanlinna-Pick problem, *Mathematical Theory of Networks and Systems,* Proceedings MTNS Conference, Delft, The Netherlands (Ed. P. Dewilde Western Periodicals, North Hollywood 1979) 207-211.

6. A. Bultheel, P. Gonzalez-Vera, E. Hendriksen and O. Njåstad, A Szegö theory for rational functions, Technical Report TW-131, K.U. Leuven, Dept. of Computer Science, May 1990.

7. A. Bultheel, P. Gonzalez-Vera, E. Hendriksen and O. Njåstad, The computation of orthogonal rational functions and their interpolating properties, *Numerical Algorithms* 2 (1992) 85-114.

8. A. Bultheel, P. Gonzalez-Vera, E. Hendriksen and O. Njåstad, A Favard theorem for orthogonal rational functions on the unit circle, *Numerical Algorithms 3* (1992) 81-90.

9. A. Bultheel, P. Gonzalez-Vera, E. Hendriksen and O. Njåstad, Moment problems and orthogonal functions, *J. Comp. Appl. Math.,* to appear.

10. A. Bultheel, P. Gonzalez-Vera, E. Hendriksen and O. Njåstad, Orthogonal rational functions with poles on the unit circle, *J. Math. Anal. Appl.,* to appear.

11. T.S. Chihara, *Introduction to Orthogonal Polynomials,* Mathematics and Its Applications Series, Gordon & Breach, New York/London/Paris, 1978.

12. P. Delsarte, Y. Genin and Y. Kamp, On the role of the Nevanlinna-Pick problem in circuit and system theory, *Int. J. Circuit Th. Appl.* 9 (1981) 177-187.

13. P. Dewilde and H. Dym, Schur recursion, error formulas, and convergence of rational estimators for stationary stochastic sequences, *IEEE Trans. on Information Theory,* IT 27 (1981) 446-461.

14. M.M. Djrbashian, A survey on the theory of orthogonal systems and some open problems, *Orthogonal Polynomials: Theory and Practice* (Ed. P. Nevai, Kluwer, Dordrecht/Boston/London 1990) 135-146.

15. Ya. L. Geronimus, *Polynomials orthogonal on a circle and their applications,* Amer. Math. Soc. Translations 104 (Providence, 1954).

16. E. Hendriksen and O. Njåstad, A Favard theorem for rational functions, *J. Math. Anal. Appl.* 142 (1989) 508-520.

17. E. Hendriksen and O. Njåstad, Positive multipoint Padé continued fractions, *Proc. Edinburgh Math. Soc.* 32 (1989) 261-269.

18. E. Hendriksen and H. van Rossum, Orthogonal Laurent polynomials, *Indag. Math.* (Ser. A) 89 (1986), 17-36.

19. W.B. Jones, Schur's algorithm extended and Schur continued fractions, *Nonlinear Numerical Methods and Rational Approximation* (Ed. A. Cuyt, D. Reidel Publ. Comp., Dordrecht 1988) 281-298.

20. W.B. Jones, O. Njåstad and W.J. Thron, Orthogonal Laurent polynomials and the strong Hamburger moment problem, *J. Math. Anal. Appl.* 98 (1984) 528-554.

21. W.B. Jones, O. Njåstad, and W.J. Thron, Two-point Padé expansions for a family of analytic functions, *J. Comput. Appl. Math.* 9 (1983), 105-123.

22. W.B. Jones, O. Njåstad and W.J. Thron, Continued fractions associated with the trigonometric and other strong moment problems, *Constr. Approx.* 2 (1986) 197-211.

23. W.B. Jones, O. Njåstad and W.J. Thron, Moment theory, orthogonal polynomials, quadrature, and continued fractions associated with the unit circle, *Bull. Lond. Math. Soc.* 21 (1989) 113-152.

24. W.B. Jones and W.J. Thron, *Continued fractions: Analytic theory and applications,* Encyclopedia of Mathematics and its Applications, No. 11, Addison-Wesley, Reading, Mass., 1980. (Now distributed by Cambridge Univ. Press.).

25. W.B. Jones and W.J. Thron, Orhogonal Laurent polynomials and Gaussian quadrature, *Quantum Mechanics in Mathematics, Chemistry and Physics* (Ed. K.E. Gustafson and W.P. Reinhardt, Plenum, New York/London 1981) 449-455.

26. M.G. Krein and A.A. Nudelman, The Markov moment problem and extremal problems, *Transl. Math. Monographs, American Mathematical Society* No. 50 (Providence 1977).

27. R. Nevanlinna, Über beschränkte Functionen die in gegebenen Punkten vorgeschriebene Werte annehmen, *Ann. Acad. Sci. Fenn.* Sec. A, 13 (1919) No. 1

28. O. Njåstad, An extended Hamburger moment problem, *Proc. Edinburgh Math. Soc.* Ser. II, 28 (1985) 167-183.

29. O. Njåstad, Unique solvability of an extended Hamburger moment problem, *J. Math. Anal. Appl.* 124 (1987) 502-519.

30. O. Njåstad, Orthogonal rational functions with poles in a finite subset of R, *Orthogonal Polynomials and their Applications,* Springer Lecture Notes in Mathematics 1329 (Ed. Alfaro et al., Berlin 1988) 300-307.

31. O. Njåstad, Multipoint Padé approximation and orthogonal rational functions, *Nonlinear Numerical Methods and Rational Approximation* (Ed. A. Cuyt, D. Reidel Publ. Comp., Dordrecht 1988) 259-270.

32. O. Njåstad, Solution of the strong Hamburger moment problem by Laurent continued fractions, *Appl. Numer. Math.* 4 (1988), 351-360.

33. O. Njåstad and W.J. Thron, The theory of sequences of orthogonal L-polynomials, in *Padé Approximants and Continued Fractions* (Ed. H. Waadeland and H. Wallin, Det Kongelige Norske Videnskabers Selskab, Skrifter, 1983).

34. G. Pick, Über die Besrchänkungen analytischer Funktionen welche durch vorgegebene Funktionswerte bewirkt werden, *Math. Ann.* 77 (1916) 7-23.

35. G. Szegö, *Orthogonal polynomials* (Amer. Math. Soc. Coll. Publ. XXIII, 4.ed. Providence, R.I., 1975).

3

Orthogonal Laurent Polynomials on the Real Line

LYLE COCHRAN, Department of Pure and Applied Mathematics, Washington State University, Pullman, WA 99164-3113, USA

S. CLEMENT COOPER[1],Department of Pure and Applied Mathematics, Washington State University, Pullman, WA 99164-3113, USA

Dedicated to the memory of Arne Magnus

1 INTRODUCTION

A new area of mathematics was opened in 1980 through a paper entitled *A Strong Stieltjes Moment Problem*, written by William B. Jones, W. J. Thron and Haakon Waadeland. The study led to a strong moment problem in which orthogonal Laurent polynomials (L-polynomials) made their appearance [6]. The theory has developed rapidly and already there have been two survey articles[5,10], but the authors felt it was time for another. The approach adopted here roughly parallels that used in T. S. Chihara's *Introduction to Orthogonal Polynomials* [1] and, as such, is different from that appearing in the other articles. The main intent of this paper is to provide a comprehensive summary of the fundamental results concerning orthogonal Laurent polynomials. One hope is that it will serve as an introduction for nonspecialists, and in that spirit, proofs are provided for all of the results. On the other hand, new results of varying degrees of importance do appear throughout, as well as some new proofs of known results, so the hope is that fellow researchers in this field also will find something of interest in the article.

This general field relates many topics in a beautiful, sometimes surprising, yet natural way. Related to one another are strong moment problems, orthogonal Laurent polynomials, Gaussian quadrature formulae, continued fractions and distribution functions. Strong moment problems on the real line, as well as on the unit circle have been studied in the

[1] Research supported in part by the U.S. National Science Foundation under Grant No. DMS- 9109095.

literature. Here we restrict ourselves to strong moment functionals on the real line. The classical moment problem and orthogonal polynomials have a rich, well-developed theory. Moreover, many of the classical orthogonal polynomials (e.g. Tchebycheff, Legendre and Hermite polynomials) are valuable tools in approximating complicated functions as well as solutions to integral and differential equations. As indicated earlier, orthogonal L-polynomials were discovered very recently. Research to date indicates that orthogonal polynomials and orthogonal L-polynomials share many similar properties, yet perhaps as importantly, differences arise due to the rational nature of the L-polynomials, adding an intriguing flavor to the study. Many standard techniques from orthogonal polynomials have been adapted to the study of L-polynomials, but due to the differences, some new approaches have also been developed.

This article will begin from the perspective of a strong moment functional defined on the real line. Basic results for orthogonal L-polynomials will be obtained, including existence criteria, recurrence relations, a Favard theorem, Christoffel-Darboux identities, quadrature rules, integral representations and a spectral analysis of distribution functions.

In order to facilitate the presentation, we begin with some basic terminology and results. Most of the terms introduced here are modifications of definitions appearing in the literature and have analogues in the polynomial setting.

A **Laurent polynomial** or **L-polynomial** is a function of a nonzero, real variable x with the form, $R(x) = \sum_{j=m}^{n} r_j x^j$, where $m, n \in \mathbb{Z}$ with $m \leq n$ and $r_j \in \mathbb{C}$ for $j = m, \ldots, n$. The set of all Laurent polynomials will be denoted by $\mathcal{R}$ while $\mathcal{R}_{m,n}$ will represent the set of all Laurent polynomials contained in the span of $\{x^j\}_{j=m}^{n}$. Two classes of L-polynomials that are particularly important in this study are

$$\mathcal{R}_{2n} = \{R \in \mathcal{R}_{-n,n} : \text{ the coefficient of } x^n \text{ is nonzero}\}$$

and

$$\mathcal{R}_{2n+1} = \{R \in \mathcal{R}_{-n-1,n} : \text{ the coefficient of } x^{-n-1} \text{ is nonzero}\}$$

for all $n \in \mathbb{Z}_0^+$.

The first result will lead to the analogue of the degree of a polynomial.

Theorem 1.1. *For each $R(x) \in \mathcal{R}$, there exists a unique integer $k \in \mathbb{Z}_0^+$ such that $R(x) \in \mathcal{R}_k$.*

Proof: Let $R(x) = \sum_{j=m}^{n} r_j x^j$ where $r_m r_n \neq 0$. Either $|m| \leq n$ or $|m| > n$. If $|m| \leq n$, then $R(x) \in \mathcal{R}_{2n}$ and $R(x) \notin \mathcal{R}_k$ for $k \neq 2n$. If $|m| > n$, then $R(x) \in \mathcal{R}_{2|m|-1}$ and $R(x) \notin \mathcal{R}_k$ for any $k \neq 2|m| - 1$. ∎

A Laurent polynomial is said to be of **L-degree** m if it is a member of $\mathcal{R}_m$, where $m \in \mathbb{Z}_0^+$. While the *degree* of a polynomial is equal to the exponent corresponding to the term with the largest power of x, this is not the case with the *L-degree* of a nonconstant Laurent polynomial. In fact, the L-degree of a polynomial is twice its degree. If $R(x) \in \mathcal{R}_{2n}$ for some $n \in \mathbb{Z}_0^+$, then the coefficient of x^n is called the **L-leading coefficient** and the coefficient of x^{-n} is called the **L-trailing coefficient**. On the other hand, if $R(x) \in \mathcal{R}_{2n+1}$ for some $n \in \mathbb{Z}_0^+$, then the coefficient of x^{-n-1} is called the **L-leading coefficient** and the coefficient of x^n is called the **L-trailing coefficient**. Note that the L-leading coefficient of an L-polynomial $R(x)$ is nonzero, but no such restriction

applies to the L-trailing coefficient of $R(x)$. However, having the L-trailing coefficient nonzero as well, turns out to be an important property. Thus, a Laurent polynomial $R(x) \in \mathcal{R}_k$ is said to be **regular** if the L-trailing coefficient of $R(x)$ is nonzero. If $\{R_n(x)\}_{n=0}^{\infty}$ is a sequence of Laurent polynomials such that $R_k(x)$ is a regular Laurent polynomial of L-degree k for each $k \in \mathbb{Z}_0^+$, then $\{R_n(x)\}_{n=0}^{\infty}$ is a **regular sequence** of Laurent polynomials.

Two more terms commonly defined for polynomials have analogues for Laurent polynomials. If the L-leading coefficient of a Laurent polynomial $R(x)$ is equal to one, then $R(x)$ is said to be **monic**. A **real** Laurent polynomial, $R(x)$, is an element of $\mathcal{R}$ satisfying the condition that all of its coefficients are real. The set of all real L-polynomials is denoted by $\mathcal{R}^{\mathbb{R}}$.

The first result is common knowledge and is mentioned in [3] without proof.

Theorem 1.2. *If $\{R_n(x)\}_{n=0}^{\infty}$ is a sequence of Laurent polynomials such that $R_k \in \mathcal{R}_k$ for each $k \in \mathbb{Z}_0^+$, then*

$$\text{span}\{x^{-m}, x^{-m+1}, \ldots, x^m\} = \text{span}\{R_0(x), R_1(x), \ldots, R_{2m}(x)\}$$

and

$$\text{span}\{x^{-m-1}, x^{-m}, \ldots, x^m\} = \text{span}\{R_0(x), R_1(x), \ldots, R_{2m+1}(x)\}$$

for all $m \in \mathbb{Z}_0^+$. Therefore $\{R_n(x)\}_{n=0}^{\infty}$ is a basis for $\mathcal{R}$.

Proof: Clearly, the elements $x^{-m}, x^{-m+1}, \ldots, x^m$ are linearly independent and

$$\text{span}\{R_0(x), R_1(x), \ldots, R_{2m}(x)\} \subseteq \text{span}\{x^{-m}, x^{-m+1}, \ldots, x^m\}$$

for all $m \in \mathbb{Z}_0^+$. Now consider $\{R_0(x), \ldots, R_{2m}(x)\}$. Suppose

$$a_0 R_0(x) + a_1 R_1(x) + \ldots + a_{2m} R_{2m}(x) = 0.$$

Since $R_{2m}(x)$ is the only term containing an element of degree m, $a_{2m} = 0$. Similarly, $a_0 = a_1 = \ldots = a_{2m-1} = 0$. Therefore, $\{R_0(x), \ldots, R_{2m}(x)\}$ is linearly independent. Both linearly independent sets have $2m + 1$ elements, so their spans are equal.

The other case is proved analogously. ∎

The concept of orthogonality is now introduced with respect to certain linear functionals $\mathcal{L} : \mathcal{R} \to \mathbb{C}$. Let $\{\mu_n\}_{-\infty}^{\infty}$ be a bisequence of complex numbers and let $\mathcal{L}$ be a complex valued function defined on the vector space $\mathcal{R}$ by

$$\mathcal{L}[R(x)] = \sum_{j=m}^{n} r_j \mu_j, \ m, n \in \mathbb{Z}$$

for $R(x) = \sum_{j=m}^{n} r_j x^j$ in $\mathcal{R}$. Note that $\mathcal{L}[x^n] = \mu_n$ for all $n \in \mathbb{Z}$. The linear functional $\mathcal{L}$ will be called the **strong moment functional** (abbreviated SMF) determined by the bisequence $\{\mu_n\}_{-\infty}^{\infty}$.

Chihara [1] defines a moment functional, whose domain is the set of all polynomials, as a linear functional which is determined by a sequence $\{\mu_n\}_{n=0}^{\infty}$. The term *strong* is used

to emphasize the fact that in the L-polynomial setting, $\mathcal{L}$ is determined by a bisequence $\{\mu_n\}_{-\infty}^{\infty}$, not just a sequence $\{\mu_n\}_{n=0}^{\infty}$.

In [8], the authors are motivated by the strong Hamburger moment problem and hence restrict themselves to strong moment functionals, $\mathcal{L}$, which give rise to inner products. Specifically, for $R, S \in \mathcal{R}^{\mathbb{R}}$, $\mathcal{L}[R(x)S(x)]$ is an inner product. Njåstad and Thron [10] also focus on the case where $\mathcal{L}$ determines an inner product. As will be seen shortly, $\mathcal{L}$ is defined more generally here and as a consequence, $\mathcal{L}$ does not necessarily determine an inner product.

A sequence $\{R_n(x)\}_{n=0}^{\infty}$ is called an **orthogonal Laurent polynomial sequence** corresponding to a SMF $\mathcal{L}$ provided $R_k(x) \in \mathcal{R}_k$ for each $k \in \mathbb{Z}_0^+$ and

$$\mathcal{L}[R_m(x)R_n(x)] = K_n\delta_{m,n}$$

for all $m, n \in \mathbb{Z}_0^+$, where $K_n \neq 0$ for all $n \in \mathbb{Z}_0^+$ and $\delta_{m,n}$ is Kronecker's delta function. The phrase *orthogonal Laurent polynomial sequence* will be abbreviated OLPS and $\{R_n(x)\}_{n=0}^{\infty}$ will be called an OLPS with respect to $\mathcal{L}$. If $K_n = 1$ for all $n \in \mathbb{Z}_0^+$, $\{R_n(x)\}_{n=0}^{\infty}$ is an **orthonormal** Laurent polynomial sequence corresponding to $\mathcal{L}$.

For a given SMF, $\mathcal{L}$, determined by a given bisequence $\{\mu_n\}_{-\infty}^{\infty}$, it may not be possible to find an OLPS for $\mathcal{L}$. Two examples which demonstrate this point are given now.

Let $\mu_0 = 0$ and suppose $\{R_n(x)\}_{n=0}^{\infty}$ is an OLPS for $\mathcal{L}$ with elements denoted as follows,

$$R_{2n}(x) = \sum_{j=-n}^{n} r_{2n,j}x^j, \quad r_{2n,n} \neq 0 \tag{1.2}$$

and

$$R_{2n+1}(x) = \sum_{j=-n-1}^{\infty} r_{2n+1,j}x^j, \quad r_{2n+1,-n-1} \neq 0. \tag{1.3}$$

Then

$$\mathcal{L}[R_0^2(x)] = r_{0,0}^2\mathcal{L}(x^0) = r_{0,0}^2\mu_0 = 0,$$

which is impossible by (1.1).

Another example is given by a bisequence $\{\mu_n\}_{-\infty}^{\infty}$ in which

$$\mu_{-2} = \mu_0 = 1 \text{ and } \mu_{-1} = -1.$$

Assume $\{R_n(x)\}_{n=0}^{\infty}$ is an OLPS corresponding to $\mathcal{L}$ whose elements are of the form given in equations (1.2) and (1.3). Then

$$0 \neq \mathcal{L}[R_0^2(x)] = r_{0,0}^2$$

and therefore

$$R_0(x) = r_{0,0} \neq 0.$$

Since

$$R_1(x) = r_{1,-1}\frac{1}{x} + r_{1,0} \text{ and } r_{1,-1} \neq 0,$$

$$0 = \mathcal{L}[R_0(x)R_1(x)] = \mathcal{L}[r_{0,0}r_{1,-1}\frac{1}{x} + r_{0,0}r_{1,0}] = -r_{0,0}r_{1,-1} + r_{0,0}r_{1,0}.$$

Thus $r_{1,0} = r_{1,-1}$ and so

$$\mathcal{L}[R_1^2(x)] = \mathcal{L}[r_{1,-1}^2 \frac{1}{x^2} + 2r_{1,0}r_{1,-1}\frac{1}{x} + r_{1,0}^2]$$

$$= r_{1,-1}^2 - 2r_{1,0}r_{1,-1} + r_{1,0}^2 = r_{1,-1}^2 - 2r_{1,-1}^2 + r_{1,-1}^2 = 0$$

yielding a contradiction to (1.1).

The question of the existence of an OLPS for a given SMF will be discussed in the next section. For now, a few more results will be given which will prove useful later. The following theorem gives equivalent ways of defining an OLPS for $\mathcal{L}$.

Theorem 1.3. Let $\mathcal{L}$ be a SMF and let $\{R_n(x)\}_{n=0}^{\infty}$ be a sequence of Laurent polynomials such that $R_n(x)$ is of L-degree n for each $n \in \mathbb{Z}_0^+$. Then the following are equivalent:
(a) $\{R_n(x)\}_{n=0}^{\infty}$ is an OLPS for $\mathcal{L}$;
(b) $\mathcal{L}[R(x)R_n(x)] = 0$ for every L-polynomial $R(x)$ of L-degree $n - 1$ or less,
    where $n \in \mathbb{Z}^+$, while $\mathcal{L}[R(x)R_n(x)] \neq 0$, where $R(x) \in \mathcal{R}_n$, $n \in \mathbb{Z}_0^+$;
(c) $\mathcal{L}[x^m R_{2n}(x)] = K_{2n}\delta_{m,n}$, where $K_{2n} \neq 0$, $n \in \mathbb{Z}_0^+$ and $m = 0, \pm 1, \pm 2, \ldots, \pm n$;
and $\mathcal{L}[x^m R_{2n+1}(x)] = K_{2n+1}\delta_{m,-(n+1)}$, where $K_{2n+1} \neq 0$, $n \in \mathbb{Z}_0^+$ and
$m = 0, \pm 1, \pm 2, \ldots, \pm n, -(n + 1)$.

Proof: It will be shown that $(a) \Rightarrow (b) \Rightarrow (c) \Rightarrow (a)$. Suppose $\{R_n(x)\}_{-\infty}^{\infty}$ is an OLPS for $\mathcal{L}$. Let $n \in \mathbb{Z}^+$ and $k \in \{0, \ldots, n - 1\}$ be given. If $R(x) \in \mathcal{R}_k$, then by Theorem 1.2.

$$R(x) = \sum_{j=0}^{k} a_j R_j(x)$$

where $a_j \in \mathbb{C}$, $j = 0, 1, \ldots, k$. By the linearity of $\mathcal{L}$ and (1.1)

$$\mathcal{L}[R(x)R_n(x)] = \sum_{j=0}^{k} a_j \mathcal{L}[R_n(x)R_j(x)] = 0.$$

If $R(x) \in \mathcal{R}_n$, then

$$R(x) = \sum_{j=0}^{n} b_j R_j(x),$$

where $b_j \in \mathbb{C}$, $j = 0, 1, \ldots, n$ with $b_n \neq 0$. Thus

$$\mathcal{L}[R(x)R_n(x)] = \sum_{j=0}^{n} b_j \mathcal{L}[R_j(x)R_n(x)] = b_n \mathcal{L}[R_n^2(x)] \neq 0$$

by (1.1). So $(a) \Rightarrow (b)$.

Now assume (b) holds. Fix an arbitrary $n \in \mathbb{Z}_0^+$ and let $m \in \{0, \pm 1, \pm 2, \ldots, \pm n\}$ be given. If $m \neq n$, then $x^m \in \mathcal{R}_k$ for $k < n$ and by part (b), $\mathcal{L}[x^m R_{2n}(x)] = 0$. If $m = n$, then $x^m \in \mathcal{R}_{2n}$ and hence $\mathcal{L}[x^m R_{2n}(x)] \neq 0$ by part (b). Now suppose $m \in \{0, \pm 1, \ldots, \pm n, -(n + 1)\}$ is given. If $m \neq -(n + 1)$, $x^m \in \mathcal{R}_k$, $k < 2n + 1$, implying

$\mathcal{L}[x^m R_{2n+1}(x)] = 0$ and if $m = -(n+1)$, $x^m \in \mathcal{R}_{2n+1}$ and therefore $\mathcal{L}[x^m R_{2n+1}(x)] \neq 0$. Hence $(b) \Rightarrow (c)$.

Finally, assume (c) holds. It will be shown that

$$\mathcal{L}[R_m(x)R_n(x)] = K_n^* \delta_{mn}, \tag{1.4}$$

where $K_n^* \neq 0$ for all $n \in \mathbb{Z}^+$. We know that $R_{2n}(x)$ and $R_{2n+1}(x)$ have the forms given in (1.2) and (1.3). Assume $m \leq n$. Then

$$\mathcal{L}[R_{2m}(x)R_{2n}(x)] = \sum_{j=-m}^{m} r_{2m,j}\mathcal{L}[x^j R_{2n}(x)] = r_{2m,m}K_{2n}\delta_{mn}$$

and

$$\mathcal{L}[R_{2m+1}(x)R_{2n+1}(x)] = \sum_{j=-(m+1)}^{m} r_{2m+1,j}\mathcal{L}[x^j R_{2n+1}(x)] = r_{2m+1,-(m+1)}K_{2n+1}\delta_{m,n}.$$

If $m < n$,

$$\mathcal{L}[R_{2m+1}(x)R_{2n}(x)] = \sum_{j=-(m+1)}^{m} r_{2m+1,j}\mathcal{L}[x^j R_{2n}(x)] = 0$$

and if $m \geq n$

$$\mathcal{L}[R_{2m+1}(x)R_{2n}(x)] = \sum_{j=-n}^{n} r_{2n,j}\mathcal{L}[x^j R_{2m+1}(x)] = 0.$$

Letting $K_{2n}^* = r_{2n,n}K_{2n}$ and $K_{2n+1}^* = r_{2n+1,-(n+1)}K_{2n+1}$, it follows that (1.4) holds and hence (c) implies (a). ∎

A simple, but useful corollary is an immediate consequence of Theorem 1.3.(c).

Corollary 1.4. *If $\{R_n(x)\}_{n=0}^{\infty}$ is an OLPS corresponding to a SMF $\mathcal{L}$, then*

$$\mathcal{L}[R_n(x)] = 0$$

for all $n \in \mathbb{Z}^+$.

According to Theorem 1.2., an L-polynomial can be expressed as a linear combination of a finite subset of $\{R_n(x)\}_{n=0}^{\infty}$. The following theorem gives the values of the coefficients in such a linear combination.

Theorem 1.5. *Let $\{R_n(x)\}_{n=0}^{\infty}$ be an OLPS with respect to $\mathcal{L}$. Then for every L-polynomial $R(x) \in \mathcal{R}_m$,*

$$R(x) = \sum_{k=0}^{m} c_k R_k(x)$$

where

$$c_k = \frac{\mathcal{L}[R(x)R_k(x)]}{\mathcal{L}[R_k^2(x)]}$$

for $k = 0, 1, \ldots, m$.

Proof: Since $R(x) \in \mathcal{R}_m$, according to Theorem 1.2., there exist constants c_k, $k = 0, 1, \ldots, m$ such that

$$R(x) = \sum_{k=0}^{m} c_k R_k(x). \tag{1.5}$$

Let $n \in \{0, 1, \ldots, m\}$ be given and apply $\mathcal{L}$ to equation (1.5) after multiplying both sides by $R_n(x)$ to see that

$$\mathcal{L}[R_n(x)R(x)] = \sum_{k=0}^{m} c_k \mathcal{L}[R_n(x)R_k(x)] = c_n \mathcal{L}[R_n^2(x)].$$

Thus

$$c_n = \frac{\mathcal{L}[R_n(x)R(x)]}{\mathcal{L}[R_n^2(x)]},$$

for $n = 0, 1, \ldots, m$. ∎

Using Theorem 1.5., it can be shown that an OLPS is unique in the sense given in the next corollary.

Corollary 1.6. If $\{R_n(x)\}_{n=0}^{\infty}$ is an OLPS for $\mathcal{L}$, then each $R_n(x)$ is uniquely determined up to an arbitrary nonzero factor. That is, if $\{Q_n(x)\}_{n=0}^{\infty}$ is also an OLPS for $\mathcal{L}$, then there exist constants $c_n \neq 0$ such that

$$Q_n(x) = c_n R_n(x),$$

for all $n \in \mathbb{Z}_0^+$.

Proof: By Theorem 1.3., $\mathcal{L}[R_k(x)Q_n(x)] = 0$ for $k < n$ where $n \in \mathbb{Z}^+$. Letting $Q_n(x)$ play the role of $R(x)$ in Theorem 1.5., it follows that

$$c_k = \frac{\mathcal{L}[Q_n(x)R_k(x)]}{\mathcal{L}[R_k^2(x)]} = 0$$

for $k < n$ and

$$c_n = \frac{\mathcal{L}[Q_n(x)R_n(x)]}{\mathcal{L}[R_n^2(x)]} \neq 0$$

by Theorem 1.3. (b). So by Theorem 1.5.,

$$Q_n(x) = \sum_{k=0}^{n} c_k R_k(x) = c_n R_n(x). \quad ∎$$

If $\{R_n(x)\}$ is an OLPS for $\mathcal{L}$, then it is obvious that $\{c_n R_n(x)\}$ is also an OLPS for $\mathcal{L}$ for every sequence $\{c_n\}_{n=0}^{\infty}$ of nonzero numbers. Therefore, by the corollary, the monic OLPS for $\mathcal{L}$ is unique. If

$$Q_n(x) = \frac{1}{\sqrt{\mathcal{L}[R_n^2(x)]}} R_n(x),$$

then

$$\mathcal{L}[Q_n^2(x)] = \frac{1}{\mathcal{L}[R_n^2(x)]}\mathcal{L}[R_n^2(x)] = 1.$$

Thus $\{Q_n(x)\}_{n=0}^{\infty}$ is the unique orthonormal L-polynomial sequence with respect to $\mathcal{L}$.
We conclude the introduction with the following theorem.

Theorem 1.7. *If* $\{R_n(x)\}_{n=0}^{\infty}$ *is an OLPS for* $\mathcal{L}$, *then* $\{R_n(x)\}_{n=0}^{\infty}$ *is also an OLPS for every SMF,* $\mathcal{L}^*$, *satisfying*

$$\mathcal{L}^*(x^n) = c\mathcal{L}(x^n)$$

where $c \in \mathbb{C}\backslash\{0\}$.

Proof: Suppose $\mathcal{L}$ is the SMF determined by the bisequence $\{\mu_n\}_{-\infty}^{\infty}$. Let $R(x)$ and $Q(x)$ be arbitrary L-polynomials. Then for α, $\beta \in \mathbb{C}$,

$$\mathcal{L}^*[\alpha R(x) + \beta Q(x)] = c\mathcal{L}[\alpha R(x) + \beta Q(x)]$$

$$= c\left(\alpha\mathcal{L}[R(x)] + \beta\mathcal{L}[Q(x)]\right) = \alpha\left(c\mathcal{L}[R(x)]\right) + \beta\left(c\mathcal{L}[Q(x)]\right)$$

$$= \alpha\mathcal{L}^*[R(x)] + \beta\mathcal{L}^*[Q(x)].$$

Hence $\mathcal{L}^*$ is indeed linear and it immediately follows that $\mathcal{L}^*$ is a SMF determined by the bisequence $\{c\mu_n\}_{-\infty}^{\infty}$. Also, since

$$\mathcal{L}^*[R_m(x)R_n(x)] = c\mathcal{L}[R_m(x)R_n(x)] = cK_n\delta_{mn},$$

where $cK_n \neq 0$, $\{R_n(x)\}_{n=0}^{\infty}$ is an OLPS for $\mathcal{L}^*$. ∎

2 EXISTENCE CRITERIA AND DETERMINANT REPRESENTATIONS FOR AN OLPS

As was illustrated in the introduction, even though a given bisequence can be used to define a strong moment functional $\mathcal{L}$, we are not guaranteed that an OLPS exists with respect to $\mathcal{L}$.

In order to discuss necessary and sufficient conditions for the existence of an OLPS, we define the Hankel determinants, $H_k^{(n)}$, associated with the given bisequence $\{\mu_n\}_{-\infty}^{\infty}$ by

$$H_0^{(n)} := 1, \quad H_k^{(n)} := \begin{vmatrix} \mu_n & \mu_{n+1} & \cdots & \mu_{n+k-1} \\ \mu_{n+1} & \mu_{n+2} & \cdots & \mu_{n+k} \\ \vdots & \vdots & \ddots & \vdots \\ \mu_{n+k-1} & \mu_{n+k} & \cdots & \mu_{n+2k-2} \end{vmatrix}$$

for all $n \in \mathbb{Z}$ and $k \in \mathbb{Z}^+$. The proof of the existence theorem is facilitated by the following lemma.

Lemma 2.1. *Let $\mathcal{L}$ be a SMF for which an OLPS exists. Let $\{D_n\}_{n=0}^{\infty}$ be a given sequence of nonzero numbers. Then there exists a unique OLPS, $\{R_n(x)\}_{n=0}^{\infty}$, corresponding to $\mathcal{L}$ such that*

$$\mathcal{L}[x^n R_{2n}(x)] = D_{2n} \quad \text{and} \quad \mathcal{L}[x^{-(n+1)} R_{2n+1}(x)] = D_{2n+1} \quad \text{for all } n \in \mathbb{Z}_0^+. \tag{2.1}$$

Proof: Let $\{Q_n(x)\}_{n=0}^{\infty}$ be a given OLPS corresponding to $\mathcal{L}$. By Theorem 1.3.(c) there exists a sequence $\{K_n\}_{n=0}^{\infty}$ of nonzero numbers such that

$$\mathcal{L}[x^m Q_{2m}(x)] = K_{2m} \text{ and } \mathcal{L}[x^{-(m+1)} Q_{2m+1}(x)] = K_{2m+1}$$

for all $m \in \mathbb{Z}_0^+$. For each $n \in \mathbb{Z}_0^+$, let $R_n(x) = \frac{D_n}{K_n} Q_n(x)$. Then $\{R_n(x)\}_{n=0}^{\infty}$ satisfies satisfies (2.1). ∎

Theorem 2.2. (Existence Theorem) *Let $\mathcal{L}$ be a SMF with the moment bisequence $\{\mu_n\}_{-\infty}^{\infty}$. Then there exists an OLPS for $\mathcal{L}$ if and only if*

$$H_{2m}^{(-2m)} \neq 0 \quad \text{and} \quad H_{2m+1}^{(-2m)} \neq 0 \text{ for all } m \in \mathbb{Z}_0^+. \tag{2.2}$$

Proof: Suppose an OLPS $\{R_n(x)\}_{n=0}^{\infty}$ exists for $\mathcal{L}$. Then the elements of $\{R_n(x)\}_{n=0}^{\infty}$ have the form

$$R_{2m}(x) = \sum_{j=-m}^{m} r_{2m,j} x^j \quad \text{where} \quad r_{2m,m} \neq 0 \quad \text{and} \quad r_{2m,j} \in \mathbb{C} \tag{2.3}$$

for $j = -m, -m+1, \ldots, m$ while

$$R_{2m+1}(x) = \sum_{j=-(m+1)}^{m} r_{2m+1,j} x^j \quad \text{where} \quad r_{2m+1,-(m+1)} \neq 0, r_{2m+1,j} \in \mathbb{C}, \tag{2.4}$$

for $j = -(m+1), -m, \ldots, m$. By Theorem 1.3.(c), there exists a sequence $\{K_n\}_{n=0}^{\infty}$ of nonzero constants such that

$$\mathcal{L}[x^m R_{2n}(x)] = K_{2n}\delta_{m,n}, \ m = 0, \pm 1, \ldots, \pm n \tag{2.5}$$

and

$$\mathcal{L}[x^m R_{2n+1}(x)] = K_{2n+1}\delta_{m,-(n+1)}, \ m = 0, \pm 1, \ldots, \pm n, -(n+1) \tag{2.6}$$

for $n \in \mathbb{Z}_0^+$. These in turn can be written as

$$r_{2n,-n}\mu_{m-n} + r_{2n,-n+1}\mu_{m-n+1} + \cdots + r_{2n,n}\mu_{m+n} = K_{2n}\delta_{m,n} \tag{2.7}$$

for $m = 0, \pm 1, \ldots, \pm n$ and

$$r_{2n+1,-(n+1)}\mu_{m-n-1} + r_{2n+1,-n}\mu_{m-n} + \cdots + r_{2n+1,n}\mu_{m+n} = K_{2n+1}\delta_{m,-(n+1)} \tag{2.8}$$

for $m = 0, \pm 1, \ldots, \pm n, -(n+1)$, where $n \in \mathbb{Z}_0^+$. Finally, (2.7) and (2.8) can be written as

$$
\begin{bmatrix}
\mu_{-2n} & \mu_{-2n+1} & \cdots & \mu_0 \\
\mu_{-2n+1} & \mu_{-2n+2} & \cdots & \mu_1 \\
\vdots & \vdots & \ddots & \vdots \\
\mu_0 & \mu_1 & \cdots & \mu_{2n}
\end{bmatrix}
\begin{bmatrix}
r_{2n,-n} \\
r_{2n,-n+1} \\
\vdots \\
r_{2n,n}
\end{bmatrix}
=
\begin{bmatrix}
0 \\
0 \\
\vdots \\
K_{2n}
\end{bmatrix}
\tag{2.9}
$$

and

$$
\begin{bmatrix}
\mu_{-2n-2} & \mu_{-2n-1} & \cdots & \mu_{-1} \\
\mu_{-2n-1} & \mu_{-2n} & \cdots & \mu_0 \\
\vdots & \vdots & \ddots & \vdots \\
\mu_{-1} & \mu_0 & \cdots & \mu_{2n}
\end{bmatrix}
\begin{bmatrix}
r_{2n+1,-(n+1)} \\
r_{2n+1,-n} \\
\vdots \\
r_{2n+1,n}
\end{bmatrix}
=
\begin{bmatrix}
K_{2n+1} \\
0 \\
\vdots \\
0
\end{bmatrix}
\tag{2.10}
$$

for all $n \in \mathbb{Z}_0^+$. According to Lemma 2.1., for the fixed sequence, $\{K_n\}_{n=0}^{\infty}$, $\{R_n(x)\}_{n=0}^{\infty}$ is the unique OLPS satisfying (2.5) and (2.6) This is equivalent to saying that the matrix equations (2.9) and (2.10) have unique solutions for all $n \in \mathbb{Z}_0^+$, which holds exactly when $H_{2n+1}^{(-2n)} \neq 0$ and $H_{2n+2}^{-(2n+2)} \neq 0$ for all $n \in \mathbb{Z}_0^+$. Or equivalently, since $H_0^{(0)} = 1$, when

$$
H_{2n+1}^{(-2n)} \neq 0 \quad \text{and} \quad H_{2n}^{(-2n)} \neq 0 \quad \text{for all } n \in \mathbb{Z}_0^+.
\tag{2.11}
$$

Conversely, if (2.2) holds, then there exist unique solutions to the matrix equations in (2.9) and (2.10) for all $n \in \mathbb{Z}_0^+$ and therefore (2.5) and (2.6) hold for all $n \in \mathbb{Z}_0^+$. The elements of $\{R_n(x)\}_{n=0}^{\infty}$ are of the form given in (2.3) and (2.4) if it can be shown that $r_{2n,-n} \neq 0$ and $r_{2n+1,-n-1} \neq 0$ for all $n \in \mathbb{Z}_0^+$. By Cramer's rule,

$$
r_{2n+1,-(n+1)} = \frac{1}{H_{2n+2}^{-(2n+2)}}
\begin{vmatrix}
K_{2n+1} & \mu_{-(2n+1)} & \cdots & \mu_{-2} & \mu_{-1} \\
0 & \mu_{-2n} & \cdots & \mu_{-1} & \mu_0 \\
\vdots & \vdots & \ddots & \vdots & \cdots \\
0 & \mu_1 & \cdots & \mu_{2n-1} & \mu_{2n}
\end{vmatrix}
$$

$$
= \frac{K_{2n+1}}{H_{2n+2}^{-(2n+2)}}
\begin{vmatrix}
\mu_{-2n} & \mu_{-2n+1} & \cdots & \mu_0 \\
\mu_{-2n+1} & \mu_{-2n+2} & \cdots & \mu_1 \\
\vdots & \vdots & \ddots & \vdots \\
\mu_1 & \mu_2 & \cdots & \mu_{2n}
\end{vmatrix}
= K_{2n+1} \frac{H_{2n+1}^{(-2n)}}{H_{2n+2}^{-(2n+2)}} \neq 0,
$$

implying $K_{2n+1} \neq 0$ for all $n \in \mathbb{Z}_0^+$. Using Cramer's rule on (2.10) to solve for $r_{2n,n}$ will yield

$$
r_{2n,n} = \frac{K_{2n} H_{2n}^{(-2n)}}{H_{2n+1}^{(-2n)}},
$$

implying that $K_{2n} \neq 0$. ∎

Theorem 2.3. *Suppose $\{R_n(x)\}_{n=0}^{\infty}$ is an OLPS for the SMF $\mathcal{L}$. Then for any L-polynomial, $R(x)$, of L-degree $2m$, where $m \in \mathbb{Z}_0^+$,*

$$
\mathcal{L}[R(x)R_{2m}(x)] = \frac{a_m r_{2m,m} H_{2m+1}^{(-2m)}}{H_{2m}^{(-2m)}}
\tag{2.12}
$$

where a_m is the coefficient of x^m in $R(x)$ and $r_{2m,m}$ is the coefficient of x^m in $R_{2m}(x)$. For any L-polynomial $Q(x)$ of L-degree $2m + 1$, where $m \in \mathbb{Z}_0^+$,

$$\mathcal{L}[Q(x)R_{2m+1}(x)] = \frac{b_{-(m+1)}r_{2m+1,-(m+1)}H_{2m+2}^{-(2m+2)}}{H_{2m+1}^{(-2m)}} \tag{2.13}$$

where $r_{2m,-(m+1)}$ is the coefficient of $x^{-(m+1)}$ in $R_{2m+1}(x)$ and $b_{-(m+1)}$ is the coefficient of $x^{-(m+1)}$ in $Q(x)$.

Proof: Write $R(x) = a_m x^m + \sum_{j=-m}^{m-1} a_j x^j$ and $Q(x) = b_{-(m+1)} x^{-(m+1)} + \sum_{j=-m}^{m} b_j x^j$. Then

$$\mathcal{L}[R(x)R_{2m}(x)] = a_m \mathcal{L}[x^m R_{2m}(x)] + \sum_{j=-m}^{m-1} a_j \mathcal{L}[x^j R_{2m}(x)]$$

$$= a_m \mathcal{L}[x^m R_{2m}(x)] = a_m K_{2m},$$

where $K_{2m} \neq 0$ for all $m \in \mathbb{Z}_0^+$. In the proof of Theorem 2.2., we saw that $r_{2m,m} = \frac{H_{2m}^{(-2m)}}{H_{2m+1}^{(-2m)}}$ from which (2.12) follows. A similar argument shows that (2.13) holds. ■

The following corollary follows immediately from Theorem 2.3.

Corollary 2.4. Suppose $\{R_n(x)\}_{n=0}^{\infty}$ is a monic OLPS for the SMF $\mathcal{L}$. Then

$$\mathcal{L}[R_{2m}^2(x)] = \frac{H_{2m+1}^{(-2m)}}{H_{2m}^{(-2m)}} \tag{2.14}$$

and

$$\mathcal{L}[R_{2m+1}^2(x)] = \frac{H_{2m+2}^{-(2m+2)}}{H_{2m+1}^{(-2m)}} \tag{2.15}$$

for all $m \in \mathbb{Z}_0^+$.

A strong moment functional, $\mathcal{L}$, for which an OLPS exists will be called **quasi-definite**.

Determinant expressions for OLPS's first appeared in [7] The monic OLPS presented here is normalized differently from the one in [7].

Theorem 2.5. Suppose $\mathcal{L}$ is a quasi-definite strong SMF and $\{R_n(x)\}_{n=0}^{\infty}$ is the corresponding monic OLPS for $\mathcal{L}$, then

$$R_{2m}(x) = \frac{1}{H_{2m}^{(-2m)}} \begin{vmatrix} \mu_{-2m} & \cdots & \mu_{-1} & x^{-m} \\ \vdots & \ddots & \vdots & \vdots \\ \mu_{-1} & \cdots & \mu_{2m-2} & x^{m-1} \\ \mu_0 & \cdots & \mu_{2m-1} & x^m \end{vmatrix}$$

and

$$R_{2m+1}(x) = \frac{-1}{H_{2m+1}^{(-2m)}} \begin{vmatrix} \mu_{-(2m+1)} & \cdots & \mu_{-1} & x^{-(m+1)} \\ \mu_{(-2m)} & \cdots & \mu_0 & x^{-m} \\ \vdots & \ddots & \vdots & \vdots \\ \mu_0 & \cdots & \mu_{2m} & x^m \end{vmatrix}.$$

Proof: Let

$$Q_{2m}(x) = \frac{1}{H_{2m}^{(-2m)}} \begin{vmatrix} \mu_{-2m} & \cdots & \mu_{-1} & x^{-m} \\ \vdots & \ddots & \vdots & \vdots \\ \mu_{-1} & \cdots & \mu_{2m-2} & x^{m-1} \\ \mu_0 & \cdots & \mu_{2m-1} & x^m \end{vmatrix}$$

and

$$Q_{2m+1}(x) = \frac{-1}{H_{2m+1}^{(-2m)}} \begin{vmatrix} \mu_{-(2m+1)} & \cdots & \mu_{-1} & x^{-(m+1)} \\ \mu_{(-2m)} & \cdots & \mu_0 & x^{-m} \\ \vdots & \ddots & \vdots & \vdots \\ \mu_0 & \cdots & \mu_{2m} & x^m \end{vmatrix}.$$

Then

$$\mathcal{L}[x^k Q_{2m}(x)] = \frac{1}{H_{2m}^{(-2m)}} \begin{vmatrix} \mu_{-2m} & \cdots & \mu_{-1} & \mu_{k-m} \\ \vdots & \ddots & \vdots & \vdots \\ \mu_{-1} & \cdots & \mu_{2m-2} & \mu_{k+m-1} \\ \mu_0 & \cdots & \mu_{2m-1} & \mu_{k+m} \end{vmatrix} = \frac{H_{2m+1}^{(-2m)}}{H_{2m}^{(-2m)}} \delta_{k,m}$$

for $k \in \{-m, -m+1, \ldots, m\}$. Similarly,

$$\mathcal{L}[x^k Q_{2m+1}(x)] = \frac{H_{2m+2}^{-(2m+2)}}{H_{2m+1}^{(-2m)}} \delta_{k,-(m+1)}$$

for $k \in \{-(m+1), -m, \ldots, m\}$. Thus $\{Q_n(x)\}_{n=0}^{\infty}$ is an OLPS for $\mathcal{L}$ and it is clear that $Q_n(x)$ is monic for all $n \in \mathbb{Z}_0^+$. Thus $Q_n(x) = R_n(x)$ for all $n \in \mathbb{Z}_0^+$, which completes the proof. ∎

From this theorem it follows that if $\mathcal{L}$ is quasi-definite, then $\{R_n(x)\}_{n=0}^{\infty}$ is regular if and only if

$$H_{2m}^{(-2m+1)} \neq 0 \quad \text{and} \quad H_{2m+1}^{-(2m+1)} \neq 0$$

for all $m \in \mathbb{Z}_0^+$.

3 POSITIVE-DEFINITE STRONG MOMENT FUNCTIONALS

A class of strong moment functionals that is of special interest in the study of orthogonal L-polynomials is the one consisting of functionals which can be represented by a Riemann-Stieltjes integral of the form

$$\mathcal{L}[x^n] = \int_a^b x^n d\psi(x), \quad -\infty \leq a < b \leq \infty$$

for all $n \in \mathbb{Z}$ and $\psi \in \Psi$, where Ψ represents the set of all bounded, real-valued, non-decreasing functions on (a, b) such that the set, $\mathcal{S}(\psi)$, of all points of increase is an infinite set. For this particular type of SMF, given $R \in \mathcal{R}$ such that $R(x) \geq 0$ with $R(x) \not\equiv 0$ for all $x \in \mathbb{R}\backslash\{0\}$, it is easy to see that $\mathcal{L}[R(x)] > 0$. In general, any SMF $\mathcal{L}$ is said to

be **positive-definite** if $\mathcal{L}[R(x)] > 0$ for all $R \in \mathcal{R}$ such that $R(x)$ is not identically zero and $R(x) \geq 0$ for all $x \in \mathbb{R}\backslash\{0\}$.

Theorem 3.1. *If $\mathcal{L}$ is a positive-definite SMF, then all of its moments are real.*

Proof: For each integer n, $x^{2n} \geq 0$ for all $x \in \mathbb{R}$ and $x^{2n} \not\equiv 0$ implying that

$$\mu_{2n} = \mathcal{L}[x^{2n}] > 0.$$

By the binomial theorem,

$$(x+1)^{2n} = \sum_{k=0}^{2n} \binom{2n}{k} x^{2n-k},$$

for all $n \in \mathbb{Z}_0^+$. For $n = 1$, $0 < \mathcal{L}[(x+1)^2] = \mu_2 + 2\mu_1 + \mu_0$. Since $\mu_2 + \mu_0 > 0$, μ_1 is real. Assume μ_{2n+1} is real for all $n \in \mathbb{Z}_0^+$ such that $0 \leq n \leq N$. Now μ_{2N+3} is real since

$$0 < \mathcal{L}\left((x+1)^{2N+4}\right) = \mathcal{L}\left[\sum_{k=0}^{2N+4} \binom{2N+4}{k} x^{2N+4-k}\right] = \sum_{k=0}^{2N+4} \binom{2N+4}{k} \mu_{2N+4-k}$$

$$= \mu_{2N+4} + (2N+4)\mu_{2N+3} + \sum_{k=2}^{2N+4} \binom{2N+4}{k} \mu_{2N+4-k}.$$

By expanding powers of $\left(\frac{1}{x} + 1\right)$, an argument similar to the one above will complete the proof. ∎

Theorem 3.2. *If the SMF $\mathcal{L}$ is positive-definite, then there exists a real, monic OLPS corresponding to $\mathcal{L}$.*

Proof: A real OLPS will be constructed using a Gram-Schmidt process. Let $R_0(x) = 1$. Then $\mathcal{L}[R_0^2(x)] = \mathcal{L}(1) = \mu_0$. Let $R_1(x) = \frac{1}{x} + aR_0(x)$ and let $a = -\frac{\mu_{-1}}{\mu_0}$ so that

$$\mathcal{L}[R_0(x)R_1(x)] = \mathcal{L}[R_1(x)] = \mu_{-1} + a\mu_0 = 0.$$

Since $\mu_{-1} \in \mathbb{R}$ and $\mu_0 > 0$, a is real. Thus $R_1(x)$ is a real L-polynomial. Also note that $\mathcal{L}[R_1^2(x)] > 0$ since $\mathcal{L}$ is positive-definite, while $R_0(x)$ and $R_1(x)$ are both monic. Suppose $\{R_0(x), R_1(x), \ldots, R_{2N-1}(x)\}$ have been constructed such that $R_j(x)$ is a real, monic L-polynomial of L-degree j for $j = 0, 1, \ldots, 2N - 1$, where $N \geq 1$ and

$$\mathcal{L}[R_m(x)R_n(x)] = K_n \delta_{m,n}$$

where $K_n \neq 0$ and $m, n = 0, 1, 2, \ldots, 2N - 1$. Define $R_{2N}(x)$ by

$$R_{2N}(x) = x^N + \sum_{k=0}^{2N-1} a_{2N,k} R_k(x).$$

Let $a_{2N,j} = \frac{-\mathcal{L}[x^N R_j(x)]}{\mathcal{L}[R_j^2(x)]}$ so that

$$\mathcal{L}[R_j(x)R_{2N}(x)] = \mathcal{L}[x^N R_j(x)] + a_{2N,j}\mathcal{L}[R_j^2(x)] = 0$$

for $j = 0, 1, \ldots, 2N - 1$. It follows that $R_{2N}(x)$ is a monic and real L-polynomial since $\mathcal{L}[R_j^2(x)] > 0$, $x^N R_j(x)$ is a real L-polynomial for $j = 0, 1, \ldots, 2N - 1$, and $\{\mu_n\}_{-\infty}^{\infty}$ is a real bisequence.

By defining

$$R_{2N+1}(x) = x^{-(N+1)} + \sum_{k=0}^{2N} a_{2N+1,k}R_k(x), \quad a_{2N+1,j} = \frac{-\mathcal{L}[x^{-(N+1)}R_j(x)]}{\mathcal{L}[R_j^2(x)]}, \quad j = 0, \ldots, 2N$$

it readily follows that $R_{2N+1}(x)$ is monic and real. ■

The next lemma concerning L-polynomials is a modification of a well-known result for polynomials [2] stating that any polynomial that is nonnegative on the real line can be expressed as the sum of the squares of two polynomials.

Lemma 3.3. Let $R \in \mathcal{R}$ be given such that $R(x) \geq 0$ for all $x \in \mathbb{R}\backslash\{0\}$. Then there exist two real L-polynomials $S(x)$ and $Q(x)$ such that

$$R(x) = S^2(x) + Q^2(x). \tag{3.1}$$

Proof: There exists a nonnegative integer m such that $x^{2m}R(x)$ is a polynomial. Define $P(x) = x^{2m}R(x)$. Since $R(x) \geq 0$ for all $x \in \mathbb{R}\backslash\{0\}$, $P(x) \geq 0$ for all $x \in \mathbb{R}$. Therefore the real roots of $P(x)$ are of even multiplicity. Since complex roots appear in conjugate pairs,

$$P(x) = r^2(x) \prod_{k=1}^{s} (x - \alpha_k - \beta_k i)(x - \alpha_k + \beta_k i),$$

where $s \in \mathbb{Z}_0^+$ and $r(x)$ is a real polynomial. If $s = 0$, define

$$\prod_{k=1}^{s} (x - \alpha_k - \beta_k i)(x - \alpha_k + \beta_k i) = 1.$$

Then there exist real polynomials $A(x)$ and $B(x)$ such that

$$\prod_{k=1}^{s} (x - \alpha_k - \beta_k i) = A(x) + iB(x)$$

and $\prod_{k=1}^{s}(x - \alpha_k - \beta_k i)(x - \alpha_k + \beta_k i) = A^2(x) + B^2(x)$. Thus

$$R(x) = \frac{P(x)}{x^{2m}} = \frac{r^2(x)A^2(x) + r^2(x)B^2(x)}{x^{2m}} = \left(\frac{r(x)A(x)}{x^m}\right)^2 + \left(\frac{r(x)B(x)}{x^m}\right)^2. \quad ■$$

The following result appears in [8] with a proof based upon quadratic forms. An alternate proof is presented here.

Theorem 3.4. *The SMF $\mathcal{L}$ is positive-definite if and only if it moments are all real, and*

$$H_{2m}^{-(2m)} > 0, \quad \text{and} \quad H_{2m+1}^{-(2m)} > 0 \quad \text{for all } m \in \mathbb{Z}_0^+. \tag{3.2}$$

Proof: If $\mathcal{L}$ is positive-definite, then all its moments are real by Theorem 3.1. Let $\{R_n(x)\}_{n=0}^\infty$ be the monic OLPS constructed by the method given in the proof of Theorem 3.2. Then by Corollary 2.4.,

$$\mathcal{L}[R_{2m}^2(x)] = \frac{H_{2m+1}^{(-2m)}}{H_{2m}^{(-2m)}} \quad \text{and} \quad \mathcal{L}[R_{2m+1}^2(x)] = \frac{H_{2m+2}^{-(2m+2)}}{H_{2m+1}^{(-2m)}} \tag{3.3}$$

for all $m \in \mathbb{Z}_0^+$. Multiplying $\mathcal{L}[R_{2m}^2(x)]$ by $\mathcal{L}[R_{2m+1}^2(x)]$ it follows that

$$\frac{H_{2m+2}^{-(2m+2)}}{H_{2m}^{(-2m)}} > 0 \quad \text{for all} \quad m \in \mathbb{Z}_0^+,$$

and since $H_0^{(0)} = 1 > 0$, $H_{2m}^{(-2m)} > 0$ for all $m \in \mathbb{Z}_0^+$. This in turn implies that $H_{2m+1}^{(-2m)} > 0$ for all $m \in \mathbb{Z}_0^+$.

Now suppose (3.2) is satisfied and all the moments are real. By Theorem 2.2., a monic OLPS exists for $\mathcal{L}$. By Corollary 2.4.,

$$\mathcal{L}[R_{2m}^2(x)] = \frac{H_{2m+1}^{(-2m)}}{H_{2m}^{(-2m)}} \quad \text{and} \quad \mathcal{L}[R_{2m+1}^2(x)] = \frac{H_{2m+2}^{-(2m+2)}}{H_{2m+1}^{(-2m)}}$$

for all $m \in \mathbb{Z}_0^+$, and by (3.2), we conclude that $\mathcal{L}[R_{2m}^2(x)] > 0$ and $\mathcal{L}[R_{2m+1}^2(x)] > 0$ for all $m \in \mathbb{Z}_0^+$. Since $\{R_n(x)\}_{n=0}^\infty$ satisfies (2.5) and (2.6), the coefficients of $R_{2m}(x)$ and $R_{2m+1}(x)$ form solutions to (2.9) and (2.10) and these solutions are real, as a consequence of the fact that K_{2m}, K_{2m+1} and the matrices whose determinants are $H_{2n+1}^{(-2n)}$ and $H_{2n+2}^{-(2n+2)}$ are all real. The fact that K_{2m} and K_{2m+1} are real follows from

$$\frac{H_{2m+1}^{(-2m)}}{H_{2m}^{-(2m)}} = \mathcal{L}[x^m R_{2m}(x)] = K_{2m}$$

and

$$\frac{H_{2m+2}^{-(2m+2)}}{H_{2m+1}^{(-2m)}} = \mathcal{L}[x^{-(m+1)} R_{2m+1}(x)] = K_{2m+1}.$$

Now suppose $R(x)$ is an L-polynomial such that $R(x) \neq 0$ and $R(x) \geq 0$ for all $x \in \mathbb{R}\backslash\{0\}$. By Lemma 3.3., there exist real L-polynomials $S(x)$ and $Q(x)$ such that $R(x) = S^2(x) + Q^2(x)$. There also exist a non-negative integer N and real constants $\alpha_0, \alpha_1, \ldots, \alpha_N$ such that

$$S(x) = \sum_{j=0}^N \alpha_j R_j(x).$$

Thus

$$\mathcal{L}[S^2(x)] = \mathcal{L}\left[\sum_{i,j=0}^{N} \alpha_i \alpha_j R_i(x) R_j(x)\right] = \sum_{i,j=0}^{N} \alpha_i \alpha_j \mathcal{L}[R_i(x) R_j(x)] = \sum_{j=0}^{N} \alpha_j^2 \mathcal{L}[R_j^2(x)] > 0.$$

By a parallel argument, $\mathcal{L}[Q^2(x)] \geq 0$. Thus $\mathcal{L}[R(x)] = \mathcal{L}[S^2(x)] + \mathcal{L}[Q^2(x)] \geq 0$. ∎

In [8] the function $\langle S, Q \rangle = \mathcal{L}[S(x)Q(x)]$ is defined for all $S, Q \in \mathcal{R}^{\mathbb{R}}$. Furthermore, $\langle \cdot, \cdot \rangle$ is an inner product if and only if all the moments are real and (3.2) is satisfied, or equivalently if and only if $\mathcal{L}$ is positive-definite. As noted before, the proof of this given in [8] depends upon the well-known fact that (3.2) is satisfied if and only if the quadratic form

$$\sum_{i,j=-n}^{n} c_{i+j} \mu_i \mu_j$$

is positive-definite for all n [for example, 12]. As stated in [1], the use of the terminology *positive-definite* goes back to positive-definite quadratic forms.

Corollary 3.5. Let $\{R_n(x)\}_{n=0}^{\infty}$ be the monic OLPS for $\mathcal{L}$. If $R_n(x)$ is a real L-polynomial for all $n \in \mathbb{Z}_0^+$, and $\mathcal{L}[R_n^2(x)] > 0$ for all $n \in \mathbb{Z}_0^+$, then $\mathcal{L}$ is positive-definite.

Proof: By hypothesis,

$$\mathcal{L}[R_{2m}^2(x)] = \frac{H_{2m+1}^{(-2m)}}{H_{2m}^{(-2m)}} > 0 \quad \text{and} \quad \mathcal{L}[R_{2m+1}^2(x)] = \frac{H_{2m+2}^{-(2m+2)}}{H_{2m+1}^{(-2m)}} > 0$$

implying that (3.2) holds. Since $R_0(x)$ and $R_1(x)$ are real, μ_0 is real and

$$0 = \mathcal{L}[R_1(x)] = \mathcal{L}\left[\frac{1}{x} + r_{1,0}\right] = \mu_{-1} + r_{1,0}\mu_0$$

implying that μ_{-1} is real. Assume $\mu_{-(n+1)}, \mu_{-n}, \ldots, \mu_n$ are real. Then

$$0 = \mathcal{L}[R_{2n+2}(x)] = \sum_{j=-(n+1)}^{n} r_{2n+2,j}\mu_j + \mathcal{L}(x^{n+1})$$

and since $\sum_{j=-(n+1)}^{n} r_{2n+2,j}\mu_j$ is real, μ_{n+1} is real. Similarly, $\mu_{-(n+2)}$ is real. So by Theorem 3.4., $\mathcal{L}$ is positive-definite. ∎

The following results can be deduced from the preceding comments made about the inner-product, but a self-contained proof is given here, based upon the results proved thus far.

Theorem 3.6. *Suppose $\mathcal{L}$ is a SMF such that an OLPS exists for $\mathcal{L}$. If $\langle R(x), S(x) \rangle = \mathcal{L}[R(x)S(x)]$ for all $R(x), S(x) \in \mathcal{R}$, then the following are equivalent.*
(a) $\mathcal{L}$ is positive-definite.

(b) $H_{2n}^{(-2n)} > 0$ and $H_{2n+1}^{(-2n)} > 0$, $m \in \mathbb{Z}_0^+$ and all the moments are real.
(c) $\langle \cdot, \cdot \rangle$ is an inner product on $\mathcal{R}^{\mathbb{R}} \times \mathcal{R}^{\mathbb{R}}$ over the field $\mathbb{R}$.

Proof: It is clear that $(a) \Leftrightarrow (b)$ by Theorem 3.4. Supposing (a) holds, it will now be shown that (c) holds. Let $R, S, Q \in \mathcal{R}^{\mathbb{R}}$ be given. By the linearity of $\mathcal{L}$,

$$\langle R + S, Q \rangle = \mathcal{L}[(R + S)(x)Q(x)] = \mathcal{L}[R(x)Q(x)] + \mathcal{L}[S(x)Q(x)] = \langle R, Q \rangle + \langle S, Q \rangle$$

and

$$\langle \alpha R, Q \rangle = \mathcal{L}[\alpha R(x)Q(x)] = \alpha \mathcal{L}[R(x)Q(x)] = \alpha \langle R, Q \rangle, \quad \alpha \in \mathbb{R}.$$

Since the moments of $\mathcal{L}$ are real and the coefficients of $R(x)$ and $Q(x)$ are real, $\mathcal{L}[Q(x)R(x)]$ is real and so $\langle R, Q \rangle = \overline{\langle R, Q \rangle} = \overline{\langle Q, R \rangle}$. The positive-definiteness of $\mathcal{L}$ implies that $\langle R, R \rangle = \mathcal{L}[R^2(x)] > 0$ if $R(x) \not\equiv 0$. If $R(x) \equiv 0$,

$$\langle R, R \rangle = \mathcal{L}(0) = 0.$$

Thus $\langle R, R \rangle \geq 0$ and $\langle R, R \rangle = 0$ if and only if $R(x) \equiv 0$. Therefore $(a) \Rightarrow (c)$. If (c) holds, let $R \in \mathcal{R}$ be given such that $R(x) \geq 0$ for all $x \in \mathbb{R} \backslash \{0\}$ and such that $R(x) \not\equiv 0$. Then by Lemma 3.3., there exist two real L-polynomials $S(x)$ and $T(x)$ such that

$$R(x) = S^2(x) + T^2(x).$$

Therefore

$$\mathcal{L}[R(x)] = \mathcal{L}[S^2(x)] + \mathcal{L}[T^2(x)] = \langle S(x), S(x) \rangle + \langle T(x), T(x) \rangle > 0.$$

Thus (a) holds. So $(a) \Leftrightarrow (c)$ and therefore $(a) \Leftrightarrow (b) \Leftrightarrow (c)$. ∎

4 THE THREE AND FIVE TERM RECURRENCE FORMULAE

The material presented so far in this paper has closely paralleled the development of the theory of orthogonal polynomials in [1]. But now some significant differences start to appear, one of which concerns the fundamental recurrence relations. A sequence of orthogonal polynomials always satisfies a *three* term recurrence formula, but this is not necessarily true of orthogonal L-polynomials. As will be shown, a three term recurrence relation can only be guaranteed if the given OLPS is *regular*.

The three term recurrence relation will be developed in the first part of this section. This recurrence relation is already known [3,5,6,7,10]. In [5,6,7,10], the SMF is positive-definite and in [3], the SMF is only required to be quasi-definite. The new information provided in this paper, regarding the three term recurrence relations, will have to do with the necessary and sufficient conditions that the recurrence formula coefficients must satisfy in order for the SMF to be positive-definite.

4.1 The Three Term Recurrence Relations

Here it will be shown that the elements of a *regular* OLPS satisfy a three term recurrence relation. Furthermore, necessary and sufficient conditions for the corresponding SMF to be positive-definite will be expressed in terms of the coefficients of the recurrence relations. The following lemma proves useful in obtaining recurrence relations.

Lemma 4.1. *If $\mathcal{L}$ is a quasi-definite SMF and $\{R_n(x)\}_{n=0}^{\infty}$ is the corresponding monic OLPS, then*

$$\mathcal{L}[x^{-(m+1)}R_{2m}(x)] = -r_{2m+1,m}\mathcal{L}[R_{2m}^2(x)] = \frac{H_{2m+1}^{-(2m+1)}}{H_{2m}^{(-2m)}}$$

and

$$\mathcal{L}[x^{(m+1)}R_{2m+1}(x)] = -r_{2m+2,-(m+1)}\mathcal{L}[R_{2m+1}^2(x)] = \frac{-H_{2m+2}^{-(2m+1)}}{H_{2m+1}^{(-2m)}}$$

for all $m \in \mathbb{Z}_0^+$.

Proof: Writing

$$x^{-(m+1)} = R_{2m+1}(x) - \sum_{j=-m}^{m} r_{2m+1,j}x^j$$

and applying Theorem 2.3. yields

$$\mathcal{L}[x^{-(m+1)}R_{2m}(x)] = \mathcal{L}[\sum_{j=-m}^{m} r_{2m+1,j}x^j R_{2m}(x)]$$

$$= -r_{2m+1,m}\mathcal{L}[x^m R_{2m}(x)] = -r_{2m+1,m}\mathcal{L}[R_{2m}^2(x)] = -r_{2m+1,m}\frac{H_{2m+1}^{(-2m)}}{H_{2m}^{(-2m)}}.$$

By Theorem 2.5.,

$$r_{2m+1,m} = \frac{-H_{2m+1}^{-(2m+1)}}{H_{2m+1}^{(-2m)}} \tag{4.1}$$

and therefore

$$\mathcal{L}[x^{-(m+1)}R_{2m}(x)] = \frac{H_{2m+1}^{-(2m+1)}}{H_{2m}^{(-2m)}}.$$

Similarly, since

$$x^{m+1} = R_{2m+2}(x) - \sum_{j=-(m+1)}^{m} r_{2m+2,j}x^j,$$

Theorems 2.3. and 2.5. can be used to show

$$r_{2m+2,-(m+1)} = \frac{H_{2m+2}^{-(2m+1)}}{H_{2m+2}^{-(2m+2)}} \tag{4.2}$$

and hence

$$\mathcal{L}[x^{m+1}R_{2m+1}(x)] = \frac{-H_{2m+2}^{-(2m+1)}}{H_{2m+1}^{(-2m)}}. \; \blacksquare$$

In preparation for developing recurrence relations, some facts about the elements of $\{R_n(x)\}_{n=0}^{\infty}$ are established. Suppose $\mathcal{L}$ is a quasi-definite SMF with the corresponding regular, monic OLPS, $\{R_n(x)\}_{n=0}^{\infty}$, of the form

$$R_{2m}(x) = \sum_{j=-m}^{m} r_{2m,j} x^j, \quad \text{where} \quad r_{2m,m} = 1, \quad r_{2m,-m} \neq 0$$

and

$$R_{2m+1}(x) = \sum_{j=-(m+1)}^{m} r_{2m+1,j} x^j, \quad \text{where} \quad r_{2m+1,-(m+1)} = 1, \quad r_{2m+1,m} \neq 0$$

for all $m \in \mathbb{Z}_0^+$. Then

$$x R_{2m+1}(x) = \sum_{j=-(m+1)}^{m} r_{2m+1,j} x^{j+1} = x^{-m} + r_{2m+1,-m} x^{-m+1} + \cdots + r_{2m+1,m} x^{m+1},$$
$$(4.3)$$

$$x^{-1} R_{2m+1}(x) = \sum_{j=-(m+1)}^{m} r_{2m+1,j} x^{j-1} = x^{-m-2} + r_{2m+1,-m} x^{-m-1} + \cdots + r_{2m+1,m} x^{m-1},$$
$$(4.4)$$

$$x R_{2m}(x) = \sum_{j=-m}^{m} r_{2m,j} x^{j+1} = r_{2m,-m} x^{-m+1} + \cdots + r_{2m,m-1} x^m + x^{m+1}, \quad (4.5)$$

and

$$x^{-1} R_{2m}(x) = \sum_{j=-m}^{m} r_{2m,j} x^{j-1} = r_{2m,-m} x^{-m-1} + \cdots + r_{2m,m-1} x^{m-2} + x^{m-1} \quad (4.6)$$

for all $m \in \mathbb{Z}_0^+$. By (4.3) and (4.5), it follows that $x R_{2k+1}(x) \in \mathcal{R}_{2k+2}$ and $x R_{2k}(x) \in \mathcal{R}_{2k+2}$ implying that

$$\mathcal{L}[R_{2m+1}(x) x R_{2k}(x)] = 0 \quad \text{and} \quad \mathcal{L}[R_{2m+1}(x) x R_{2k+1}(x)] = 0$$

for $k = 0, 1, \ldots, m-1$ and $m \in \mathbb{Z}^+$. Thus

$$\mathcal{L}[R_{2m+1}(x) x R_k(x)] = 0 \quad \text{for} \quad k = 0, 1, \ldots, 2m-1 \quad \text{and} \quad m \in \mathbb{Z}^+. \quad (4.7)$$

From (4.4), $x^{-1} R_{2k+1}(x) \in \mathcal{R}_{2k+3}$ and (4.6), $x^{-1} R_{2k}(x) \in \mathcal{R}_{2k+1}$ implying

$$\mathcal{L}[R_{2m}(x) x^{-1} R_{2k+1}(x)] = 0 \quad \text{for} \quad k = 0, 1, \ldots, m-2, \quad m \in \mathbb{Z}^+, \quad m \geq 2$$

and

$$\mathcal{L}[R_{2m}(x) x^{-1} R_{2k}(x)] = 0 \quad \text{for} \quad k = 0, 1, \ldots, m-1, \quad m \in \mathbb{Z}^+.$$

Therefore,

$$\mathcal{L}[R_2(x)x^{-1}R_0(x)] = 0 \quad \text{and} \quad \mathcal{L}[R_{2m}(x)x^{-1}R_k(x)] = 0 \tag{4.8}$$

for $k = 0, 1, \ldots, 2m-2$ and $m \in \mathbb{Z}^+$, $m \geq 2$. Equations (4.7) and (4.8) will be used to prove the following theorem which gives a three term recurrence formula for a regular OLPS.

Theorem 4.2. *Suppose $\mathcal{L}$ is a quasi-definite SMF whose monic OLPS, $\{R_n(x)\}_{n=0}^{\infty}$, is regular with elements of the form*

$$R_{2m}(x) = \sum_{j=-m}^{m} r_{2m,j}x^j \quad \text{and} \quad R_{2m+1}(x) = \sum_{j=-(m+1)}^{m} r_{2m+1,j}x^j$$

for all $m \in \mathbb{Z}_0^+$. Then

$$R_{2m+1}(x) = \left(\frac{x^{-1}}{\alpha_{2m+1}} + \beta_{2m+1} \right) R_{2m}(x) + \lambda_{2m+1}R_{2m-1}(x) \tag{4.9}$$

and

$$R_{2m+2}(x) = \left(\frac{x}{\alpha_{2m+2}} + \beta_{2m+2} \right) R_{2m+1}(x) + \lambda_{2m+2}R_{2m}(x) \tag{4.10}$$

where

$$\alpha_{2m+1} = \frac{H_{2m}^{-(2m-1)}}{H_{2m}^{-(2m)}}, \quad \alpha_{2m+2} = -\frac{H_{2m+1}^{-(2m+1)}}{H_{2m+1}^{-(2m)}}, \tag{4.11}$$

$$\beta_{m+1} = \alpha_{m+2} \tag{4.12}$$

and

$$\lambda_{2m+2} = \frac{-H_{2m+2}^{-(2m+1)}H_{2m}^{-(2m)}}{H_{2m+1}^{-(2m)}H_{2m+1}^{-(2m+1)}}, \quad \lambda_{2m+3} = \frac{-H_{2m+3}^{-(2m+3)}H_{2m+1}^{-(2m)}}{H_{2m+2}^{-(2m+2)}H_{2m+2}^{-(2m+1)}} \tag{4.13}$$

for all $m \in \mathbb{Z}_0^+$ with $R_{-1}(x) \equiv 0$.

Comment: In [3], a three term recurrence relation is developed for an OLPS which is not monic. The method of proof used here in establishing the three term recurrence relations for the monic OLPS is similar to that in [3]. Also, it will be shown that (4.9) and (4.10) hold where

$$\alpha_{2m+1} = r_{2m,-m} \quad \text{and} \quad \alpha_{2m+2} = r_{2m+1,m}, \quad m \in \mathbb{Z}_0^+. \tag{4.14}$$

The determinant formulae given for the recurrence coefficients in (4.11) and (4.12) follow immediately from (4.1), (4.2), and (4.14).

Proof of Theorem 4.2.: Since

$$R_{2m+2}(x) - \frac{x}{r_{2m+1,m}}R_{2m+1}(x) = \sum_{j=-m-1}^{m+1} r_{2m+2,j}x^j - \frac{1}{r_{2m+1,m}} \sum_{j=-m-1}^{m} r_{2m+1,j}x^{j+1}$$

$$= \sum_{j=-m-1}^{m} r_{2m+2,j}x^j - \frac{1}{r_{2m+1,m}} \sum_{j=-m-1}^{m-1} r_{2m+1,j}x^{j+1},$$

$$R_{2m+2}(x) - \frac{x}{r_{2m+1,m}} R_{2m+1}(x) \in \mathcal{R}_{2m+1} \quad \text{for all} \quad m \in \mathbb{Z}_0^+. \tag{4.15}$$

By (4.7),

$$\mathcal{L}\left[\left(R_{2m+2}(x) - \frac{x}{r_{2m+1,m}}R_{2m+1}(x)\right)R_k(x)\right] = \frac{-1}{r_{2m+1,m}}\mathcal{L}[xR_{2m+1}(x)R_k(x)] = 0 \tag{4.16}$$

for $k = 0, 1, \ldots, 2m - 1$ and $m \in \mathbb{Z}^+$. By (4.15), there exist constants $\alpha_{2m+1,j}$ for $j = 0, 1, \ldots, 2m + 1$ such that

$$R_{2m+2}(x) - \frac{x}{r_{2m+1,m}}R_{2m+1}(x) = \sum_{j=0}^{2m+1} \alpha_{2m+1,j}R_j(x)$$

and by (4.16),

$$0 = \mathcal{L}\left[\sum_{j=0}^{2m+1} \alpha_{2m+1,j}R_j(x)R_k(x)\right] = \alpha_{2m+1,k}\mathcal{L}[R_k^2(x)]$$

for $k = 0, 1, \ldots, 2m - 1$. From (4.16) and the fact that $\mathcal{L}[R_k^2(x)] \neq 0$, $\alpha_{2m+1,k} = 0$ for $k = 0, 1, \ldots, 2m - 1$. Let $\beta_{2m+2} = \alpha_{2m+1,2m+1}$ and $\lambda_{2m+2} = \alpha_{2m+1,2m}$ for all $m \in \mathbb{Z}^+$. Then

$$R_{2m+2}(x) - \frac{x}{r_{2m+1,m}}R_{2m+1}(x) = \lambda_{2m+2}R_{2m}(x) + \beta_{2m+2}R_{2m+1}(x)$$

for all $m \in \mathbb{Z}^+$. So

$$R_{2m+2}(x) = \left(\frac{x}{r_{2m+1,m}} + \beta_{2m+2}\right)R_{2m+1}(x) + \lambda_{2m+2}R_{2m}(x) \quad \text{for all} \quad m \in \mathbb{Z}^+. \tag{4.17}$$

To show that (4.10) also holds for $m = 0$, note that there exist constants a and b such that

$$R_2(x) = x + aR_1(x) + bR_0(x).$$

Since

$$\frac{xR_1(x)}{r_{1,0}} = \frac{x}{r_{1,0}}(x^{-1} + r_{1,0}) = \frac{1}{r_{1,0}} + x,$$

$$x = \frac{xR_1(x)}{r_{1,0}} - \frac{1}{r_{1,0}}.$$

Therefore,

$$R_2(x) = \frac{xR_1(0)}{r_{1,0}} - \frac{1}{r_{1,0}} + aR_1(x) + bR_0(x)$$

$$= (\frac{x}{r_{1,0}} + a)R_1(x) + (b - \frac{1}{r_{1,0}})R_0(x).$$

Letting $\lambda_2 = b - \dfrac{1}{r_{1,0}}$, and observing that $a = r_{2,-1}$ it follows that (4.17) holds for all $x \in \mathbb{Z}_0^+$. Therefore (4.10) follows by letting $\alpha_{2m+2} = r_{2m+1,m}$ for all $m \in \mathbb{Z}_0^+$.

Similarly,

$$R_{2m+1}(x) - \frac{x^{-1}}{r_{2m,-m}} R_{2m}(x) \in \mathcal{R}_{2m}. \tag{4.18}$$

So by (4.8),

$$\mathcal{L}[(R_{2m+1}(x) - \frac{x^{-1}}{r_{2m,-m}} R_{2m}(x)) R_k(x)] = \frac{-1}{r_{2m,-m}} \mathcal{L}[x^{-1} R_{2m}(x) R_k(x)] = 0 \tag{4.19}$$

for $k = 0, 1, \ldots, 2m - 2$, $m \geq 2$ and

$$\mathcal{L}[(R_3(x) - \frac{x^{-1}}{r_{2,-1}} R_2(x)) R_0(x)] = \frac{-1}{r_{2,-1}} \mathcal{L}[x^{-1} R_2(x) R_0(x)] = 0. \tag{4.20}$$

By (4.18), there exist constants $\alpha_{2m,j}$ for $j = 0, 1, \ldots, 2m$ such that

$$R_{2m+1}(x) - \frac{x^{-1}}{r_{2m,-m}} R_{2m}(x) = \sum_{j=0}^{2m} \alpha_{2m,j} R_j(x)$$

for all $m \in \mathbb{Z}^+$. Hence by (4.19),

$$0 = \mathcal{L}[\sum_{j=0}^{2m} \alpha_{2m,j} R_j(x) R_k(x)] = \alpha_{2m,k} \mathcal{L}[R_k^2(x)]$$

for $k = 0, 1, \ldots, 2m - 2$ and by (4.20),

$$0 = \mathcal{L}[\sum_{j=0}^{2} \alpha_{2,j} R_j(x) R_0(x)] = \alpha_{2,0} \mathcal{L}[R_0^2(x)]$$

implying that $\alpha_{2m,k} = 0$ for $k = 0, 1, \ldots, 2m - 2$, $m \geq 1$. Letting $\beta_{2m+1} = \alpha_{2m,2m}$ and $\lambda_{2m+1} = \alpha_{2m,2m-1}$, it follows that

$$R_{2m+1}(x) = \left(\frac{x^{-1}}{r_{2m,-m}} + \beta_{2m+1} \right) R_{2m}(x) + \lambda_{2m+1} R_{2m-1}(x) \tag{4.21}$$

holds for all $m \in \mathbb{Z}^+$. It is not difficult to show that (4.21) holds for all $m \in \mathbb{Z}_0^+$. Thus (4.9) follows by letting $\alpha_{2m+1} = r_{2m,-m}$. By comparing coefficients of x^{-m-1} in (4.10), one can see that $\beta_{2m+2} = r_{2m+2,-m-1}$. Hence $\beta_{2m+2} = \alpha_{2m+3}$ for all $m \in \mathbb{Z}_0^+$. Similarly, by comparing coefficients of x^m in (4.9), it is evident that $\beta_{2m+1} = r_{2m+1,m}$ and therefore $\beta_{2m+1} = \alpha_{2m+2}$. Hence (4.12) holds for all $m \in \mathbb{Z}_0^+$.

It remains to show that the λ_k's are given by the formulae in (4.13) for $k \in \mathbb{Z}_0^+$. Multiplying equation (4.9) by x^{-m} and applying $\mathcal{L}$ yields

$$\lambda_{2m+1} = \frac{-\mathcal{L}[x^{-m-1} R_{2m}(x)]}{r_{2m,-m} \mathcal{L}[R_{2m-1}^2(x)]} \quad \text{for all} \quad m \in \mathbb{Z}^+.$$

Applying Lemma 4.1., it follows that

$$\lambda_{2m+1} = \frac{r_{2m+1,m}\mathcal{L}[R_{2m}^2(x)]}{r_{2m,-m}\mathcal{L}[R_{2m-1}^2(x)]} = \frac{-H_{2m+1}^{-(2m+1)}H_{2m-1}^{-(2m-2)}}{H_{2m}^{(-2m)}H_{2m}^{-(2m-1)}} \tag{4.22}$$

for all $m \in \mathbb{Z}^+$.

Similarly, multiplying equation (4.10) by x^m and applying $\mathcal{L}$ yields

$$\lambda_{2m+2} = \frac{-\mathcal{L}[x^{m+1}R_{2m+1}(x)]}{r_{2m+1,m}\mathcal{L}[R_{2m}^2(x)]}.$$

and by Lemma 4.1.,

$$\lambda_{2m+2} = \frac{r_{2m+2,-m-1}\mathcal{L}[R_{2m+1}^2(x)]}{r_{2m+1,m}\mathcal{L}[R_{2m}^2(x)]} = \frac{-H_{2m+2}^{-(2m+1)}H_{2m}^{(-2m)}}{H_{2m+1}^{(-2m)}H_{2m+1}^{-(2m+1)}} \tag{4.23}$$

for all $m \in \mathbb{Z}_0^+$. ∎

Theorem 4.3. *Suppose $\mathcal{L}$ is a quasi-definite SMF whose monic OLPS, $\{R_n(x)\}_{n=0}^\infty$, is regular with elements of the form*

$$R_{2m}(x) = \sum_{j=-m}^{m} r_{2m,j}x^j \quad and \quad R_{2m+1}(x) = \sum_{j=-(m+1)}^{m} r_{2m+1,j}x^j$$

for all $m \in \mathbb{Z}_0^+$ and let $\{R_n(x)\}_{n=0}^\infty$ satisfy equations (4.9) through (4.13) for all $m \in \mathbb{Z}_0^+$. If $\lambda_1 = \mu_0 r_{1,0}$, then $\mathcal{L}$ is positive-definite if and only if

$$\frac{\lambda_n \alpha_n}{\beta_n} > 0 \quad for \ all \quad n \in \mathbb{Z}^+ \tag{4.24}$$

and $\{R_n(x)\}_{n=0}^\infty$ is a real OLPS.

Proof: First consider $\lambda_1 = \mu_0 r_{1,0}$. Note that

$$\frac{r_{0,0}}{r_{1,0}}\lambda_1 = \mu_0 \frac{r_{1,0}}{r_{1,0}} = \mathcal{L}[R_0^2(x)] = \mu_0 > 0.$$

By (4.22) and (4.23),

$$\frac{\alpha_{2m+1}}{\beta_{2m+1}}\lambda_{2m+1} = \frac{\mathcal{L}[R_{2m}^2(x)]}{\mathcal{L}[R_{2m-1}^2(x)]} \quad and \quad \frac{\alpha_{2m}}{\beta_{2m}}\lambda_{2m} = \frac{\mathcal{L}[R_{2m-1}^2(x)]}{\mathcal{L}[R_{2m-2}^2(x)]} \tag{4.25}$$

for all $m \in \mathbb{Z}^+$. So if $\mathcal{L}$ is positive-definite, then by (4.25) it is clear that

$$\frac{\alpha_{2m}}{\beta_{2m}}\lambda_{2m} > 0 \quad and \quad \frac{\alpha_{2m-1}}{\beta_{2m-1}}\lambda_{2m-1} > 0 \quad for \ all \quad m \in \mathbb{Z}^+.$$

By Theorem 2.5. and Theorem 3.1., $\{R_n(x)\}_{n=0}^\infty$ is a real OLPS.

Next assume that (4.24) holds and $\{R_n(x)\}_{n=0}^{\infty}$ is a real OLPS. Then by (4.25),

$$\frac{\mathcal{L}[R_{2m}^2(x)]}{\mathcal{L}[R_{2m-1}^2(x)]} > 0 \quad \text{and} \quad \frac{\mathcal{L}[R_{2m-1}^2(x)]}{\mathcal{L}[R_{2m-2}^2(x)]} > 0 \quad \text{for all} \quad m \in \mathbb{Z}^+. \tag{4.26}$$

By (4.24),

$$0 < \frac{\alpha_1}{\beta_1}\lambda_1 = \frac{\lambda_1}{r_{1,0}} = \mu_0$$

implying

$$\mathcal{L}[R_0^2(x)] > 0$$

and therefore

$$\mathcal{L}[R_n^2(x)] > 0 \quad \text{for all} \quad n \in \mathbb{Z}_0^+.$$

So by Corollary 3.5., $\mathcal{L}$ is positive-definite. ∎

The following theorem gives another set of equations for λ_n and formulae for $\prod_{i=1}^{n}\lambda_i$.

Theorem 4.4. *Suppose $\mathcal{L}$ is a quasi-definite SMF whose monic OLPS, $\{R_n(x)\}_{n=0}^{\infty}$, is regular with elements of the form*

$$R_{2m}(x) = \sum_{j=-m}^{m} r_{2m,j}x^j \quad and \quad R_{2m+1}(x) = \sum_{j=-(m+1)}^{m} r_{2m+1,j}x^j$$

for all $m \in \mathbb{Z}_0^+$ and let $\{R_n(x)\}_{n=0}^{\infty}$ satisfy equations (4.9) through (4.13) for all $m \in \mathbb{Z}_0^+$. If $\lambda_1 = r_{1,0}\mu_0$ then the following hold for all $m \in \mathbb{Z}^+$.

(a) $\lambda_{2m} = \dfrac{\beta_{2m}\mathcal{L}[R_{2m-1}^2(x)]}{\alpha_{2m}\mathcal{L}[R_{2m-2}^2(x)]}$;

(b) $\lambda_{2m+1} = \dfrac{\beta_{2m+1}\mathcal{L}[R_{2m}^2(x)]}{\alpha_{2m+1}\mathcal{L}[R_{2m-1}^2(x)]}$;

(c) $\displaystyle\prod_{j=1}^{2m-1} \lambda_j = \beta_{2m-1}\mathcal{L}[R_{2m-2}^2(x)] = \dfrac{-H_{2m-1}^{-(2m-1)}}{H_{2m-2}^{-(2m-2)}}$; and

(d) $\displaystyle\prod_{j=1}^{2m} \lambda_j = \beta_{2m}\mathcal{L}[R_{2m-1}^2(x)] = \dfrac{H_{2m}^{-(2m-1)}}{H_{2m-1}^{-(2m-2)}}$.

Proof: Parts (a) and (b) hold by (4.25) and parts (c) and (d) follow from parts (a) and (b) together with (4.22) and (4.23). ∎

4.2 The Five Term Recurrence Formulae

Now we will drop the assumption that a given OLPS is regular. In this case, a five term recurrence relation will hold.

Theorem 4.5. *Suppose $\mathcal{L}$ is a quasi-definite SMF with a corresponding monic OLPS $\{R_n(x)\}_{n=0}^{\infty}$. For each $n \in \mathbb{Z}_0^+$,*

$$R_{2n+2}(x) = \gamma_{2n+2,2n+1}R_{2n+1}(x) + (x + \gamma_{2n+2,2n})R_{2n}(x) \tag{4.27}$$

$$+\gamma_{2n+2,2n-1}R_{2n-1}(x) + \gamma_{2n+2,2n-2}R_{2n-2}(x)$$

and

$$R_{2n+3}(x) = \gamma_{2n+3,2n+2}R_{2n+2}(x) + (x^{-1} + \gamma_{2n+3,2n+1})R_{2n+1}(x) \tag{4.28}$$

$$+\gamma_{2n+3,2n}R_{2n}(x) + \gamma_{2n+3,2n-1}R_{2n-1}(x)$$

where

$$\gamma_{2n+2,k} = \frac{-\mathcal{L}[xR_k(x)R_{2n}(x)]}{\mathcal{L}[R_k{}^2(x)]} \quad for \quad k = 2n-2, \ldots, 2n+1, \quad k \geq 0. \tag{4.29}$$

and

$$\gamma_{2n+3,k} = \frac{-\mathcal{L}[x^{-1}R_k(x)R_{2n-1}(x)]}{\mathcal{L}[R_k{}^2(x)]} \quad for \quad k = 2n-1, \ldots, 2n+2, \quad k \geq 0. \tag{4.30}$$

If $k < 0$, then $\gamma_{m,k} = 0$, $m \geq 2$.

Proof: Equations (4.28) and (4.30) will be verified ((4.27) and (4.29) follow by similar arguments).

By equation (4.4), $x^{-1}R_{2n+1}(x)$ is a monic element of $\mathcal{R}_{2n+3}$ for all $n \in \mathbb{Z}_0^+$. Hence,

$$x^{-1}R_{2n+1}(x) = R_{2n+3}(x) - \sum_{j=0}^{2n+2} \gamma_{2n+3,j}R_j(x) \tag{4.31}$$

where

$$\gamma_{2n+3,k} = \frac{-\mathcal{L}[x^{-1}R_k(x)R_{2n+1}(x)]}{\mathcal{L}[R_k{}^2(x)]}, \quad k = 0, 1, \ldots, 2n+2, \quad n \in \mathbb{Z}_0^+.$$

Since $x^{-1}R_{2m+1}(x) \in \mathcal{R}_{2m+3}$, $\gamma_{2n+3,2m+1} = 0$ for $m = 0, \ldots, n-2$ where $n \geq 2$. Equation (4.6) implies $x^{-1}R_{2m}(x)$ is of L-degree $2m+1$ or less and therefore $\gamma_{2n+3,2m} = 0$ for $m = 0, \ldots, n-1$ where $n \geq 1$. Hence (4.31) reduces to

$$x^{-1}R_{2n+1}(x) = R_{2n+3}(x) - \sum_{k=2n-1}^{2n+2} \gamma_{2n+3,k}R_k(x), \quad n \geq 1. \tag{4.32}$$

For $n = 0$, (4.32) reduces to

$$x^{-1}R_1(x) = R_3(x) - \sum_{j=0}^{2} \gamma_{3,j}R_j(x)$$

which implies that

$$R_3(x) = \gamma_{3,2}R_2(x) + (x^{-1} - \gamma_{3,1})R_1(x) - \gamma_{3,0}R_0(x).$$

Hence (4.28) holds for all $n \in \mathbb{Z}_0^+$. ∎

5 A FAVARD THEOREM

In this section, a Favard Theorem for Laurent polynomials will be derived. A special case of Favard's Theorem for L-polynomials is given in [5]. In [3], a more general Favard's Theorem for L-polynomials is developed. The general Favard's Theorem given here includes necessary and sufficient conditions for determining when the corresponding SMF is positive-definite.

Theorem 5.1. *Let $\{\alpha_n\}_{n=1}^{\infty}$ and $\{\beta_n\}_{n=1}^{\infty}$ be sequences of nonzero complex numbers related by $\beta_n = \alpha_{n+1}$ for $n \in \mathbb{Z}^+$ (where $\alpha_1 = 1$) and let $\{\lambda_n\}_{n=1}^{\infty}$ be a sequence of complex numbers. Define $\{R_n(z)\}_{n=1}^{\infty}$ by letting $R_{-1}(z) \equiv 0$, $R_0(z) \equiv 1$,*

$$R_{2m+1}(z) = (\frac{z^{-1}}{\alpha_{2m+1}} + \beta_{2m+1})R_{2m}(z) + \lambda_{2m+1}R_{2m-1}(z) \tag{5.1}$$

and

$$R_{2m+2}(z) = (\frac{z}{\alpha_{2m+2}} + \beta_{2m+2})R_{2m+1}(z) + \lambda_{2m+2}R_{2m}(z) \tag{5.2}$$

for all $m \in \mathbb{Z}_0^+$. Then $\{R_n(x)\}_{n=0}^{\infty}$ is regular with the L-trailing coefficients of $R_{2m+1}(z)$ and $R_{2m+2}(z)$, given by β_{2m+1} and β_{2m+2}, respectively, while the respective L-leading coefficients are equal to one for all $m \in \mathbb{Z}_0^+$. There also exists a unique SMF $\mathcal{L}$ for which $\{R_n(x)\}_{n=0}^{\infty}$ is the corresponding monic OLPS if and only if $\lambda_n \neq 0$ for all $n \in \mathbb{Z}^+$. The SMF $\mathcal{L}$ is positive-definite if and only if

$$\frac{\lambda_n \alpha_n}{\beta_n} > 0 \quad for \ all \quad n \in \mathbb{Z}^+ \tag{5.3}$$

and $\{R_n(x)\}_{n=0}^{\infty}$ is a real OLPS for $\mathcal{L}$.

Comment: *Notice that the recurrence relations are determined by the two sequences $\{\alpha_n\}_{n=1}^{\infty}$ and $\{\lambda_n\}_{n=1}^{\infty}$ since $\beta_{m+1} = \alpha_{m+2}$ for all $m \in \mathbb{Z}_0^+$.*

Proof of Theorem 5.1: By (5.1) and (5.2),

$$R_1(z) = \frac{z^{-1}}{\alpha_1} + \beta_1 \quad \text{and} \quad R_2(z) = (\frac{z}{\alpha_2} + \beta_2)(\frac{z^{-1}}{\alpha_1} + \beta_1) + \lambda_2$$

or

$$R_1(z) = z^{-1} + \beta_1 \quad \text{and} \quad R_2(z) = \beta_2 z^{-1} + (\frac{1}{\alpha_2} + \beta_1\beta_2 + \lambda_2) + z$$

implying that $R_1(z)$ and $R_2(z)$ are regular, monic L-polynomials of L-degree 1 and 2, respectively. Also, β_1 and β_2 are the L-trailing coefficients of $R_1(z)$ and $R_2(z)$ respectively.

Assume $\{R_n(z)\}_{n=0}^{2m}$ is regular and monic with β_{2n-1} and β_{2n} being the L-trailing coefficients of $R_{2n-1}(z)$ and $R_{2n}(z)$ respectively, for some $n \in \mathbf{Z}^+$. By (5.1), the L-trailing coefficient of $R_{2m+1}(z)$ is β_{2m+1} and the L-leading coefficient is $\frac{\beta_{2m}}{\alpha_{2m+1}} = 1$. By (5.2), it is clear that the L-leading coefficient of $R_{2m+2}(z)$ is 1 and the L-trailing coefficient of $R_{2m}(z)$ is β_{2m+2}.

Next, define the functional $\mathcal{L}$ by

$$\mathcal{L}[R_n(z)] = 0 \quad \text{for all} \quad n \in \mathbf{Z}^+$$

and

$$\mathcal{L}(1) = \mu_0, \quad \text{where} \quad \mu_0 = \frac{\lambda_1}{\beta_1}.$$

If $\mu_n = \mathcal{L}(z^n)$ for $n \in \mathbf{Z}$, then

$$0 = \mathcal{L}[R_1(z)] = \mathcal{L}\left[\frac{1}{z} + \beta_1\right]$$

implying

$$\mu_{-1} = \mathcal{L}\left[\frac{1}{z}\right] = -\lambda_1.$$

Furthermore,

$$0 = \mathcal{L}[R_2(z)] = \mathcal{L}(z) + \left(\frac{1}{\alpha_2} + \beta_1\beta_2 + \lambda_2\right)\mu_0 + \beta_2\mathcal{L}\left(\frac{1}{z}\right)$$

implying

$$\mu_1 = \mathcal{L}(z) = -\beta_2\mathcal{L}\left[\frac{1}{z}\right] - \left(\frac{1}{\alpha_2} + \beta_1\beta_2 + \lambda_2\right)\mu_0.$$

Continuing in this fashion, a unique bisequence $\{\mu_n\}_{-\infty}^{\infty}$ is obtained implying that $\mathcal{L}$ is indeed a SMF. From (5.2), it follows that

$$\frac{z}{\alpha_{2m+2}}R_{2m+1}(z) = R_{2m+2}(z) - \beta_{2m+2}R_{2m+1}(z) - \lambda_{2m+2}R_{2m}(z)$$

for all $m \in \mathbf{Z}_0^+$. Replacing m by $m-1$ yields

$$\frac{z}{\alpha_{2m}}R_{2m-1}(z) = R_{2m}(z) - \beta_{2m}R_{2m-1}(z) - \lambda_{2m}R_{2m-2}(z), \quad m \in \mathbf{Z}^+. \tag{5.4}$$

From (5.1), it immediately follows that

$$\beta_{2m+1}R_{2m}(z) = R_{2m+1}(z) - \frac{z^{-1}}{\alpha_{2m+1}}R_{2m}(z) - \lambda_{2m+1}R_{2m-1}(z), \quad m \in \mathbf{Z}_0^+. \tag{5.5}$$

By applying an inductive argument to k, it will be shown that

$$\mathcal{L}[z^k R_{2m}(z)] = \mathcal{L}[z^k R_{2m-1}(z)] = 0 \tag{5.6}$$

for $k = 0, 1, \ldots, m - 1$ where $m \in \mathbb{Z}^+$. The fact that

$$\mathcal{L}[z^{-k} R_{2m}(z)] = \mathcal{L}[z^{-k} R_{2m+1}(z)] = 0 \qquad (5.7)$$

for $k = 1, \ldots, m$ where $m \in \mathbb{Z}^+$, will follow by a similar inductive argument. From the hypothesis, we deduce that (5.6) holds for $k = 0$ and $m \in \mathbb{Z}^+$. Suppose $k = 1$. By applying $\mathcal{L}$ to (5.4), it is clear that

$$\mathcal{L}[z R_{2m-1}(z)] = 0 \quad \text{for} \quad m \geq 2.$$

Then multiplying (5.5) by z and applying $\mathcal{L}$ yields

$$\mathcal{L}[z R_{2m}(z)] = 0 \quad \text{for} \quad m \geq 2.$$

Now assume that there exists an $N \in \mathbb{Z}^+$ such that (5.6) holds for each positive k satisfying $k \leq N$ where $N \leq m - 1$. We will show that (5.6) holds for $k = N + 1$ provided $N + 1 \leq m - 1$ or equivalently $N \leq m - 2$. Multiply (5.4) by z^N and apply $\mathcal{L}$ to see that

$$\mathcal{L}[z^{N+1} R_{2m-1}(z)] = 0 \quad \text{for} \quad N + 1 \leq m - 1.$$

Thus $\mathcal{L}[z^k R_{2m-1}(z)] = 0$ for any k provided $0 \leq k \leq m-1$. Similarly, by multiplying (5.5) by z^{N+1} and applying $\mathcal{L}$, it follows that

$$\mathcal{L}[z^{N+1} R_{2m}(z)] = 0 \quad \text{for} \quad N + 1 \leq m - 1.$$

Thus $\mathcal{L}[z^k R_{2m}(z)] = 0$ for any k provided $0 \leq k \leq m - 1$. Therefore, by induction on k, (5.6) holds.

From (5.5),

$$\frac{z^{-1}}{\alpha_{2m+1}} R_{2m}(z) = R_{2m+1}(z) - \beta_{2m+1} R_{2m}(z) - \lambda_{2m+1} R_{2m-1}(z), \quad m \in \mathbb{Z}_0^+. \qquad (5.8)$$

and from (5.4),

$$\beta_{2m} R_{2m-1}(z) = R_{2m}(z) - \frac{z}{\alpha_{2m}} R_{2m-1}(z) - \lambda_{2m} R_{2m-2}(z), \quad m \in \mathbb{Z}^+. \qquad (5.9)$$

Using an inductive argument together with equations (5.8) and (5.9), equation (5.7) can be shown to hold for $k = 1, 2, \ldots, m$ where $m \in \mathbb{Z}^+$.

From (5.6) and (5.7),

$$\mathcal{L}[z^k R_{2n}(z)] = 0 \quad \text{for} \quad k = -n, -n + 1, \ldots, n - 1, \quad n \in \mathbb{Z}^+$$

and

$$\mathcal{L}[z^k R_{2n+1}(z)] = 0 \quad \text{for} \quad k = -n, \ldots, n, \quad n \in \mathbb{Z}_0^+.$$

By the linearity of $\mathcal{L}$ it follows that

$$\mathcal{L}[R_m(z) R_n(z)] = 0 \quad \text{for all} \quad m \neq n, \quad m, n \in \mathbb{Z}_0^+.$$

Multiply equation (5.5) by z^m and apply $\mathcal{L}$ to obtain the equation

$$\beta_{2m+1}\mathcal{L}[z^m R_{2m}(z)] = -\lambda_{2m+1}\mathcal{L}[z^m R_{2m-1}(z)], \quad \text{for} \quad m \in \mathbb{Z}_0^+.$$

Similarly,

$$\frac{1}{\alpha_{2m}}\mathcal{L}[z^m R_{2m-1}(z)] = -\lambda_{2m}\mathcal{L}[z^{m-1}R_{2m-2}(z)], \quad m \in \mathbb{Z}^+$$

is obtained from (5.4). Therefore,

$$\mathcal{L}[z^m R_{2m}(z)] = \frac{\alpha_{2m}\lambda_{2m}\lambda_{2m+1}}{\beta_{2m+1}}\mathcal{L}[z^{m-1}R_{2m-2}(z)]$$

$$= \frac{\alpha_{2m-2}\alpha_{2m}\lambda_{2m-2}\lambda_{2m-1}\lambda_{2m}\lambda_{2m+1}}{\beta_{2m-1}\beta_{2m+1}}\mathcal{L}[z^{m-2}R_{2m-4}(z)]$$

or

$$\mathcal{L}[z^m R_{2m}(z)] = \frac{\alpha_{2m-2}}{\beta_{2m+1}}\lambda_{2m-2}\lambda_{2m-1}\lambda_{2m}\lambda_{2m+1}\mathcal{L}[z^{m-2}R_{2m-4}(z)]$$

or

$$\mathcal{L}[z^m R_{2m}(z)] = \frac{\alpha_4}{\beta_{2m+1}}\left(\prod_{k=4}^{2m+1}\lambda_k\right)\mathcal{L}[zR_2(z)] = \left(\frac{\alpha_4}{\beta_{2m+1}}\prod_{k=4}^{2m+1}\lambda_k\right)\frac{-\lambda_3}{\beta_3}\mathcal{L}[zR_1(z)]$$

$$= \left(\frac{-1}{\beta_{2m+1}}\prod_{k=3}^{2m+1}\lambda_k\right)\mathcal{L}[zR_1(z)] = \left(\frac{-1}{\beta_{2m+1}}\prod_{k=3}^{2m+1}\lambda_k\right)\alpha_2(-\lambda_2)\frac{\lambda_1}{\beta_1} = \frac{1}{\beta_{2m+1}}\prod_{k=1}^{2m+1}\lambda_k.$$

Therefore,

$$\mathcal{L}[z^m R_{2m}(z)] = \frac{1}{\beta_{2m+1}}\prod_{k=1}^{2m+1}\lambda_k, \tag{5.10}$$

for all $m \in \mathbb{Z}_0^+$.

Multiplying (5.9) by z^{-m} and applying $\mathcal{L}$ yields

$$\beta_{2m}\mathcal{L}[z^{-m}R_{2m-1}(z)] = -\lambda_{2m}\mathcal{L}[z^{-m}R_{2m-2}(z)], \quad \text{for} \quad m \in \mathbb{Z}^+.$$

Replacing m by $m-1$ in (5.8) produces

$$\frac{z^{-1}}{\alpha_{2m-1}}R_{2m-2}(z) = R_{2m-1}(z) - \beta_{2m-1}R_{2m-2}(z) - \lambda_{2m-1}R_{2m-3}(z), \quad m \in \mathbb{Z}^+.$$

Thus

$$\mathcal{L}[z^{-m}R_{2m-2}(z)] = -\alpha_{2m-1}\lambda_{2m-1}\mathcal{L}[z^{-m+1}R_{2m-3}(z)], \quad m \in \mathbb{Z}^+.$$

By an argument similar to the one used in establishing (5.10),

$$\mathcal{L}[z^{-m}R_{2m-1}(z)] = \frac{1}{\beta_{2m}}\prod_{k=1}^{2m}\lambda_k \tag{5.11}$$

for all $m \in \mathbb{Z}^+$. Since $\beta_n \neq 0$, $\mathcal{L}$ is quasi-definite if and only if $\lambda_n \neq 0$ for all $n \in \mathbb{Z}^+$.

Suppose $\mathcal{L}$ is positive-definite. Then by (5.10) and (5.11),

$$\frac{1}{\beta_{2m-1}} \prod_{k=1}^{2m-1} \lambda_k > 0 \quad \text{and} \quad \frac{1}{\beta_{2m}} \prod_{k=1}^{2m} \lambda_k > 0 \quad \text{for all} \quad m \in \mathbb{Z}^+.$$

Therefore,

$$\frac{\beta_{2m-1} \prod_{k=1}^{2m} \lambda_k}{\beta_{2m} \prod_{k=1}^{2m-1} \lambda_k} = \frac{\alpha_{2m} \lambda_{2m}}{\beta_{2m}} > 0$$

for all $m \in \mathbb{Z}^+$ and

$$\frac{\beta_{2m-2} \prod_{k=1}^{2m-1} \lambda_k}{\beta_{2m-1} \prod_{k=1}^{2m-2} \lambda_k} = \frac{\alpha_{2m-1} \lambda_{2m-1}}{\beta_{2m-1}} > 0$$

for all $m \in \mathbb{Z}^+$ which proves that (5.3) holds.

Now assume that (5.3) holds and $\{R_n(x)\}_{n=0}^{\infty}$ is a real OLPS. Then

$$\mathcal{L}(1) = \frac{\alpha_1}{\beta_1} \lambda_1 > 0.$$

By (5.10) and (5.11),

$$\frac{\mathcal{L}[R_{2m}^2(z)]}{\mathcal{L}[R_{2m-1}^2(z)]} = \frac{\alpha_{2m+1}}{\beta_{2m+1}} \lambda_{2m+1} > 0 \qquad (5.12)$$

and

$$\frac{\mathcal{L}[R_{2m-1}^2(z)]}{\mathcal{L}[R_{2m-2}^2(z)]} = \frac{\alpha_{2m}}{\beta_{2m}} \lambda_{2m} > 0 \qquad (5.13)$$

for all $m \in \mathbb{Z}^+$. Since $\mathcal{L}[R_0^2(z)] > 0$, then (5.12) and (5.13) imply $\mathcal{L}[R_n^2(z)] > 0$ for all $n \in \mathbb{Z}_0^+$. The result follows from Corollary 3.5. ∎

6 A CHRISTOFFEL-DARBOUX IDENTITY

As in the previous sections, the identities given here were motivated by the orthogonal polynomial analogues. Other Christoffel identities involving L-polynomials can be found in [10].

Theorem 6.1. *Suppose a sequence of L-polynomials, $\{R_n(x)\}_{n=0}^{\infty}$, satisfies the three term recurrence formulae given in Theorem 4.2., where α_{2m+1}, α_{2m+2}, and λ_{2m+1} are all nonzero for each $m \in \mathbb{Z}_0^+$. Then, with $\alpha_0 = 1$ and $R_{-1}(x) \equiv 0$,*

$$\sum_{m=0}^{n} \left[\frac{(x^{-1} - u^{-1})R_{2m}(x)R_{2m}(u)}{\alpha_{2m+1} \prod\limits_{i=1}^{2m+1} \lambda_i} - \frac{(x - u)R_{2m-1}(x)R_{2m-1}(u)}{\alpha_{2m} \prod\limits_{i=1}^{2m} \lambda_i} \right] \qquad (6.1)$$

$$= \frac{R_{2n+1}(x)R_{2n}(u) - R_{2n+1}(u)R_{2n}(x)}{\prod\limits_{i=1}^{2n+1} \lambda_i}$$

and

$$\sum_{m=0}^{n}\left[\frac{(x-u)R_{2m+1}(x)R_{2m+1}(u)}{\alpha_{2m+2}\prod_{i=1}^{2m+2}\lambda_i}-\frac{(x^{-1}-u^{-1})R_{2m}(x)R_{2m}(u)}{\alpha_{2m+1}\prod_{i=1}^{2m+1}\lambda_i}\right] \qquad (6.2)$$

$$=\frac{R_{2n+2}(x)R_{2n+1}(u)-R_{2n+2}(u)R_{2n+1}(x)}{\prod_{i=1}^{2n+2}\lambda_i}$$

for all $n \in \mathbb{Z}_0^+$, provided $\prod_{i=1}^{0}\lambda_i = 1$.

Proof: In (4.10), replace m by $m-1$ to see that

$$\frac{x}{\alpha_{2m}}R_{2m-1}(x) = R_{2m}(x) - \beta_{2m}R_{2m-1}(x) - \lambda_{2m}R_{2m-2}(x).$$

Then

$$\frac{x}{\alpha_{2m}}R_{2m-1}(x)R_{2m-1}(u) = R_{2m}(x)R_{2m-1}(u) - \beta_{2m}R_{2m-1}(x)R_{2m-1}(u) \qquad (6.3)$$

$$-\lambda_{2m}R_{2m-2}(x)R_{2m-1}(u)$$

and by interchanging u and x it immediately follows that

$$\frac{u}{\alpha_{2m}}R_{2m-1}(x)R_{2m-1}(u) = R_{2m}(u)R_{2m-1}(x) - \beta_{2m}R_{2m-1}(u)R_{2m-1}(x) \qquad (6.4)$$

$$-\lambda_{2m}R_{2m-2}(u)R_{2m-1}(x)$$

for all $m \in \mathbb{Z}^+$. Subtracting equations (6.3) and (6.4) shows that

$$\frac{x-u}{\alpha_{2m}}R_{2m-1}(x)R_{2m-1}(u) = [R_{2m}(x)R_{2m-1}(u) - R_{2m}(u)R_{2m-1}(x)]$$

$$-\lambda_{2m}[R_{2m-2}(x)R_{2m-1}(u) - R_{2m-2}(u)R_{2m-1}(x)].$$

Divide both sides of the equation by $\prod_{i=1}^{2m}\lambda_i$ to obtain

$$\frac{(x-u)R_{2m-1}(x)R_{2m-1}(u)}{\alpha_{2m}\prod_{i=1}^{2m}\lambda_i} = \frac{R_{2m}(x)R_{2m-1}(u) - R_{2m}(u)R_{2m-1}(x)}{\prod_{i=1}^{2m}\lambda_i} \qquad (6.5)$$

$$+\frac{R_{2m-1}(x)R_{2m-2}(u) - R_{2m-1}(u)R_{2m-2}(x)}{\prod_{i=1}^{2m-1}\lambda_i}$$

for all $m \in \mathbb{Z}^+$.

Define

$$F_n(x, u) = \frac{R_n(x)R_{n-1}(u) - R_n(u)R_{n-1}(x)}{\displaystyle\prod_{i=1}^{n} \lambda_i}, \quad n \in \mathbb{Z}_0^+,$$

and let $F_{-1}(x) = 0$. Then (6.5) reduces to

$$\frac{(x - u)R_{2m-1}(x)R_{2m-1}(u)}{\displaystyle\alpha_{2m}\prod_{i=1}^{2m} \lambda_i} = F_{2m}(x, u) + F_{2m-1}(x, u), \quad m \in \mathbb{Z}_0^+. \qquad (6.6)$$

The fact that (6.6) also holds for $m = 0$ follows since $R_{-1}(x) = F_{-1}(x) = 0$. Therefore, for $n \in \mathbb{Z}_0^+$,

$$\sum_{m=0}^{n} \left[\frac{(x - u)R_{2m+1}(x)R_{2m+1}(u)}{\displaystyle\alpha_{2m+2}\prod_{i=1}^{2m+2} \lambda_i} - \frac{(x^{-1} - u^{-1})R_{2m}(x)R_{2m}(u)}{\displaystyle\alpha_{2m+1}\prod_{i=1}^{2m+1} \lambda_i} \right]$$

$$= \sum_{m=0}^{n} [F_{2m+2}(x, u) - F_{2m}(x, u)] = F_{2n+2}(x, u).$$

The proof of equation (6.2) is completely analogous to the proof of (6.1). ∎

The following corollary to Theorem 6.1. will be useful in proving a separation of zeros property later.

Corollary 6.2. *Suppose a sequence of L-polynomials, $\{R_n(x)\}_{n=0}^{\infty}$, satisfies the three term recurrence formulae given in Theorem 4.2., where α_{2m+1}, α_{2m+2}, and λ_{2m+1} are all nonzero for each $m \in \mathbb{Z}_0^+$. Then, with $\alpha_0 = 1$ and $R_{-1}(x) \equiv 0$,*

$$\sum_{m=0}^{n} \left[\frac{-R_{2m}^2(x)}{x^2 \alpha_{2m+1} \displaystyle\prod_{i=1}^{2m+1} \lambda_i} - \frac{R_{2m-1}^2(x)}{\alpha_{2m} \displaystyle\prod_{i=1}^{2m} \lambda_{2m}} \right] = \frac{R'_{2n+1}(x)R_{2n}(x) - R'_{2n}(x)R_{2n+1}(x)}{\displaystyle\prod_{i=1}^{2n+1} \lambda_i} \qquad (6.7)$$

and

$$\sum_{m=0}^{n} \left[\frac{R_{2m+1}^2(x)}{\alpha_{2m+2} \displaystyle\prod_{i=1}^{2m+2} \lambda_i} + \frac{R_{2m}^2(x)}{x^2 \alpha_{2m+1} \displaystyle\prod_{i=1}^{2m+1} \lambda_i} \right] = \frac{R'_{2n+2}(x)R_{2n+1}(x) - R'_{2n+1}(x)R_{2n+2}(x)}{\displaystyle\prod_{i=1}^{2n+2} \lambda_i}.$$

$$(6.8)$$

for each $n \in \mathbb{Z}_0^+$.

Proof: Dividing equation (6.1) by $x - u$ and rewriting

$$R_{2n+1}(x)R_{2n}(u) - R_{2n+1}(u)R_{2n}(x)$$

as

$$[R_{2n+1}(x) - R_{2n+1}(u)]R_{2n}(x) - [R_{2n}(x) - R_{2n}(u)]R_{2n+1}(x),$$

yields

$$\sum_{m=0}^{n} \left[\frac{-R_{2m}(x)R_{2m}(u)}{ux\alpha_{2m+1}\prod_{i=1}^{2m+1}\lambda_i} - \frac{R_{2m-1}(x)R_{2m-1}(u)}{\alpha_{2m}\prod_{i=1}^{2m}\lambda_i} \right]$$

$$= \frac{[R_{2n+1}(x) - R_{2n+1}(u)]R_{2n}(x) - [R_{2n}(x) - R_{2n}(u)]R_{2n+1}(x)}{(x-u)\prod_{i=1}^{2n+1}\lambda_i}.$$

Equation (6.7) immediately follows by letting $u \to x$. The proof of (6.8) is similar. ∎

7 POSITIVE-DEFINITENESS ON SUBSETS OF $\mathbb{R}$

In this section, a generalization of the definition of a positive-definite SMF is given which will lead to some new results regarding the zeros of elements of an OLPS corresponding to a given SMF. Let $E \subseteq \mathbb{R}\backslash\{0\}$ be given. A SMF $\mathcal{L}$ is said to be **positive-definite on E** if $\mathcal{L}[R(x)] > 0$ for every real L-polynomial $R(x)$ that does not vanish identically on E and is non-negative on E.

By comparing the two definitions for positive-definiteness, one can see that $\mathcal{L}$ is positive-definite if and only if $\mathcal{L}$ is positive-definite on $(-\infty, 0) \cup (0, \infty)$. The special case where $\mathcal{L}$ is positive-definite on $(0, \infty)$ will be the main focus of the rest of the paper.

Note that Chihara [1] gives a similar definition for a positive-definite moment functional defined on the set of all polynomials. The definition differs here in that the set E excludes zero since Laurent polynomials may be undefined at $x = 0$.

Theorem 7.1. *Suppose E is an infinite set with $E \subseteq \mathbb{R}\backslash\{0\}$ and $\mathcal{L}$ is a SMF which is positive-definite on E. Then*

(i) $\mathcal{L}$ is positive-definite on every set $S \subseteq \mathbb{R}\backslash\{0\}$ such that $E \subseteq S$, and

(ii) $\mathcal{L}$ is positive-definite on every dense subset of E.

Proof: Suppose the hypothesis holds with $S \subseteq \mathbb{R}\backslash\{0\}$ and $E \subseteq S$. Let $R \in \mathcal{R}^{\mathbb{R}}$ be given satisfying the conditions that $R(x)$ does not vanish on S and $R(x) \geq 0$ for all $x \in S$. Then $R(x) \geq 0$ on E. Combining the fact that E is an infinite set with the result that $R(x)$ has only a finite number of zeros, it follows that $R(x)$ does not vanish on E. Therefore $\mathcal{L}[R(x)] > 0$.

Now let $F \subseteq E$ be given such that F is dense in E and consider a real L-polynomial $R(x)$ such that $R(x)$ does not vanish identically on F and $R(x) \geq 0$ for all $x \in F$. Since $R(x)$ is continuous on E and F is dense in E, $R(x) \geq 0$ for all $x \in E$. ∎

From Theorem 7.1. it follows that if $\mathcal{L}$ is positive-definite on a nonempty open interval not containing 0, then $\mathcal{L}$ is positive-definite.

Part of the next theorem is found in [8], but the extended definition of positive-definiteness given here leads to a stronger result.

Theorem 7.2. *If $\mathcal{L}$ is a SMF which is positive-definite on an open interval $I \subseteq \mathbb{R}\backslash\{0\}$, where I is bounded or unbounded, and $\{R_n(x)\}_{n=0}^{\infty}$ is the corresponding monic OLPS for $\mathcal{L}$, then $\{R_n(x)\}_{n=0}^{\infty}$ is a regular sequence of L-polynomials and all the zeros of $R_n(x)$ are simple and located in I for each $n \in \mathbb{Z}^+$.*

Proof: Suppose a given SMF $\mathcal{L}$ is positive-definite on an open interval I not containing zero. Then either $I \subseteq (-\infty, 0)$ or $I \subseteq (0, \infty)$. Let $n \in \mathbb{Z}^+$ be given. Suppose that ν_n represents the number of zeros of $R_n(x)$ with odd multiplicity contained in I. Let the zeros of $R_{2n-1}(x)$ be denoted by $x_{2n-1,1} < x_{2n-1,2} < \cdots < x_{2n-1,\nu_{2n-1}}$. Define

$$T_{2n-1}(x) = \frac{\displaystyle\prod_{i=1}^{\nu_{2n-1}} (x - x_{2n-1,i})}{x^{n-1}},$$

where $\displaystyle\prod_{i=1}^{\nu_{2n-1}} (x - x_{2n,i}) = 1$ if $\nu_{2n-1} = 0$, and recall that

$$R_{2n-1}(x) = \frac{P_{2n-1}(x)}{x^n}$$

where $P_{2n-1}(x)$ is a polynomial of degree $2n-1$ or less having the same roots as $R_{2n-1}(x)$. Then

$$T_{2n-1}(x)R_{2n-1}(x) = \frac{\displaystyle\prod_{i=1}^{\nu_{2n-1}} (x - x_{2n-1,i})P_{2n-1}(x)}{x^{2n-1}}.$$

Since $\displaystyle\prod_{i=1}^{\nu_{2n-1}} (x - x_{2n-1,i})P_{2n-1}(x)$ is a polynomial having only zeros with even multiplicity, either

$$T_{2n-1}(x)R_{2n-1}(x) \geq 0 \quad \text{for all} \quad x \in I \tag{7.1}$$

or

$$T_{2n-1}(x)R_{2n-1}(x) \leq 0 \quad \text{for all} \quad x \in I. \tag{7.2}$$

Inequality (7.1) will hold if the leading coefficient of $P_{2n-1}(x)$ is negative and $I \subseteq (-\infty, 0)$ or if the leading coefficient of $P_{2n-1}(x)$ is positive and $I \subseteq (0, \infty)$. The second inequality (7.2) will hold if the leading coefficient of $P_{2n-1}(x)$ is positive and $I \subseteq (-\infty, 0)$ or if the leading coefficient of $P_{2n-1}(x)$ is negative and $I \subseteq (0, \infty)$. So in any case, either $\mathcal{L}[T_{2n-1}(x)R_{2n-1}(x)] < 0$ for all $x \in I$ or $\mathcal{L}[T_{2n-1}(x)R_{2n-1}(x)] > 0$ for all $x \in I$. If $\nu_{2n-1} < 2n - 1$, then $T_{2n-1}(x)$ is of L-degree $2n - 2$ or $2n - 3$ which in turn implies $\mathcal{L}[T_{2n-1}(x)R_{2n-1}(x)] = 0$, a contradiction. It follows that $\nu_{2n-1} = 2n - 1$ and hence all $2n - 1$ zeros of $R_{2n-1}(x)$ are simple and located in I.

A similar argument will show that $\nu_{2n} = 2n$ and hence all the zeros of $R_{2n}(x)$ are simple and located in I.

Since the number of simple zeros of $R_n(x)$ is n for each $n \in \mathbb{Z}_0^+$, it is clear that $\{R_n(x)\}_{n=0}^{\infty}$ is a regular sequence. ∎

The following corollary, which is an immediate consequence of Theorem 7.2., will be useful since the special case where $I = (0, \infty)$ will be examined in detail.

Corollary 7.3. If a SMF $\mathcal{L}$ is positive-definite on $(0, \infty)$ and $\{R_n(x)\}_{n=0}^{\infty}$ is the corresponding monic OLPS for $\mathcal{L}$, then $\{R_n(x)\}_{n=0}^{\infty}$ is a regular sequence of L-polynomials and for each $n \in \mathbb{Z}^+$, $R_n(x)$ has n simple zeros contained in $(0, \infty)$.

Suppose $\{R_n(x)\}_{n=0}^{\infty}$ is the monic OLPS corresponding to a SMF $\mathcal{L}$ which is positive-definite on $(0, \infty)$. Since $\{R_n(x)\}_{n=0}^{\infty}$ is regular, it satisfies the following three term recurrence relations

$$R_{2n+1}(x) = \left(\frac{x^{-1}}{r_{2n,-n}} + r_{2n+1,n} \right) R_{2n}(x) + \lambda_{2n+1} R_{2n-1}(x) \tag{7.3}$$

and

$$R_{2n+2}(x) = \left(\frac{x}{r_{2n+1,n}} + r_{2n+2,-n-1} \right) R_{2n+1}(x) + \lambda_{2n+2} R_{2n}(x) \tag{7.4}$$

for all $n \in \mathbb{Z}_0^+$, where $R_{-1}(x) \equiv 0$ and λ_1 is defined to be $\mu_0 r_{1,0}$. Further, $r_{2n,-n}$ and $r_{2n+1,n}$ are the L-trailing coefficients of $R_{2n}(x)$ and $R_{2n+1,n}(x)$, respectively. Since $R_{2n}(x)$ is monic and has $2n$ real, simple zeros in $(0, \infty)$, $\lim_{x \to \infty} R_{2n}(x) = \infty$ and $\lim_{x \to 0^+} R_{2n}(x) = \infty$. Therefore, the L-trailing coefficient of $R_{2n}(x)$ is positive. Next, $\lim_{x \to 0^+} R_{2n-1}(x) = \infty$, since $R_{2n-1}(x)$ is monic, implying $\lim_{x \to \infty} R_{2n-1}(x) = -\infty$. Hence, the L-trailing coefficient of $R_{2n-1}(x)$ is negative and the following lemma holds.

Lemma 7.4. If $\{R_n(x)\}_{n=0}^{\infty}$ is a monic OLPS corresponding to a SMF $\mathcal{L}$ that is positive-definite on $(0, \infty)$, then the sign of the L-trailing coefficient of $R_n(x)$ equals $(-1)^n$.

This lemma will be used to prove the next theorem which gives a necessary condition that the Hankel determinants must satisfy when a SMF is positive-definite on $(0, \infty)$.

The Hankel determinants involved in the next result have been discussed before, with respect to the necessary and sufficient conditions for the existence of a solution to the strong Stieltjes moment problem, in [5,8,9]. The proof of Theorem 7.5. given here is new and the theorem is stated without any explicit reference to the strong Stieltjes moment problem.

Theorem 7.5. If $\mathcal{L}$ is a SMF for a given bisequence $\{\mu_n\}_{-\infty}^{\infty}$, which is positive-definite on $(0, \infty)$, then

$$H_{2n-1}^{(-2n+1)} > 0 \quad and \quad H_{2n}^{(-2n+1)} > 0$$

for all $n \in \mathbb{Z}^+$.

Proof: Let $\{R_n(x)\}_{n=0}^{\infty}$ be the OLPS corresponding to $\mathcal{L}$, where $\mathcal{L}$ is positive-definite on $(0, \infty)$. By Theorem 2.5., for each $n \in \mathbb{Z}^+$,

$$r_{2n,-n} = \frac{H_{2n}^{(-2n+1)}}{H_{2n}^{(-2n)}} \quad and \quad r_{2n-1,n-1} = -\frac{H_{2n-1}^{(-2n+1)}}{H_{2n-1}^{(-2n+2)}}$$

for each $n \in \mathbb{Z}^+$. Since $\mathcal{L}$ is positive-definite,

$$H_{2n}^{(-2n)} > 0 \quad \text{and} \quad H_{2n-1}^{(-2n+2)} > 0$$

for all $n \in \mathbb{Z}_0^+$ according to Theorem 3.4.. Hence, by Lemma 7.4.,

$$H_{2n}^{(-2n+1)} > 0 \quad \text{and} \quad H_{2n-1}^{(-2n+1)} > 0$$

for all $n \in \mathbb{Z}^+$. ∎

Theorem 7.6. *Let $\mathcal{L}$ be a SMF which is positive-definite on $(0, \infty)$ and let $\{R_n(x)\}_{n=0}^{\infty}$ be the monic OLPS corresponding to $\mathcal{L}$. Then the elements of $\{R_n(x)\}_{n=0}^{\infty}$ satisfy equations (7.3) and (7.4) for all $n \in \mathbb{Z}_0^+$ with $\lambda_n < 0$ for all $n \in \mathbb{Z}^+$.*

Proof: Since $\mathcal{L}$ is positive-definite on $(0, \infty)$, then $\{R_n(x)\}_{n=0}^{\infty}$ is regular and therefore satisfies (7.3) and (7.4). By Lemma 7.4,

$$\frac{r_{2m-1,m-1}}{r_{2m,-m}} < 0 \quad \text{and} \quad \frac{r_{2m-2,-(m-1)}}{r_{2m-1,m-1}} < 0$$

for all $m \in \mathbb{Z}^+$. So by equation (4.24), $\lambda_n < 0$ for all $n \in \mathbb{Z}^+$. ∎

A separation of zeros property for elements of an OLPS corresponding to a SMF which is positive-definite on $(0, \infty)$ can now by stated. This theorem will lay the groundwork for examining the integral representations for a SMF which is positive-definite on $(0, \infty)$.

Theorem 7.7. *Let $\mathcal{L}$ be a SMF that is positive-definite on $(0, \infty)$ and suppose $\{R_n(x)\}_{n=0}^{\infty}$ is the corresponding monic OLPS. Let the zeros of $R_n(x)$ be denoted by $x_{n,1} < x_{n,2} < \cdots < x_{n,n}$ for all $n \in \mathbb{Z}^+$. Then*

$$x_{n+1,i} < x_{n,i} < x_{n+1,i+1}$$

for $i = 1, 2, \ldots, n$ and each $n \in \mathbb{Z}^+$.

Proof: By (6.7),

$$\frac{R'_{2n+1}(x_{2n+1,i})R_{2n}(x_{2n+1,i})}{\lambda_1 \lambda_2 \cdots \lambda_{2n+1}} > 0$$

or

$$R'_{2n+1}(x_{2n+1,i})R_{2n}(x_{2n+1,i}) < 0 \tag{7.5}$$

for $i = 1, 2, \ldots, 2n + 1$, where $n \in \mathbb{Z}^+$. Since all the zeros of $R_{2n+1}(x)$ are simple and $\lim\limits_{x \to \infty} R_{2n+1}(x) = -\infty$,

$$\text{sign } R'_{2n+1}(x_{2n+1,i}) = (-1)^i$$

for $i = 1, 2, \ldots, 2n + 1$. So by equation (7.5),

$$\text{sign } R_{2n}(x_{2n+1,i}) = (-1)^{i-1}$$

for $i = 1, 2, \ldots, 2n + 1$. Therefore, $R_{2n}(x)$ contains exactly one zero on the interval $(x_{2n+1,i}, x_{2n+1,i+1})$ for $i = 1, 2, \ldots, 2n$, where $n \in \mathbb{Z}^+$. Thus

$$x_{2n+1,i} < x_{2n,i} < x_{2n+1,i+1}$$

for $i = 1, 2, \ldots, 2n$, $n \in \mathbb{Z}^+$.

Using (6.8), a similar argument shows that

$$x_{2n+2,i} < x_{2n+1,i} < x_{2n+2,i+1}$$

for $i = 1, 2, \ldots, 2n + 1$, $n \in \mathbb{Z}_0^+$. ∎

8 GAUSSIAN QUADRATURE AND REPRESENTATIVE FUNCTIONS

A SMF $\mathcal{L}$ which is positive-definite on $(0, \infty)$, has a Riemann-Stieltjes integral representation. In order to establish this result, some quadrature rules will be established first. The formulae given here actually hold in the setting of positive-definite on $\mathbb{R}\backslash\{0\}$ and can be found in [5,6,7,8,10].

Let $\{R_n(x)\}_{n=0}^{\infty}$ be the monic OLPS corresponding to a SMF $\mathcal{L}$ which is positive-definite on $(0, \infty)$ and let $n \in \mathbb{Z}^+$ be given. Suppose $x_{n,1}, x_{n,2}, \ldots, x_{n,n}$, where $x_{n,1} < x_{n,2} < \ldots < x_{n,n}$, represent the zeros of $R_n(x)$ for each $n \in \mathbb{Z}^+$. Define

$$l_{2n,j}(x) = \frac{R_{2n}(x)}{(x - x_{2n,j})R'_{2n}(x_{2n,j})}$$

for $j = 1, 2, \ldots, 2n$ and

$$l_{2n+1,j}(x) = \frac{x R_{2n+1}(x)}{x_{2n+1,j}(x - x_{2n+1,j})R'_{2n+1}(x_{2n+1,j})}$$

for $j = 1, 2, \ldots, 2n+1$. Since $R_{2n}(x)$ is a regular, monic L-polynomial of L-degree $2n$ with $2n$ positive, real, simple zeros, $R_{2n}(x) = \dfrac{P_{2n}(x)}{x^n}$, where $P_{2n}(x)$ is a monic polynomial of exact degree $2n$ with a nonzero constant term which in turn implies $\frac{P_{2n}(x)}{x - x_{2n,j}}$ is a monic polynomial of exact degree $2n - 1$ with a nonzero constant term for $j = 1, 2, \ldots, 2n$. Therefore $l_{2n,j}(x)$ is of L-degree $2n - 1$. Furthermore,

$$l_{2n,j}(x_{2n,j}) = \lim_{x \to x_{2n,j}} \frac{R_{2n}(x) - R_{2n}(x_{2n,j})}{(x - x_{2n,j})R'_{2n}(x_{2n,j})} = \frac{R'_{2n}(x_{2n,j})}{R'_{2n}(x_{2n,j})} = 1$$

and hence $l_{2n,j}(x_{2n,k}) = \delta_{j,k}$ for $j, k = 1, 2, \ldots, 2n$. By a similar argument, we can show that $l_{2n+1,j}(x)$ is of L-degree $2n$, $l_{2n+1,j}(x_{2n+1,j}) = 1$, and $l_{2n+1,j}(x_{2n+1,k}) = \delta_{j,k}$ for $j, k = 1, 2, \ldots, 2n+1$. So for each $n \in \mathbb{Z}^+$, $l_{n,j}(x)$ is of L-degree $n-1$ with $l_{n,j}(x_{n,k}) = \delta_{j,k}$ for $j, k = 1, 2, \ldots, n$.

Next, define

$$L_n(x) = \sum_{k=1}^{n} y_k l_{n,k}(x)$$

for each $n \in \mathbb{Z}^+$, where $\{y_1, \ldots, y_n\}$ is an arbitrary set of complex numbers. Thus, $L_n(x)$ is of L-degree at most $n - 1$ and $L_n(x_{n,j}) = y_j$ for $j = 1, 2, \ldots, n$.

The function $L_n(x)$ is called a **Lagrange interpolating L-polynomial** corresponding to the nodes $\{x_{n,1}, x_{n,2}, \ldots, x_{n,n}\}$ and the ordinates $\{y_1, y_2, \ldots, y_n\}$. With these results in hand, we are ready for the following theorem.

Theorem 8.1. *Suppose $\mathcal{L}$ is a SMF which is positive-definite on $(0, \infty)$ and is determined by the bisequence $\{\mu_n\}_{-\infty}^{\infty}$. Let $\{R_n(x)\}_{n=0}^{\infty}$ represent the corresponding monic OLPS for $\mathcal{L}$. For each $n \in \mathbb{Z}^+$, if $x_{n,1}, x_{n,2}, \ldots, x_{n,n}$ represent the zeros of $R_n(x)$ then*

$$\mathcal{L}[R(x)] = \sum_{k=1}^{n} A_{n,k} R(x_{n,k}) \tag{8.1}$$

for every L-polynomial, $R(x)$, of L-degree $2n - 1$ or less. Furthermore, $A_{n,k} > 0$ for $k = 1, 2, \ldots, n$ and $\sum_{k=1}^{n} A_{n,k} = \mu_0$.

Proof: Fix $n \in \mathbb{Z}^+$ and define

$$L_n(x) = \sum_{j=1}^{n} R(x_{n,j}) l_{n,j}(x)$$

so that $L_n(x)$ is the Lagrange interpolating L-polynomial corresponding to the nodes $\{x_{n,j}\}_{j=1}^{n}$ and the ordinates $\{R_n(x_{n,j})\}_{j=1}^{n}$. Note that $L_n(x)$ is of L-degree at most $n - 1$ since $l_{n,j}(x)$ is of L-degree $n - 1$ for $j = 1, 2, \ldots, n$. Next, let $Q(x) = L_n(x) - R(x)$ and note that $Q(x)$ is of L-degree at most $2n - 1$ with $Q(x_{n,k}) = L_n(x_{n,k}) - R(x_{n,k}) = 0$. Hence,

$$Q(x) = R_n(x) S(x)$$

where $S(x)$ is an L-polynomial of L-degree at most $n - 1$. Then due to orthogonality, $\mathcal{L}[Q(x)] = 0$ and it follows that

$$\mathcal{L}[R(x)] = \mathcal{L}[L_n(z) - Q(z)] = \mathcal{L}[L_n(x)] = \sum_{j=1}^{n} R(x_{n,j}) \mathcal{L}[l_{n,j}(x)].$$

Let $A_{n,j} = \mathcal{L}[l_{n,j}(x)]$ for $j = 1, \ldots, n$ so that (8.1) holds. To show that $A_{n,j} > 0$ for $j = 1, \ldots, n$, let $R(x) = l_{n,k}^2(x)$ for $j = 1, \ldots, n$. Since $\mathcal{L}$ is positive-definite,

$$0 < \mathcal{L}[l_{n,k}^2(x)] = \mathcal{L}\left[\sum_{j=1}^{n} l_{n,k}^2(x_{n,j}) l_{n,j}(x)\right] = \mathcal{L}[l_{n,k}(x)] = A_{n,k}$$

or $A_{n,k} > 0$ for $k = 1, \ldots, n$. If $R(x) = 1$, then $\mu_0 = \mathcal{L}(1) = \sum_{j=1}^{n} A_{n,j}$ completing the proof. ∎

By Theorem 8.1.,

$$\mu_j = \mathcal{L}[x^j] = \sum_{k=1}^{n} A_{n,k} x_{n,k}^j$$

for $j = -n, \ldots, n-1$ where $n \in \mathbb{Z}^+$. Define

$$\psi_n(x) = \begin{cases} 0, & x < x_{n,1} \\ A_{n,1} + \cdots + A_{n,p}, & x_{n,p} \leq x < x_{n,p+1}, \quad 1 \leq p \leq n-1 \\ \mu_0, & x \geq x_{n,n} \end{cases} \tag{8.2}$$

It immediately follows that

$$\mu_k = \int_0^\infty x^k d\psi_n(x)$$

for $k = -n, \ldots, n-1$. To proceed, the following two theorems, known as *Helly's Selection Theorems*, will be used. These theorems will be stated here without proof. The theorems and their respective proofs can be found in [1].

Theorem 8.2. *Suppose* $\{\psi_n(x)\}_{n=1}^\infty$ *is a sequence of uniformly bounded, nondecreasing functions on* $(-\infty, \infty)$. *Then there exists a subsequence* $\{\psi_{n_k}(x)\}_{k=1}^\infty$ *of* $\{\psi_n(x)\}_{n=1}^\infty$ *which converges on* $(-\infty, \infty)$ *to a bounded, nondecreasing function* $\psi(x)$.

Theorem 8.3. *If* $\{\phi_n(x)\}_{n=1}^\infty$ *is a sequence of uniformly bounded, nondecreasing functions on a compact interval* $[a, b]$ *which converges on* $[a, b]$ *to some limit function* $\phi(x)$, *then for every real function* $f(x)$ *which is continuous on* $[a, b]$,

$$\lim_{n \to \infty} \int_a^b f(x) d\phi_n(x) = \int_a^b f(x) d\phi(x).$$

By Theorem 8.2., there exists a subsequence $\{\psi_{n_k}(x)\}_{k=1}^\infty$ of $\{\psi_n(x)\}_{n=1}^\infty$ such that $\psi_{n_k}(x) \to \psi(x)$ as $k \to \infty$, where $\psi(x)$ is a bounded, nondecreasing function. It is claimed that

$$\int_0^\infty x^k d\psi(x) = \mu_k$$

for all $k \in \mathbb{Z}$. To prove this, the case where $k \in \mathbb{Z}_0^+$ is considered first. Let $\alpha > 0$ be given. Then

$$\left| \int_0^\alpha x^k d\psi(x) - \mu_k \right| = \left| \int_0^\alpha x^k d\psi(x) - \int_0^\infty x^k d\psi_{n_i}(x) \right|$$

$$\leq \left| \int_0^\alpha x^k d\psi(z) - \int_0^\alpha x^k d\psi_{n_i}(x) \right| + \left| \int_\alpha^\infty x^k d\psi_{n_i}(x) \right|$$

for $n_i \geq k+1$. Since

$$\left| \int_\alpha^\infty x^k d\psi_{n_i}(x) \right| = \left| \int_\alpha^\infty \frac{x^{k+1}}{x} d\psi_{n_i}(x) \right| \leq \frac{1}{\alpha} \left| \int_\alpha^\infty x^{k+1} d\psi_{n_i}(x) \right|$$

$$\leq \frac{1}{\alpha} \left| \int_0^\infty x^{k+1} d\psi_{n_i}(x) \right| = \frac{1}{\alpha} \mu_{k+1}$$

for $k+1 \leq n_i - 1$ (or $n_i \geq k+2$),

$$\left| \int_0^\alpha x^k d\psi(x) - \mu_k \right| \leq \left| \int_0^\alpha x^k d\psi(x) - \int_0^\alpha x^k d\psi_{n_i}(x) \right| + \frac{1}{\alpha} \mu_{k+1}$$

for $n_i \geq k + 2$. Let $i \to \infty$ and apply Theorem 8.3. to see that

$$\left| \int_0^\alpha x^k d\psi(x) - \mu_k \right| \leq \frac{1}{\alpha} \mu_{k+1}$$

and then let $\alpha \to \infty$ to conclude

$$\int_0^\infty x^k d\psi(x) = \mu_k, \quad \text{for} \quad k \geq 0.$$

Now the case where $k \in \mathbb{Z}^-$ will be considered. Let α and β be given such that $0 < \alpha < 1 < \beta < \infty$. Then

$$\left| \int_\alpha^\beta x^k d\psi(x) - \mu_k \right| = \left| \int_\alpha^\beta x^k d\psi(x) - \int_0^\infty x^k d\psi_{n_i}(x) \right|$$

$$\leq \left| \int_\alpha^\beta x^k d\psi(x) - \int_\alpha^\beta x^k d\psi_{n_i}(x) \right| + \left| \int_0^\alpha x^k d\psi_{n_i}(x) \right| + \left| \int_\beta^\infty x^k d\psi_{n_i}(x) \right|$$

for $n_i \geq -k$ and

$$\left| \int_0^\alpha x^k d\psi_{n_i}(x) \right| = \int_0^\alpha x^k d\psi_{n_i}(x) = \int_0^\alpha x^{k-1} x d\psi_{n_i}(x)$$

$$\leq \alpha \int_0^\alpha x^{k-1} d\psi_{n_i}(x) \leq \alpha \mu_{k-1}$$

for $n_i \geq -k + 1$. Also

$$\int_\beta^\infty x^k d\psi_{n_i}(x) \leq \mu_0 \beta^k.$$

Thus

$$\left| \int_\alpha^\beta x^k d\psi_{(x)} - \mu_k \right| \leq \left| \int_\alpha^\beta x^k d\psi(x) - \int_\alpha^\beta x^k d\psi_{n_i}(x) \right|$$

$$+ \alpha \mu_{k-1} + \beta^k \mu_0.$$

Let $i \to \infty$, then let $\alpha \to 0$ and $\beta \to \infty$ to conclude

$$\int_0^\infty x^k d\psi(x) = \mu_k, \quad \text{for} \quad k \geq 0.$$

Suppose $\psi(x)$ has only a finite number of points of increase, say at $x = x_i$ for $i = 1, 2, \ldots, j$ where $j \geq 1$. Define $P(x) = \prod_{i=1}^j (x - x_i)^2$. Since $\mathcal{L}$ is positive-definite on $(0, \infty)$ and $P(x) \geq 0$ with $P(x) \not\equiv 0$ on $(0, \infty)$, $\mathcal{L}[P(x)] > 0$. But $\mathcal{L}[P(x)] = \int_0^\infty P(x) d\psi(x) = 0$ since $P(x)$ vanishes at every point of increase of $\psi(x)$. Hence a contradiction has been reached. Therefore every SMF which is positive-definite on $(0, \infty)$ can be expressed as a Riemann-Stieltjes integral and the following theorem has been proved.

Theorem 8.4. *Let $\mathcal{L}$ be a SMF which is positive-definite on $(0, \infty)$ and define $\psi_n(x)$ as in (8.2) for each $n \in \mathbf{Z}^+$. Then there exists a subsequence $\{\psi_{n_k}(x)\}_{k=1}^{\infty}$ of $\{\psi_n(x)\}_{n=1}^{\infty}$ which converges to a bounded, nondecreasing function $\psi(x)$ with infinitely many points of increase such that*

$$\mathcal{L}(x^n) = \mu_n = \int_0^{\infty} x^n d\psi(x)$$

for each $n \in \mathbf{Z}$.

9 RELATED CONTINUED FRACTIONS

The purpose of this section is first to show the relationship between T-fractions and orthogonal L-polynomials and then to show how the quadrature weights introduced in the last section can be used to obtain an integral representation for a specific T-fraction that is closely related to the corresponding SMF.

Recall [4,11,12] that a continued fraction

$$\frac{a_1}{b_1} + \frac{a_2}{b_2} + \frac{a_3}{b_3} + \cdots$$

has an nth approximant C_n satisfying

$$C_n = \frac{A_n}{B_n}, \quad n \in \mathbf{Z}^+$$

where

$$A_n = b_n A_{n-1} + a_n A_{n-2}, \quad A_{-1} = 1, \quad A_0 = 0 \tag{9.1}$$

and

$$B_n = b_n B_{n-1} + a_n B_{n-2}, \quad B_{-1} = 0, \quad B_0 = 1 \tag{9.2}$$

for all $n \in \mathbf{Z}^+$. Furthermore,

$$A_n B_{n-1} - A_{n-1} B_n = (-1)^{n+1} a_1 a_2 \cdots a_n \tag{9.3}$$

for all $n \in \mathbf{Z}^+$.

Suppose $\{R_n(x)\}_{n=0}^{\infty}$ is a regular OLPS of the form given in Theorem 4.2 satisfying equations (4.9) through (4.13), for all $m \in \mathbf{Z}_0^+$, with $R_{-1}(x) \equiv 0$. Define $\{Q_n(x)\}_{n=0}^{\infty}$ such that

$$Q_{-1}(x) \equiv 1, \quad Q_0(x) \equiv 0, \tag{9.4}$$

$$Q_{2m+1}(x) = \left(\frac{x^{-1}}{\alpha_{2m+1}} + \beta_{2m+1} \right) Q_{2m}(x) + \lambda_{2m+1} Q_{2m-1}(x), \tag{9.5}$$

and

$$Q_{2m+2}(x) = \left(\frac{x}{\alpha_{2m+2}} + \beta_{2m+2} \right) Q_{2m+1}(x) + \lambda_{2m+2} Q_{2m}(x) \tag{9.6}$$

for all $m \in \mathbf{Z}_0^+$. The recurrence coefficient, λ_1, will be defined to be $\lambda_1 = \mu_0 \alpha_2 = \mu_0 \beta_1$ as in Theorem 4.3. Then, for each $n \in \mathbf{Z}^+$, $Q_n(x)$ and $R_n(x)$ represent the nth partial numerator and denominator, respectively, of the continued fraction

$$\frac{\lambda_1}{\frac{x^{-1}}{\alpha_1} + \beta_1} + \frac{\lambda_2}{\frac{x}{\alpha_2} + \beta_2} + \frac{\lambda_3}{\frac{x^{-1}}{\alpha_3} + \beta_3} + \frac{\lambda_4}{\frac{x}{\alpha_4} + \beta_4} + \cdots \tag{9.7}$$

for all $n \in \mathbf{Z}^+$. I.e., if $C_n(x)$ represents the nth approximant of (9.7), then

$$C_n(x) = \frac{Q_n(x)}{R_n(x)}$$

for all $n \in \mathbf{Z}^+$. It is not difficult to show that (9.7) is a T-fraction with the equivalent form

$$\frac{F_1 x}{1 + G_1 x} + \frac{F_2 x}{1 + G_2 x} + \frac{F_3 x}{1 + G_3 x} + \frac{F_4 x}{1 + G_4 x} + \cdots \qquad (9.8)$$

where

$$F_1 = \alpha_1 \lambda_1, \quad F_{2n} = \frac{\alpha_{2n-1} \lambda_{2n}}{\beta_{2n}}, \quad F_{2n+1} = \lambda_{2n+1}$$

and

$$G_{2n-1} = \alpha_{2n-1} \beta_{2n-1}, \quad G_{2n} = \frac{1}{\alpha_{2n} \beta_{2n}}$$

for all $n \in \mathbf{Z}^+$.

It is claimed that

$$Q_{2n-1}(x) = \sum_{j=-(n-1)}^{n-1} q_{2n-1,j} x^j$$

and

$$Q_{2n}(x) = \sum_{j=-(n-1)}^{n} q_{2n,j} x^j$$

for all $n \in \mathbf{Z}^+$ where $q_{2n-1,j} \in \mathbb{C}$ for $j = 1, 2, \ldots, 2n-1$ with $q_{2n-1,n-1} = \mu_0 \beta_{2n-1}$ and $q_{2n,j} \in \mathbb{C}$ for $j = 1, 2, \ldots, 2n$ with $q_{2n,n} = \mu_0$. To see this, note that for $n = 1$, (9.5) implies $Q_1(x) = \lambda_1$ so that $q_{1,0} = \lambda_1 = \beta_1 \frac{\lambda_1}{\beta_1} = \beta_1 \mu_0$. Then (9.6) yields $Q_2(x) = \left(\frac{x}{\alpha_2} + \beta_2\right) \lambda_1$ implying $q_{2,0} = \beta_2 \lambda_1$ and $q_{2,1} = \frac{\lambda_1}{\alpha_2} = \mu_0$. Now assume that (9.5) and (9.6) hold for $n = m$ where $q_{2m-1,m-1} = \mu_0 \beta_{2m-1}$ and $q_{2m,m} = \mu_0$. Then by (9.5), the coefficient corresponding to the highest power of x in $Q_{2m+1}(x)$ is the coefficient of x^m in $\beta_{2m+1} Q_{2m}(x)$. Therefore, this coefficient is $q_{2m+1,m} = \beta_{2m+1} q_{2m,m} = \beta_{2m+1} \mu_0$. The coefficient corresponding to the lowest power of x in $Q_{2m+1}(x)$ is the coefficient of x^{-m} in $\frac{x^{-1}}{\alpha_{2m+1}} Q_{2m}(x)$. This coefficient will be called $q_{2m+1,-m}$. One can then use (9.6) to complete the claim.

Let $T_{2n-2}(x) = x^{n-1} Q_{2n-1}(x)$ and $T_{2n-1}(x) = x^{n-1} Q_{2n}(x)$ for all $n \in \mathbf{Z}^+$. Then the leading coefficients of $T_{2n-2}(x)$ and $T_{2n-1}(x)$ are $\mu_0 \beta_{2n-1}$ and μ_0, respectively, where $T_n(x)$ is a polynomial of degree at most n, for each $n \in \mathbf{Z}^+$. Recall that $R_{2n-1}(x) = \frac{P_{2n-1}(x)}{x^n}$ where $P_{2n-1}(x)$ is a polynomial of degree $2n-1$ with a leading coefficient of $r_{2n-1,n} = \alpha_{2n}$ and $R_{2n}(x) = \frac{P_{2n}(x)}{x^n}$ where $P_{2n}(x)$ is a polynomial of degree $2n$ with a leading coefficient of one. It follows that

$$C_n(x) = \frac{x T_{n-1}(x)}{P_n(x)}, \quad n \in \mathbf{Z}^+.$$

Now make the further assumption that $\{R_n(x)\}_{n=0}^{\infty}$ corresponds to a SMF, $\mathcal{L}$, which is positive-definite on $(0, \infty)$. Then, by Theorem 7.6., $\lambda_n < 0$ for all $n \in \mathbb{Z}^+$. By equation (9.3),

$$Q_{2m+1}(x)R_{2m}(x) - Q_{2m}(x)R_{2m+1}(x) = \prod_{i=1}^{2m+1} \lambda_i < 0$$

and

$$Q_{2m+2}(x)R_{2m+1}(x) - Q_{2m+1}(x)R_{2m+2}(x) = - \prod_{i=1}^{2m+2} \lambda_i < 0$$

for all $m \in \mathbb{Z}_0^+$ so that

$$Q_{n+1}(x)R_n(x) - Q_n(x)R_{n+1}(x) < 0 \tag{9.9}$$

for all $n \in \mathbb{Z}^+$. Recall that $R_n(x)$ has a set, $\{x_{n,i}\}_{i=1}^{n}$, of n simple zeros satisfying

$$0 < x_{n,1} < x_{n,2} < \cdots < x_{n,n}.$$

Then (9.9) implies

$$Q_n(x_{n,i})R_{n+1}(x_{n,i}) > 0, \quad i = 1, 2, \ldots, n, \quad n \in \mathbb{Z}^+.$$

Since all the zeros of both $R_n(x)$ and $R_{n+1}(x)$ are simple and mutually separate each other, $R_{n+1}(x_{n,i})$ and $R_{n+1}(x_{n,i+1})$ have opposite signs. Hence, $Q_n(x_{n,i})$ and $Q_n(x_{n,i+1})$ also have opposite signs for $i = 1, 2, \ldots, n-1$. Therefore, in between consecutive zeros of $R_n(x)$ lies at least one zero of $Q_n(x)$. Since the nonzero roots of $Q_n(x)$ are the nonzero roots of $T_{n-1}(x)$ for $n > 2$ and $T_{n-1}(x)$ has at most $n-1$ simple zeros, there is exactly one zero of $Q_n(x)$ in between consecutive roots of $R_n(x)$. Therefore $Q_n(x)$ has exactly $n-1$ simple zeros of the form $y_{n,i}$, satisfying

$$x_{n,i} < y_{n,i} < x_{n,i+1}$$

for $i = 1, 2, \ldots, n-1$ where $n \geq 2$. The work completed so far in this section is summarized in the following theorem.

Theorem 9.1. Suppose $\{R_n(x)\}_{n=0}^{\infty}$ is an OLPS of the form given in Theorem 4.2. satisfying equations (4.9) through (4.13) with $R_{-1}(x) \equiv 0$ and $\lambda_1 = \mu_0 \alpha_2 = \mu_0 \beta_1$. If $\{Q_n(x)\}_{n=0}^{\infty}$ is defined as in (9.4) through (9.6), then

$$C_n(x) = \frac{Q_n(x)}{R_n(x)}, \quad n \in \mathbb{Z}^+$$

where $C_n(x)$ is the nth approximant of (9.7). Furthermore, if $\{R_n(x)\}_{n=0}^{\infty}$ corresponds to a SMF which is positive-definite on $(0, \infty)$, then the zeros of $Q_n(x)$ separate the zeros of $R_n(x)$. In other words, for each integer $n \geq 2$,

$$x_{n,i} < y_{n,i} < x_{n,i+1}, \quad i = 1, \ldots, n-1$$

where $\{y_{n,i}\}_{i=1}^{n}$ are the zeros of $Q_n(x)$ and $\{x_{n,i}\}_{i=1}^{n}$ are the zeros of $R_n(x)$.

The zeros of $T_{n-1}(x)$ are identically the zeros of $Q_n(x)$ and the zeros of $P_n(x)$ are exactly those of $R_n(x)$. Therefore, the zeros of $T_{2n-1}(x)$ separate the zeros of $P_{2n}(x)$ and this together with the fact that the degree of $T_{2n-1}(x)$ is one less than $P_{2n}(x)$, implies that there exist constants $D_{2n,1}, D_{2n,2}, \ldots, D_{2n,2n}$ such that

$$\frac{T_{2n-1}(x)}{P_{2n}(x)} = \frac{D_{2n,1}}{(x - x_{2n,1})} + \cdots + \frac{D_{2n,2n}}{(x - x_{2n,2n})}.$$

Therefore,

$$T_{2n-1}(x) = \frac{D_{2n,1}P_{2n}(x)}{(x - x_{2n,1})} + \cdots + \frac{D_{2n,2n}P_{2n}(x)}{(x - x_{2n,2n})}$$

implying that

$$\lim_{x \to x_{2n,j}} T_{2n-1}(x) = D_{2n,j}P'_{2n}(x_{2n,j})$$

and hence

$$D_{2n,j} = \frac{T_{2n-1}(x_{2n,j})}{P'_{2n}(x_{2n,j})}$$

for $j = 1, 2, \ldots, 2n$. Also

$$\sum_{j=1}^{2n} D_{2n,j} = \sum_{j=1}^{2n} \lim_{x \to \infty} \frac{x}{x - x_{2n,j}} D_{2n,j} = \lim_{x \to \infty} x \sum_{j=1}^{2n} \frac{D_{2n,j}}{x - x_{2n,j}} = \lim_{x \to \infty} \frac{x T_{2n-1}(x)}{P_{2n}(x)} = \mu_0.$$

In a similar fashion, there exist $D_{2n-1,1}, D_{2n-1,2}, \ldots, D_{2n-1,2n-1}$ such that

$$\frac{T_{2n-2}(x)}{P_{2n-1}(x)} = \frac{D_{2n-1,1}}{(x - x_{2n-1,1})} + \cdots + \frac{D_{2n-1,2n-1}}{(x - x_{2n-1,2n-1})}.$$

It can be shown that

$$D_{2n-1,j} = \frac{T_{2n-2}(x_{2n-1,j})}{P'_{2n-1}(x_{2n-1,j})}$$

for $j = 1, 2, \ldots, 2n - 1$ and

$$\sum_{j=1}^{2n-1} D_{2n-1,j} = \mu_0.$$

Recall the quadrature weights, $A_{n,k}$, first given in Theorem 8.1. It is claimed that

$$A_{n,j} = D_{n,j}$$

for all $n \in \mathbb{Z}^+$ and $j = 1, 2, \ldots, n$. Let $n \in \mathbb{Z}^+$ be given. Recall that

$$A_{2n-1,j} = \frac{1}{x_{2n-1,j} R'_{2n-1,j}(x_{2n-1,j})} \mathcal{L}\left[\frac{x R_{2n-1}(x)}{x - x_{2n-1,j}}\right]$$

and

$$D_{2n-1,j} = \frac{T_{2n-2}(x_{2n-1,j})}{P'_{2n-1}(x_{2n-1,j})}$$

for $j = 1, 2, \ldots, 2n - 1$ while

$$A_{2n,j} = \frac{1}{R'_{2n,j}(x_{2n,j})} \mathcal{L}\left[\frac{R_{2n}(x)}{x - x_{2n,j}}\right]$$

and

$$D_{2n,j} = \frac{T_{2n-1}(x_{2n,j})}{P'_{2n}(x_{2n,j})}$$

for $j = 1, 2, \ldots, 2n$. Since $P_{2n-1}(x) = x^n R_{2n-1}(x)$,

$$P'_{2n-1}(x) = nx^{n-1} R_{2n-1}(x) + x^n R'_{2n-1}(x).$$

Hence

$$P'_{2n-1}(x_{2n-1,j}) = x_{2n-1,j}^n R'_{2n-1}(x_{2n-1,j}), \quad j = 1, 2, \ldots, 2n - 1.$$

Similarly,

$$P'_{2n}(x_{2n,j}) = x_{2n,j}^n R'_{2n}(x_{2n,j}), \quad j = 1, 2, \ldots, 2n.$$

Also,

$$T_{2n-2}(x_{2n-1,j}) = x_{2n-1,j}^{n-1} Q_{2n-1}(x_{2n-1,j}), \quad j = 1, 2, \ldots, 2n - 1$$

and

$$T_{2n-1}(x_{2n,j}) = x_{2n,j}^{n-1} Q_{2n}(x_{2n,j}), \quad j = 1, 2, \ldots, 2n.$$

Hence

$$D_{2n-1,j} = \frac{Q_{2n-1}(x_{2n-1,j})}{x_{2n-1,j} R'_{2n-1}(x_{2n-1,j})}, \quad j = 1, 2, \ldots, 2n - 1$$

and

$$D_{2n,j} = \frac{Q_{2n}(x_{2n,j})}{x_{2n,j} R'_{2n}(x_{2n,j})}, \quad j = 1, 2, \ldots, 2n.$$

So in order to show that $A_{n,j} = D_{n,j}$, for $j = 1, 2, \ldots, n$, it suffices to show

$$Q_{2n-1}(x_{2n,j}) = \mathcal{L}\left[\frac{xR_{2n-1}(x)}{x - x_{2n-1,j}}\right], \quad j = 1, 2, \ldots, 2n - 1$$

and

$$\frac{Q_{2n}(x_{2n,j})}{x_{2n,j}} = \mathcal{L}\left[\frac{R_{2n}(x)}{x - x_{2n,j}}\right], \quad j = 1, 2, \ldots, 2n.$$

More generally it will be shown that

$$Q_{2n-1}(t) = \mathcal{L}\left[\frac{zR_{2n-1}(x) - tR_{2n-1}(t)}{x - t}\right] \tag{9.10}$$

and

$$\frac{Q_{2n}(t)}{t} = \mathcal{L}\left[\frac{R_{2n}(x) - R_{2n-1}(t)}{x - t}\right] \tag{9.11}$$

for all $n \in \mathbb{Z}^+$ by an argument used in [7]. For $n = 1$, $Q_1(t) = \lambda_1$ and

$$\mathcal{L}\left[\frac{xR_1(x) - tR_1(t)}{x - t}\right] = \mathcal{L}\left[\frac{\beta_1(x - t)}{x - t}\right] = \beta_1 \mathcal{L}[1] = \frac{\beta_1 \lambda_1}{\beta_1} = \lambda_1.$$

Therefore (9.10) holds for $n = 1$. Similarly, (9.11) holds for $n = 1$. Now assume (9.10) and (9.11) hold for $n = 1, 2, \ldots, m$.

Then, using (4.9) and (4.10),

$$\mathcal{L}\left[\frac{xR_{2m+1}(x) - tR_{2m+1}(t)}{x - t}\right] =$$

$$\mathcal{L}\left[\frac{\lambda_{2m+1}[xR_{2m-1}(x) - tR_{2m-1}(t)] + x[\frac{x^{-1}}{\alpha_{2m+1}} + \beta_{2m+1}]R_{2m}(x)}{x - t}\right.$$

$$\left.\frac{-t[\frac{t^{-1}}{\alpha_{2m+1}} + \beta_{2m+1}]R_{2m}(t)}{x - t}\right]$$

$$= \mathcal{L}\left[\frac{\lambda_{2m+1}[xR_{2m-1}(x) - tR_{2m-1}(t)]}{x - t}\right] + \mathcal{L}\left[\frac{\frac{1}{\alpha_{2m+1}}[R_{2m}(x) - R_{2m}(t)]}{x - t}\right]$$

$$+ \mathcal{L}\left[\frac{\beta_{2m+1}[xR_{2m}(x) - tR_{2m}(t)]}{x - t}\right]$$

$$= \lambda_{2m+1}Q_{2m-1}(t) + \frac{Q_{2m}(t)}{\alpha_{2m+1}t} + \mathcal{L}\left[\frac{\beta_{2m+1}[xR_{2m}(x) - tR_{2m}(x) + tR_{2m}(x) - tR_{2m}(t)]}{x - t}\right]$$

$$= \lambda_{2m+1}Q_{2m-1}(t) + \frac{Q_{2m}(t)}{\alpha_{2m+1}t} + \beta_{2m+1}Q_{2m}(t) = Q_{2m+1}(t).$$

Thus (9.10) holds for $n = m + 1$. Using this result along with the inductive step (9.11), it can be shown to hold for $n = m + 1$ in a similar fashion. Therefore

$$A_{n,j} = D_{n,j}, \quad n \in \mathbb{Z}^+, \quad j = 1, \ldots, n.$$

We will use the following theorem known as Grommer's Selection Theorem [9].

Theorem 9.2. *Let $\{\psi_n(t)\}$ be a sequence of real-valued nondecreasing functions defined on $-\infty < t < \infty$, such that $c \le \psi_n(t) \le C$ for all $-\infty < t < \infty$ where $n \in \mathbb{Z}^+$. Then there exists a real-valued nondecreasing function $\psi(t)$ defined on $-\infty < t < \infty$ such that $c \le \psi(t) \le C$ for all $-\infty < t < \infty$, and there exists a subsequence, $\{n_k\}$, of positive integers such that $\lim_{k \to \infty} \psi_{n_k}(t) = \psi(t)$ for $-\infty < t < \infty$. Moreover, if $g(t)$ is a continuous complex-valued function of the real variable t such that $\lim_{t \to \pm\infty} g(t) = 0$, then*

$$\lim_{k \to \infty} \int_{-\infty}^{\infty} g(t)d\psi_{n_k}(t) = \int_{-\infty}^{\infty} g(t)d\psi(t).$$

Recall that $A_{n,j} > 0$ for all $n \in \mathbb{Z}^+$ and $j = 1, \ldots, n$. Define

$$\psi_n(x) = \begin{cases} 0 & x < x_{n,1} \\ \sum_{j=1}^{p} A_{n,j} & x_{n,p} \le x < x_{n,p+1} \\ \mu_0 & x \ge x_{n,n} \end{cases} \tag{9.12}$$

so that

$$C_n(x) = \int_0^\infty \frac{x}{x-t} d\psi_n(t).$$

Note that $\{\psi_n(t)\}_{n=1}^\infty$ is a sequence of real, nondecreasing functions defined on $(-\infty, \infty)$ with $0 \le \psi_n(t) \le \mu_0$ for all $t \in (-\infty, \infty)$, for each $n \in \mathbb{Z}^+$. By Grommer's Selection Theorem, there exists a real-valued, nondecreasing function $\psi(t)$ defined on $(-\infty, \infty)$ such that $0 \le \psi(t) \le \mu_0$ for each $n \in \mathbb{Z}^+$ and a subsequence $\{\psi_{n_k}(t)\}_{k=1}^\infty$ of $\{\psi_n(t)\}_{n=1}^\infty$ such that

$$\lim_{k \to \infty} \psi_{n_k}(t) = \psi(t)$$

for $-\infty < t < \infty$.

Suppose $x \in \mathbb{R}$, but $x \notin (0, \infty)$ and $\mathcal{L}$ is positive-definite on $(0, \infty)$. Let $g(t) = \frac{x}{x-t}$. Note that $g(t)$ is continuous everywhere on $\mathbb{R}$ except at $t = x$. But this is not a problem since $\psi(t)$ is constant off of $(0, \infty)$. Choose $\varepsilon > 0$ but small enough such that $(x - \varepsilon, x + \varepsilon) \cap (0, \infty) = \emptyset$. Let $\hat{g}(t) = g(t)$ for $t \notin (x - \varepsilon, x + \varepsilon)$ and let $\hat{g}(t)$ be a straight line on $[x - \varepsilon, x + \varepsilon]$ satisfying $\hat{g}(x + \varepsilon) = g(x + \varepsilon)$ and $\hat{g}(x - \varepsilon) = g(x - \varepsilon)$ so that $\hat{g}(t)$ is a continuous, real-valued function of the real variable t with $\lim_{t \to \pm\infty} \hat{g}(t) = 0$. Therefore

$$\int_0^\infty \frac{x}{x-t} d\psi_{n_k}(t) = \int_0^\infty \hat{g}(t) d\psi_{n_k}(t)$$

and so by Grommer's Selection Theorem

$$\lim_{k \to \infty} C_{n_k}(x) = \lim_{k \to \infty} \int_0^\infty \hat{g}(t) d\psi_{n_k}(t) = \int_0^\infty \hat{g}(t) d\psi(t) = \int_0^\infty \frac{x}{x-t} d\psi(t).$$

The following theorem has been proved.

Theorem 9.3. *Suppose $\{R_n(x)\}_{n=0}^\infty$ is an OLPS corresponding to a SMF $\mathcal{L}$, which is positive-definite on $(a, b] \subseteq (0, \infty)$ and satisfies equations (4.9) through (4.13) with $\lambda_1 = \mu_0 \alpha_2 = \mu_0 \beta_1$. Let $\{Q_n(x)\}_{n=0}^\infty$ be defined as in (9.4) through (9.6). Then the quadrature weights, $A_{n,k}$, first given in Theorem 8.1., satisfy*

$$A_{n,k} = \frac{Q_n(x_{n,j})}{x_{n,j} R_n'(x_{n,j})}, \quad j = 1, \ldots, n, \quad n \in \mathbb{Z}^+$$

where $\{x_{n,j}\}_{j=1}^n$ are the roots of $R_n(x)$. Furthermore if the elements of $\{\psi_n(t)\}_{n=1}^\infty$ are defined as in (9.12), then

$$C_n(x) = \int_0^\infty \frac{x}{x-t} d\psi_n(t)$$

and there exists a subsequence $\{\psi_{n_k}(t)\}_{k=1}^\infty$ of $\{\psi_n(t)\}_{n=1}^\infty$ such that for every nonzero $x \in \mathbb{R} \backslash [0, \infty]$,

$$\lim_{k \to \infty} C_{n_k}(x) = \int_0^\infty \frac{x}{x-t} d\psi(t).$$

10 SPECTRAL ANALYSIS

A nondecreasing function $\psi(x)$ with infinitely many points of increase such that

$$\mathcal{L}(x^n) = \mu_n = \int_0^\infty x^n \, d\psi(x)$$

for a SMF $\mathcal{L}$ will be called a **representative function** for $\mathcal{L}$ and if $\psi(x)$ is the limit of a subsequence of step functions $\{\psi_{n_k}\}_{k=1}^\infty$, as defined in (8.2), then $\psi(x)$ will be called a **natural representative** for $\mathcal{L}$.

In this section, the zeros of the elements of an OLPS that correspond to a SMF which is positive-definite on $(0, \infty)$, are related to the spectra of natural representatives. The terminology and notation used in this section are motivated by Chihara's spectral analysis of the zeros of orthogonal polynomials [1]. Similarities between the theorems in this section and in [1] can be seen clearly, but there are also some subtle differences.

Suppose $\psi(x)$ is a real-valued, nondecreasing function of the real variable x. The **spectrum**, $\mathcal{S}(\psi)$, is defined as

$$\mathcal{S}(\psi) = \{x \in (-\infty, \infty) : \psi(x + \varepsilon) - \psi(x - \varepsilon) > 0 \quad \text{for all} \quad \varepsilon > 0\}.$$

An element of $\mathcal{S}(\psi)$ is called a **spectral point** of $\psi(x)$. A real-valued, non-decreasing function, $\psi(x)$ of the real variable x is called a **strong distribution function** if $\mathcal{S}(\psi)$ is an infinite set and

$$\int_{-\infty}^\infty x^n \, d\psi(x) < \infty$$

for all $n \in \mathbb{Z}$.

The next theorem states that if a SMF $\mathcal{L}$ has a representative function $\psi(x)$, then $\mathcal{L}$ is positive-definite on the spectrum of $\psi(x)$.

Theorem 10.1. *Suppose $\psi(x)$ is a strong distribution function and let*

$$\mu_n = \int_{-\infty}^\infty x^n \, d\psi(x)$$

for all $n \in \mathbb{Z}$. If $\mathcal{L}$ is the SMF determined by the sequence $\{\mu_n\}_{-\infty}^\infty$, then $\mathcal{S}(\psi)$ is a supporting set for $\mathcal{L}$ (i.e. $\mathcal{L}$ is a positive-definite SMF on $\mathcal{S}(\psi)$).

Proof: Suppose $S(x) \in \mathcal{R}^{\mathbb{R}}$ is given such that $S(x) > 0$ for all $x \in \mathcal{S}(\psi)$ and suppose $S(x)$ does not vanish on $\mathcal{S}(\psi)$. Since $\psi(x)$ is non-decreasing, either $\psi(x)$ consists entirely of jump discontinuities or $\psi(x)$ contains an interval of the real line of the form (a, b) where $-\infty < a < b < \infty$.

First, assume $\psi(x)$ consists entirely of jump discontinuities. Since $S(x)$ only has a finite number of zeros, there exists $y \in \mathcal{S}(\psi)$ such that $\psi(y) > 0$. Thus

$$\int_{-\infty}^\infty S(x) d\psi(x) \geq LS(y)$$

where

$$L = \lim_{\varepsilon \to 0} [\psi(y + \varepsilon) - \psi(y - \varepsilon)].$$

Now $L > 0$ since $S(x)$ has a jump discontinuity at $x = y$ and hence $\mathcal{L}[S(x)] > 0$ since $\mathcal{L}[S(x)] = \int_{-\infty}^{\infty} S(x)d\psi(x) > 0$.

Next, suppose $\mathcal{S}(\psi)$ contains an open interval of the form (a, b) where $-\infty < a < b < \infty$. Choose c and d such that $a < c < d < b$. Then $S(x) > 0$ for all $x \in [c, d]$. Let $m = \min\{S(x) : c \le x \le d\}$. Then

$$\int_{-\infty}^{\infty} S(x) \ge \int_c^d S(x) \ge m[S(d) - S(c)] > 0$$

and so once again it follows that $\mathcal{L}[S(x)] > 0$. ∎

For the remainder of this section, it will be assumed that $\mathcal{L}$ is a SMF which is positive-definite on $(0, \infty)$ with a corresponding monic OLPS denoted by $\{R_n(x)\}_{n=0}^{\infty}$. Let $x_{n,1} < x_{n,2} < \cdots < x_{n,n}$ denote the zeros of $R_n(x)$ for all $n \in \mathbb{Z}^+$. By Theorem 7.7.,

$$x_{n+1,i} < x_{n,i} < x_{n+1,i+1} \tag{10.1}$$

for $i = 1, \ldots, n$ and $n \in \mathbb{Z}^+$.

For each $i \in \mathbb{Z}^+$ and $j \in \mathbb{Z}^+$, define

$$\xi_i = \lim_{n \to \infty} x_{n,i} \quad \text{and} \quad \eta_j = \lim_{n \to \infty} x_{n,n-j+1}.$$

By (10.1), it is clear that $\{x_{n,i}\}_{n=i}^{\infty}$ is a decreasing sequence for all $i \in \mathbb{Z}^+$ and hence $\xi_i < \infty$ for all $i \in \mathbb{Z}^+$. However $\{x_{n,n-j+1}\}_{n=j}^{\infty}$ is an increasing sequence and therefore it may be possible that $\eta_j = \infty$ for some $j \in \mathbb{Z}^+$. The following diagrams illustrate these definitions.

$$
\begin{array}{c}
x_{1,1} \\
x_{2,1} \quad x_{2,2} \\
x_{3,1} \quad x_{3,2} \quad x_{3,3} \\
x_{4,1} \quad x_{4,2} \quad x_{4,3} \quad x_{4,4} \\
x_{5,1} \quad x_{5,2} \quad x_{5,3} \quad x_{5,4} \quad x_{5,5} \\
x_{6,1} \quad x_{6,2} \quad x_{6,3} \quad x_{6,4} \quad x_{6,5} \quad x_{6,6} \\
\swarrow \quad \swarrow \quad \swarrow \quad \swarrow \quad \swarrow \quad \swarrow \\
\xi_1 \quad \xi_2 \quad \xi_3 \quad \xi_4 \quad \xi_5 \quad \xi_6 \quad \cdots
\end{array}
$$

$$
\begin{array}{c}
x_{1,1} \\
x_{2,1} \quad x_{2,2} \\
x_{3,1} \quad x_{3,2} \quad x_{3,3} \\
x_{4,1} \quad x_{4,2} \quad x_{4,3} \quad x_{4,4} \\
x_{5,1} \quad x_{5,2} \quad x_{5,3} \quad x_{5,4} \quad x_{5,5} \\
x_{6,1} \quad x_{6,2} \quad x_{6,3} \quad x_{6,4} \quad x_{6,5} \quad x_{6,6} \\
\searrow \quad \searrow \quad \searrow \quad \searrow \quad \searrow \quad \searrow \\
\cdots \quad \eta_6 \quad \eta_5 \quad \eta_4 \quad \eta_3 \quad \eta_2 \quad \eta_1
\end{array}
$$

Lemma 10.2. For each $i \in \mathbb{Z}^+$, $\xi_i < \eta_i$.

Proof: Let $i \in \mathbb{Z}^+$ be given. As stated earlier, $\{x_{n,i}\}_{n=i}^{\infty}$ is a decreasing sequence converging to ξ_i and $\{x_{n,n-i+1}\}_{n=i}^{\infty}$ is an increasing sequence converging to η_i for each

$i \in \mathbf{Z}^+$. Combining this with the fact that the ith element of each sequence is $x_{2i-1,i}$ yields $\xi_i < x_{2i-1,i} < \eta_i$. ∎

Next, let

$$\sigma = \lim_{i \to \infty} \xi_i \quad \text{and} \quad \tau = \lim_{j \to \infty} \eta_j$$

where either or both of σ and τ may equal ∞. Then Lemma 10.2. implies

$$0 \leq \xi_1 \leq \xi_2 \leq \cdots \leq \sigma \leq \tau \leq \cdots \leq \eta_2 \leq \eta_1 \leq \infty.$$

From the next theorem we conclude that the smallest closed interval which is a supporting set for $\mathcal{L}$ is $[\xi_1, \eta_1]$.

Theorem 10.3. *Let $\mathcal{L}$ be a SMF which is positive-definite on $(0, \infty)$. Further, let $\psi(x)$ be a natural representative function of $\mathcal{L}$ and let $\phi(x)$ denote an arbitrary representative function of $\mathcal{L}$. Then each of the following hold:*

(a) $\mathcal{S}(\psi) \subseteq [\xi_1, \eta_1]$; and

(b) if $\mathcal{S}(\phi) \subseteq [a, b]$, where $-\infty \leq a < b \leq \infty$, then $[\xi_1, \eta_1] \subseteq [a, b]$.

Proof: Let $\psi(x)$ be a natural representative of $\mathcal{L}$. Then $\psi(x)$ is the limit of a subsequence of $\{\psi_n(x)\}_{n=1}^{\infty}$, where $\psi_n(x)$ is defined as in (8.2) for all $n \in \mathbf{Z}^+$. Since $\xi_1 < x_{n,1}$ for all $n \in \mathbf{Z}^+$, $\psi_n(x) = 0$ for all $x < \xi_1$ and $n \in \mathbf{Z}^+$. Therefore $\psi(x) = 0$ for all $x < \xi_1$. Similarly, $\psi_n(x) = \mu_0$ for all $x > \eta_1$, since $\eta_1 > x_{n,n}$ for all $n \in \mathbf{Z}^+$, and hence $\psi(x) = \mu_0$ for all $x > \eta_1$. Thus the proof of part (a) is completed.

Let $\phi(x)$ be any representative of $\mathcal{L}$ and suppose $-\infty \leq a < b \leq \infty$ with $\mathcal{S}(\phi) \subseteq [a, b]$. By Theorem 10.1., $\mathcal{S}(\phi)$ is a supporting set for $\mathcal{L}$ and hence, by Theorem 7.1., $[a, b]$ is a supporting set for $\mathcal{L}$. Theorem 7.2. implies that both $x_{n,1} \in [a, b]$ and $x_{n,n} \in [a, b]$ for all $n \in \mathbf{Z}^+$ and hence $[\xi_1, \eta_1] \subseteq [a, b]$. ∎

Since $[\xi_1, \eta_1]$ is the smallest closed interval which is also a supporting set for $\mathcal{L}$, it is natural to define the closed set $[\xi_1, \eta_1]$ be the **true interval of orthogonality** for $\mathcal{L}$.

Theorem 10.4. *Let $\phi(x)$ be an arbitrary representative of a SMF $\mathcal{L}$ which is positive-definite on $(0, \infty)$. Then*

$$\mathcal{S}(\phi) \cap (x_{n,i}, x_{n,i+1}) \neq \emptyset$$

for $i = 1, \ldots, n-1$ and integers $n > 1$.

Proof: Suppose there exist $n \in \mathbf{Z}^+$ and $i \in \{1, \ldots, 2n-1\}$ such that $\mathcal{S}(\phi) \cap (x_{2n,i}, x_{2n,i+1}) = \emptyset$. Define

$$S(x) = \frac{R_{2n}(x)}{(x - x_{2n,i})(x - x_{2n,i+1})}.$$

Then $S(x) \in \mathcal{R}_{2n-1}$ which implies $\mathcal{L}[S(x)R_{2n}(x)] = 0$. But

$$S(x)R_{2n}(x) = \frac{R_{2n}^2(x)}{(x - x_{2n,i})(x - x_{2n,i+1})} \geq 0$$

for all nonzero, real $x \notin (x_{2n,i}, x_{2n,i+1})$ and therefore $S(x)R_{2n}(x) \geq 0$ for all $x \in S(\phi)$. Since $S(\phi)$ is a supporting set for $\mathcal{L}$, $\mathcal{L}[S(x)R_{2n}(x)] > 0$ and hence a contradiction has been reached.

A similar argument shows that $S \cap (x_{2n+1,i}, x_{2n+1,i+1}) \neq \emptyset$ for $n \in \mathbb{Z}^+$ and $i \in \{1, \ldots, 2n\}$. ∎

Theorem 10.5. *Let $\mathcal{L}$ be a SMF which is positive-definite on $(0, \infty)$ and suppose $\{R_n(x)\}_{n=0}^\infty$ is the corresponding monic OLPS. Suppose G is an open set with the property that $R_n(x)$ does not contain any zeros in G for integers $n \geq N$ for some fixed $N \in \mathbb{Z}^+$. Then for any natural representative function $\psi(x)$ of $\mathcal{L}$, $S(\psi) \cap G = \emptyset$.*

Proof: Let I be any open interval contained in G. There exists an $N \in \mathbb{Z}^+$ such that $R_n(x) \neq 0$ for all $x \in I$ and $n \geq N$. Assume $\psi(x)$ is a subsequential limit of the sequence $\{\psi_n(x)\}_{n=1}^\infty$ whose elements are of the form given in (8.2). Let $x, y \in I$ be given. Then $\psi_n(x) - \psi_n(y) = 0$ for all $n \geq N$ and hence $\psi(x) - \psi(y) = 0$ and so $\psi(x)$ is constant on I implying $S(\psi) \cap I = \emptyset$. Hence $S(\psi) \cap G = \emptyset$. ∎

Corollary 10.6. *Let $\psi(x)$ be a natural representative function for a SMF $\mathcal{L}$ which is positive-definite on $(0, \infty)$ and let $s \in S(\psi)$ be given. Then every neighborhood of s contains a zero of $R_n(x)$ for infinitely many values of $n \in \mathbb{Z}^+$.*

Proof: Let $\varepsilon > 0$ be given so that $(s - \varepsilon, s + \varepsilon)$ is an arbitrary neighborhood of s. Suppose there exists an $N \in \mathbb{Z}^+$ such that $R_n(x)$ contains no zeros in $(s - \varepsilon, s + \varepsilon)$ for all $n \geq N$. Then $(s - \varepsilon, s + \varepsilon) \cap S(\psi) = \emptyset$, which is a contradiction. ∎

A simple, but useful lemma is

Lemma 10.7. *The spectrum, $S(\phi)$, of a non-decreasing, real-valued function $\phi(x)$ is closed.*

Proof: Suppose x is a limit point of $S(\phi)$. Let $\varepsilon > 0$ be given. Then there exists an $s \in S(\phi)$ such that $s \in (x - \frac{\varepsilon}{2}, x + \frac{\varepsilon}{2})$. Then $x + \varepsilon \geq s + \frac{\varepsilon}{2}$ and $x - \varepsilon \leq s - \frac{\varepsilon}{2}$ and hence

$$\phi(x + \varepsilon) - \phi(x - \varepsilon) \geq \phi\left(s + \frac{\varepsilon}{2}\right) - \phi\left(s - \frac{\varepsilon}{2}\right) > 0.$$

Thus $\phi(x + \varepsilon) - \phi(x - \varepsilon) > 0$ for all $\varepsilon > 0$ which implies $x \in S(\phi)$. ∎

Theorem 10.8. *Let $\mathcal{L}$ be a SMF that is positive-definite on $(0, \infty)$ and suppose $\phi(x)$ is an arbitrary representative function of $\mathcal{L}$. If $\xi_i = \xi_{i+1}$ for some $i \in \mathbb{Z}^+$, then ξ_i is a limit point of $S(\phi)$ and hence $\xi_i \in S(\phi)$.*

Proof: Let $\varepsilon > 0$ be given. Then there exists an $N \in \mathbb{Z}^+$ such that $(x_{n,i}, x_{n,i+1}) \subseteq (\xi_i, \xi_i + \varepsilon)$ for $n \geq N$. Therefore by Theorem 10.4., $S(\phi) \cap (\xi_i, \xi_i + \varepsilon) \neq \emptyset$. Hence every neighborhood of ξ_i contains an element of $S(\phi)$ not equal to ξ_i implying that ξ_i is a limit point of $S(\phi)$ and since $S(\phi)$ is closed, $\xi_i \in S(\phi)$. ∎

Theorem 10.9. *Let $\mathcal{L}$ be a SMF which is positive-definite on $(0, \infty)$ and suppose $\psi(x)$ is a natural representative of $\mathcal{L}$. Then $\xi_i \in S(\psi)$ for all $i \in \mathbb{Z}^+$.*

Proof: Suppose $\xi_1 \notin S(\psi)$. Then there exists an $\varepsilon > 0$ such that $\psi(\xi_1 + \varepsilon) - \psi(\xi_1 - \varepsilon) = 0$ and so by Theorem 10.3 (a), $S(\psi) \subseteq [\xi_1 + \varepsilon, \eta_1]$. But from Theorem 10.3.(b), $[\xi_1, \eta_1] \subseteq [\xi_1 + \varepsilon, \eta_1]$, a contradiction. Thus $\xi_1 \in S(\psi)$.

Now let $i \in \mathbb{Z}^+\backslash 1$. If $\xi_{i-1} = \xi_i$, then by Theorem 10.8., $\xi_i \in \mathcal{S}(\psi)$. Suppose $\xi_{i-1} < \xi_i$ and let $\varepsilon > 0$. Then there exists an $N \in \mathbb{Z}^+$ such that $(x_{n,i-1}, x_{n,i}) \subseteq (\xi_{i-1}, \xi_i + \varepsilon)$ for $n \geq N$. So by Theorem 10.4., $\mathcal{S}(\psi) \cap (\xi_{i-1}, \xi_i + \varepsilon) \neq \emptyset$. Choose α such that $\xi_{i-1} < \alpha < \xi_i$. Since $\xi_i < x_{n,i}$ for all $n \in \mathbb{Z}^+$ and $\xi_{i-1} < x_{n,i-1} < \alpha$ for sufficiently large n, (α, ξ_i) contains no zeros of $R_n(x)$ for sufficiently large n. Hence Theorem 10.5. implies that $\mathcal{S}(\psi) \cap (\alpha, \xi_i) = \emptyset$. Thus $\mathcal{S}(\psi) \cap (\xi_{i-1}, \xi_i) = \emptyset$ and therefore $\mathcal{S}(\psi) \cap [\xi_i, \xi_i + \varepsilon) \neq \emptyset$. So either $\xi_i \in \mathcal{S}(\psi)$ or ξ_i is a limit point of $\mathcal{S}(\psi)$. Either way, $\xi_i \in \mathcal{S}(\psi)$ since $\mathcal{S}(\psi)$ is closed.

∎

Theorem 10.10. *If $\mathcal{L}$ is a SMF which is positive-definite on $(0, \infty)$ and $\psi(x)$ is a natural representative of $\mathcal{L}$, then the following hold:*

a) $\sigma \in \mathcal{S}(\psi)$;

b) if $x \in \mathcal{S}(\psi)$ and $x < \sigma$, then $x = \xi_i$ for some $i \in \mathbb{Z}^+$;

c) if $\xi_i = \xi_{i+1}$ for some $i \in \mathbb{Z}^+$, then $\xi_p = \sigma$ for all $p \geq i$.

Proof: (a) Suppose $\xi_i < \xi_{i+1}$ for all $i \in \mathbb{Z}^+$. By Theorem 10.9., $\xi_i \in \mathcal{S}(\psi)$ and since $\lim_{i \to \infty} \xi_i = \sigma$, σ is a limit point of $\mathcal{S}(\psi)$ and hence $\sigma \in \mathcal{S}(\psi)$. If $\xi_i = \xi_{i+1}$ for some $i \in \mathbb{Z}^+$, the result still holds since $\xi_i \in \mathcal{S}(\psi)$.

(b) Let $x \in \mathcal{S}(\psi)$ and $x < \sigma$. Then by Theorem 10.3.(a), $\xi_1 \leq x$. In order to reach a contradiction, suppose $x \neq \xi_i$ for any $i \in \mathbb{Z}^+$. Then there exists a $j \in \mathbb{Z}^+$ such that $x \in (\xi_j, \xi_{j+1})$. But $(\xi_j, \xi_{j+1}) \cap \mathcal{S}(\psi) = \emptyset$ by the proof of Theorem 10.9. and hence $x = \xi_i$ for some $i \in \mathbb{Z}^+$.

(c) Suppose there exists an $i \in \mathbb{Z}^+$ such that $\xi_i = \xi_{i+1}$. It is claimed that $\xi_i = \sigma$. To show this, suppose instead that $\xi_i < \sigma$. Let a sufficiently small $\varepsilon > 0$ be given satisfying $\xi_i + \varepsilon < \sigma$. Since $\lim_{n \to \infty} \xi_n = \sigma$ and since the ξ_n's are the only elements of $\mathcal{S}(\psi)$ which are possibly smaller than σ, there are only finitely many elements of $\mathcal{S}(\psi)$ contained in $(\xi_i - \varepsilon, \xi_i + \varepsilon)$ and therefore ξ_i is not a limit point of $\mathcal{S}(\psi)$ which contradicts Theorem 10.8. Thus $\xi_i = \sigma$ and hence (c) follows since $\xi_n \leq \sigma$ for all $n \in \mathbb{Z}^+$. ∎

From Theorem 10.10. it is deduced that

$$0 \leq \xi_1 < \xi_2 < \xi_3 < \cdots < \sigma$$

or

$$0 \leq \xi_1 < \cdots < \xi_p = \sigma$$

for some $p \in \mathbb{Z}^+$.

An analysis similar to the one just completed for $\{\xi_i\}_{i=1}^{\infty}$ will now be given for $\{\eta_j\}_{j=1}^{\infty}$. Some minor differences will arise since it is possible for $\eta_i = \infty$ for some $i \in \mathbb{Z}^+$.

Theorem 10.11. *Let $\mathcal{L}$ be a SMF which is positive-definite on $(0, \infty)$, let $\phi(x)$ be an arbitrary representative function of $\mathcal{L}$ and let $\psi(x)$ be any natural representative function of $\mathcal{L}$. If $\eta_1 < \infty$, then $\eta_1 \in \mathcal{S}(\phi)$ and if $\eta_1 = \infty$, then $\eta_1 = \eta_2$ and η_1 is a limit point of $\mathcal{S}(\psi)$.*

Proof: First suppose that $\eta_1 < \infty$ and assume that $\eta_1 \notin \mathcal{S}(\phi)$. Then there exists an $\varepsilon > 0$ such that $\xi_1 < \eta_1 - \varepsilon$ and $\phi(\eta_1 + \varepsilon) - \phi(\eta_1 - \varepsilon) = 0$ implying $\mathcal{S}(\phi) \subseteq [\xi_1, \eta_1 - \varepsilon]$. By Theorem 10.3.(b), $[\xi_1, \eta_1] \subseteq [\xi_1, \eta_1 - \varepsilon]$, a contradiction. Thus $\eta_1 \in \mathcal{S}(\phi)$ if $\eta_1 < \infty$.

Next, suppose $\eta_1 = \infty$. In search of a contradiction, suppose $\eta_1 \neq \eta_2$. Choose α such that $\eta_2 < \alpha < \infty$. For sufficiently large n, $R_n(x)$ contains no zeros in (η_2, α) and so, by Theorem 10.5., $(\eta_2, \alpha) \cap S(\psi) = \emptyset$. Therefore, $S(\psi) \subseteq [\xi_1, \eta_2]$ which is a contradiction to Theorem 10.3.(b). Thus, $\eta_1 = \eta_2$ when $\eta_1 = \infty$. Furthermore, for any $N > 0$, $(x_{n,n-1}, x_{n,n}) \subseteq (N, \infty)$ for a sufficiently large n implying $S(\psi) \cap (N, \infty) \neq \emptyset$ for any $N > 0$ and hence $\eta_1 = \infty$ is a limit point of $S(\psi)$. ∎

Theorem 10.12. *Let $\mathcal{L}$ be a SMF which is positive-definite on $(0, \infty)$ and let $\psi(x)$ be a natural representative function of $\mathcal{L}$. Then $(\eta_1, \infty) \cap S(\psi) = \emptyset$ and $(\eta_{k+1}, \eta_k) \cap S(\psi) = \emptyset$.*

Proof: The fact that $(\eta_1, \infty) \cap S(\psi) = \emptyset$ holds since all the points of increase of a natural representative lie in $[\xi_1, \eta_1]$. Let $k \in \mathbb{Z}^+$ be given. If $\eta_k = \eta_{k+1}$, the result holds vacuously. If $\eta_{k+1} < \eta_k$ then choose α such that $\eta_{k+1} < \alpha < \eta_k$. The interval (η_{k+1}, α) does not contain any zeros of $R_n(x)$ for any sufficiently large n, so by Theorem 10.5., $S(\psi) \cap (\eta_{k+1}, \alpha) = \emptyset$ and since $\alpha \in (\eta_{k+1}, \eta_k)$ was arbitrarily chosen, $S(\psi) \cap (\eta_{k+1}, \eta_k) = \emptyset$. ∎

Theorem 10.13. *Let $\mathcal{L}$ be a SMF which is positive-definite on $(0, \infty)$ and let $\psi(x)$ be a natural representative function of $\mathcal{L}$. For every $k \in \mathbb{Z}^+ \backslash 1$, η_k is a limit point of $S(\psi)$ and $\eta_k \in S(\psi)$ whenever $\eta_k < \infty$.*

Proof: Let $k \in \mathbb{Z}^+ \backslash 1$ and suppose $\eta_k < \infty$. Choose an arbitrary $\varepsilon > 0$. Then $(x_{n,n-k+1}, x_{n,n-k+2}) \subseteq (\eta_k - \varepsilon, \eta_{k-1})$ for all sufficiently large n. By Theorem 10.4., $S(\psi) \cap (\eta_k - \varepsilon, \eta_{k-1}) \neq \emptyset$ and hence by Theorem 10.12., $S(\psi) \cap (\eta_k - \varepsilon, \eta_k) \neq \emptyset$. Thus η_k is a limit point of $S(\psi)$. Since $S(\psi)$ is closed, $\eta_k \in S(\psi)$. Suppose instead that $\eta_k = \infty$. Let $N > 0$ be given. For sufficiently large n, $(x_{n,n-k+1}, x_{n,n-k+2}) \subseteq (N, \infty)$ and hence $S(\psi) \cap (N, \infty) \neq \emptyset$ implying that $\eta_k = \infty$ is a limit point of $S(\psi)$. ∎

A theorem analogous to Theorem 10.10. is

Theorem 10.14. *Let $\mathcal{L}$ be a SMF which is positive-definite on $(0, \infty)$ and let $\psi(x)$ be a natural representative function of $\mathcal{L}$. Then the following hold:*

(a) τ is a limit point of $S(\psi)$;

(b) $S(\psi)$ contains no points larger than τ other than

each finite η_j for $j \in \mathbb{Z}^+$; and

(c) if $\eta_{j+1} = \eta_j$ for some $j \in \mathbb{Z}^+$, then $\tau = \eta_j$.

Proof: (a) If $\tau = \infty$, then $\infty = \eta_j = \tau$ for all $j \in \mathbb{Z}^+$ and so τ is a limit point of $S(\psi)$ according to Theorem 10.13.. If $\tau < \infty$, then eventually $\eta_j < \infty$. Since $\lim_{j \to \infty} \eta_j = \tau$ and since $\eta_j \in S(\psi)$ whenever $\eta_j < \infty$ for $j \in \mathbb{Z}^+$, τ is a limit point of $S(\psi)$.

(b) This part of the proof follows immediately from Theorem 10.12..

(c) Suppose $\eta_{j+1} = \eta_j$ for some $j \in \mathbb{Z}^+$, but $\tau < \eta_j$. Since $\lim_{j \to \infty} \eta_j = \tau$ and since the only possible elements in $S(\psi)$ larger than τ are the η_j's, η_{j+1} is not a limit point of $S(\psi)$ contradicting Theorem 10.13.. Thus $\tau = \eta_j$. ∎

So either

$$\tau < \cdots < \eta_3 < \eta_2 < \eta_1 < \infty,$$

$$\tau = \eta_p < \cdots < \eta_1 < \infty$$

for some $p \in \mathbb{Z}^+\backslash 1$ or

$$\tau = \eta_p = \infty$$

for all $p \in \mathbb{Z}^+$.

Therefore $\sigma = \infty$ implies that $\tau = \infty$ and hence $\mathcal{S}(\psi) = \{\xi_i\}_{i=1}^{\infty}$. If $\sigma < \infty$ and $\tau = \infty$, then $\mathcal{S}(\psi) \subseteq \{\xi_i\}_{i=1}^{\infty} \cup [\sigma, \infty)$. If both σ and τ are finite, then

$$\mathcal{S}(\psi) \subseteq \{\xi_i\}_{i=1}^{\infty} \cup [\sigma, \tau] \cup \{\eta_i\}_{i=1}^{\infty}.$$

Finally, if $\tau = 0$, it follows that $\mathcal{S}(\psi) = \{\eta_i\}_{i=1}^{\infty} \cup \{0\}$.

REFERENCES

[1] T. S. CHIHARA, *An Introduction to Orthogonal Polynomials*, Gordon and Breach, 1978.

[2] GEZA FREUD, *Orthogonal Polynomials*, Pergamon Press, Elmsford, New York, 1971.

[3] E. HENDRIKSON AND H. VAN ROSSUM, Orthogonal Laurent polynomials, *Proceedings of the Koninklijke Nederlandse Akademie von Wetenschappen*, Proceedings **A 89(1)** (1986), 17-36.

[4] WILLIAM B. JONES AND W. J. THRON, *Continued Fractions: Analytic Theory and Applications, Encyclopedia of Mathematics and Its Applications 11*, Addison- Wesley Publ. Co., Reading, MA, 1980, (distributed now by Cambridge Univ. Press, NY).

[5] WILLIAM B. JONES AND W. J. THRON, Survey of continued fraction methods of solving moment problems and related topics, *W. B. Jones, W.J.Thron and H. Waadeland(eds.), Analytic Theory of Continued Fractions, Proceedings Loen, Norway 1981, Lecture Notes in Mathematics, Springer-Verlag* **932** (1982), 4-37.

[6] WILLIAM B. JONES AND W. J. THRON, Orthogonal Laurent polynomials and Gaussian quadrature, *(Karl E. Gustafson and William P. Reinhardt, (ed.s)), Quantum Mechanics in Mathematics, Chemistry, and Physics, Plenum Publ. Corp.* (1981), 449-445.

[7] WILLIAM B. JONES, OLAV NJÅSTAD, AND W. J. THRON, Two- point Padé expansions for a family of analytic functions, *JCAM* **9** (1983), 105-123.

[8] WILLIAM B. JONES, OLAV NJÅSTAD, AND W. J. THRON, Orthogonal Laurent polynomials and the strong Hamburger moment problem, *J. Math. Anal. and Appl.* **98** (1984), 528-554.

[9] WILLIAM B. JONES, W. J. THRON AND H. WAADELAND, A Strong Stieltjes Moment Problem, *Trans. of the AMS* **261** (1980), 503-528.

[10] O. NJÅSTAD AND W. J. THRON, The theory of sequences of orthogonal L- polynomials, *Det Kongelige Norske Videnskabers Selskab* **1** (1983), 54-91.

[11] O. PERRON, *Die Lehre von den Kettenbrüchen II*, Teubner, Stuttgart, 1957.

[12] H. S. WALL, *Analytic Theory of Continued Fractions*, Van Nostrand, New York, 1948.

4

Separate Convergence for Log-Normal Modified *S*-Fractions

S. CLEMENT COOPER[1],Department of Pure and Applied Mathematics, Washington State University, Pullman, WA 99164-3113

WILLIAM B. JONES[2],Department of Mathematics, Box 395, University of Colorado, Boulder, CO 80309-0395

W.J. THRON[3], Department of Mathematics, Box 395, University of Colorado, Boulder, CO 80309-0395

Dedicated to the memory of Arne Magnus

1. INTRODUCTION

Log- normal distributions $d\phi(t)$ defined by

$$\phi'(t) := \frac{q^{\frac{1}{2}}}{2\kappa\sqrt{\pi}} e^{-(\frac{\log t}{2\kappa})^2}, \quad 0 < q < 1, \quad q = e^{-2\kappa^2}, \quad 0 < t < \infty \qquad (1.1)$$

have received considerable recent attention due to their applications to geophysical sciences, economics and business (see, e.g. [6]). A special case of (1.1) with $\kappa = \frac{1}{2}$ and $q = e^{-\frac{1}{2}}$ was used by Stieltjes [12] for an example of an indeterminate moment problem. Wigert [13] investigated orthogonal polynomials with respect to log- normal distributions and obtained explicit expressions for the orthogonal polynomials and for the limit of this sequence of polynomials. Attention should also be paid to a paper by Moak [9] in which convergence results similar to ours appear. Orthogonal Laurent polynomials with respect to (1.1) have been studied more recently by [10], [3], [4], [7].

---

[1] Research supported in part by the U.S. National Science Foundation under Grant No. DMS- 9109095

[2] Research supported in part by the United States Educational Foundation in Norway(Fulbright Grant), the Norwegian Research Council(NAVF) and the U.S. National Science Foundation under Grant No. DMS- 9103141

[3] Research supported in part by the U.S. National Science Foundation under Grant No. DMS- 9103141

The present paper deals with separate convergence of the polynomial sequences $\{A_{2n}(z)\}$, $\{B_{2n}(z)\}$, $\{A_{2n+1}(z)\}$, and $\{B_{2n+1}(z)\}$ where $A_n(z)$ and $B_n(z)$ denote the nth numerator and denominator, respectively, of the continued fractions

$$\frac{a_1}{1} + \frac{a_2}{z} + \frac{a_3}{1} + \frac{a_4}{z} + \cdots, \qquad a_n > 0, \qquad (1.2a)$$

where

$$a_1 := 1, \quad a_{2n} := q^{-2n+\frac{1}{2}}, \quad a_{2n+1} := q^{-2n-\frac{1}{2}}(1-q^n), \quad n = 1, 2, 3, \ldots. \qquad (1.2b)$$

The continued fractions (1.2) are called *log- normal modified S- fractions* because of the correspondence property

$$\frac{A_n(z)}{B_n(z)} - L_0(z) = O\left(\left(\frac{1}{z}\right)^n\right), \quad n = 1, 2, \ldots, \qquad (1.3)$$

where

$$L_0(z) := c_0 + c_1 z^{-1} + c_2 z^{-2} + \cdots \qquad (1.4)$$

and where c_n denotes the nth moment

$$c_n := \int_0^\infty (-t)^n \, d\phi(t) = (-1)^n q^{-\frac{n^2}{2}-n}, \quad n = 0, 1, 2, \ldots \qquad (1.5)$$

for the log- normal distribution function $\phi(t)$. Explicit expressions for the polynomials $A_n(z)$ and $B_n(z)$ are obtained in Section 3. The limits

$$A^{(\nu)}(z) := \lim_{n\to\infty} A_{2n+\nu}(z) \quad \text{and} \quad B^{(\nu)}(z) := \lim_{n\to\infty} B_{2n+\nu}(z), \quad \nu = 0, 1 \qquad (1.6)$$

are shown (Theorem 3.1) to exist (locally uniformly on $\mathbb{C}$) as entire functions satisfying $A^{(1)}(z)B^{(0)}(z) - A^{(0)}(z)B^{(1)}(z) \equiv 1$. This ensures divergence (by oscillation) of the continued fraction (1.2). Explicit power series expressions for $A^{(0)}(z), A^{(1)}(z), B^{(0)}(z)$, $B^{(1)}(z)$ are also given (Theorem 3.2). To obtain the separate convergence results we prove a lemma (Theorem 2.1) making use of ideas of von Koch (see, e.g., [11, Satz 2.6] and [8, Theorem 4.14]) together with the additional information and structure for the continued fractions.

The results of this paper are closely related to log- normal orthogonal polynomials. The sequence $\{B_{2n}(z)\}$ of denominators of (1.2) is the monic orthogonal polynomial sequence with respect to $\phi(t)$ in (1.1). These are, up to multiplicative factors, the polynomials studied by Wigert [13]. $\{B_{2n+1}(z)\}$ is the kernel polynomial sequence orthogonal with respect to the distribution $t\,d\phi(t)$ [2, pp.35-40]. The numerator sequences $\{A_{2n}(z)\}$ and $\{A_{2n+1}(z)\}$ are the associated polynomials for the sequences $\{B_{2n}(z)\}$ and $\{B_{2n+1}(z)\}$, respectively.

The remainder of this introduction contains background material for use in later sections and for relating the present paper to topics such as moment theory, asymptotic series and integral representations. In 1894 [12] T. J. Stieltjes posed and solved the following Stieltjes moment problem (SMP): Given a sequence of real numbers $\{\mu_n\}_{n=0}^\infty$,

does there exist a distribution function $\psi(t)$ (bounded, non-decreasing with infinitely many points of increase) on $(0, \infty)$ such that

$$\mu_n = \int_0^\infty (-t)^n d\psi(t), \quad n = 0, 1, 2, \ldots? \tag{1.7}$$

Such a function $\psi(t)$ is called a *solution* to the SMP and μ_n is called the nth *moment* for ψ. Associated with $\{\mu_n\}$ are the *Hankel determinants*

$$H_0^{(m)} := 1, \quad H_k^{(m)} := det(\mu_{m+i+j})_{i,j=0}^{k-1}, \quad m = 0, 1, 2, \ldots, \quad k = 1, 2, 3, \ldots. \tag{1.8}$$

Stieltjes proved the equivalence of the following three statements.
- The SMP for $\{\mu_n\}_{n=0}^\infty$ has a solution. $\quad$ (1.9a)
- $H_n^{(0)} > 0$ and $(-1)^n H_n^{(1)} > 0$, $n = 0, 1, 2, \ldots$. $\quad$ (1.9b)
- There exists a modified S-fraction (1.2a) corresponding to $\quad$ (1.9c)
  $L(z) = \sum_{k=0}^\infty \mu_k z^{-k}$ in the sense that

$$\frac{A_n(z)}{B_n(z)} - L(z) = O\left(\left(\frac{1}{z}\right)^n\right), \quad n = 1, 2, 3, \ldots.$$

In proving this result Stieltjes showed: (a) if (1.9b) holds, then the coefficients a_n of (1.2a) are given by

$$a_1 = H_1^{(0)}, \quad a_{2m} = \frac{-H_{m-1}^{(0)} H_m^{(1)}}{H_m^{(0)} H_{m-1}^{(1)}}, \quad a_{2m+1} = \frac{-H_{m+1}^{(0)} H_{m-1}^{(1)}}{H_m^{(0)} H_m^{(1)}}, \quad m = 1, 2, 3, \ldots; \tag{1.10}$$

(b) the sequences of approximants $\{\frac{A_{2n}(z)}{B_{2n}(z)}\}$ and $\{\frac{A_{2n+1}(z)}{B_{2n+1}(z)}\}$ converge to analytic functions $F_0(z)$ and $F_1(z)$, respectively, in the region $S_\pi := [z : |\arg z| < \pi]$; and (c) these functions have integral representations of the form

$$F_\nu(z) := \lim_{n \to \infty} \frac{A_{2n+\nu}(z)}{B_{2n+\nu}(z)} = \int_0^\infty \frac{z d\psi_\nu(t)}{z + t}, \quad z \in S_\pi, \quad \nu = 0, 1, \tag{1.11}$$

for some distribution functions $\psi_0(t)$ and $\psi_1(t)$. Regarding uniqueness, Stieltjes proved the equivalence of the following three statements.
- A solvable SMP for $\{\mu_n\}_{n=0}^\infty$ has a unique solution. $\quad$ (1.12a)
- At least one of the following series is divergent $\quad$ (1.12b)

$$\sum_{n=1}^\infty \left| \frac{a_1 a_3 \cdots a_{2n-1}}{a_2 a_4 \cdots a_{2n}} \right|, \quad \sum_{n=1}^\infty \left| \frac{a_2 a_4 \cdots a_{2n}}{a_1 a_3 \cdots a_{2n+1}} \right|.$$

- The modified S-fraction (1.2a) corresponding to $\quad$ (1.12c)
  $L(z) := \sum_{k=0}^\infty \mu_k z^{-k}$ in the sense of (1.9c) is convergent for $z \in S_\pi$.

Finally, Stieltjes proved that the series $L(z) = \sum_{k=0}^\infty \mu_k z^{-k}$ is the asymptotic expansion of the functions $F_0(z)$ and $F_1(z)$ for $z \in S_\alpha := [z : |\arg z| < \alpha]$, $0 \le \alpha < \pi$ as $z \to \infty$. Thus, he obtained (his primary motivation for studying this problem) a method for "summing" certain divergent series.

To see that the even denominators $\{B_{2n}(z)\}$ of the log- normal modified S- fraction (1.2) form the sequence of monic polynomials orthogonal with respect to $\phi(t)$, it suffices to note that the even part of (1.2) is a real J- fraction (multiplied by $a_1 z$)

$$\frac{a_1 z}{a_2 + z} - \frac{a_2 a_3}{a_3 + a_4 + z} - \frac{a_4 a_5}{a_5 + a_6 + z} - \frac{a_6 a_7}{a_7 + a_8 + z} - \cdots, \tag{1.13}$$

and then apply well known results from the theory of continued fractions and orthogonal polynomials (e.g., [2], [8, Sect. 7.2]). In fact it can be readily shown that

$$U_n(z) = A_{2n}(z) \quad \text{and} \quad V_n(z) = B_{2n}(z), \quad n = 1, 2, 3, \ldots, \tag{1.14}$$

where $U_n(z)$ and $V_n(z)$ denote the nth numerator and denominator, respectively, of (1.13). The formulas (1.2b) for the elements a_n can be derived from the moments c_n in (1.5) either by use of the quotient- difference algorithm [8, Sect. 7.1] or directly from the determinant formulas (1.10) and the known result for Hankel determinants associated with the log- normal moments c_n

$$H_k^{(m)} = (-1)^{-km} \left[q^{\frac{-(m+k-1)^2}{2} - (m+k-1)} \right]^k q^{\frac{-k(k^2-1)}{6}} \sum_{j=1}^{k-1} (1 - q^j)^{k-j}. \tag{1.15}$$

From (1.2b) one can readily establish the convergence of the series (1.12b) and hence deduce the fact that the SMP for $\{c_n\}$ is indeterminate. Another distribution function $\psi(t)$ that generates the log- normal moments (1.5) can be obtained by use of a beta- type integral introduced by Ramanujan (see, e.g., [1]). It is not known whether the distribution functions $\psi_0(t)$ and $\psi_1(t)$ in (1.11) overlap with either the log- normal distribution function $\phi(t)$ or with the one obtained from the beta integral. The inequalities (2.13) in Section 2 give truncation error bounds for the sequences $\{\frac{P_{2n+\nu}(z)}{Q_{2n+\nu}(z)}\}_{n=0}^{\infty}, \nu = 0, 1,$ and hence might be useful in the problem of determining the limit functions $F_\nu(z)$ and associated distribution functions $\psi_\nu(t), \nu = 0, 1$.

We conclude this section by stating the difference equations that define the polynomials $A_n(z)$ and $B_n(z)$ for a modified S- fraction (1.2a):

$$A_{-1}(z) := 1, \quad A_0(z) := 0, \quad B_{-1}(z) := 0, \quad B_0(z) := 1, \tag{1.16a}$$

$$\begin{pmatrix} A_{2n}(z) \\ B_{2n}(z) \end{pmatrix} = z \begin{pmatrix} A_{2n-1}(z) \\ B_{2n-1}(z) \end{pmatrix} + a_{2n} \begin{pmatrix} A_{2n-2}(z) \\ B_{2n-2}(z) \end{pmatrix}, \quad n = 1, 2, \ldots, \tag{1.16b}$$

$$\begin{pmatrix} A_{2n+1}(z) \\ B_{2n+1}(z) \end{pmatrix} = \begin{pmatrix} A_{2n}(z) \\ B_{2n}(z) \end{pmatrix} + a_{2n+1} \begin{pmatrix} A_{2n-1}(z) \\ B_{2n-1}(z) \end{pmatrix}, \quad n = 1, 2, \ldots. \tag{1.16c}$$

2. SEPARATE CONVERGENCE FOR DIVERGENT CONTINUED FRACTIONS

In this section we consider continued fractions of the form

$$\mathop{\mathrm{K}}_{n=1}^{\infty} \left(\frac{1}{b_n(z)} \right) = \frac{1}{b_1(z)} + \frac{1}{b_2(z)} + \frac{1}{b_3(z)} + \cdots, \tag{2.1}$$

where the partial denominators $b_n(z)$ are functions of a complex variable z of the form

$$b_n(z) = \varepsilon_n f_n(z), \quad n = 1, 2, 3, \dots, \tag{2.2}$$

where

$$E := \sum_{n=1}^{\infty} |\varepsilon_n| < \infty, \quad \varepsilon_n \in \mathbb{C} \text{ for } n = 1, 2, 3, \dots, \tag{2.3}$$

and the functions $f_n(z)$ are holomorphic in a region D and uniformly bounded on every compact subset K of D; that is, for each compact set $K \subseteq D$, there exists a number $M(K) > 0$ such that

$$|f_n(z)| \le M(K) \quad \text{for all} \quad z \in K, \quad n = 1, 2, 3, \dots. \tag{2.4}$$

The nth numerator $P_n(z)$ and denominator $Q_n(z)$ of (2.1) are defined by difference equations

$$P_{-1}(z) := 1, \quad P_0(z) := 0, \quad Q_{-1}(z) := 0, \quad Q_0(z) := 1, \tag{2.5a}$$

$$\begin{pmatrix} P_n(z) \\ Q_n(z) \end{pmatrix} := b_n(z) \begin{pmatrix} P_{n-1}(z) \\ Q_{n-1}(z) \end{pmatrix} + \begin{pmatrix} P_{n-2}(z) \\ Q_{n-2}(z) \end{pmatrix}, \quad n = 1, 2, 3, \dots. \tag{2.5b}$$

Theorem 2.1. *Let* $\mathrm{K}(\frac{1}{b_n(z)})$ *be a continued fraction of the form described by (2.1) to (2.5). Then:*

(A.) *The sequences* $\{P_{2n}(z)\}$, $\{Q_{2n}(z)\}$, $\{P_{2n+1}(z)\}$, $\{Q_{2n+1}(z)\}$ *converge locally uniformly on the region* D *to holomorphic functions* $P^{(0)}(z)$, $Q^{(0)}(z)$, $P^{(1)}(z)$, $Q^{(1)}(z)$, *respectively.*

(B.) *For all* $z \in D$,

$$P^{(1)}(z)Q^{(0)}(z) - P^{(0)}(z)Q^{(1)}(z) = 1 \tag{2.6}$$

and hence

$$\frac{P^{(0)}(z)}{Q^{(0)}(z)} := \lim_{n \to \infty} \frac{P_{2n}(z)}{Q_{2n}(z)} \ne \lim_{n \to \infty} \frac{P_{2n+1}(z)}{Q_{2n+1}(z)} =: \frac{P^{(1)}(z)}{Q^{(1)}(z)}, \quad z \in D. \tag{2.7}$$

(C.) *For each compact subset* $K \subseteq D$ *there exists a constant* $M(K)$ *such that for all* $z \in K$,

$$|P^{(\nu)}(z) - P_{2n+\nu}(z)| \le M(K)e^{EM(K)} \sum_{k=n+1}^{\infty} |\varepsilon_{2k+\nu}|, \quad \nu = 0, 1, \tag{2.8a}$$

and

$$|Q^{(\nu)}(z) - Q_{2n+\nu}(z)| \le M(K)e^{EM(K)} \sum_{k=n+1}^{\infty} |\varepsilon_{2k+\nu}|, \quad \nu = 0, 1, \tag{2.8b}$$

where E *is defined by (2.3).*

Proof. (A.): Using the difference equations (2.5) one can successively derive the following relations, with $\varepsilon_0 := 0$ and $n \ge 1 - \nu$, $\nu = 0, 1$:

$$P_{2n+\nu}(z) = \nu + \sum_{k=1}^{n} \varepsilon_{2k+\nu} f_{2k+\nu}(z) P_{2k+\nu-1}(z), \quad \nu = 0, 1 \tag{2.9a}$$

$$Q_{2n+\nu}(z) = 1 - \nu + \sum_{k=0}^{n} \varepsilon_{2k+\nu} f_{2k+\nu}(z) Q_{2k+\nu-1}(z), \quad \nu = 0, 1 \tag{2.9b}$$

and the bounds

$$|P_n(z)| \leq \prod_{k=2}^{n} \left(1 + |\varepsilon_k f_k(z)|\right), |Q_n(z)| \leq \prod_{k=1}^{n} \left(1 + |\varepsilon_k f_k(z)|\right) \tag{2.10}$$

valid for all $z \in D$ and $n \geq 1$. Let K be a given compact subset of D and let $M(K)$ satisfy (2.4). It follows from (2.10) that $|P_n(z)|$ and $|Q_n(z)|$ are both bounded above for all $z \in K$ and $n \geq 1$ by

$$\prod_{k=1}^{\infty} \left(1 + \varepsilon_k M(K)\right) \leq e^{EM(K)}. \tag{2.11}$$

Therefore the infinite series

$$\nu + \sum_{k=1}^{\infty} \varepsilon_{2k+\nu} f_{2k+\nu}(z) P_{2k+\nu-1}(z) \quad \text{and} \quad 1 - \nu + \sum_{k=0}^{\infty} \varepsilon_{2k+\nu} f_{2k+\nu}(z) Q_{2k+\nu-1}(z)$$

converge uniformly and absolutely on K for $\nu = 0, 1$. Hence by (2.9) we have for all $z \in D$, $\nu = 0, 1$,

$$P^{(\nu)}(z) := \lim_{n \to \infty} P_{2n+\nu}(z) = \nu + \sum_{k=1}^{\infty} \varepsilon_{2k+\nu} f_{2k+\nu}(z) P_{2k+\nu-1}(z), \tag{2.12a}$$

and

$$Q^{(\nu)}(z) := \lim_{n \to \infty} Q_{2n+\nu}(z) = 1 - \nu + \sum_{k=0}^{\infty} \varepsilon_{2k+\nu} f_{2k+\nu}(z) Q_{2k+\nu-1}(z). \tag{2.12b}$$

(B.) follows immediately from the determinant formulas for continued fractions [8, p.20]. (C.) is a consequence of (2.9), (2.11), and (2.12). ∎

The truncation error bounds (2.8) for $A_{2n+\nu}(z)$ and $B_{2n+\nu}(z)$ provide the following truncation error bounds for the approximants $\frac{P_{2n+\nu}(z)}{Q_{2n+\nu}(z)}$ of (2.1), $\nu = 0, 1$. For that purpose we write

$$W_n(K) := M(K) e^{EM(K)} \sum_{k=n+1}^{\infty} |\varepsilon_{2k+\nu}|, \quad z \in K \subseteq D, \quad n \geq 1, \quad \nu = 0, 1.$$

By (2.8) and the triangle inequality

$$|P^{(\nu)}(z)| \leq |P_{2n+\nu}(z)| + |W_n(K)| \quad \text{and} \quad |Q^{(\nu)}(z)| \geq |Q_{2n+\nu}(z)| - W_n(K),$$

and hence for all $z \in K \subseteq D$, $n \geq 1$, $\nu = 0, 1$,

$$\begin{aligned}
\left| \frac{P^{(\nu)}(z)}{Q^{(\nu)}(z)} - \frac{P_{2n+\nu}(z)}{Q_{2n+\nu}(z)} \right| &\leq \frac{|P^{(\nu)}(z)||Q_{2n+\nu}(z) - Q^{(\nu)}(z)|}{|Q^{(\nu)}(z)||Q_{2n+\nu}(z)|} + \frac{|P^{(\nu)}(z) - P_{2n+\nu}(z)|}{|Q_{2n+\nu}(z)|} \\
&\leq \frac{W_n(K)}{|Q_{2n+\nu}(z)|} \left[1 + \frac{B|P_{2n+\nu}(z)| + W_n(K)}{|Q_{2n+\nu}(z)| - W_n(K)} \right].
\end{aligned} \tag{2.13}$$

3. REPRESENTATION THEOREMS

Throughout this section we let $A_n(z)$ and $B_n(z)$ denote the nth numerator and denominator, respectively, of a log- normal modified S- fraction (1.2). We apply Theorem 2.1 to obtain the separate convergence of the sequences $\{A_{2n+\nu(z)}\}$ and $\{B_{2n+\nu(z)}\}$, $\nu = 0, 1$ (Theorem 3.2). We then derive infinite power series representations for the limit functions of these four sequences (Theorem 3.3). We begin by giving explicit expressions for the polynomials $A_n(z)$ and $B_n(z)$.

Theorem 3.1. *(A.)*

$$A_{2n}(z) = \sum_{j=1}^{n} a_{2n,j} z^j, \quad a_{2n,n} = 1, \quad n = 1, 2, \ldots, \tag{3.1a}$$

where for $j = 1, 2, \ldots, n-1$,

$$a_{2n,j} = q^{-n^2 - \frac{n}{2} + j^2 + \frac{i}{2}} \sum_{i=0}^{n-j} (-1)^i q^{\frac{i^2}{2} + i(2j - \frac{1}{2})} \prod_{k=1}^{i+j} \left(\frac{1 - q^{n-k+1}}{1 - q^k} \right) \tag{3.1b}$$

and

$$A_{2n+1}(z) = \sum_{j=0}^{n} a_{2n+1,j} z^j, \quad a_{2n+1,n} = 1, \quad n = 0, 1, 2, \ldots, \tag{3.1c}$$

where for $j = 0, 1, \ldots, n-1$,

$$a_{2n+1,j} = q^{-n^2 - \frac{3n}{2} + j^2 + \frac{3j}{2}} \sum_{i=0}^{n-j} (-1)^i q^{\frac{i^2}{2} + i(2j + \frac{1}{2})} \prod_{k=1}^{i+j} \left(\frac{1 - q^{n-k+1}}{1 - q^k} \right). \tag{3.1d}$$

(B.) $B_1(z) = 1$ *and for* $n = 1, 2, 3, \ldots$,

$$B_{2n}(z) = q^{-n^2 - \frac{n}{2}} \sum_{j=0}^{n} q^{j^2 + \frac{i}{2}} \prod_{i=1}^{j} \left(\frac{1 - q^{n-i+1}}{1 - q^i} \right) z^j, \tag{3.2a}$$

and

$$B_{2n+1}(z) = q^{-n^2 - \frac{3n}{2}} \sum_{j=0}^{n} q^{j^2 + \frac{3j}{2}} \prod_{i=1}^{j} \left(\frac{1 - q^{n-i+1}}{1 - q^i} \right) z^j. \tag{3.2b}$$

Proof: (B.) Formulas (3.2) can be verified by induction using (1.2b) and the difference equations (1.16). (A.) Using (3.2) with the notation $B_{2n+\nu}(z) = \sum_{j=0}^{n} b_{2n+\nu,j} z^j$, where $b_{2n+\nu,n} = 1$ and the correspondence property (1.3), we have

$$A_{2n+\nu}(z) - L_0(z) B_{2n+\nu}(z) = O\left(\left(\frac{1}{z} \right)^{n+\nu} \right), \quad \nu = 0, 1. \tag{3.3}$$

Thus, $a_{2n+\nu,j} = \sum_{i=0}^{n-j} c_i b_{2n+\nu,i+j}$, $j = 1 - \nu, \ldots, n$, $n \geq 1$. ∎

In order to apply Theorem 2.1, we will use equivalence transformations to rewrite the log- normal modified S- fraction (1.2) in the form $K\left(\frac{1}{b_n^*(z)}\right)$. Thus, we seek a sequence $\{r_n\}_{n=0}^{\infty}$ such that $r_0 = 1$, $r_{n-1}r_n a_n = 1$, $b_{2n}^* = r_{2n}z$ and $b_{2n+1}^* = r_{2n+1}$ for $n = 1, 2, \ldots$. With the convention that an empty product has the value 1, we find that

$$r_{2n} = q^{n+\frac{1}{2}} \prod_{k=1}^{n-1}(1 - q^k) \quad \text{and} \quad r_{2n+1} = \frac{q^n}{\prod_{k=1}^{n}(1 - q^k)} \tag{3.4}$$

for $n = 1, 2, \ldots$. Denote the nth numerator and denominator of $K\left(\frac{1}{b_n^*(z)}\right)$ by $A_n^*(z)$ and $B_n^*(z)$, respectively, and the relationship between the numerators and denominators of the two continued fractions is given by

$$\begin{pmatrix} A_n^*(z) \\ B_n^*(z) \end{pmatrix} = \left(\prod_{k=0}^{n} r_k\right) \begin{pmatrix} A_n(z) \\ B_n(z) \end{pmatrix}. \tag{3.5}$$

Thus, $B_1^*(z) = B_1(z)$

$$B_{2n}^*(z) = \sum_{j=0}^{n} q^{j^2+\frac{i}{2}} \prod_{i=1}^{j}\left(\frac{1 - q^{n-i+1}}{1 - q^i}\right) z^j \tag{3.6a}$$

$$B_{2n+1}^*(z) = \frac{1}{\prod_{k=1}^{n}(1 - q^k)} \left[\sum_{j=0}^{n} q^{j^2+\frac{3j}{2}} \prod_{i=1}^{j}\left(\frac{1 - q^{n-i+1}}{1 - q^i}\right) z^j\right] \tag{3.6b}$$

for $n = 1, 2, \ldots$. Similarly, $A_1^*(z) = A_1(z)$ and

$$A_{2n}^*(z) = \sum_{j=1}^{n} a_{2n,j}^* z^j \tag{3.7a}$$

where for $j = 1, 2, \ldots, n$

$$a_{2n,j}^* = q^{j^2+\frac{i}{2}} \sum_{i=0}^{n-j}(-1)^i q^{\frac{i^2}{2}+i(2j-\frac{1}{2})} \prod_{k=1}^{i+j}\left(\frac{1 - q^{n-k+1}}{1 - q^k}\right) \tag{3.7b}$$

and

$$A_{2n+1}^*(z) = \sum_{j=0}^{n} a_{2n+1,j}^* z^j \tag{3.7c}$$

where for $j = 0, 1, \ldots, n$

$$a_{2n+1,j}^* = \frac{1}{\prod_{k=1}^{n}(1 - q^k)} q^{j^2+\frac{3j}{2}} \sum_{i=0}^{n-j}(-1)^i q^{\frac{i^2}{2}+i(2j+\frac{1}{2})} \prod_{k=1}^{i+j}\left(\frac{1 - q^{n-k+1}}{1 - q^k}\right) \tag{3.7d}$$

valid for $n = 1, 2, \ldots$.

Theorem 3.2. *Consider* $\mathrm{K}\left(\frac{1}{b_n^*(z)}\right)$ *with*

$$b_{2n}^*(z) = q^{n+\frac{1}{2}}\left[\prod_{k=1}^{n-1}(1-q^k)\right]z, \quad b_{2n-1}^*(z) = \frac{q^{n-1}}{\prod_{k=1}^{n-1}(1-q^k)}, \quad n = 1, 2, \ldots \quad (3.8)$$

where $0 < q < 1$. *Let* $A_n^*(z)$ *and* $B_n^*(z)$ *be the* n^{th} *numerator and denominator, respectively of this continued fraction. Then:*

A.) *There exist entire functions* $A^{(0)}(z)$, $A^{(1)}(z)$, $B^{(0)}(z)$, *and* $B^{(1)}(z)$, *such that the sequences* $\{A_{2n}^*(z)\}_{n=1}^\infty$, $\{A_{2n+1}^*(z)\}_{n=0}^\infty$, $\{B_{2n}^*(z)\}_{n=1}^\infty$ *and* $\{B_{2n+1}^*(z)\}_{n=0}^\infty$, *respectively, converge to them. The convergence is uniform on each set* $K_R = \{z : |z| \leq R, \; R > 0\}$.

B.) *For all* z,

$$A^{(1)}(z)B^{(0)}(z) - A^{(0)}(z)B^{(1)}(z) = 1 \quad (3.9)$$

and hence

$$\frac{A^{(0)}(z)}{B^{(0)}(z)} := \lim_{n\to\infty}\frac{A_{2n}^*(z)}{B_{2n}^*(z)} \neq \lim_{n\to\infty}\frac{A_{2n+1}^*(z)}{B_{2n+1}^*(z)} =: \frac{A^{(1)}(z)}{B^{(1)}(z)}. \quad (3.10)$$

C.) *Let* $\lambda = q^{\frac{1}{2}}$, $\mu = \frac{1}{\prod_{k=1}^\infty(1-q^k)}$ *and* $N(R) = exp\{R\frac{\mu+\lambda q}{1-q}\}$. *Then*

$$|A^{(\nu)}(z) - A_{2n+\nu}^*(z)| \leq (\lambda + \mu)R \cdot N(R)\frac{q^{n+1}}{1-q} \quad (3.11a)$$

and

$$|B^{(\nu)}(z) - B_{2n+\nu}^*(z)| \leq (\lambda + \mu)R \cdot N(R)\frac{q^{n+1}}{1-q} \quad (3.11b)$$

for $\nu = 0, 1$ *and* $z \in \mathbb{C}$.

Proof: Let $\varepsilon_{2n} = q^{n+\frac{1}{2}}\prod_{k=1}^{n-1}(1-q^k)$, $\varepsilon_{2n+1} = \frac{q^n}{\prod_{k=1}^{n-1}(1-q^k)}$, $f_{2n+\nu}(z) = \nu + (1-\nu)z$. Then $|\varepsilon_{2n}| \leq \lambda q^n$ and $|\varepsilon_{2n+1}| \leq \mu q^n$. Note that $\mu < \infty$ since $\sum_{k=1}^\infty q^k < \infty$. Thus, $\sum_{n=1}^\infty|\varepsilon_n| = \sum_{n=1}^\infty|\varepsilon_{2n}| + \sum_{n=0}^\infty|\varepsilon_{2n+1}| \leq \lambda\sum_{n=1}^\infty q^n + \mu\sum_{n=0}^\infty q^n = \frac{\mu+\lambda q}{1-q} < \infty$. For each $R > 0$, $|f_n(z)| \leq R + 1$ for all z such that $|z| \leq R$. The results now easily follow from Theorem 2.1. ∎

With the separate convergence established for the sequences, we conclude by deriving explicit power series representations for the limit functions.

Theorem 3.3. *Consider the continued fraction* $\mathrm{K}\left(\frac{1}{b_n^*(z)}\right)$ *where* $b_n^*(z)$ *is given by (3.8). The limit functions* $A^{(0)}(z)$, $A^{(1)}(z)$, $B^{(0)}(z)$, *and* $B^{(1)}(z)$ *whose existence is confirmed by Theorem 3.2 have the following power series representations*

$$A^{(0)}(z) = \sum_{j=1}^\infty a_j^{(0)}z^j \quad \text{where} \quad a_j^{(0)} = q^{j^2+\frac{j}{2}}\sum_{i=0}^\infty(-1)^i\frac{q^{\frac{i^2}{2}+2ij-\frac{i}{2}}}{\prod_{k=1}^{i+j}(1-q^k)}, \quad (3.12)$$

$$A^{(1)}(z) = \sum_{j=0}^\infty a_j^{(1)}z^j \quad \text{where} \quad a_j^{(1)} = \frac{q^{j^2+\frac{3j}{2}}}{\prod_{k=1}^\infty(1-q^k)}\sum_{i=0}^\infty(-1)^i\frac{q^{\frac{i^2}{2}+2ij+\frac{i}{2}}}{\prod_{k=1}^{i+j}(1-q^k)}, \quad (3.13)$$

$$B^{(0)}(z) = \sum_{j=0}^{\infty} q^{j^2+\frac{i}{2}} \left[\prod_{i=1}^{j} \frac{1}{(1-q^i)} \right] z^j, \tag{3.14}$$

$$B^{(1)}(z) = \frac{1}{\prod_{k=1}^{\infty}(1-q^k)} \sum_{j=0}^{\infty} q^{j^2+\frac{3j}{2}} \left[\prod_{i=1}^{j} \frac{1}{(1-q^i)} \right] z^j. \tag{3.15}$$

Proof: For $\nu = 0, 1$, let $a_{2n+\nu,j}^*$, $b_{2n+\nu,j}^*$, $\tilde{a}_j^{(\nu)}$, $\tilde{b}_j^{(\nu)}$ be defined by

$$A_{2n+\nu}^*(z) =: \sum_{j=0}^{n} a_{2n+\nu,j}^* z^j, \quad B_{2n+\nu}^*(z) =: \sum_{j=0}^{n} b_{2n+\nu,j}^* z^j, \quad n = 0, 1, \ldots,$$

$$A^{(\nu)}(z) =: \sum_{j=0}^{\infty} \tilde{a}_j^{(\nu)} z^j, \quad B^{(\nu)}(z) =: \sum_{j=0}^{\infty} \tilde{b}_j^{(\nu)} z^j.$$

It follows that

$$\tilde{a}_j^{(\nu)} = \lim_{n \to \infty} a_{2n+\nu,j}^* \quad \text{and} \quad \tilde{b}_j^{(\nu)} = \lim_{n \to \infty} b_{2n+\nu,j}, \quad \nu = 0, 1, \quad j = 0, 1, 2, \ldots, \tag{3.16}$$

which is readily verified by use of the formulas for Taylor series coefficients:

$$\left| \tilde{a}_j^{(\nu)} - a_{2n+\nu,j}^* \right| = \frac{1}{2\pi} \left| \int_C \frac{A^{(\nu)}(z) - A_{2n+\nu}^*(z)}{z^{j+1}} dz \right|$$
$$\leq R^{-j} \max_{z \in C} \left| A^{(\nu)}(z) - A_{2n+\nu}^*(z) \right|,$$

where C denotes the circle $|z| = R$, $R > 0$. The right hand side tends to zero as $n \to \infty$ by uniform convergence of $\{A_{2n+\nu}^*(z)\}_{n=0}^{\infty}$. Similar inequalities hold for $|\tilde{b}_j^{(\nu)} - b_{2n+\nu,j}^*|$. Equations (3.14) and (3.15) follow immediately from (3.6) and (3.16).

The assertions (3.12) and (3.13) can be deduced from Tannery's Theorem (see for example [5]). We include an independent proof here for completeness. To verify (3.12) and (3.13) it suffices to show that

$$\tilde{a}_j^{(\nu)} = a_j^{(\nu)}, \quad \nu = 0, 1, \quad j = 0, 1, \ldots. \tag{3.17}$$

We establish (3.17) for $\nu = 0$. An analogous argument holds for $\nu = 1$ and hence is omitted. Recall

$$a_{2n,j}^* = q^{j^2+\frac{i}{2}} \sum_{i=0}^{n-j} (-1)^i q^{\frac{i^2}{2}+2ij-\frac{i}{2}} \prod_{k=1}^{i+j} \left(\frac{1-q^{n-k+1}}{1-q^k} \right)$$

$$= q^{j^2+\frac{i}{2}} \prod_{k=1}^{n} (1-q^k) \sum_{i=0}^{n-j} (-1)^i \frac{q^{\frac{i^2}{2}+2ij-\frac{i}{2}}}{\prod_{k=1}^{i+j}(1-q^k) \prod_{k=1}^{n-i-j}(1-q^k)}.$$

We claim that

$$\lim_{n \to \infty} \sum_{i=0}^{n-j} (-1)^i \frac{q^{\frac{i^2}{2}+2ij-\frac{i}{2}}}{\prod_{k=1}^{i+j}(1-q^k) \prod_{k=1}^{n-i-j}(1-q^k)} = \sum_{i=0}^{\infty} (-1)^i \frac{q^{\frac{i^2}{2}+2ij-\frac{i}{2}}}{\prod_{k=1}^{i+j}(1-q^k) \prod_{k=1}^{\infty}(1-q^k)},$$

from which the result will easily follow. Let $\epsilon > 0$. Consider

$$\left| \sum_{i=0}^{n-j} (-1)^i \frac{q^{\frac{i^2}{2}+2ij-\frac{i}{2}}}{\prod_{k=1}^{i+j}(1-q^k)\prod_{k=1}^{n-i-j}(1-q^k)} - \sum_{i=0}^{\infty} (-1)^i \frac{q^{\frac{i^2}{2}+2ij-\frac{i}{2}}}{\prod_{k=1}^{i+j}(1-q^k)\prod_{k=1}^{\infty}(1-q^k)} \right|$$

$$\leq \underbrace{\left| \sum_{i=0}^{n-j} (-1)^i \frac{q^{\frac{i^2}{2}+2ij-\frac{i}{2}}}{\prod_{k=1}^{i+j}(1-q^k)} \left[\frac{\prod_{k=1}^{\infty}(1-q^k) - \prod_{k=1}^{n-i-j}(1-q^k)}{\prod_{k=1}^{n-i-j}(1-q^k)\prod_{k=1}^{\infty}(1-q^k)} \right] \right|}_{(1)}$$

$$+ \underbrace{\left| \sum_{i=n-j+1}^{\infty} (-1)^i \frac{q^{\frac{i^2}{2}+2ij-\frac{i}{2}}}{\prod_{k=1}^{i+j}(1-q^k)\prod_{k=1}^{\infty}(1-q^k)} \right|}_{(2)}.$$

The ratio test applied to the series appearing in (2) yields $\dfrac{q^{2(i+j)+\frac{1}{2}}}{1-q^{i+j+1}}$ which approaches zero as i goes to infinity and hence there exists an N_1 such that for all $n \geq N_1$

$$\left| \sum_{i=n-j+1}^{\infty} (-1)^i \frac{q^{\frac{i^2}{2}+2ij-\frac{i}{2}}}{\prod_{k=1}^{i+j}(1-q^k)\prod_{k=1}^{\infty}(1-q^k)} \right| < \frac{\epsilon}{2}.$$

Now concentrate on (1) and note that $[\![x]\!]$ denotes the greatest integer in x.

$$\left| \sum_{i=0}^{n-j} (-1)^i \frac{q^{\frac{i^2}{2}+2ij-\frac{i}{2}}}{\prod_{k=1}^{i+j}(1-q^k)} \left[\frac{\prod_{k=1}^{\infty}(1-q^k) - \prod_{k=1}^{n-i-j}(1-q^k)}{\prod_{k=1}^{n-i-j}(1-q^k)\prod_{k=1}^{\infty}(1-q^k)} \right] \right|$$

$$= \left| \sum_{i=0}^{n-j} (-1)^i \frac{q^{\frac{i^2}{2}+2ij-\frac{i}{2}}}{\prod_{k=1}^{i+j}(1-q^k)} \left[\frac{\prod_{k=n-i-j+1}^{\infty}(1-q^k) - 1}{\prod_{k=1}^{\infty}(1-q^k)} \right] \right|$$

$$\leq \underbrace{\left| \sum_{i=0}^{[\frac{n-j}{2}]} (-1)^i \frac{q^{\frac{i^2}{2}+2ij-\frac{i}{2}}}{\prod_{k=1}^{i+j}(1-q^k)} \left[\frac{\prod_{k=n-i-j+1}^{\infty}(1-q^k) - 1}{\prod_{k=1}^{\infty}(1-q^k)} \right] \right|}_{(3)}$$

$$+ \underbrace{\left| \sum_{i=[\frac{n-j}{2}]+1}^{n-j} (-1)^i \frac{q^{\frac{i^2}{2}+2ij-\frac{i}{2}}}{\prod_{k=1}^{i+j}(1-q^k)} \left[\frac{\prod_{k=n-i-j+1}^{\infty}(1-q^k) - 1}{\prod_{k=1}^{\infty}(1-q^k)} \right] \right|}_{(4)}$$

First, we will show that (3) is arbitrarily small for sufficiently large n. We begin by observing that

$$\prod_{k=n-i-j+1}^{\infty} (1-q^k) \leq \prod_{[\frac{n-j}{2}]+2}^{\infty} (1-q^k).$$

Then applying the ratio test to

$$\sum_{i=0}^{\infty} (-1)^i \frac{q^{\frac{i^2}{2}+2ij-\frac{i}{2}}}{\prod_{k=1}^{i+j}(1-q^k)}$$

it is readily evident that the series in question converges to, say s. Thus, there exists an N_2 such that for all $n \geq N_2$,

$$\left| \sum_{i=0}^{[\frac{n-j}{2}]} (-1)^i \frac{q^{\frac{i^2}{2}+2ij-\frac{i}{2}}}{\prod_{k=1}^{i+j}(1-q^k)} \right| \leq |s| + 1.$$

Also, since $0 < q < 1$, $\sum_{n=1}^{\infty}(-q)^n$ converges and hence there exists an $0 < M < \infty$ such that $\prod_{k=1}^{\infty}(1-q^k) = M$. Thus, there exists an N_3 such that for all $n \geq N_3$,

$$\prod_{[\frac{n-j}{2}]+2}^{\infty} (1-q^k) < 1 + \frac{\varepsilon M}{4(|s|+1)}.$$

Therefore define $N_4 = \max\{N_2, N_3\}$ and for all $n \geq N_4$,

$$\left| \sum_{i=0}^{[\frac{n-j}{2}]} (-1)^i \frac{q^{\frac{i^2}{2}+2ij-\frac{i}{2}}}{\prod_{k=1}^{i+j}(1-q^k)} \left[\frac{\prod_{k=n-i-j+1}^{\infty}(1-q^k)-1}{\prod_{k=1}^{\infty}(1-q^k)} \right] \right|$$

$$\leq \left| \frac{\prod_{k=[\frac{n-j}{2}]+2}^{\infty}(1-q^k)-1}{\prod_{k=1}^{\infty}(1-q^k)} \right| \left| \sum_{i=0}^{[\frac{n-j}{2}]} (-1)^i \frac{q^{\frac{i^2}{2}+2ij-\frac{i}{2}}}{\prod_{k=1}^{i+j}(1-q^k)} \right|$$

$$\leq \frac{\varepsilon}{4(|s|+1)}(|s|+1) = \frac{\varepsilon}{4}.$$

Finally, we must consider (4). Note that

$$\left| \sum_{i=[\frac{n-j}{2}]+1}^{n-j} (-1)^i \frac{q^{\frac{i^2}{2}+2ij-\frac{i}{2}}}{\prod_{k=1}^{i+j}(1-q^k)} \left[\frac{\prod_{k=n-i-j+1}^{\infty}(1-q^k)-1}{\prod_{k=1}^{\infty}(1-q^k)} \right] \right|$$

$$\leq \left| \frac{\prod_{k=1}^{\infty}(1-q^k)-1}{\prod_{k=1}^{\infty}(1-q^k)} \right| \left| \sum_{i=[\frac{n-j}{2}]+1}^{n-j} (-1)^i \frac{q^{\frac{i^2}{2}+2ij-\frac{i}{2}}}{\prod_{k=1}^{i+j}(1-q^k)} \right|.$$

Since $\sum_{i=0}^{\infty}(-1)^i \dfrac{q^{\frac{i^2}{2}+2ij-\frac{i}{2}}}{\prod_{k=1}^{i+j}(1-q^k)}$ converges there exists an N_5 such that for all $n \geq N_5$

$$\left| \sum_{i=[\frac{n-j}{2}]+1}^{n-j} (-1)^i \frac{q^{\frac{i^2}{2}+2ij-\frac{i}{2}}}{\prod_{k=1}^{i+j}(1-q^k)} \right| < \frac{\varepsilon}{4}\left| \frac{M}{M-1} \right|.$$

Therefore, let $n = \max\{N_1, N_4, N_5\}$ and for all $n \geq N$,

$$\left| \sum_{i=0}^{n-j}(-1)^i \frac{q^{\frac{i^2}{2}+2ij-\frac{i}{2}}}{\prod_{k=1}^{i+j}(1-q^k)\prod_{k=1}^{n-i-j}(1-q^k)} - \sum_{i=0}^{\infty}(-1)^i \frac{q^{\frac{i^2}{2}+2ij-\frac{i}{2}}}{\prod_{k=1}^{i+j}(1-q^k)\prod_{k=1}^{\infty}(1-q^k)} \right| < \varepsilon$$

which verifies our claim.

REFERENCES

[1] R. ASKEY, Beta integrals and q- extensions, *Proc. Ramanujan Centennial International Conference, Annamaalainagar*, (1987).

[2] T. S. CHIHARA, *An Introduction to Orthogonal Polynomials*, Gordon and Breach, 1978.

[3] S. CLEMENT COOPER, WILLIAM B. JONES, AND W.J.THRON, Orthogonal Laurent Polynomials and Continued Fractions Associated with Log- Normal Distributions, *Journal CAM*, **32** (1990), 39-46.

[4] S. CLEMENT COOPER, WILLIAM B. JONES, AND W.J.THRON, Asymptotics of Orthogonal L- Polynomials for Log- Normal Distributions, *Constr. Approx*, **8** (1992), 59-67.

[5] E. T. COPSON, *Theory of Functions of a Complex Variable*, University Press, Oxford, 1935.

[6] EDWIN L. CROW AND KUNIO SHIMIZU (EDS.), *Log- normal Distributions, Theory and Applications*, Marcel Dekker, Inc., New York, 1988.

[7] WILLIAM B. JONES, ARNE MAGNUS, AND W. J. THRON, PC- fractions and Orthogonal Laurent Polynomials for Log- normal Distributions, *J. Math. Anal. Appl.*, to appear.

[8] WILLIAM B. JONES AND W. J. THRON, *Continued Fractions: Analytic Theory and Applications, Encyclopedia of Mathematics and Its Applications 11*, Addison- Wesley Publ. Co., Reading, MA, 1980, (distributed now by Cambridge Univ. Press, NY).

[9] DANIEL S. MOAK, The *q*-Analogue of the Laguerre Polynomials, *J. Math. Anal. and Appl.*, **81** (1981), 20-47.

[10] P. I. PASTRO, Orthogonal polynomials and some *q*- beta integrals of Ramanujan, *J. Math. Anal. and Appl.*, **112** (1985), 517-540.

[11] O. PERRON, *Die Lehre von den Kettenbrüchen II*, Teubner, Stuttgart, 1957.

[12] T. J. STIELTJES, Recherches sur les fractions continues, *Ann. Fac. Sci. Toulouse Math.*, **8** (1894), 1-122; **9** (1895), 5-47.

[13] S. WIGERT, Sur les polynômes orthogonaux et l'approximation des fonctions continues, *Ark. Mat. Astr. Fysik*, **17** (1923).

5

Best Truncation Error Bounds for Continued Fractions $K(1/b_n)$, $\lim\limits_{n \to \infty} b_n = \infty$

C. M. CRAVIOTTO, Department of Mathematics, University of Colorado, Boulder, CO 80309–0395, U.S.A.

WILLIAM B. JONES Department of Mathematics, University of Colorado, Boulder, CO 80309–0395, U.S.A.

W. J. THRON Department of Mathematics, University of Colorado, Boulder, CO 80309–0395, U.S.A.

Dedicated to the memory of Arne Magnus

1   INTRODUCTION

In this paper we focus our attention on continued fractions of the form

$$K(1/b_n) = \cfrac{1}{b_1} + \cfrac{1}{b_2} + \cfrac{1}{b_3} + \cdots \tag{1.1}$$

where

$$\lim_{n\to\infty} b_n = \infty, \quad b_n \in \mathbb{C} \quad \text{for} \quad n \geq 1. \tag{1.2}$$

The condition $\lim\limits_{n\to\infty} b_n = \infty$ is sufficient to ensure that $K(1/b_n)$ converges to f in the extended complex plane [3, Theorem 4.35]. We are interested in sharp bounds for the truncation error $|f - f_n|$ obtained when $f = K(1/b_n) \neq \infty$ is replaced by its nth approximant f_n.

Special consideration is given to a class of continued fractions denoted by $LP[k, \{\rho_n\}]$ and defined as follows. Let k be a fixed non-negative integer and $\{\rho_n\}_{n=0}^{\infty}$ be a sequence of positive numbers such that

$$\lim_{n\to\infty} \rho_n = 0. \tag{1.3a}$$

Let

$$E_n := \left[w : |w| \geq \rho_n + \frac{1}{\rho_{n-1}} \right]. \tag{1.3b}$$

Research supported in part by the U.S. National Science Foundation under Grants DMS–9103141 and INT–9113400.

Then

$$LP[k, \{\rho_n\}] := [K(1/b_n) : b_n \in E_n \text{ and } \rho_n + \frac{1}{\rho_{n-1}} \geq 2 \text{ for } n = k+1, k+2, \ldots]. \quad (1.3c)$$

Remark. If $K(1/b_n) \in LP[k, \{\rho_n\}]$, then $\lim b_n = \infty$ and hence $K(1/b_n)$ converges, possibly to ∞. In most cases of interest in the present paper, the value f of $K(1/b_n)$ is finite.

Let

$$V_n := [v : |v| \leq \rho_n], \quad n = 0, 1, 2, \ldots . \quad (1.4)$$

One can easily show that

$$\frac{1}{b_n + V_n} \subseteq V_{n-1} \quad \text{whenever} \quad b_n \in E_n, \ n = 1, 2, 3, \ldots . \quad (1.5)$$

In Section 3 we develop two types of error bounds: (a) a priori bounds expressed in terms of the complex elements b_n or parameters related to them; (b) a posteriori bounds for f_n obtained only after one has computed $f_1, f_2, \ldots, f_n$ or comparable quantities. In Section 4 (Theorem 4.2) we show that the bounds given by Theorem 3.2 are best with respect to associated element regions. In Section 5 we provide a numerical example.

The following formulas and notation are used throughout the paper (see [3] for further details). Linear fractional transformations s_n and $S_n^{(m)}$ associated with the continued fraction (1.1) are defined by

$$s_n(w) := \frac{1}{b_n + w} \qquad n = 1, 2, 3, \ldots, \quad (1.6)$$

and

$$S_0^{(m)}(w) := w, \quad S_n^{(m)}(w) := S_{n-1}^{(m)}(s_{m+n}(w)), \quad m = 0, 1, 2, \ldots, \ n = 1, 2, 3, \ldots . \quad (1.7)$$

For $m = 0, 1, 2, \ldots$, the mth tail $f^{(m)}$ of (1.1) is the continued fraction

$$f^{(m)} := \frac{1}{b_{m+1}} + \frac{1}{b_{m+2}} + \cdots \quad (1.8)$$

The nth approximant $f_n^{(m)}$ of $f^{(m)}$ is given by

$$f_n^{(m)} := S_n^{(m)}(0) = \frac{1}{b_{m+1}} + \frac{1}{b_{m+2}} + \cdots + \frac{1}{b_{m+n}}, \quad (1.9)$$

where we adopt the notation

$$f = f^{(0)}, \quad f_n = f_n^{(0)}, \quad \text{and} \quad S_n = S_n^{(0)}. \quad (1.10)$$

The nth numerator A_n and nth denominator B_n of (1.1) are defined by the second order linear difference equations

$$A_{-1} := 1 \quad A_0 := 0 \quad B_{-1} := 0 \quad B_0 := 1 \tag{1.11a}$$

$$A_n := b_n A_{n-1} + A_{n-2} \quad n = 1, 2, 3, \ldots \tag{1.11b}$$

$$B_n := b_n B_{n-1} + B_{n-2} \quad n = 1, 2, 3, \ldots . \tag{1.11c}$$

These A_n and B_n satisfy the determinant formula

$$A_n B_{n-1} - A_{n-1} B_n = (-1)^{n-1} \quad n = 1, 2, 3, \ldots , \tag{1.12}$$

and also

$$S_n(w) = \frac{A_n + w A_{n-1}}{B_n + w B_{n-1}} \quad n = 0, 1, 2, \ldots . \tag{1.13}$$

Thus

$$f_n := S_n(0) = \frac{A_n}{B_n} \quad \text{for} \quad n = 0, 1, 2, \ldots . \tag{1.14}$$

The ratios $h_n := \dfrac{B_n}{B_{n-1}}$, $n = 0, 1, 2, \ldots$ satisfy

$$h_0 = \infty, \quad h_1 = b_1 \tag{1.15a}$$

$$h_n = b_n + \frac{1}{h_{n-1}} = b_n + \frac{1}{b_{n-1}} + \frac{1}{b_{n-2}} + \cdots + \frac{1}{b_1}, \quad n = 2, 3, 4, \ldots , \tag{1.15b}$$

and

$$S_n(-h_n) = \infty. \tag{1.15c}$$

Some useful properties of h_n are discussed in Section 2. We let $\mathbb{C}$ denote the complex plane. The boundary, interior, closure and diameter of an arbitrary subset A of $\mathbb{C}$ are denoted by $\partial A, \mathrm{Int}\,(A), c(A)$, and $\mathrm{diam}\,(A)$, respectively. By $\mathrm{rad}\,(A)$ and $\mathrm{center}\,(A)$ we denote, respectively, the radius and center of a disk A.

2   PRELIMINARIES

In this section we establish bounds for $|h_n|, |f_n^{(m)}|$ and $|f^{(m)}|$ and limiting values for $\{h_n\}$ that are subsequently employed.

THEOREM 2.1 *Let $K(1/b_n)$ be a given continued fraction.*

(A) *If $|h_{m-1}| \geq 2 \left| \dfrac{1}{b_m} \right|$, for some integer $m \geq 1$, then $\left| \dfrac{h_m}{b_m} - 1 \right| \leq \dfrac{1}{2}$.*

(B) *If $\left| \dfrac{h_{m-1}}{b_{m-1}} \right| \geq \dfrac{1}{2}$, for some integer $m \geq 2$, then $\left| \dfrac{h_m}{b_m} - 1 \right| \leq 2 \left| \dfrac{1}{b_m b_{m-1}} \right|$.*

(C) *If* $|b_m| \geq 2$, *for all* $m \geq 1$, *then for all* $m = 2, 3, \ldots,$

$$\left|\frac{h_m}{b_m} - 1\right| \leq 2\left|\frac{1}{b_{m-1}b_m}\right| \leq \frac{1}{2} \quad \text{and} \quad \left|\frac{h_m}{b_m}\right| \geq \frac{1}{2}. \tag{2.1}$$

Proof: (A) and (B) follow from (1.15b) since

$$\left|\frac{h_m}{b_m} - 1\right| = \left|\frac{1}{h_{m-1}b_m}\right| \leq \frac{1}{2}.$$

To prove (C), we use induction on m. Since $h_2 = b_2 + \dfrac{1}{b_1}$, we obtain

$$\left|\frac{h_2}{b_2} - 1\right| = \left|\frac{1}{b_1 b_2}\right| \leq 2\left|\frac{1}{b_1 b_2}\right| \leq \frac{1}{2},$$

and

$$\left|\frac{h_2}{b_2}\right| = \left|1 + \frac{1}{b_1 b_2}\right| \geq \left|1 - \left|\frac{1}{b_1 b_2}\right|\right| \geq \frac{3}{4} \geq \frac{1}{2}.$$

Hence (2.1) holds for $m = 2$. If we assume that (2.1) holds for some $m \geq 2$, then (B) implies that (2.1) holds with m replaced by $m + 1$. $\square$

THEOREM 2.2 *Let* $K(1/b_n)$ *be a given continued fraction such that*

$$\lim_{n \to \infty} b_n = \infty. \tag{2.2}$$

Let f *denote the value of* $K(1/b_n)$.
    (A) *If* $f = \infty$, *then* $\lim\limits_{n \to \infty} h_n = 0$.
    (B) *If* $f \neq \infty$, *then* $\lim\limits_{n \to \infty} h_n = \infty$.

Our proof of this theorem makes use of the following two lemmas.

LEMMA 2.3 *Let* $K(1/b_n)$ *be a given continued fraction such that for some non-negative integer* k,

$$|b_n| \geq 2 \qquad n = k + 1, k + 2, \ldots \tag{2.3}$$

Then

$$|f_n^{(k)}| < 2\left|\frac{1}{b_{k+1}}\right| \leq 1, \quad \text{for} \quad n = 1, 2, 3, \ldots, \tag{2.4a}$$

and

$$|f^{(m)}| \leq 2\left|\frac{1}{b_{m+1}}\right| \leq 1, \quad \text{for} \quad m = k, k + 1, k + 2, \ldots. \tag{2.4b}$$

Proof: Without loss of generality we may assume $k = 0$. Let $n \geq 1$ be fixed. We now show by induction on j that, for $j = 1, 2, \ldots, n$,

$$|f_j^{(n-j)}| < 2\left|\frac{1}{b_{n-j+1}}\right| \leq 1. \tag{2.5}$$

For $j = 1$, it follows from (1.9) that $|f_1^{(n-1)}| = \left|\dfrac{1}{b_n}\right| < 2\left|\dfrac{1}{b_n}\right| \leq 1$, so that (2.5) holds for $j = 1$. If we assume that (2.5) holds for some j such that $1 \leq j \leq n - 1$, then by (1.9) and (2.5)

$$|f_{j+1}^{(n-j-1)}| = \left| \frac{1}{b_{n-j} + f_j^{(n-j)}} \right| \leq \frac{1}{|b_{n-j}|} \frac{1}{1 - \frac{1}{|b_{n-j}|}} < 2\left|\frac{1}{b_{n-j}}\right| \leq 1.$$

Letting $j = n$ in (2.5) gives $|f_n^{(0)}| < 2\left|\dfrac{1}{b_1}\right| \leq 1$, which establishes (2.4a). It follows that (2.4b) holds when $m = k$. If $m > k$, then $|b_n| \geq 2$ for $n = m + 1, m + 2, \ldots$ and thus (2.4b) follows. $\square$

LEMMA 2.4 *Let $K(1/b_n)$ be a given continued fraction and let k be a non-negative integer such that (2.3) holds. Then for each $n = k + 1, k + 2, \ldots$, either*

$$|h_n| \geq \frac{1}{2} |b_n| \tag{2.6}$$

or

$$|\varepsilon_n| \geq |\varepsilon_k| \prod_{j=k+1}^{n} |b_{j-1}| \left(\frac{|b_{j-1}|}{2} - \frac{1}{|b_{j+1}|} \right) \tag{2.7}$$

where

$$\varepsilon_m := \frac{1}{b_m} (h_m + f^{(m)}), \qquad m = 1, 2, 3, \ldots .$$

If (2.6) holds for some positive integer n, then it holds for all subsequent values of n.

Proof: First we note that $0 \neq |f^{(m)}| \leq 1$ for all $m \geq k$. The inequality $|f^{(m)}| \leq 1$ follows from Lemma 2.3. Moreover, $f^{(m)} \neq 0$ follows from $f^{(m)} = \dfrac{1}{b_m + f^{(m+1)}} \neq 0$ since $b_m \neq \infty$ and $f^{(m+1)} \neq \infty$. Thus $\varepsilon_m = \infty$ if and only if $h_m = \infty$ (a possibility that can occur).

Suppose that $|h_m| \geq \frac{1}{2} |b_m|$ holds for some $m \geq k$. Then by Theorem 2.1(B) and (2.3) we have

$$\left| \frac{h_{m+1}}{b_{m+1}} - 1 \right| \leq 2 \left| \frac{1}{b_{m+1} b_m} \right| \leq \frac{1}{2}.$$

Hence $|h_{m+1}| \geq \frac{1}{2} |b_{m+1}|$. It follows by induction on n that $|h_n| \geq \frac{1}{2} |b_n|$ for all $n \geq m$. This proves the validity of the last sentence in Lemma 2.4. We will make subsequent use of the following two observations.

Observation 1: If $|h_m| \geq 2\left|\dfrac{1}{b_{m+1}}\right|$ holds for some $m \geq k$, then $|h_n| \geq \frac{1}{2} |b_n|$ holds for all $n \geq m + 1$. To verify this, we use Theorem 2.1(A) to conclude that $\left| \dfrac{h_{m+1}}{b_{m+1}} - 1 \right| \leq \dfrac{1}{2}$;

hence $|h_{m+1}| \geq \frac{1}{2}|b_{m+1}|$. The observation then follows from the last sentence of Lemma 2.4.

Observation 2: If $0 \leq |h_m| < 2\left|\dfrac{1}{b_{m+1}}\right|$ holds for some $m \geq k$, then

$$|\varepsilon_{m+1}| > |\varepsilon_m b_m|\left(\frac{|b_{m+1}|}{2} - \frac{1}{|b_{m+2}|}\right).$$

To verify this we use (1.15b) and the relation $f^{(m)} = \dfrac{1}{b_{m+1} + f^{(m+1)}}$ to obtain (assuming $h_m \neq 0$)

$$\begin{aligned}
|\varepsilon_{m+1}| &= \left|\frac{1}{b_{m+1}}\right||h_{m+1} + f^{(m+1)}| \\
&= \left|\frac{1}{b_{m+1}}\right|\left|\frac{1}{h_m} + \frac{1}{f^{(m)}}\right| \\
&= \left|\frac{1}{b_{m+1}}\right|\left|\frac{f^{(m)} + h_m}{h_m f^{(m)}}\right| \\
&= |\varepsilon_m||b_m|\left|\frac{1}{b_{m+1}h_m f^{(m)}}\right| \\
&= |\varepsilon_m||b_m|\left|\frac{1 + f^{(m+1)}/b_{m+1}}{h_m}\right| \\
&> |\varepsilon_m||b_m|\left|\frac{b_{m+1}}{2} + \frac{f^{(m+1)}}{2}\right| \\
&\geq |\varepsilon_m b_m|\left|\frac{|b_{m+1}|}{2} - \frac{|f^{(m+1)}|}{2}\right|.
\end{aligned}$$

The observation follows from this since, by Lemma 2.3, $|f^{(m+1)}| \leq 2\left|\dfrac{1}{b_{m+2}}\right|$. If $h_m = 0$, then $h_{m+1} = \infty$, $\varepsilon_m = \dfrac{1}{b_m}f^{(m)} \neq \infty$ and $\varepsilon_{m+1} = \dfrac{1}{b_{m+1}}(h_{m+1} + f^{(m+1)}) = \infty$. Thus the assertion of Observation 2 holds.

Now let S denote the set

$$S := \left[m : m \text{ is an integer, } m \geq k \text{ and } |h_m| \geq 2\left|\frac{1}{b_{m+1}}\right|\right].$$

Case 1: Suppose S is not the empty set. Let m_0 be the least element of S. It follows from Observation 1 that $|h_n| \geq \frac{1}{2}|b_n|$ for all $n \geq m_0 + 1$. Thus if $m_0 = k$, the assertions of the lemma follow. Suppose that $m_0 > k$. Then since

$$0 \leq |h_m| < 2\left|\frac{1}{b_{m+1}}\right| \quad \text{for} \quad k \leq m < m_0$$

it follows from successive applications of Observation 2 that (2.7) holds for $n = k$, $k+1, \ldots, m_0$. By Observation 1, (2.6) holds for $n = m_0 + 1, m_0 + 2, \ldots$. Hence the assertions of the lemma hold.

Case 2: Suppose S is the null set. Thus

$$0 \le |h_m| < 2 \left| \frac{1}{b_{m+1}} \right|, \qquad m = k, k+1, \ldots$$

Then by successively applying Observation 2 we obtain (2.7) for all $n = k+1, k+2, \ldots$ Hence the assertions of the Lemma hold. $\square$

Proof of Theorem 2.2: That $K(1/b_n)$ converges to a limit f (finite or infinite) follows from the Śleszyński–Pringsheim condition (i.e., $|b_n| \ge 2$ for all n implies $K(1/b_n)$ is convergent), and the remark following equation (1.3c). (A): Suppose $f = \infty$. Then for $n \ge 1$, $S_n(-h_n) = \infty = f = S_n(f^{(n)})$ and hence $-h_n = f^{(n)}$. By Lemma 2.3, for sufficiently large n,

$$|h_n| = |f^{(n)}| \le 2 \left| \frac{1}{b_{n+1}} \right|$$

from which assertion (A) follows, since $b_n \to \infty$.

(B): Suppose $f \ne \infty$. Then $f^{(n)} = S_n^{-1}(f) \ne S_n^{-1}(\infty) = -h_n$ which implies $f^{(n)} \ne -h_n$ for all n. Let $k \ge 2$ be chosen such that $|b_n| \ge 3$, for $n = k, k+1, k+2, \ldots$ Then by Lemma 2.3

$$|f^{(k+m)}| \le 2 \left| \frac{1}{b_{k+m-1}} \right| \le \frac{2}{3} \quad \text{for} \quad m \ge 1.$$

Suppose there exists an integer $m \ge 1$ such that $|h_{k+m}| \ge \frac{1}{2}|b_{k+m}|$. Then by Lemma 2.4, $\lim\limits_{n \to \infty} |h_n| \ge \lim\limits_{n \to \infty} \frac{1}{2}|b_n| = \infty$ and hence (B) holds. Suppose $|h_{k+m}| < \frac{1}{2}|b_{k+m}|$ for all $m = 1, 2, \ldots$. By Lemma 2.4

$$|\varepsilon_{k+m}| \ge |\varepsilon_k| \prod_{j=k+1}^{k+m} |b_{j-1}| \left(\frac{|b_j|}{2} - \frac{1}{|b_{j+1}|} \right)$$

$$\ge |\varepsilon_k| \prod_{j=k+1}^{k+m} 3 \left(\frac{3}{2} - \frac{1}{3} \right) = |\varepsilon_k| \prod_{j=k+1}^{k+m} \left(\frac{7}{2} \right)$$

$$\ge |\varepsilon_k| 3^m \quad \text{for} \quad m = 1, 2, \ldots \ .$$

Clearly $\varepsilon_k \ne 0$ since $\varepsilon_k = \frac{1}{b_k}(h_k + f^{(k)})$ and $f^{(m)} \ne -h_m$ for all m. For $m \ge 1$,

$$|f^{(k+m)}| \le \frac{2}{3} \quad \text{and} \quad |\varepsilon_{k+m}| \ge 3^m |\varepsilon_k|$$

which implies

$$\left|\frac{h_{k+m}}{b_{k+m}}\right| \geq |\varepsilon_{k+m}| - \left|\frac{f^{k+m}}{b_{k+m}}\right|$$

$$\geq 3^m |\varepsilon_k| - \frac{2}{9}.$$

Hence there exists an n_0 such that $\left|\dfrac{h_{k+n}}{b_{k+n}}\right| > \dfrac{1}{2}$ whenever $n \geq n_0$. Thus $\lim\limits_{n\to\infty} |h_n| \geq \lim\limits_{n\to\infty} \frac{1}{2}|b_n| = \infty$. $\square$

COROLLARY 2.5 *If $K(1/b_n)$ is a continued fraction converging to a finite value f and if $\lim\limits_{n\to\infty} b_n = \infty$, then*

$$\lim_{n\to\infty} (h_n - b_n) = 0 \tag{2.8}$$

and

$$\lim_{n\to\infty} \frac{h_n}{b_n} = 1. \tag{2.9}$$

Proof: It follows from (1.15b) that (2.8) holds, from which (2.9) follows. $\square$

3 BOUNDS

In this section we develop a posteriori and a priori bounds for the truncation error.

THEOREM 3.1 *Let $K(1/b_n) \in LP[k, \{\rho_n\}]$ be given and let f_n denote the nth approximant of $K(1/b_n)$ and $f = \lim\limits_{n\to\infty} f_n$.*

(A) *If there exists a $k_0 \geq k$ such that $|h_{k_0}| > \rho_{k_0}$, then*

$$|h_n| > \rho_n \qquad n = k_0, k_0 + 1, \dots, \tag{3.1}$$

$$K(1/b_n) \text{ converges to a finite value } f, \tag{3.2}$$

and

$$f \in S_n(V_n), \qquad n = k_0, k_0 + 1, \dots, \tag{3.3}$$

where S_n and V_n are defined by (1.4), (1.6) and (1.7).

(B) *If $K(1/b_n)$ converges to a finite value f, then there exists a $k_0 \geq k$ such that $|h_{k_0}| > \rho_{k_0}$ and hence the assertions of (A) hold.*

Proof: (A): Since $|h_{k_0}| > \rho_{k_0}$, $-h_{k_0} \notin V_{k_0}$, and thus from (1.15c), $\infty \notin S_{k_0}(V_{k_0})$. Hence $S_{k_0}(V_{k_0})$ is a bounded circular disk. By (1.6), (1.7) and (1.5),

$$S_{k_0+1}(V_{k_0+1}) = S_{k_0}\left(\frac{1}{b_{k_0+1} + V_{k_0+1}}\right) \subseteq S_{k_0}(V_{k_0})$$

and hence $S_{k_0+1}(V_{k_0+1})$ is also a bounded disk, which implies $-h_{k_0+1} \notin V_{k_0+1}$. This proves (3.1) for $n = k_0 + 1$. By induction one can verify that (3.1) holds for all $n \geq k_0 + 1$, and that $S_n(V_n) \subseteq S_{k_0}(V_{k_0})$ for all $n \geq k_0$. Hence $\{S_n(V_n)\}_{n=k_0}^{\infty}$ is a nested sequence of bounded circular disks and

$$f_{n+m} := S_{n+m}(0) \in S_{n+m}(V_{n+m}) \subseteq S_n(V_n) \subseteq S_{k_0}(V_{k_0}), \text{ for } n \geq k_0 \text{ and } m \geq 0. \quad (3.4)$$

Since $f = \lim f_n$ exists, it follows from (3.4) that f is finite. (3.3) also follows from (3.4).

(B): By Theorem 2.2B, $\lim_{n \to \infty} h_n = \infty$. Since $\lim_{n \to \infty} \rho_n = 0$, there exists $k_0 \geq k$ such that $|h_{k_0}| \geq \rho_{k_0}$, and thus (3.1), (3.2) and (3.3) hold. $\square$

In the following theorem an a posteriori bound is established.

THEOREM 3.2 *Let $K(1/b_n) \in LP[k, \{\rho_n\}]$ be given, let f_n denote the nth approximant of $K(1/b_n)$, and let $f = \lim f_n$.*

(A) If there exists a $k_0 \geq k$ such that $|h_{k_0}| > \rho_{k_0}$, then $K(1/b_n)$ converges to a finite value f and, for $n \geq k_0 + 1$,

$$|f - f_n| \leq \sup[|S_n(v) - f_n| : v \in V_n] = \frac{\rho_n}{|B_{n-1}|^2 |h_n|(|h_n| - \rho_n)}. \quad (3.5)$$

(B) If $K(1/b_n)$ converges to a finite value f, then there exists a $k_0 \geq k$ such that $|h_{k_0}| > \rho_{k_0}$ and hence the assertions of (A) hold.

Proof (A): It follows from (3.4) and the fact that $S_{k_0}(V_{k_0})$ is a bounded disk that $f_{n-1} \neq \infty$ for $n \geq k_0 + 1$. Hence by (1.14) $B_{n-1} \neq 0$ for $n \geq k_0 + 1$.

By Theorem 3.1A, $f \neq \infty$ and $f \in S_n(V_n)$, $n \geq k_0$, from which the inequality in (3.5) follows. By (1.14), (1.13), and (1.12),

$$|S_n(v) - f_n| = |S_n(v) - S_n(0)| = \left| \frac{A_n + v A_{n-1}}{B_n + v B_{n-1}} - \frac{A_n}{B_n} \right| = \frac{|v|}{|B_{n-1}|^2 |h_n + v| |h_n|}. \quad (3.6)$$

By Theorem 3.1A, $h_n \notin V_n$ for $n \geq k_0$. Hence $\min[|h_n + v| : v \in V_n]$ is attained at the unique point lying at the intersection of ∂V_n and the segment joining 0 to h_n. Clearly $\max[|v| : v \in V_n] = \rho_n$. Therefore the equality in (3.5) is a direct consequence of (3.6).

(B): This result follows immediately from Lemma 3.1B. $\square$

The following result obtained from [2, Theorem 2.1] is stated for later reference.

THEOREM 3.3 *Let $K(1/b_n) \in LP[0, \{\rho_n\}]$. Then*

$$|f - f_n| \leq 2\rho_0 \prod_{j=2}^{n} \frac{\rho_j \rho_{j-1}\left(1 + \frac{\rho_{j-1}}{|h_{j-1}|}\right)}{1 + 2\rho_j \rho_{j-1} - \frac{\rho_{j-1}}{|h_{j-1}|}}, \text{ for } n \geq 2. \quad (3.7)$$

THEOREM 3.4 Let $K(1/b_n) \in LP[0, \{\rho_n\}]$. Then

$$|f - f_n| \leq 2\rho_0 \prod_{j=2}^{n} \rho_j \rho_{j-1} \left(\frac{1 + 3\rho_{j-1}\rho_{j-2}}{2\rho_j(\rho_{j-1})^2 \rho_{j-2} + 2\rho_j \rho_{j-1} - \rho_{j-1}\rho_{j-2} + 1} \right). \quad (3.8)$$

Proof: From Theorem 3.3 we have

$$|f - f_n| \leq 2\rho_0 \prod_{j=2}^{n} \rho_j \rho_{j-1} f_j \left(\frac{1}{|h_{j-1}|} \right) \quad (3.9)$$

where

$$f_j(x) = \frac{1 + \rho_{j-1}x}{1 + 2\rho_j\rho_{j-1} - \rho_{j-1}x}.$$

It follows from Theorem 2.1C and (1.3) that

$$\frac{1}{|h_{j-1}|} \leq \frac{2}{|b_{j-1}|} \leq \frac{2\rho_{j-2}}{\rho_{j-1}\rho_{j-2} + 1}. \quad (3.10)$$

Since $f_j(x)$ is an increasing function of x,

$$f_j \left(\frac{1}{|h_{j-1}|} \right) \leq f_j \left(\frac{2\rho_{j-2}}{\rho_{j-1}\rho_{j-2} + 1} \right)$$

and it follows from (3.9) that (3.8) holds. $\square$

4   BEST TRUNCATION ERROR BOUNDS

A sequence $E := \{E_n\}_{n=1}^{\infty}$ of non-empty subsets of $\mathbb{C}$ is called a *sequence of element regions* for a continued fraction of the form

$$K \left(\frac{1}{b_n} \right) = \frac{1}{b_1} + \frac{1}{b_2} + \frac{1}{b_3} + \cdots \quad (4.1a)$$

if

$$b_n \in E_n \quad \text{for} \quad n = 1, 2, 3, \ldots . \quad (4.1b)$$

A sequence $V = \{V_n\}_{n=0}^{\infty}$ of subsets of $\mathbb{C}$ is called a *sequence of value regions corresponding to* $\{E_n\}$ if

$$\frac{1}{E_n} \subseteq V_{n-1} \quad \text{and} \quad \frac{1}{E_n + V_n} \subseteq V_{n-1} \quad \text{for} \quad n = 1, 2, 3, \ldots \quad (4.2)$$

where

$$\frac{1}{E_n} := \left[\frac{1}{b} : b \in E_n \right] \quad \text{and} \quad \frac{1}{E_n + V_n} := \left[\frac{1}{b+v} : b \in E_n, \, v \in V_n \right]$$

We denote by $\mathcal{V}(E)$ the family of all sequences of value regions $V = \{V_n\}$ corresponding to the sequence E of element regions. The following result is easily verified.

THEOREM 4.1 *Let $E = \{E_n\}$ be a given sequence of non-empty subsets of $\mathbb{C}$.*
    (A) If $V = \{V_n\}_{n=0}^{\infty}$ satisfies

$$0 \in V_{n-1} \quad \text{and} \quad \frac{1}{E_n + V_n} \subseteq V_{n-1} \quad \text{for} \quad n = 1, 2, 3, \ldots, \tag{4.3}$$

then $V \in \mathcal{V}(E)$.
    (B) If $V = \{V_n\} \in \mathcal{V}(E)$, then $\{c(V_n)\} \in \mathcal{V}(E)$.

In addition, if $0 \in V_{n-1}$ for $n \geq 1$, then by (1.9) and (4.2)

$$f_n = S_n(0) \in S_n(V_n) = S_{n-1}(s_n(V_n)) \subseteq S_{n-1}(V_{n-1}), \quad n \geq 1. \tag{4.4}$$

It follows that $\{S_n(V_n)\}$ is a nested sequence of non-empty subsets of $\mathbb{C}$ and

$$f_{n+m} \in S_n(V_n) \quad \text{for} \quad n \geq 1, \ m \geq 0. \tag{4.5}$$

Hence

$$|f_{n+m} - f_n| \leq \sup[|S_n(v) - f_n| : v \in V_n] \tag{4.6}$$

and in particular

$$|f - f_n| \leq \sup[|S_n(v) - f_n| : v \in V_n] \tag{4.7}$$

whenever $\{V_n\}$ is a sequence of closed value regions corresponding to $\{E_n\}_{n=1}^{\infty}$. If

$$U = \{U_N\}_{n=0}^{\infty} \in \mathcal{V}(E) \text{ and } U_n \subseteq V_n \text{ for } n = 0, 1, 2, \ldots \text{ for every } \{V_n\}_{n=0}^{\infty} \in \mathcal{V}(E), \tag{4.8}$$

then U is called the *best sequence of value regions corresponding to* E, which we denote by $U(E) = \{U_n(E)\}_{n=0}^{\infty}$. The existence and uniqueness of $U(E)$ follows from [3, Theorem 4.1]; in particular one has

$$U_n(E) = \left[v : v = \frac{1}{b_{n+1}} + \frac{1}{b_{n+2}} + \cdots + \frac{1}{b_{n+m}} : b_k \in E_k, \right.$$
$$\left. n+1 \leq k \leq n+m, \ m = 1, 2, 3, \ldots \right]. \tag{4.9}$$

Let $K = K(1/b_n)$ be a given continued fraction with nth approximant f_n and with elements b_n satisfying (4.1b) for some fixed sequence $E = \{E_n\}$ of element regions. For each $n = 1, 2, \ldots$, let

$$I_n(K, E) := [f_n] \cup S_n(c(U_n(E))) \tag{4.10}$$

which consists of f_n and all approximants of the form

$$\frac{1}{b_1} + \cdots + \frac{1}{b_n} + \frac{1}{\widehat{b}_{n+1}} + \cdots + \frac{1}{\widehat{b}_{n+m}} \tag{4.11a}$$

such that

$$\hat{b}_j \in E_j \quad \text{for} \quad n+1 \le j \le n+m \quad \text{for} \quad m = 1, 2, \ldots, \tag{4.11b}$$

and their limits. Since $I_n(K, E)$ contains only those terminating continued fractions whose elements satisfy (4.16) and whose first n elements agree with those of K, $I_n(K, E)$ is called the *best nth inclusion region for the continued fraction K with respect to the sequence of element regions E.* From (4.5) it is clear that $f_{n+m} \in I_n(K, E)$ for all $n \ge 1$, $m \ge 0$. Therefore if K converges to a finite limit f, then for $n = 1, 2, \ldots,$

$$|f - f_n| \le B_n(K, E) \tag{4.12}$$

where

$$B_n(K, E) := \sup[|\hat{f} - f_n| : \hat{f} \in I_n(K, E)]. \tag{4.13}$$

$B_n(K, E)$ is called the *best truncation error bound for the nth approximant f_n with respect to E.*

The term *best* is justified here since (4.13) is the sharpest bound that can be found subject to the limited information available; here we assume that the only things known about K are: (a) the first n elements $b_1, b_2, \ldots, b_n$, and (b) remaining elements satisfy (4.11b). From (4.13) and (4.8) we see that best truncation error bounds can be obtained from best value regions. We turn now to the problem of determining when a given sequence of value regions is best.

THEOREM 4.2 *Let $K(1/b_n) \in LP[k, \{\rho_n\}]$ be given. If there exists $k_0 \ge k$ such that $|h_{k_0}| > \rho_{k_0}$, then for $n \ge k_0 + 1$ the truncation error bounds (3.5) of Theorem 3.2 are best with respect to the element regions* (1.3b).

Proof: Let $\{V_n\}_{n=0}^{\infty}$ be defined by (1.4). It follows from (1.5) that $\dfrac{1}{E_n + V_n} \subseteq V_{n-1}$ for all integers $n \ge k + 1$. By Theorem 4.1, $\{V_n\}_{n=k}^{\infty}$ and $\{c(V_n)\}_{n=k}^{\infty}$ are sequences of value regions corresponding to $\{E_n\}_{n=k}^{\infty}$. To see that $V_{n-1} - \{0\} \subseteq \dfrac{1}{E_n + V_n}$ for $n \ge k$, fix $n \ge k$ and let $v \in V_{n-1} - \{0\}$. Then $v = \rho e^{i\theta}$ where $0 < \rho \le \rho_{n-1}$. Set $b_n = \left(\rho_n + \dfrac{1}{\rho}\right) e^{-i\theta}$ and $v_n = -\rho_n e^{-i\theta}$. Then $b_n \in E_n$ and $v_n \in V_n$. Furthermore $v = \dfrac{1}{b_n + v_n} \subseteq \dfrac{1}{E_n + V_n}$. Therefore $c\left(\dfrac{1}{E_n + V_n}\right) = V_{n-1}$ for $n \ge k$.

Let $\{U_m\}_{m=0}^{\infty}$ be the sequence of best value regions corresponding to $\{E_n\}_{n=1}^{\infty}$. By (4.8) and (1.5), $c(U_m) \subseteq V_m$ for all integers $m \ge 0$. We now show $V_m \subseteq c(U_m)$, for all integers $m \ge k$. Fix $m \ge k$ and choose $v_m \in V_m - \{0\}$.

Case 1. Suppose there exist $b_{m+1}, b_{m+2}, \ldots, b_{m+p}$ such that $b_{m+j} \in E_{m+j}$, $1 \le j \le p$, and

$$v_m = \cfrac{1}{b_{m+1}} + \cfrac{1}{b_{m+2}} + \cdots + \cfrac{1}{b_{m+p}}.$$

Then by (4.9) we have $v_m \in U_m$ and hence $v_m \in c(U_m)$.

Case 2. Suppose

$$v_m \neq \cfrac{1}{b_{m+1} + \cdots + b_{m+p}} \quad \text{for every choice of } b_{m+j} \in E_{m+j}, \ 1 \leq j \leq p. \qquad (4.14)$$

Then by (4.4) there exists $b_{m+1} \in E_{m+1}$, $v_{m+1} \in V_{m+1}$ such that $v = \cfrac{1}{b_{m+1} + v_{m+1}}$ and by (4.14) $v_{m+1} \neq 0$. Continuing in this manner there exist sequences $\{b_j\}_{j=m+1}^{\infty}$ and $\{v_j\}_{j=m+1}^{\infty}$ such that

$$v_m = \cfrac{1}{b_{m+1}} + \cfrac{1}{b_{m+2}} + \cdots + \cfrac{1}{b_{m+n} + v_{m+n}} \qquad (4.15)$$

where $b_{m+j} \in E_{m+j}$, $1 \leq j \leq n$ and $v_{m+n} \in V_{m+n} - \{0\}$ for $n = 2, 3, \ldots$. Let $\{S_n^{(m)}\}_{n=1}^{\infty}$ and $\{s_n\}_{n=1}^{\infty}$ be the linear fractional transformations associated with (4.15). Then by Theorem 3.3,

$$\text{Rad } S_n^{(m)}(V_{n+m}) \leq 2\rho_m \prod_{j=m+2}^{m+n} \frac{\rho_j \rho_{j-1}\left(1 + \frac{\rho_{j-1}}{|h_{j-1}|}\right)}{1 + 2\rho_j \rho_{j-1} - \frac{\rho_{j-1}}{|h_{j-1}|}}.$$

It follows from (1.3a) and Theorem 2.2B, that

$$\lim_{n \to \infty} \sup_{b_j \in E_j} \text{Rad } S_n^{(m)}(V_{n+m}) = 0. \qquad (4.16)$$

Since $v \in S_n(V_{n+m})$ for $m \geq 1$,

$$\cfrac{1}{b_{m+1}} + \cdots + \cfrac{1}{b_{m+n}} \in S_n(V_{n+m}), \quad m \geq 1. \qquad (4.17)$$

By (4.16) and (4.17), we have

$$v = \cfrac{1}{b_{m+1}} + \cfrac{1}{b_{m+2}} + \cfrac{1}{b_{m+3}} + \cdots.$$

It follows from (4.9) that $v \in c(U_m)$. $\square$

5 APPLICATION

We now consider a numerical example of the best truncation error bound for the continued fraction expansion of $f(z) := \tan z$ for $z = 2e^{i\pi/4}$. It is well known that (see [3]),

$$\begin{aligned}
\tan z &= \cfrac{z}{1} - \cfrac{z^2}{3} - \cfrac{z^2}{5} - \cfrac{z^2}{7} - \cfrac{z^2}{9} - \cdots \\
&= \cfrac{1}{1/z} + \cfrac{1}{-3/z} + \cfrac{1}{5/z} + \cfrac{1}{-7/z} + \cfrac{1}{9/z} + \cdots.
\end{aligned} \qquad (5.1)$$

Thus $\tan z = K(1/b_n)$ where $b_n = \dfrac{(-1)^{n+1}(2n-1)}{z}$, $n = 1, 2, 3, \ldots$. Set $\rho_n = \dfrac{|z|}{2n-1}$ for $n = 1, 2, 3, \ldots$. For a fixed $z \in \mathbb{C}$, choose k such that $\rho_n + \dfrac{1}{\rho_{n-1}} \geq 2$ for $n \geq k+1$, and $|z^2| \leq 4n - 2$ for $n \geq k+1$. It follows from (1.3) that $K(1/b_n) \in LP[k, \{\rho_n\}]$. In particular, when $z = 2e^{i\pi/4}$ we set $k = 2$. Then $K(1/b_n) \in LP[2, \{\rho_n\}]$ where $b_n = \dfrac{(-1)^{n+1}(2n-1)}{2e^{i\pi/4}}$. Furthermore $|h_2| = 10/4 > 2$ and hence the best truncation error bound given by (3.5) holds for $n \geq 3$. The following table provides a comparison of the actual error $|f - f_n|$, with the best truncation error bound (3.5) of $f(z) := \tan z$, for $z = 2e^{i\pi/4}$.

n	$|f - f_n|$	best truncation error bound (3.5)
3	$0.223(-1)$	$0.320(-1)$
6	$0.155(-5)$	$0.184(-5)$
9	$0.617(-11)$	$0.690(-11)$
12	$0.256(-17)$	$0.387(-17)$
15	$0.480(-24)$	$0.513(-24)$

REFERENCES

1. C. Baltus and W. B. Jones, Truncation error bounds for limit-periodic continued fractions $K(a_n/1)$ with $\lim a_n = 0$, *Numer. Math.* **46** (1985), 541–569.
2. D. A. Field and W. B. Jones, A priori estimates for truncation error of continued fractions $K(1/b_n)$, *Numer. Math.* **19** (1972), 283–302.
3. W. B. Jones and W. J. Thron, *Continued fractions: Analytic theory and applications*, in Encyclopedia of Mathematics and Its Applications, Addison–Wesley, Reading, MA, 1980.

6

Sequences of Linear Fractional Transformations and Reverse Continued Fractions

JOHN GILL, Department of Mathematics, University of Southern Colorado, Pueblo, CO 81001-4901 USA

Dedicated to the memory of Arne Magnus

1. INTRODUCTION

In the study of continued fractions the modified n^{th} approximant of $K[a_n/b_n]$ can be expressed as

$$\phi_n(z) = \frac{a_1}{b_1} + \frac{a_2}{b_2} + \cdots + \frac{a_n}{b_n + z}$$

where $\phi_n(z) := f_1 \circ f_2 \circ \ldots \circ f_n(z)$ and $f_j(z) := a_j/(b_j + z), j = 1, 2, \ldots$. $\phi_n(z)$ is an example of an *inner composition* of complex functions $\{f_j\}$. Inner compositions of linear fractional transformations (LFT's) have been studied extensively, particularly with regard to the convergence and divergence of continued fractions whose partial numerators and denominators are complex numbers. See, e.g., Magnus & Mandell[9], Gill[2], Thron & Jones[7], and Thron & Waadeland[10]. Investigations of inner compositions of more general analytic functions include Lorentzen[8] and Gill[3].

The study of the analytic theory of continued fractions at times involves an associated expansion that can be called a *reverse* continued fraction, with suggested notation $RK[a_n/b_n]$

$$\cdots + \frac{a_n}{b_n} + \frac{a_{n-1}}{b_{n-1}} + \cdots + \frac{a_1}{b_1}.$$

For example, if the nth approximant of the continued fraction $\boldsymbol{K}[a_n/1]$ is written A_n/B_n, then

$$\frac{B_n}{B_{n-1}} = 1 + \frac{a_n}{1} + \frac{a_{n-1}}{1} + \cdots + \frac{a_2}{1}$$

which may be interpreted as $1 + F_n(0)$, where $F_n(z) := f_n \circ f_{n-1} \circ \ldots \circ f_2(z)$ and $f_j(z) := a_j/(1+z), j = 2, 3, \ldots$. $F_n(z)$ is an *outer composition* of complex functions $\{f_j\}$, in this case LFT's. The general theory of such compositional forms has been explored in two recent papers by the author [4], [6]. The following result may be found there (in two parts).

Theorem 1.1. *Let $\{g_n\}$ be a sequence of functions analytic on a simply connected region S and continuous on the closure S' of S. Suppose there exists a compact set D contained in S such that $D \supset g_n(S)$ for all n. Then $G_n(z) := g_n \circ g_{n-1} \circ \ldots \circ g_1(z) \to \alpha$, a point of D, uniformly on S' if and only if the sequence of fixed points $\{\alpha_n\}$ of $\{g_n\}$ (in D) converges to α.*

The author, in [4], shows how this theorem can be used to compute fixed points of certain functions defined by continued fractions.

Theorem 1.2. *Let $F(z) := \boldsymbol{K}[a_n(z)/b_n(z)]$, where $a_n(z), b_n(z)$ are analytic in $S := \{z : |z| < R\}$ and continuous on S'. Set $h_n(w) := a_n(z)/(b_n(z) + w)$, $g_n(z) := h_1 \circ \ldots \circ h_n(z)$, and $G_n(z) := g_n \circ g_{n-1} \circ \ldots \circ g_1(z)$. If $|h_n(w)| \leq r < R$ for $|w| \leq r$ and $|h_n(0)| \leq r$ for $z \in S'$, then the solution of $F(z) = z$ in S is given iteratively by $\{G_n(z)\}$.*

In [6] the author shows how reverse continued fractions might be used as sequence modifiers:

Theorem 1.3. *Let $S := \{z : |z - \alpha| < r\}$ and suppose $|\alpha_n - \alpha| < \varepsilon_n \to 0$. Set $r_n := r + \varepsilon_n$. Then the modified reverse continued fraction*

$$\frac{\alpha_n \beta_n}{\alpha_n + \beta_n} - \frac{\alpha_{n-1} \beta_{n-1}}{\alpha_{n-1} + \beta_{n-1}} - \cdots - \frac{\alpha_1 \beta_1}{\alpha_1 + \beta_1 - z}$$

converges to α provided $|\beta_n| \geq r_n(1 + |\alpha_n|/(p - \varepsilon_n))$, where $p < r$.

Certain time-dependent dynamical systems also involve iterative structures of the form $\{F_n(z)\}$. There appears to be justification, then, for further exploring the convergence behavior of outer compositions, particularly of LFT's.

Let us begin with a review of the simplest cases, iterations $(F_n(z) = f^n(z))$ of single LFT's. Subsequent theory developed in this paper parallels these elementary dynamical systems.

An *LFT* of the form $f(z) = (az + b)/(cz + d)$, normalized so that $ad - bc = 1$, is called *parabolic* if $a + d = 2$ or -2. It is a simple consequence of this condition that $f(z)$ has but one fixed point α in the complex plane C (non-parabolic, normalized *LFT*'s have two fixed points, α and β). It is possible to write $f(z)$ implicitly in the following way (Ford, [1]):

$$\frac{1}{f(z) - \alpha} = \frac{1}{z - \alpha} + c \tag{1.1}$$

provided $a + d = 2$ (if $a + d = -2$ then c is replaced by $-c$).

Now, α is a "neutral fixed point" of f, that is to say $f'(\alpha) = 1$, and dynamic behavior at such points frequently is difficult to determine; however (1.1) clearly shows that for this special transformation the iterated function $f^n(z) \to \alpha$ as $n \to \infty$ for all $z \in C$. In fact, (1.1) gives

$$f^n(z) - \alpha = (z - \alpha)(nc(z - \alpha) + 1)^{-1} \tag{1.2}$$

so that convergence is of a simple arithmetic nature (and hence may be quite slow).

When f is not parabolic, it can be shown [1] to have two distinct fixed points α and β and can be written implicitly as

$$\frac{f(z) - \alpha}{f(z) - \beta} = K\frac{z - \alpha}{z - \beta}, \quad \text{where } K = \frac{a - c\alpha}{a - c\beta}. \tag{1.3}$$

When the constants a and d of the normalized $f(z)$ are such that $a + d$ is real and $|a + d| < 2$, f is called *elliptic*. In this case it can be shown that $|K| = 1$ but $K \neq 1$.

(1.3) is particularly suitable for an iteration of the function f. In fact, iterating and manipulating the resulting expression, one finds that for a fixed value of z

$$f^n(z) = \beta + (K^n C_1 + C_2)^{-1}, \tag{1.4}$$

where $C_1 = (z - \alpha)/D(z - \beta)$ and $C_2 = -1/D$, for $D = \beta - \alpha$. Thus in the elliptic case there is oscillatory divergence of $\{f^n(z)\}$ except when $z = \alpha$ or β.

There are two remaining classifications of f (not including the identity function): the *hyperbolic* $(a + d \in R, |a + d| > 2)$ and *loxodromic* $(a + d \notin R)$ cases. In each of these the fixed points are unequal in absolute value and are usually labelled so that $|K| < 1$. (1.3) and (1.4) then show that $\{f^n(z)\}$ converges to α, the *attracting* fixed point of f, for all $z \neq \beta$, the *repelling* fixed point of f.

A first step beyond simple iteration of a single *LFT* is the so-called *limit periodic* case (the expression comes from the theory of continued fractions). For inner compositions this means $\phi_n(z) = f_1 \circ \ldots \circ f_n(z)$ with $f_n \to f$. Magnus & Mandell [9] and the author [2] previously described conditions on the rate at which $f_n \to f$ that are sufficient to insure that convergence behavior of $\{\phi_n(z)\}$ is similar to if not exactly the same as that of $\{f^n(z)\}$. Such results shed light on the convergence/divergence of *limit periodic* continued fractions.

In the present paper and in one previous monograph (Gill, [5]) the author presents several results for outer composition structures of LFT's $\{f_n\}$, where $f_n \to f$. Conditions on the rate of convergence of $\{f_n\}$ are given which imply that the resulting convergence behavior of $\{F_n(z)\}$ roughly parallels that of $\{f^n(z)\}$ - except in the special case $f(z) \equiv z$. The results are not exactly the same as for inner compositions, and the proofs, after initial restructuring of the composition chains, differ from those found in [4] and [2]. For the case $f(z) \equiv z$, a simple observation reveals that any LFT F (or constant) might be the limit of the sequence $\{F_n(z)\}$: set $f_1 := F$, and $f_j \equiv z$ for $j > 1$. This case is not particularly relevant to the study of continued fractions and is not treated in this paper.

In [5] the author describes the case when the limit function f is hyperbolic or loxodromic. It is proven that the convergence behavior of $\{F_n(z)\}$ always parallels that of the simple iteration $\{f^n(z)\}$:

Theorem 1.4. *If f and $\{f_n\}$ are all hyperbolic/loxodromic LFT's with $f_n \to f$, then $\lim_{n\to\infty} F_n(z) = \lim_{n\to\infty} f_n \circ \ldots \circ f_1(z) = \alpha$, the attracting fixed point of f, for all $z \in C$ except possibly one value z_0. In this case $\lim_{n\to\infty} F_n(z_0) = \beta$, the repelling fixed point of f.*

A corollary for reverse continued fractions follows easily.

Corollary 1.5. *If $\lim a_n = a \in C - \{z : z \leq -1/4\}$, then for all $z \in C$*

$$\lim_{n\to\infty} \left[\frac{a_n}{1} + \frac{a_{n-1}}{1} + \cdots + \frac{a_1}{1+z} \right] = \alpha$$

where $\alpha = (-1 + \sqrt{(1+4a)})/2$ is the branch of the root chosen to minimize $|\alpha|$.

The remaining elliptic and parabolic limit periodic cases are now developed, leading to a relatively complete general picture of the outer composition limit periodic variation of simple iteration of a non-singular LFT.

2. THE PARABOLIC CASE

The right side of (1.2) tends to zero as n goes to infinity. In order to approximate this behavior in the more general context of $\{F_n\}$ it is convenient to decompose each f_n in a conjugate format, and to reassociate the parts of the composition chain $F_n(z) = f_n \circ f_{n-1} \circ \ldots \circ f_1(z)$. After this initial step has been taken the proofs of the inner composition parabolic theorem in [2] and the present outer composition theorem develop along different lines.

Since the individual f_n's may or may not be parabolic one can algebraically manipulate (1.3) to produce

$$\frac{1}{f_n(z) - \alpha_n} = \frac{K_n}{z - \alpha_n} + q_n \tag{2.1}$$

where $K_n := 1$ if f_n is parabolic, and $K_n := (a_n - c_n\beta_n)/(a_n - c_n\alpha_n)$ otherwise, and $q_n := c_n$ if f_n is parabolic, and $q_n := (K_n - 1)/(\alpha_n - \beta_n) = c_n/(a_n - c_n\beta_n)$ otherwise (this last expression reduces to c_n if f_n is parabolic) [2]. α_n and β_n are chosen so that $|K_n| \geq 1$.

Next, set $H_n(z) := 1/(z - \alpha_n)$, $Q_n(z) := z + q_n$, and $K_n(z) := K_n z$. With $H_n^{-1}(z) = \alpha_n + 1/z$, write

$$F_n(z) = H_n^{-1} \circ Q_n \circ K_n \circ H_n \circ H_{n-1}^{-1} \circ Q_{n-1} \circ K_{n-1} \circ H_{n-1} \circ \ldots \circ H_1^{-1} \circ Q_1 \circ K_1 \circ H_1(z). \tag{2.2}$$

Now, set

$$G_n(z) := Q_n \circ K_n \circ H_n \circ H_{n-1}^{-1}(z) = q_n + z(K_n/(\varepsilon_n z + 1)), \tag{2.3}$$

$\varepsilon_n := \alpha_{n-1} - \alpha_n$ for $n = 2, 3, \ldots$. Observe that when $f_n \equiv f$, $G_n(z) \equiv z$. The idea behind the basic convergence theorem of this section is that for large n, $G_n(z) \approx z$. Thus, for large n and $W_n(z) := Q_n \circ K_n \circ H_n(F_{n-1}(z))$, it appears that

$$
\begin{aligned}
F_{n+p}(z) &= H_{n+p}^{-1} \circ G_{n+p} \circ G_{n+p-1} \circ \ldots \circ G_{n+1} \circ W_n(z) \\
&\approx H_{n+p}^{-1}(W_n(z) + q_n + \ldots + q_{n+p}) \\
&= \alpha_n + 1/(W_n(z) + q_n + \ldots + q_{n+p}) \to \alpha \text{ as } p \to \infty.
\end{aligned}
$$

Now when f_n is elliptic and $f_n \equiv f$, an infinite iteration of this single function $\{f^n(z)\}$ gives a divergent sequence provided $z \neq \alpha, \beta$. So that, in the more general scenario, if $f_n(\text{elliptic}) \to f$ (parabolic) very slowly, it seems plausible that the sequence $\{F_n(z)\}$ might diverge for some values of z. Since $|K_n| = 1$ and $K_n \neq 1$ if f_n is elliptic, it is entirely reasonable that there be a condition on the rate at which $K_n \to 1$.

Theorem 2.1. *Suppose a sequence of linear fractional transformations $\{f_n\}$, with $|K_n| := |(a_n - c_n\beta_n)/(a_n - c_n\alpha_n)| \geq 1$, converges to a parabolic transformation f in such a manner that $\Sigma|\alpha_n - \beta_n| < \infty$ and $|\alpha_n - \alpha_{n-1}| < r^n$ for some $r \in [0, 1)$. Then $\lim_{n \to \infty} F_n(z) := \lim_{n \to \infty} f_n \circ \ldots \circ f_1(z) = \alpha$, the attracting fixed point of f, for all $z \in C$.*

Proof. Suppose that it is not true that $F_n(z) \to \alpha$ for all z. Since $w_n(z) = q_n + K_n/(F_{n-1}(z) - \alpha_n)$, $q_n \to c$, and $K_n \to K$, there exists a z and an $M > 0$ such that

$$|w_n(z)| \leq M \tag{2.4}$$

for an infinite number of values of n, and, also, that $|q_n| \leq M$ for all n.

Set $Z_n := w_n(z)$ and $Z_{n+m} := G_{n+m}(Z_{n+m-1})$ for $m = 1, 2, \ldots$.

One can write (from (2.1), (2.2), and (2.3))

$$Z_{n+k} = q_{n+k} + K_{n+k}Z_{n+k-1}(1 + \varepsilon_{n+k}Z_{n+k-1}) = q_{n+k} + Z_{n+k-1}(1 + v_{n+k})(1 + u_{n+k}),$$

where $v_{n+k} = c_{n+k}(\alpha_{n+k} - \beta_{n+k})/(a_{n+k} - c_{n+k}\alpha_{n+k})$ and $u_{n+k} = -\varepsilon_{n+k}Z_{n+k-1}/(1 + \varepsilon_{n+k}Z_{n+k-}$

Thus $U_{n+k} := |u_{n+k}| \le |\varepsilon_{n+k}||Z_{n+k-1}|/(1 - |\varepsilon_{n+k}||Z_{n+k-1}|)$.

Let $V_{n+k} := |v_{n+k}|$. Set $s^2 := (r+1)/2 < 1$. Thus $0 \le r < s^2 < s < 1$. Then for sufficiently large n, $r^{n+k} \le s^{2n+2k}/(n+k)$ for $k \ge 1$. Set $R := 1/s > 1$. Then for large enough n,

$$\begin{aligned}|\varepsilon_{n+k}| &\le r^{n+k} \le s^{2n+2k}/(n+k) \le (s^{n+k})^2/k = s^{n+k}s^n(1/kR^k) \\ &\le s^{n+k}s^n(1/[1 + R + \ldots + R^{k-1}]).\end{aligned}$$

Let n be large enough to guarantee that $(1 + V_{n+k}) < c := (1 + R)/2$ for $k = 1, 2, \ldots$. This is possible since "ΣV_j converges" implies $\Pi(1 + V_j)$ converges, which is equivalent to $\Pi(1 + V_{n+k}) \approx 1$ for large fixed n as $k \to \infty$. Now, let n be large enough to insure $Ms^n/(1 - Ms^n) < \varepsilon := R/c - 1 = (R-1)/(R+1)$. Observe that $R = c(1 + \varepsilon)$.

Assume in all that follows that n is a value that satisfies (2.4) and all conditions specified above.

Recalling that $|Z_n| \le M$ and $|q_{n+k}| \le M$, for sufficiently large values of n

$$\begin{aligned}|Z_{n+1}| &\le M + (1 + V_{n+1})M(1 + |\varepsilon_{n+1}||Z_n|/(1 - |\varepsilon_{n+1}||Z_n|)) \\ &\le M + cM(1 + s^{n+1}Ms^n/(1 - s^{n+1}Ms^n)) \\ &< M + cM(1 + \varepsilon) = M(1 + R)\end{aligned}$$

$$\begin{aligned}|Z_{n+2}| &\le M + (1 + V_{n+2})M(1 + R)[1 + |\varepsilon_{n+2}|M(1 + R)/(1 - |\varepsilon_{n+2}|M(1 + R))] \\ &\le M + cM(1 + R)(1 + s^{n+2}Ms^n/(1 - s^{n+2}Ms^n)) \\ &< M + cM(1 + R)(1 + \varepsilon) = M(1 + R + R^2) \\ &\quad\vdots\end{aligned}$$

$$\begin{aligned}|Z_{n+m}| &\le M + cM(1 + R + R^2 + \ldots + R^{m-1})(1 + s^{n+m}Ms^n/(1 - s^{n+m}Ms^n)) \\ &\le M + cM(1 + R + R^2 + \ldots + R^{m-1})(1 + \varepsilon) = M(1 + R + R^2 + \ldots + R^m).\end{aligned}$$

Expanding Z_{n+m} by repeated substitution gives

$$Z_{n+m} = q_{n+m} + q_{n+m-1}P_n(m, m) + q_{n+m-2}P_n(m-1, m) + \ldots + q_{n+1}P_n(2, m) + Z_nP_n(1, m)$$

where

$$P_n(k, m) = \prod_{j=k}^{m}(1 + W_{n+j}) = \prod_{j=k}^{m}(1 + V_{n+j})\prod_{j=k}^{m}(1 + U_{n+j})$$

with $U_{n+k} \leq s^{n+k}Ms^n/(1 - s^{n+k}Ms^n)$ for $0 < s < 1$, and $V_{n+k} \leq A|\alpha_{n+k} - \beta_{n+k}|$, $A =$ constant, for large n's.

Consequently, $\Pi(1 + U_j)$ and $\Pi(1 + V_j)$ converge by virtue of the convergence of $\Sigma|U_j|$ and $\Sigma|V_j|$. Thus for sufficiently large n, $P_n(k, m) \approx 1$ for $m \geq 1$ and $1 \leq k \leq m$.

Set $d_k := P_n(k, m)q_{n+k-1} - c$, for $k = 2, \ldots, m + 1$. $(P_n(m + 1, m) := 1)$.

$$
\begin{aligned}
\text{Then } |d_k| &\leq |P_n(k, m)q_{n+k-1} - q_{n+k-1}| + |q_{n+k-1} - c| \\
&= |q_{n+k-1}||P_n(k, m) - 1| + |q_{n+k-1} - c| \\
&< |c|/2 \text{ for } n \text{ sufficiently large and all } m \geq 1, \ 1 \leq k \leq m + 1.
\end{aligned}
$$

Therefore

$$
\begin{aligned}
Z_{n+m} &= \sum_{k=2}^{m+1} P_n(k, m)q_{n+k-1} + P_n(1, m)Z_n = \sum_{k=2}^{m+1}(c + d_k) + P_n(1, m)Z_n \\
&= mc + \sum_{k=2}^{m+1} d_k + P_n(1, m)Z_n \text{ and this implies}
\end{aligned}
$$

$$
\begin{aligned}
|Z_{n+m}| &\geq m|c| - \sum_{k=2}^{m+1}|d_k| - |P_n(1, m)||Z_n| \geq m|c| - m\left|\frac{c}{2}\right| - (M + 1) \\
&= m\left|\frac{c}{2}\right| - (M + 1) \to \infty \text{ as } m \to \infty.
\end{aligned}
$$

Therefore $Z_{n+m} \to \infty$ as $m \to \infty$. Consequently

$$
F_{n+m}(z) = H_{n+m}^{-1}(Z_{n+m}) = \alpha_{n+m} + 1/Z_{n+m} \to \alpha \text{ as } m \to \infty. \ \blacksquare
$$

For the important class of reverse continued fractions $\boldsymbol{RK}[a_n/1]$, there is

Corollary 2.2. If $|a_n + 1/4| < p^n$, where $0 \leq p < 1$ and $\{|a_n + 1/4|\}(\downarrow)$, then

$$
\lim_{n\to\infty}\left(\frac{a_n}{1} + \frac{a_{n-1}}{1} + \cdots + \frac{a_1}{1 + z}\right) = -\frac{1}{2}
$$

for all $z \in C$.

Proof. Choose positive t such that $\rho < t^2 < 1$. Then $2|a_{n-1} + 1/4|^{1/2} < 2\rho^{(n-1)/2} < 2t^{n-1} < r^n$ for large enough values of $n = n(r)$ provided $t < r < 1$. Since α_n is one of $[-1 \pm \sqrt{(1 + 4a_n)}]/2$, $|\alpha_n - \alpha_{n-1}| = (1/2)|(a_n+1/4)^{1/2} - (a_{n-1}+1/4)^{1/2}| \leq 2|a_{n-1}+1/4|^{1/2} < r^n$ for large n. Also, $\Sigma|\alpha_n - \beta_n| = \Sigma 2|a_n + 1/4|^{1/2} < \Sigma 2\rho^{n/2} < \infty$.

3. THE ELLIPTIC CASE

The simple iteration $\{f^n(z)\}$ of an elliptic LFT diverges by oscillation for all $z \neq \alpha, \beta$. Clearly $f^n(\alpha) \equiv \alpha$ and $f^n(\beta) \equiv \beta$. In the outer composition scenario, however, two possibilities arise.

Theorem 3.1. *Given a sequence* $\{f_n\}$ *of* LFT*'s where* $f_n \to f$ *elliptic, define* $F_n(z) := f_n \circ f_{n-1} \circ \ldots \circ f_1(z)$, $n = 1, 2, \ldots$. *Let* $\{\alpha_n\}$ *and* $\{\beta_n\}$ *be the fixed points of* $\{f_n\}$ *and* α *and* β *be the fixed points of* f, *designated so that* $|K_n| = |(a_n - c_n\alpha_n)/(a_n - c_n\beta_n)| \leq 1$ *and* $|K| = |(a - c\alpha)/(a - c\beta)| = 1$, $k \neq 1$.

(A) *If* $\Sigma|\alpha_{n-1} - \alpha_n|$, $\Sigma|\beta_{n-1} - \beta_n|$, *and* $\Pi|K_n|$ *all converge, then, for each* $z \in C$, *either* $F_n(z) \to \alpha$ *or* $F_n(z) \to \beta$ *or* $\{F_n(z)\}$ *diverges by oscillation.*

(B) *If* $\Sigma|\alpha_{n-1} - \alpha_n|$ *and* $\Sigma|\beta_{n-1} - \beta_n|$ *both converge, and* $\Pi|K_n|$ *diverges to 0, then for each* $z \in C$, *either* $F_n(z) \to \alpha$ *or* $F_n(z) \to \beta$.

Proof: Let us assume that the f_n's are non-parabolic. This will certainly be the case for n larger than some N, and by reindexing the f_n's (set $f_j := f_{N+j}$ for $j = 1, 2, \ldots$) and letting $f_N \circ \ldots \circ f_1(z)$, a one-to-one LFT, be replaced by z, we can begin our argument at this point.

Then, from (2.1) each f_n may be written as $f_n(z) = H_n^{-1} \circ K_n \circ H_n(z)$, where $H_n(z) = (z - \alpha_n)/(z - \beta_n)$, $K_n(z) = K_n z$, $H_n^{-1}(z) = (\beta_n z - \alpha_n)/(z - 1) = \beta_n + D_n/(z - 1)$, with $D_n := \beta_n - \alpha_n$. Set $\varepsilon_n := |\alpha_{n-1} - \alpha_n|$ and $\delta_n := |\beta_{n-1} - \beta_n|$. Thus

$$\begin{aligned} F_{n+m}(z) &= H_{n+m}^{-1} \circ K_{n+m} \circ H_{n+m} \circ H_{n+m-1}^{-1} \circ \quad K_{n+m-1} \circ H_{n+m-1} \circ \ldots \\ &\qquad \ldots \circ H_{n+1}^{-1} \circ K_{n+1} \circ K_{n+1} \circ H_{n+1}(F_n(z)) \\ &= H_{n+m}^{-1} \circ G_{n+m} \circ \ldots \circ G_{n+1}(w_n(z)), \end{aligned}$$

where $G_j(z) = K_j \circ H_j \circ H_{j-1}^{-1}(z)$ and $w_n(z) = K_n(F_{n-1}(z) - \alpha_n)/(F_{n-1}(z) - \beta_n)$.

Now, it is possible that there are z's such that $F_n(z) \to \alpha$ or $F_n(z) \to \beta$. For instance, if $f_n \equiv f$, elliptic, then clearly $F_n(\alpha) \equiv \alpha$ and $F_n(\beta) \equiv \beta$. Hence, in the following exposition let us consider a complex number z where $\{F_n(z)\}$ does not converge to either α or β.

Set $Z_n := w_n(z)$ and $Z_{n+j} := G_{n+j}(Z_{n+j-1})$ for $j = 1, 2, \ldots$. It is convenient to incorporate two short lemmas into the structure of the proof of the Theorem 3.1.

Lemma 3.2. *There exist a number* $R > 0$ *(distinct from the R defined in the proof of Theorem 2.1.) and an infinite number of positive integers n such that* $|Z_n| \leq R$.

Proof: Since $\{F_n(z)\}$ does not converge to β, there exist $d > 0$ and an infinite number of n's such that $|F_n(z) - \beta| > d$. Thus, for such an n, $Z_{n+1} = K_{n+1}[1 + D_{n+1}/(F_n(z) - \beta_{n+1})]$, which gives $|Z_{n+1}| \leq 1 + (|D| + 1)/(d/2) =: R > 0$, where $D := \lim D_n$. In what follows replace such an $n + 1$ with n. Then there exists an infinite sequence of positive integers n with $|Z_n| \leq R$. ∎

Next, set $S_k := \max\{|\varepsilon_k|, |\delta_k|\}$ for $k = 1, 2, \dots$.

Lemma 3.3. If $X_k := (9R^2/|D|)S_k$, then for n sufficiently large and $j = 1, 2, \dots$

$$S_{n+j} \leq \frac{X_{n+j}|D|}{2(1 + R + \sum_{j=1}^{\infty} X_{n+j})}$$

Proof: Follows easily from the fact that $\Sigma X_k < \infty$. ∎

Returning now to $Z_{n+j} := G_{n+j}(Z_{n+j-1}) = K_{n+j}(Z_{n+j-1} + T_{n+j})$, where, after some algebra, $|T_{n+j}| \leq S_{n+j}(1 + |Z_{n+j-1}|)^2/[|D|/2 - S_{n+j}(1 + |Z_{n+j-1}|)]$.

Repeated substitution gives

$$Z_{n+m} = (K_{n+m} \dots K_{n+1})Z_n + (K_{n+m} \dots K_{n+1})T_{n+1}$$
$$+ (K_{n+m} \dots K_{n+2})T_{n+2} + \dots + K_{n+m}T_{n+m}$$

So that $|Z_{n+m}| \leq |Z_n| + |T_{n+1}| + |T_{n+2}| + \dots + |T_{n+m}|$.

Lemmas 3.2 and 3.3 are used to show that

$$|T_{n+1}| \leq S_{n+1}(1 + |Z_n|)^2/[|D|/2 - S_{n+1}(1 + |Z_n|)] \leq X_{n+1}$$

for certain sufficiently large values of n.

Assume now that for such an n, $|T_{n+j}| \leq X_{n+j}$ for $j = 1, 2, \dots, p$. Then $|Z_{n+p}| \leq R + X_{n+1} + X_{n+2} + \dots + X_{n+p}$, and

$$|T_{n+p+1}| \leq \frac{S_{n+p+1}(1 + R + X_{n+1} + \dots + X_{n+p})^2}{\frac{|D|}{2} - S_{n+p+1}(1 + R + X_{n+1} + \dots + X_{n+p})}$$

$$\leq \frac{\frac{X_{n+p+1}|D|}{2(1+R+\dots+X_{n+p})(1+R+\dots+X_{n+p+1})}(1 + R + \dots + X_{n+p})^2}{\frac{|D|}{2} - \frac{X_{n+p+1}|D|}{2(1+R+\dots+X_{n+p})(1+R+\dots+X_{n+p+1})}(1 + R + \dots + X_{n+p})} = X_{n+p+1}$$

Therefore $|Z_{n+m}| \leq R + X_{n+1} + X_{n+2} + \dots + X_{n+m}$ for $m = 1, 2, \dots$. More precisely

$$Z_{n+m} = \left(\prod_{j=1}^{m} K_{n+j}\right) Z_n + \sum_{j=1}^{m} \left(\prod_{p=j}^{m} K_{n+p}\right) T_{n+j}$$

where the sum can be made arbitrarily small (for all $m > 0$) by choosing n large enough.

That $\{Z_n\}$ is bounded away from 0 is shown by the following argument: suppose some subsequence of $\{Z_n\}$ converges to 0. Since $|Z_{n+m}| \leq |Z_n| + X_{n+1} + X_{n+2} + \ldots + X_{n+m}$, $|Z_{n+m}|$ can be made arbitrarily small (uniformly for all $m > 0$) by choosing a large enough value of n associated with this subsequence. Simple computations show that

$$F_{n+m}(z) - \alpha = H_{n+m}^{-1}(Z_{n+m}) - \alpha = \ldots = [(\alpha_{n+m} - \alpha) + Z_{n+m}(\alpha - \beta_{n+m})]/[1 - Z_{n+m}],$$

which can be made arbitrarily close to 0, uniformly for all m, by choosing large enough n. This implies $F_j(z) \to \alpha$, contradicting our original assumption that $\{F_n(z)\}$ does not converge to either α or β. Thus, for large values of n,

$$Z_{n+m} \approx \left(\prod_{j=1}^{m} K_{n+j}\right) Z_n$$

and if $\Pi|K_j|$ converges, with $K_j \to K, |K| = 1, K \neq 1$, the equation

$$F_{n+m}(z) = \beta_{n+m} + (\beta_{n+m} - \alpha_{n+m})/(Z_{n+m} - 1)$$

reveals oscillating behavior of $\{F_{n+m}(z)\}$ as $m \to \infty$.

Now, if one returns to the beginning of the proof of Theorem 3.1 and assumes that z is a value for which $\{F_n(z)\}$ does not converge to β (allowing possible convergence to α), the condition $\Pi|K_j| \to 0$ applied to the preceding equation demonstrates that, indeed, $\{F_{n+m}(z)\}$ can be made arbitrarily close to α, uniformly for all m, if n is chosen large enough. Hence $F_j(z) \to \alpha$ as $j \to \infty$.

Corollary 3.4. *Suppose $a_n \to a < -1/4$. For each n choose the branch of the square root so that $|1 - \sqrt{(1 + 4a_n)}| \leq |1 + \sqrt{(1 + 4a_n)}|$. For each $z \in C$ let the sequence of modified convergents of the reverse continued fraction $RK[a_n/1]$ be defined as*

$$F_n(z) = \frac{a_n}{1} + \frac{a_{n-1}}{1} + \cdots + \frac{a_1}{1 + z}$$

Then:

(A) *If $\Sigma|a_n - a_{n-1}|$ and $\Sigma(|1 + \sqrt{(1 + 4a_n)}| - |1 - \sqrt{(1 + 4a_n)}|)$ both converge, $\{F_n(z)\}$ either converges to $(-1 + \sqrt{(1 + 4a)})/2$ or $(-1 - \sqrt{(1 + 4a)})/2$, or diverges by oscillation, or*

(B) *If $\Sigma|a_n - a_{n-1}|$ converges and $\Sigma(|1 + \sqrt{(1 + 4a_n)}| - |1 - \sqrt{(1 + 4a_n)}|)$ diverges, $\{F_n(z)\}$ converges to either $(-1 + \sqrt{(1 + 4a)})/2$ or $(-1 - \sqrt{(1 + 4a)})/2$.*

Proof: $\alpha_n = (-1 + \sqrt{(1 + 4a_n)})/2$ and $\beta_n = (-1 - \sqrt{(1 + 4a_n)})/2$. Thus the convergence of $\Sigma|a_n - a_{n-1}|$ is easily seen to imply the convergence of both $\Sigma|\alpha_{n-1} - \alpha_n|$

and $\Sigma|\beta_{n-1} - \beta_n|$. Also, $|K_n| = |1 - \sqrt{(1 + 4a_n)}|/|1 + \sqrt{(1 + 4a_n)}| = 1 - U_n$, where $U_n = (|1 + \sqrt{(1 + 4a_n)}| - |1 - \sqrt{(1 + 4a_n)}|)/|1 + \sqrt{(1 + 4a_n)}| \geq 0$. Then $\Pi|K_n|$ converges or diverges (to 0) as ΣU_n converges or diverges.

REFERENCES

1. L.R. Ford, *Automorphic Functions*, McGraw-Hill, New York, 1929, Ch.1

2. J. Gill, Infinite Compositions of Mobius Transformations, *Trans. Amer. Math. Soc.*, Vol. 176, Feb. 1973, 479-487.

3. J. Gill, Complex Dynamics of the Limit Periodic System $F_n(z) = F_{n-1}(f_n(z))$, $f_n \to f$, *J. Comp. & Appl. Math.*, 32 (1990), 89-96.

4. J. Gill, The Use of the Sequence $f_n(z) := f_n \circ f_{n-1} \circ \ldots \circ f_1(z)$ in Computing Fixed Points of Continued Fractions, Products, and Series, *App. Num. Math.* 8(1991) 469-476.

5. J. Gill, Outer Compositions of Hyperbolic/Loxodromic Linear Fractional Transformations, *Intl. J. Math. & Math. Sci.*, Vol. 15, No. 4 (1992) 819-822.

6. J. Gill, A Note on the Dynamics of the System $F_n(z) = f_n(F_{n-1}(z))$, $f_n \to f$, *Comm. Anal. Th. Cont. Fractions*, Vol. 1 (1992) 35-40.

7. W. Jones & Thron, *Continued Fractions: Analytic Theory & Applications*, Addison-Wesley, Reading, MA (1980).

8. L. Lorentzen, Compositions of Contractions, *J. Comp. & Appl. Math.*, 32 (1990), 169-178.

9. A. Magnus & M. Mandell, On Convergence of Sequences of Linear Fractional Transformations, *Math. Z.* 115 (1970) 11-17.

10. W. Thron & Waadeland, Modifications of Continued Fractions, *Springer Lecture Notes* #932 (1982), 38-66.

7

An Alternative Way of Using Szegő Polynomials in Frequency Analysis

W. B. JONES, Department of Mathematics, University of Colorado, Boulder, CO 80309–0426, USA

O. NJÅSTAD, Department of Mathematical Sciences, University of Trondheim (NTH), N–7034 Trondheim, Norway

H. WAADELAND, Department of Mathematics and Statistics, University of Trondheim (AVH), N–7055 Dragvoll, Norway

Dedicated to the memory of Arne Magnus

1. THE PROBLEM

Given a truncated trigonometric signal

$$x_N(m) = \sum_{j=-I}^{j=I} \alpha_j e^{i\omega_j m}, \quad \text{if} \quad m = 0, 1, 2, ..., N-1, \quad \text{otherwise} \quad x_N(m) = 0. \quad (1.1a)$$

Here $x_N(0) \neq 0$, and

$$\alpha_0 \geq 0, \quad 0 \neq \alpha_{-j} = \overline{\alpha}_j \in \mathbf{C}, \quad \omega_{-j} = -\omega_j \in \mathbf{R}, \quad j = 1, 2, ..., I, \quad (1.1b)$$

where the frequencies are arranged such that $0 = \omega_0 < \omega_1 < ... < \omega_I < \pi$. We shall assume that there is no noise.

The *frequency analysis problem* is to determine the unknown frequencies $\omega_1, \omega_2, ..., \omega_I$ from signal values $x_N(m)$ with a sample of size N. This problem has recently been dealt with in the papers [2], [3], [7], [8], [9], [11] by the Wiener- Levinson method [10], [12], formulated in terms of monic Szegő polynomials, i. e. polynomials orthogonal on the unit circle with respect to a certain absolutely continuous distribution function arising from the given signal. We shall briefly outline how this goes according to [7], since the method to be presented here is a modification of the procedure there. For details we refer to [7]. We shall, for short, refer to the method in [7] as the N-process, and the modification to be discussed here as the R-process. The reason will become clear later.

2. THE N-PROCESS

The absolutely continuous distribution function in question, $\psi_N(\theta)$, is defined by

$$\psi_N'(\theta) := \frac{1}{2\pi}|X_N(e^{i\theta})|^2, \quad -\pi \leq \theta \leq \pi, \quad \text{where} \quad X_N(z) = \sum_{m=0}^{N-1} x_N(m)z^{-m}, \quad (2.1)$$

The moments for this distribution, $\mu_m^{(N)}$, are known to form a positive definite hermitian sequence $\{\mu_m^{(N)}\}$, $m = 0, \pm 1, \pm 2, ...$, which means that $\mu_0^{(N)} > 0$, $\mu_{-m}^{(N)} = \overline{\mu_m^{(N)}}$ for $m > 0$, and that certain Toeplitz determinants with the moments as elements are positive. (In our case the moments are actually real.) It is also known (and easily seen), that they can be computed as autocorrelation coefficients of the signal:

$$\mu_m^{(N)} = \sum_{k=0}^{N-m-1} x_N(k)x_N(k+m), \quad m = 0, 1, 2, ..., \quad \text{and} \quad \mu_m^{(N)} = \mu_{-m}^{(N)}. \quad (2.2)$$

The Szegö polynomials $\rho_n(\psi_N; z)$ and the *reciprocal polynomials* $\rho_n^*(\psi_N; z) :=$ $z^n \overline{\rho_n(\psi_N; 1/\bar{z})}$ may be determined in different ways: 1: By the *Levinson algorithm*. 2: By explicit determinant formulas involving Toeplitz determinants. For details we again refer to [7]. For practical purposes the Levinson algorithm is by far the best method, in fact the only useful one in most cases. But for the purpose of what will be discussed in the present note the determinant formulas are more appropriate. Since the formulas play such a central role in the arguments, we shall give a brief review of them, see [1] and [6]:

With coefficients as given in (2.2) the power series

$$L_0(z) := \mu_0^{(N)} + 2\sum_{k=1}^{\infty} \mu_k^{(N)} z^k \quad (2.3)$$

represents a *Carathéodory* function, i. e. it is holomorphic in the open unit disk and maps it into the open right half plane $\Re w > 0$, moreover, $L_0(0) > 0$ (it is *normalized*). The power series

$$L_\infty(z) := -\mu_0^{(N)} - 2\sum_{k=1}^{\infty} \mu_{-k}^{(N)} z^{-k} \quad (2.4)$$

then represents a function, which is holomorphic in the region $|z| > 1$ and maps it into the open left half plane $\Re w < 0$.

A continued fraction of the form

$$\delta_0^{(N)} - \frac{2\delta_0^{(N)}}{1} + \frac{1}{\delta_1^{(N)}z} + \frac{(1-|\delta_1^{(N)}|^2)z}{\delta_1^{(N)}} + \frac{1}{\delta_2^{(N)}z} + \frac{(1-|\delta_2^{(N)}|^2)z}{\delta_2^{(N)}} + \cdots, \quad (2.5)$$

where $\delta_0^{(N)} > 0$ and $|\delta_k^{(N)}| < 1$ for $k > 0$, is called a *positive Perron-Carathéodory continued fraction*, or a positive PC-fraction for short. Perron-Carathéodory continued fractions were introduced in [4] by Jones, Njåstad and Thron. See also [5]. (The N in

(2.5) is included for the specific needs in the present note. Furthermore, in our case it will turn out that all $\delta_k^{(N)}$ are real, so that the absolute value and conjugate symbols may be left out in (2.5).)

To every pair of power series (2.3), (2.4), where the μ-sequence is a positive definite hermitian sequence, there corresponds a unique positive PC- fraction (2.5), such that the sequence of even approximants of (2.5) corresponds to $L_0(z)$, and the sequence of odd approximants of (2.5) corresponds to $L_\infty(z)$. Moreover, the sequences of even and odd approximants converge to the functions represented by the power series $L_0(z)$ and $L_\infty(z)$ respectively, in both cases uniformly on compact subsets of their domains of definition. Conversely, given a positive PC-fraction (2.5), there is a unique corresponding pair (2.3), (2.4), representing functions with the mentioned properties. The parameters $\delta_k^{(N)}$, called *reflection coefficients*, can be expressed by means of determinants, where the elements are the moments $\mu_m^{(N)}$, but they also come out of the Levinson algorithm. The formulas mentioned above are in our case:

For the Toeplitz determinant :

$$\Delta_n^{(N)} = \begin{vmatrix} \mu_0^{(N)} & \mu_{-1}^{(N)} & \mu_{-2}^{(N)} & \cdot & \cdot & \mu_{-n}^{(N)} \\ \mu_1^{(N)} & \mu_0^{(N)} & \mu_{-1}^{(N)} & \cdot & \cdot & \mu_{-n+1}^{(N)} \\ \mu_2^{(N)} & \mu_1^{(N)} & \mu_0^{(N)} & \cdot & \cdot & \mu_{-n+2}^{(N)} \\ \cdot & \cdot & \cdot & \cdot & \cdot & \cdot \\ \mu_n^{(N)} & \mu_{n-1}^{(N)} & \mu_{n-2}^{(N)} & \cdot & \cdot & \mu_0^{(N)} \end{vmatrix} . \tag{2.6}$$

For the reflection coefficients:

$$\delta_n^{(N)} = \frac{(-1)^n}{\Delta_{n-1}^{(N)}} \begin{vmatrix} \mu_{-1}^{(N)} & \mu_0^{(N)} & \cdot & \cdot & \cdot & \mu_{n-2}^{(N)} \\ \mu_{-2}^{(N)} & \mu_{-1}^{(N)} & & & & \mu_{n-3}^{(N)} \\ \cdot & \cdot & \cdot & \cdot & \cdot & \cdot \\ \cdot & \cdot & \cdot & \cdot & \cdot & \cdot \\ \mu_{-n}^{(N)} & \mu_{-n+1}^{(N)} & \cdot & \cdot & \cdot & \mu_{-1}^{(N)} \end{vmatrix} . \tag{2.7}$$

For the Szegö polynomials

$$\rho_n(\psi_N; z) = Q_{2n+1}(\psi_N; z) = \frac{1}{\Delta_{n-1}^{(N)}} \begin{vmatrix} \mu_0^{(N)} & \mu_{-1}^{(N)} & \cdot & \cdot & \cdot & \mu_{-n}^{(N)} \\ \mu_1^{(N)} & \mu_0^{(N)} & \cdot & \cdot & \cdot & \mu_{-n+1}^{(N)} \\ \cdot & \cdot & \cdot & \cdot & \cdot & \cdot \\ \mu_{n-1}^{(N)} & \mu_{n-2}^{(N)} & \cdot & \cdot & \cdot & \mu_{-1}^{(N)} \\ 1 & z & \cdot & \cdot & \cdot & z^n \end{vmatrix} . \tag{2.8}$$

$Q_n(\psi_N; z)$ denotes the nth denominator of the PC-fraction (2.5). Further references to PC-fractions, Carathéodory functions and Szegö polynomials are for instance [6] and [2] and the references therein. For later use: Observe that all formulas remain valid with the lefthand side unchanged if all $\mu_m^{(N)}$ are replaced by $\mu_m^{(N)}/N$. The significance of this observation is that the limit of the latter as $N \to \infty$ exists and will be crucial in the discussion.

The connection to Szegö polynomials is as follows: With standard normalization the denominators of the odd order approximants are the Szegö polynomials, the denominators of the even order approximants are the reciprocal polynomials. With notation from [7] we have:

$$Q_{2n}(\psi_N; z) = \rho_n^*(\psi_N; z), \quad Q_{2n+1}(\psi_N; z) = \rho_n(\psi_N; z), \quad n = 0, 1, 2, \ldots \qquad (2.9)$$

Having determined the Szegö polynomials, one way or another, they can lead us to the unknown frequencies as follows: Let

$$n_0 = 2I + L, \quad \text{where} \quad L = 0 \quad \text{if} \quad \alpha_0 = 0 \quad \text{and} \quad L = 1 \quad \text{otherwise}. \qquad (2.10)$$

Then the following holds, uniformly on compact subsets of $\mathbf{C}$ (see e. g. [7, Lemma 3.2.C]):

$$\lim_{N \to \infty} \rho_{n_0}(\psi_N; z) = (z - 1)^L \prod_{j=1}^{I} (z - e^{i\omega_j})(z - e^{-i\omega_j}) \qquad (2.11 .)$$

The main ideas of the proof are as follows: The formula for the nth Szegö polynomial can be written as a fraction, where numerator and denominator are as in (2.8),except that both are divided by N^n, which is the same as replacing all $\mu_m^{(N)}$ by $\mu_m^{(N)}/N$. We have

$$\lim_{N \to \infty} \frac{\mu_m^{(N)}}{N} = \sum_{j=-I}^{I} |\alpha_j|^2 e^{i\omega_j m} =: \mu_m. \qquad (2.12)$$

Hence we find $\lim_{N \to \infty} \rho_n(\psi_N; z)$ by taking limits separately in numerator and denominator of (2.8), provided that the limit of the denominator is $\neq 0$. This turns out to be the case for all $n \leq n_0$, but generally not for larger n. In fact, the series (2.3), divided by N, and with $\mu_k^{(N)}/N$ replaced by μ_k, represents the Carathéodory function

$$F(z) = \sum_{j=-I}^{I} |\alpha_j|^2 \frac{e^{i\omega_j} + z}{e^{i\omega_j} - z}, \qquad (2.13)$$

which can also be represented as a terminating continued fraction of the form

$$F(z) = \delta_0 - \frac{2\delta_0}{1} + \frac{1}{\overline{\delta_1}z} + \frac{(1 - |\delta_1|^2)z}{\delta_1} + \cdots + \frac{1}{\overline{\delta_{n_0-1}}z} + \frac{(1 - |\delta_{n_0-1}|^2)z}{\delta_{n_0-1}} + \frac{1}{\overline{\delta_{n_0}}z}. \qquad (2.14)$$

Here $\delta_0 = \lim_{N \to \infty} \delta_0^{(N)}/N$ and $\delta_k = \lim_{N \to \infty} \delta_k^{(N)}$ for $k = 1, 2, \ldots, n_0$. Furthermore $|\delta_k| < 1$ for $k = 1, 2, \ldots, n_0 - 1$, whereas $|\delta_{n_0}| = 1$. $F(z)$ is thus equal to the $2n_0$th as well as the $(2n_0 + 1)$th approximant of the expansion (2.14). (Keep in mind that $(1 - |\delta_{n_0}|^2) = 0$.) The denominator of (2.13), normalized to be monic, is the polynomial in (2.11).

From the convergence of $\rho_{n_0}(\psi_N; z)$ follows (by use of Hurwitz's theorem or Rouchè's theorem) that the zeros converge [7, Lemma 3.2 C]:

If the zeros $z(j, n_0, N)$ of $\rho_{n_0}(\psi_N; z)$ are ordered appropriately, then

$$\lim z(j, n_0, N) = e^{i\omega_j}, \quad j = \pm 1, \pm 2, \ldots, \pm I. \qquad (2.15)$$

If $\alpha_0 > 0$, we have in addition $\lim_{N\to\infty} z(0, n_0, N) = e^0 = 1$.

What we have seen above is of little use, unless we know the number of frequencies. "Normally" we do not know that. If we do not know n_0, we take an n, assumed to be $> n_0$, and try out the same scheme. If we really have $n > n_0$, we run into the trouble that the denominator and numerator in the formula for $\rho_n(\psi_N; z)$ both tend to zero. Since the Szegö polynomials are monic and have all their zeros in the open unit disk (see e. g. [1]) they must, in any disk $|z| \leq K$, be bounded above by $(1 + K)^n$ (this may also be seen from the recurrence formulas for Szegö polynomials and reciprocal polynomials [1] ,[6]). Hence the family of monic Szegö polynomials of degree n is uniformly bounded on any disk $|z| \leq K$, and thus there exist subsequences for which we have convergence. Important is, that regardless of which convergent sequence of Szegö polynomials we have, the limit polynomial will always have the polynomial in (2.11) as a factor [7, Lemma 3.3 C]. This, in turn, leads to the following convergence result for the zeros [7, Theorem 3.1]:

For every $n \geq n_0$ and $N \geq 1$ the zeros $z(j, n, N)$ of $\rho_n(\psi_N; z)$ can be ordered in such a manner that we have

$$\lim_{N\to\infty} z(j, n, N) = e^{i\omega_j}, \quad j = \pm1, \pm2, ..., \pm I. \tag{2.16}$$

If $L \neq 0$, we also have $\lim_{N\to\infty} z(0, n, N) = e^{0i} = 1$. If, in addition, we have a subsequence of $\{N\}$ such that the sequence of Szegö polynomials converges, then we have an additional $n - n_0$ zeros of the limit polynomial. These zeros will normally depend upon the subsequence chosen, but, since zeros of Szegö polynomials always are located in the open unit disk, those zeros (the "uninteresting" zeros) must all have absolute value ≤ 1. We even know more: From [11] it can be extracted, that to any $n > n_0$ there is a number $K_n \in (0, 1)$, depending only upon n and the given signal, such that $|\zeta| \leq K_n$ for all "uninteresting" zeros ζ. This knowledge, that the "uninteresting" zeros stay away from the unit circle, is very useful in the process of determining the frequences.

We conclude this section by showing a very simple illustrating example.

Example 1 Assume that the signal has only two frequencies, $\pm\pi/3$, and positive amplitude. Without loss of generality we may assume the amplitude to be $= 1$. The truncated signal is then

$$x_N(m) = 2\cos(m\pi/3), \quad \text{for} \quad m = 0, 1, 2, ..., N - 1, \quad \text{otherwise} \quad x_N(m) = 0.$$

By the procedure described above we find for instance (by using MACSYMA):

$$N \equiv 0 \quad (\text{mod } 3): \quad \lim_{N\to\infty} \rho_3(\psi_N; z) = (z - e^{i\pi/3})(z - e^{-i\pi/3})(z + 2/5)$$

$$N \equiv 1 \quad (\text{mod } 3): \quad \lim_{N\to\infty} \rho_3(\psi_N; z) = (z - e^{i\pi/3})(z - e^{-i\pi/3})(z + 2/5)$$

$$N \equiv 2 \quad (\text{mod } 3): \quad \lim_{N\to\infty} \rho_3(\psi_N; z) = (z - e^{i\pi/3})(z - e^{-i\pi/3})(z + 1/7)$$

The "uninteresting" zero in the two first cases is $-2/5$, in the third case $-1/7$. As we can see it depends upon the subsequence.

3. THE R-PROCESS

Let R be a positive number < 1. The R-process is to replace, in the N- process

$$\{\mu_m^{(N)}\} \quad \text{by} \quad \{R^{|m|}\mu_m^{(N)}\},$$

which is again a positive definite hermitian sequence, as will be seen later, and proceed to the Szegö polynomials $\rho_n(\psi_N^{(R)}; z)$. Then the following holds:

THEOREM 1. *The limit*

$$\lim_{R\uparrow 1}(\lim_{N\to\infty} \rho_n(\psi_N^{(R)}; z)) = \tilde{\rho}_n(z) \tag{3.1}$$

exists, and for $n \geq n_0$ the polynomial $\tilde{\rho}_n(z)$ is of the form

$$\tilde{\rho}_n(z) = (z-1)^L \prod_{j=1}^{I}(z-e^{i\omega_j})(z-e^{-i\omega_j}) \cdot \prod_{p=1}^{n-n_0}(z-z_p^{(n)}), \quad \text{where} \quad \text{all} \quad |z_p^{(n)}| \leq 1. \tag{3.2}$$

Before proving Theorem 1 we shall show two examples, the first one related to Example 1, but slightly more general.

Example 2 For an ω with $0 < \omega < \pi$ let the signal be such that $\pm\omega$ are the only frequencies. Moreover, let the amplitude be 1. Then the signal is as in Example 1, except that the special angle $\pi/3$ is now replaced by ω. Then the R-process leads to the following results (again by MACSYMA):

$$\tilde{\rho}_2(z) = (z - e^{i\omega})(z - e^{-i\omega})$$

$$\tilde{\rho}_3(z) = (z - e^{i\omega})(z - e^{-i\omega}) \cdot (z + (\cos\omega)/2)$$

$$\tilde{\rho}_4(z) = (z - e^{i\omega})(z - e^{-i\omega}) \cdot (z - \zeta)(z - \bar{\zeta}),$$

where

$$\zeta = -\frac{t}{4-t^2}(1 - i\sqrt{3-t^2}), \quad t = \cos\omega.$$

For $t = \cos\omega = \cos(\pi/3) = 1/2$, the "uninteresting" factor in $\tilde{\rho}_3(z)$ is $(z+1/4)$. Compare to Example 1, where we had the "uninteresting" factors $(z + 1/7)$ and $(z + 2/5)$ (twice) for three different subsequences of $\{N\}$. Observe finally that for *all* permitted ω the following inequality holds:

$$|\zeta| = \frac{|t|}{\sqrt{4-t^2}} \leq \frac{1}{\sqrt{3}} < 1$$

Example 3 For an $a > 0$ and an $\omega \in (0, \pi)$ we study the signal

$$x_N(m) = 1 + 2a\cos(m\omega) \quad \text{for} \quad m = 0, 1, ..., N - 1, \quad \text{otherwise} \quad x_N(m) = 0.$$

The R-process yields for instance (by MACSYMA)

$$\tilde{\rho}_4(z) = (z - 1)(z - e^{i\omega})(z - e^{-i\omega}) \cdot (z + \frac{2a^2(\cos\omega + 1) + 2\cos\omega}{2(\cos\omega)^2 + 2a^2(\cos\omega + 2) + 1}).$$

By simple calculus one can prove that the absolute value of the "uninteresting" zero of the fourth degree polynomial in the present example has an absolute value at most $1/\sqrt{2}$, regardless of frequency and amplitude.

In the proof of Theorem 1 we shall use the following

Observation. *Given a sequence $\{\gamma_n\}_{-\infty}^{\infty}$ of complex numbers, such that*

$$\gamma_{-n} = \overline{\gamma}_n \quad for \quad n \neq 0, \quad \gamma_0 > 0,$$

and such that the function, defined by the series

$$(3.3) \qquad \gamma_0 + 2\sum_{n=1}^{\infty} \gamma_n z^n$$

is holomorphic in the open unit disk $|z| < 1$ and maps it into the open right half plane, i. e. is a Carathéodory function. Then for any $R \in (0,1)$ the sequence

$$(3.4) \qquad \{R^{|n|}\gamma_n\}_{-\infty}^{\infty}$$

is a positive definite hermitian sequence.

This observation follows from a result stated in [6], which for our purpose can be phrased as follows: If (3.3) is a normalized Carathéodory function, not of the form (2.13), then the sequence $\{\gamma_n\}_{-\infty}^{\infty}$ is a positive definite hermitian sequence. In fact, regardless of whether (3.3) is of the form (2.13) or not, the function

$$\gamma_0 + 2\sum_{n=1}^{\infty} \gamma_n R^n z^n$$

is a normalized Carathéodory function, not of the form (2.13), and hence (3.4) is a positive definite hermitian sequence.

Outline of proof of Theorem 1. From the Observation it follows in particular that with $\mu_m^{(N)}$ as in (2.2) the sequence $\{R^{|m|}\mu_m^{(N)}\}$ is a positive definite hermitian sequence for all $R \in (0,1)$. For a fixed n the corresponding Szegö polynomial, $\rho_n(\psi_N^{(R)}; z)$, can be expressed as the ratio of two determinants, the denominator determinant with elements $R^{|m|}\mu_m^{(N)}/N$, in fact the Toeplitz determinant below (with n replaced by $n-1$)

$$\begin{vmatrix} \dfrac{\mu_0^{(N)}}{N} & \dfrac{\mu_{-1}^{(N)}R}{N} & \cdot & \cdot & \dfrac{\mu_{-n}^{(N)}R^n}{N} \\ \dfrac{\mu_1^{(N)}R}{N} & \dfrac{\mu_0^{(N)}}{N} & \cdot & \cdot & \dfrac{\mu_{-n+1}^{(N)}R^{n-1}}{N} \\ \cdot & & \cdot & \cdot & \cdot \\ \cdot & \cdot & \cdot & \cdot & \cdot \\ \cdot & \cdot & \cdot & \cdot & \\ \dfrac{\mu_n^{(N)}R^n}{N} & \dfrac{\mu_{n-1}^{(N)}R^{n-1}}{N} & \cdot & \cdot & \dfrac{\mu_0^{(N)}}{N} \end{vmatrix},$$

and the numerator determinant with the same kind of elements, except for $1, z, z^2, ..., z^n$ in the bottom row. (See (2.8) and the subsequent observation.) Hence we can take the limit as N tends to infinity by taking it separately in numerator and denominator, provided that the limit of the denominator is $\neq 0$. This is the case, since the determinant

in the denominator tends to the same type of Toeplitz determinant, but with the elements $R^{|m|}\mu_m$, where the numbers μ_m are as in (2.12):

$$
\begin{vmatrix}
\mu_0 & \mu_{-1}R & . & . & . & \mu_{-n+1}R^{n-1} \\
\mu_1 R & \mu_0 & . & . & . & \mu_{-n+2}R^{n-2} \\
. & & . & . & . & . \\
. & & & . & . & . \\
. & & & & . & . \\
\mu_{n-1}R^{n-1} & \mu_{n-2}R^{n-2} & . & . & . & \mu_0
\end{vmatrix}
$$

According to the observation above the sequence $\{R^{|m|}\mu_m\}$ is a positive definite hermitian sequence for all $R \in (0,1)$, and hence the Toeplitz determinant is $\neq 0$. Let the corresponding sequence of Szegö polynomials be $\{\rho_n(\psi^{(R)}; z)\}$. We then get, for all $n \geq 1$, that

$$
\lim_{N \to \infty} \rho_n(\psi_N^{(R)}; z) = \rho_n(\psi^{(R)}; z). \tag{3.5}
$$

Before we proceed to the next step in the proof we have to fill a notational gap:

Remark On the righthand side of (3.5) and in the line above there is a (so far) undefined symbol, namely $\psi^{(R)}$. It is the unique (up to an additive constant) distribution function solving the moment problem for the sequence $\{R^{|m|}\mu_m\}$. On the other hand it is also the limit as N tends to infinity of the distribution function $\psi_N^{(R)}$, not only in the weak* sense, but in the following sense:

$$
\lim_{N \to \infty} \frac{d}{d\theta}(\psi_N^{(R)}(\theta)) = \frac{1}{2\pi} \lim_{N \to \infty} \Re(\mu_0^{(N)}/N + \frac{2}{N}\sum_{n=1}^{\infty} R^n \mu_n^{(N)} e^{ni\theta}) = \tag{3.6}
$$

$$
\frac{1}{2\pi}\Re(\mu_0 + 2\sum_{n=1}^{\infty} R^n \mu_n e^{ni\theta}) = \frac{1}{2\pi}\Re(\sum_{j=-I}^{I} |\alpha_j|^2 \frac{e^{i\omega_j} + Re^{i\theta}}{e^{i\omega_j} - Re^{i\theta}}) = \frac{d}{d\theta}(\psi^{(R)}(\theta))
$$

Here we have used the fact (likely to be folklore, but also mentioned in [13]), that if the Carathéodory function $L_0(z)$ is holomorphic in the *closed* unit disk $|z| \leq 1$ then the absolutely continuous distribution function Φ is (except for an additive constant) uniquely given by

$$
\frac{d}{d\theta}\Phi(\theta) = \frac{1}{2\pi}\Re(L_0(e^{i\theta})). \tag{3.7}
$$

The steps in (3.6) are easily justified.

The second limit, i.e. the limit when R tends to 1, is somewhat more tricky, since the denominator (for $n > n_0$) may tend to 0. We deal with it in the following way:

With $R = 1 - h$ the numerator and denominator in the formula for $\rho_n(\psi^{(R)}; z)$ are polynomials in h. Moreover, for all $R \in (0,1)$ we have $|\rho_n(\psi^{(R)}; z)| \leq (1+K)^n$ in the disk $|z| \leq K$, (as we also had in the N-process). From this it follows, that the degree of the lowest power of h in the denominator must be $\leq$ the degree of the lowest power of h in the numerator, and hence the limit as $h \downarrow 0$ exists. In the same way we can prove the existence of the same type of limits for the reflection coefficients:

Let $\delta_m^{(N,R)}, m \geq 0$, be the reflection coefficients coming from the moment sequence $\{R^{|n|}\mu_n^{(N)}/N\}_{-\infty}^{\infty}$. We have in particular $\delta_0^{(N,R)} = \mu_0^{(N)}/N$. For $m \geq 1$ the reflection

coefficient $\delta_m^{(N,R)}$ is equal to the ratio of two Toeplitz determinants where the elements are moments from that particular sequence, compare (2.7). Since that moment sequence is a positive definite hermitian sequence the determinant in the denominator (the same as in the formula for $\rho_n(\psi_N^{(R)}; z)$) is different from 0. Taking the limit as $N \to \infty$ in numerator and denominator separately we still have a determinant different from 0 in the denominator, since the moment sequence $\{R^{|n|}\mu_n\}$ is hermitian positive definite. The determinant in the denominator is

$$
\begin{vmatrix}
\mu_0 & \mu_{-1}R & \cdot & \cdot & \cdot & \mu_{-n+1}R^{n-1} \\
\mu_1 R & \mu_0 & \cdot & \cdot & \cdot & \mu_{-n+2}R^{n-2} \\
\cdot & & & & & \cdot \\
& & & & & \\
\cdot & \cdot & & & & \\
\mu_{n-1}R^{n-1} & \mu_{n-2}R^{n-2} & \cdot & \cdot & \cdot & \mu_0
\end{vmatrix}
$$

The ratio is the reflection coefficient $\delta_m(R)$ coming from that particular sequence of moments. The absolute value is at most 1, regardless of the value of $R \in (0,1)$. With $R = 1 - h$ the reflection coefficient $\delta_m(R)$ is a rational function of h, where the lowest power of h in the denominator is at most equal to the lowest power of h in the numerator, and hence the limit

$$\lim_{R \uparrow 1} \delta_m(R) =: \tilde{\delta}_m, \quad m \geq 1, \tag{3.8}$$

exists and has absolute value ≤ 1. For $m = 0$ we have

$$\delta_0(R) =: \tilde{\delta}_0 = \mu_0. \tag{3.8'}$$

In conclusion we have that the limit

$$\lim_{R \uparrow 1} \left(\lim_{N \to \infty} \delta_m^{(N,R)} \right) = \tilde{\delta}_m \tag{3.9}$$

exists for any non-negative integer m. (Since the coefficients of the Szegö polynomials are continuous functions of the reflection coefficients this gives an alternative proof of the existence of the limit in (3.1).) Combined with the uniform boundedness on compact sets of all monic Szegö polynomials of a given degree we get uniform convergence on compact sets for N-limit as well as R-limit in (3.1).

It is important that for $m \leq n_0$ we can switch the order of the limits (N-limit and R-limit), since

$$\lim_{N \to \infty} \left(\lim_{R \uparrow 1} R^{|m|} \mu_m^{(N)}/N \right) = \lim_{R \uparrow 1} \left(\lim_{N \to \infty} R^{|m|} \mu_m^{(N)}/N \right),$$

and in these cases no denominator will vanish. Hence, for $m \leq n_0$ we have

$$\tilde{\delta}_m = \delta_m, \tag{3.10}$$

in particular $\tilde{\delta}_{n_0} = \delta_{n_0} = \pm 1$. (The numbers δ_m are only defined for $0 \leq m \leq n_0$, whereas the numbers $\tilde{\delta}_m$ are defined for all non-negative integers m.) From this we get "for free" from the N-process that (3.2) holds for $n = n_0$.

Since the statement about the limit for $n > n_0$ is crucial, we shall outline the proof, although it is very close to the proof in [7].

In order to prove (3.2) we go to the continued fraction (2.5), except for replacing $\delta_k^{(N)}$ by $\delta_k(R)$. The normalized denominator of any odd order approximant (order $2n+1$) then is the Szegö polynomial $\rho_n(\psi^{(R)}; z)$. For $n > n_0$ we may express this in terms of $Q_{2n_0}(\psi^{(R)}; z)$ and $Q_{2n_0-1}(\psi^{(R)}; z)$ by repeated use of the recurrence relations for denominators (and numerators) of the continued fraction, starting with

$$Q_{2n_0+1}(\psi^{(R)}; z) = \delta_{n_0}(R)Q_{2n_0}(\psi^{(R)}; z) + (1 - (\delta_{n_0}(R))^2)zQ_{2n_0-1}(\psi^{(R)}; z).$$

We find, that $Q_{2n+1}(\psi^{(R)}; z)$ can be expressed in the form

$$Q_{2n+1}(\psi^{(R)}; z) = A_n(z, R)Q_{2n_0}(\psi^{(R)}; z) + B_n(z, R)(1 - (\delta_{n_0}(R))^2)Q_{2n_0-1}(\psi^{(R)}; z),$$
$$(3.11)$$

where A_n and B_n are polynomials. We know that the limit of each term separately when $R \uparrow 1$ exists, in particular that $(\delta_{n_0}(R))^2$ goes to 1, and also that the righthand side is a monic polynomial of degree n, and so is the limit, which is then of the form

$$\lim_{R\uparrow 1} Q_{2n+1}(\psi^{(R)}; z) = \tilde{\rho}_n(z) = A_n(z)(z-1)^L \prod_{j=1}^{I}(z - e^{i\omega_j})(z - e^{-i\omega_j}).$$

Since $\tilde{\rho}_n(z)$ is monic of degree n and $(z-1)^L \prod_{j=1}^{I}(z - e^{i\omega_j})(z - e^{-i\omega_j})$ is monic of degree n_0, $A_n(z)$ must be monic of degree $n - n_0$, and hence of the form $\prod_{p=1}^{n-n_0}(z - z_p^{(n)})$. Finally, since the polynomial (3.11) is a Szegö polynomial and hence has all its zeros in the open unit disk we must have $|z_p^{(n)}| \leq 1$.

Theorem 1 is thus proved.

Convergence of zeros is also proved essentially as in [7], or alternatively by using the theorem of Rouché . But since here *all* zeros converge, the result can be phrased as follows:

COROLLARY 2. *For $n \geq 2$ the zeros $z^{(R)}(k, n, N)$ of $\rho_n(\psi_N^{(R)}; z)$ can be arranged in such a way that for all $k \in \{1, 2, ..., n\}$ the limits*

$$\lim_{R\uparrow 1}(\lim_{N\to\infty} z^{(R)}(k, n, N)) = z_k^{(n)}$$

exist, and $|z_k^{(n)}| \leq 1$. For $n \geq n_0$ the set of limits contains the set of all $e^{i\omega}$, where ω is a frequency in the signal.

4. FINAL REMARKS

When we compare the N-process and the R-process we note some differences:

1. $\mu_m^{(N)} R^m / N$ is simpler to deal with than $\mu_m^{(N)} / N$.
2. In the R-process, limits exist without going to subsequences.

3. For a given signal and a given n (also for $n > n_0$) all zeros are unique in the R-process.
4. The R-process contains one additional parameter.
5. For the R-process no "K_n-result" is (so far) at hand.

One possible attempt to remove the additional parameter R is to relate R to N in such a way that $R \to 1$ and $N \to \infty$ simultaneously. With some "luck" the two limits (N-limit first, then R-limit) may in this way be replaced by one, namely an N-limit. Inspired by the iterated limit in the R-process it seems to be a good idea to let R tend to 1 "slowly" as compared to the speed at which $N \to \infty$.

A slightly different idea from the one introduced in Section 2 would be to throw in the factor R^m directly into the signal, i.e. to replace (1.1) by $\tilde{x}_N(m) := x_N(m)R^m$. It turns out that this is almost the same as the R-process, the only essential difference being that there will be no division by N, but in the bargain we will have to multiply by the factor $(1 - R^2)$ before letting $R \to 1$.

REFERENCES

1. U. Grenander and G. Szegö, "Toeplitz Forms and their Applications," University of California Press, Berkeley, 1958.
2. W.B. Jones and O. Njåstad, *Applications of Szegö Polynomials to Digital Signal Processing*, Rocky Mountain Journal of Mathematics 21, 387-436 (1992).
3. W.B. Jones, O. Njåstad and E.B. Saff, *Szegö Polynomials Associated with Wiener-Levinson Filters*, Journal of Computational and Applied Mathematics 32, 387-406 (1990).
4. W.B. Jones, O. Njåstad and W.J. Thron, *Continued Fractions Associated with Trigonometric and Other Strong Moment Problems*, Constructive Approximation 2, 197-211 (1986).
5. W.B. Jones, O. Njåstad and W.J. Thron, *Schur Fractions, Perron-Carathéodory Fractions, a Survey*, Lecture Notes in Mathematics, Springer-Verlag 1199, 127-158 (1986).
6. W.B. Jones, O. Njåstad and W.J. Thron, *Moment Theory, Orthogonal Polynomials, Quadrature, and Continued Fractions associated with the Unit Circle*, Bulletin of the London Mathematical Society 21, 113-152 (1989).
7. W.B. Jones, O. Njåstad, W.J. Thron and H. Waadeland, *Szegö Polynomials Applied to Frequency Analysis*, Journal of Computational and Applied Mathematics (to appear).
8. W.B. Jones, O. Njåstad and H. Waadeland, *Asymptotics for Szegö Polynomial Zeros*, Numerical Algorithms (1992) (to appear).
9. W.B. Jones, O. Njåstad and H. Waadeland, *Application of Szegö Polynomials to Frequency Analysis*, (to appear).
10. N. Levinson, *The Wiener RMS (root mean square) Error Criterion in Filter Design and Prediction*, Journal of Mathematics and Physics 25, 261-278 (1947).
11. K. Pan and E.B. Saff, *Asymptotics for Zeros of Szegö Polynomials Associated with Trigonometric Polynomial Signals*, ICM- Report Number 91-014, Department of Mathematics, University of South Florida..
12. N. Wiener, "Extrapolation, Interpolation and Smoothing of Stationary Time Series," The Technology Press of Massachusetts Institute of Technology/ John Wiley and

Sons, Inc.,New York, 1949.

13. H. Waadeland, *A Szegő Quadrature Formula for the Poisson Formula*, Computational and Applied Mathematics I – Algorithms and Theory, Proc. of the 13th IMACS World Congress, Dublin, Ireland 1991 (to appear).

8

Asymptotics of Zeros of Orthogonal and Para-Orthogonal Szegö Polynomials in Frequency Analysis

WILLIAM B. JONES*   Department of Mathematics, University of Colorado, Boulder, Colorado 80309–0395.

OLAV NJÅSTAD   Department of Mathematical Sciences, University of Trondheim (NTH), N-7034 Trondheim, Norway.

HAAKON WAADELAND   Department of Mathematics and Statistics, University of Trondheim (AVH), N-7055 Dragvoll, Norway.

Dedicated to the memory of Arne Magnus

1   INTRODUCTION

The present paper deals with asymptotics of zeros of Szegö polynomials $\rho_n(\psi_{N,I}; z)$ and para-orthogonal polynomials $B_n(\psi_{N,I}; u_n; z)$ associated with discrete time signals $x_{N,I} = \{x_{n,I}(m)\}_{m=-\infty}^{\infty}$ of the form

$$
x_{N,I}(m) = \begin{cases} \displaystyle\sum_{j=-I}^{I} \alpha_j e^{i\omega_j m}, & 0 \leq m \leq N-1, \ x_{N,I}(0) \neq 0, \\ 0 & \text{elsewhere,} \end{cases} \tag{1.1a}
$$

where

$$
1 \leq N \leq \infty, \qquad 1 \leq I \leq \infty, \tag{1.1b}
$$

$$
\omega_0 = 0, \quad 0 < \omega_j = -\omega_{-j} < \pi, \quad \omega_j \neq \omega_k \text{ for } j \neq k, \ j, k \geq 1, \tag{1.1c}
$$

$$
0 \leq \alpha_0 \leq r_0, \ 0 < |\alpha_1| \leq r_1, \ 0 \leq |\alpha_j| \leq r_j, \ \alpha_j = \bar{\alpha}_{-j}, \ r_j = r_{-j} \text{ for } j \geq 1, \tag{1.1d}
$$

$$
\sum_{j=-I}^{I} \sum_{k=-I}^{I} r_j r_k \left| \csc\left(\frac{\omega_j - \omega_k}{2} \right) \right| < \infty. \tag{1.1e}
$$

*Research supported in part by the United States Educational Foundation in Norway (Fulbright Grant), the Norwegian Research Council (NAVF) and the U.S. National Science Foundation under grants DMS–9103141 and INT–9113400.

Condition (1.1e) implies

$$\sum_{j=-I}^{I} r_j < \infty. \tag{1.2}$$

If $1 \leq I < \infty$, then (1.1e) holds trivially.

We denote by $\rho_n(\psi_{N,I}; z)$ the monic nth degree Szegö polynomial and by

$$B_n(\psi_{N,I}; u_n; z) := \rho_n(\psi_{N,I}; z) + u_n \rho_n^*(\psi_{N,I}; z) \tag{1.3}$$

the nth degree Szegö para-orthogonal polynomial all orthogonal on the unit circle with respect to the distribution function $\psi_{N,I}(\theta)$ defined as follows:

(a) If $1 \leq N < \infty$ and $1 \leq I \leq \infty$, then $\psi_{N,I}(\theta)$ is an absolutely continuous function on $[-\pi, \pi]$ such that

$$\psi'_{N,I}(\theta) := \frac{1}{2\pi} |X_{N,I}(e^{i\theta})|^2, \quad X_{N,I}(z) := \sum_{m=0}^{N-1} x_{N,I}(m) z^{-m}, \tag{1.4}$$

which is readily seen to imply symmetry

$$\psi'_{N,I}(-\theta) = \psi'_{N,I}(\theta) \quad \text{for} \quad -\pi \leq \theta \leq \pi. \tag{1.5}$$

(b) If $N = \infty$ and $1 \leq I \leq \infty$, then

$$\psi_{\infty,I}(\theta) := \sum_{\omega_j \leq \theta} |\alpha_j|^2, \quad -\pi \leq \theta \leq \pi. \tag{1.6}$$

It follows that, if $1 \leq I < \infty$, then $\psi_{\infty,I}(\theta)$ is a step function with jumps $|\alpha_j|^2$ at $\theta = \omega_j$.

Some recent results on asymptotics of zeros $z_j(n, \psi_{N,I})$ of Szegö polynomials $\rho_n(\psi_{N,I}; z)$ provide a theoretical basis for their use in solving the frequency analysis problem (FAP): From a sample of N observed values $\{x_{N,I}(m)\}_{m=0}^{N-1}$ can we determine the unknown frequencies $\omega_1, \omega_2, \ldots, \omega_I$, where $I < \infty$? In two recent papers [4] and [8] the following result has been proved:

For $1 \leq I < \infty$ and $n \geq n_o(I)$ the zeros $z_j(n, \psi_{N,I})$ can be arranged so that

$$\lim_{N \to \infty} z_j(n, \psi_{N,I}) = e^{i\omega_j} \quad \text{for all} \quad j \in J(I), \tag{1.6a}$$

where

$$n_0(I) := 2I + L, \quad L = 0 \quad \text{if} \quad \alpha_0 = 0 \quad \text{and} \quad L = 1 \quad \text{if} \quad \alpha_0 > 0, \tag{1.6b}$$

$$J(I) = J_0 \cup [\pm 1, \pm 2, \ldots, \pm I], \quad J_0 := \begin{cases} \emptyset & \text{if} \quad \alpha_0 = 0 \\ [0] & \text{if} \quad \alpha_0 > 0. \end{cases} \tag{1.6c}$$

In addition, from the proof given by [8] one can show that for $1 \leq I < \infty$ and for each $n \geq n_0(I)$ there exists a number R_n such that, for all of the remaining $n - n_0(I)$ zeros $z_j(n, \psi_{N,I})$ of $\rho_n(\psi_{N,I}; z)$ we have

$$|z_j(n, \psi_{N,I})| \leq R_n < 1 \quad \text{for all} \quad 1 \leq N < \infty. \tag{1.7}$$

These results give support for using the following well known method for approximating the critical points $e^{i\omega_j}$, $j \in J(I)$: for fixed $n \geq n_0(I)$ and sample size N we use the $n_0(I)$ zeros $z_j(n, \psi_{N,I})$ with largest moduli to approximate the $e^{i\omega_j}$, $j \in J(I)$. This method is a reformulation of one given by Wiener and Levinson based on digital filters and linear prediction (see, e.g., [7], [10]). Other methods that have been used to solve the FAP with $1 \leq I < \infty$ can be found in [1], [6], [9].

In a more recent paper [5], signals (1.1) are considered for which the number I of frequencies ω_j can be infinite, i.e., $1 \leq I \leq \infty$. We choose fixed sequences of numbers $\{\omega_j\}$ and $\{r_j\}$ satisfying (1.1c, d, e) and then allow the complex amplitudes α_j to vary subject to the conditions (1.1). With this understanding it is shown [5, Theorem 4.1] that if K satisfies $1 \leq K < I \leq \infty$, then the zeros $z_j(n_0(K), \psi_{N,I})$ can be arranged so that

$$\lim_{\substack{N \to \infty \\ \alpha(K,I) \to 0}} z_j(n_0(K), \psi_{N,I}) = e^{i\omega_j} \quad \text{for all} \quad j \in J(K) \tag{1.8a}$$

where

$$\alpha(K, I) := \sum_{j=K+1}^{I} |\alpha_j|. \tag{1.8b}$$

This asymptotic result on zeros of $\rho_n(\psi_{N,I}; z)$ for which $n = n_0(K) < n_0(I)$ suggests that the method can be used to approximate frequencies ω_j associated with $|\alpha_j|$ larger than other amplitudes. Results from a computational example were given in [5] to illustrate the method's applicability. In the present paper we extend the result (1.8) by proving (Theorem 2.1) that if $1 \leq K < I \leq \infty$, then there exists a constant C such that

$$|z_j(n_0(K), \psi_{N,I}) - e^{i\omega_j}| \leq C \left(\frac{1}{N} + \sum_{j=k+1}^{I} |\alpha_j| \right), \, j \in J(K), \, 1 \leq N < \infty. \tag{1.9}$$

In the special case for which $\alpha_j = 0$ for $K + 1 \leq j < I + 1$, (1.9) reduces to

$$z_j(n_0(K), \psi_{N,K}) = e^{i\omega_j} + O\left(\frac{1}{N}\right) \quad \text{for} \quad j \in J(K), \tag{1.10}$$

a result proved in [8].

Asymptotic results for the zeros of para-orthogonal Szegö polynomials $B_n(\psi_{N,I}; u; z)$ are given in Section 3. For $1 \leq I < \infty$, $|u| = 1$ and $n \geq n_0(I)$ it is shown (Theorem 3.1) that zeros $z_j(n, \psi_{N,I}; u)$ of $B_n(\psi_{N,I}; u_j; z)$ can be arranged so that

$$\lim_{N \to \infty} z_j(n, \psi_{N,I}; u) = e^{i\omega_j} \quad \text{for all} \quad j \in J(I).$$

This analogue of the result of [4] and [8] in (1.6) suggests that the zeros of $B_n(\psi_{N,I}; u; z)$ might also be used to solve the frequency analysis problem (FAP). However, for the zeros $z_j(n, \psi_{N,I}; u)$ we do not have a property analogous to (1.7) that separates the uninteresting zeros from the interesting ones. Thus the para-orthogonal polynomial zeros seem to be of limited value for use in the FAP. Nevertheless, we believe it is of interest to study the convergence and speed of convergence properties of the

$B_n(\psi_{N,I}; u; z)$ and of their zeros. Therefore in Theorem 3.3 we give an asymptotic result for $z_j((n_0(K), \psi_{N,I}; u)$ with $1 \leq K < I \leq \infty$ analogous to that given for $z_j(n_0(K); \psi_{N,I})$ in Theorem 2.1.

Section 4 is devoted to a discussion of results obtained from computational experiments that illustrate many of the asymptotic properties of the zeros of Szegő orthogonal and para-orthogonal polynomials.

We conclude this introduction with a summary of known properties of Szegő orthogonal and para-orthogonal polynomials that are subsequently used. Additional information can be found in the references [2], [3].

The mth moments $\mu_m^{(N,I)}$ with respect to $\psi_{N,I}$ are defined by

$$\mu_m^{(N,I)} := \int_{-\pi}^{\pi} e^{-im\theta} d\psi_{N,I}(\theta), \quad m = 0, \pm 1, \pm 2, \ldots, \quad 1 \leq N, I \leq \infty. \tag{1.11}$$

They can be computed by the following equations valid for $1 \leq I \leq \infty$:

$$\mu_m^{(N,I)} = \begin{cases} \displaystyle\sum_{k=0}^{N-m-1} x_{N,I}(k) x_{N,I}(k+m), & m \geq 0, \\ \mu_{-m}^{(N,I)}, & m < 0, \end{cases} \quad \text{for} \quad 1 \leq N < \infty, \tag{1.12a}$$

$$\mu_m^{(\infty,I)} = \sum_{j=-I}^{I} |\alpha_j|^2 e^{i\omega_j m}, \quad m = 0, \pm 1, \pm 2, \ldots, \quad \text{for } N = \infty. \tag{1.12b}$$

For $1 \leq N < \infty$, $a \leq I \leq \infty$ and for $N = \infty$, $I = \infty$, the bisequence $\{\mu_m^{(N,I)}\}_{m=-\infty}^{\infty}$ is positive definite; that is,

$$\mu_{-n}^{(N,I)} = \mu_n^{(N,I)} \quad \text{and} \quad \Delta_n^{(N,I)} := T_{N+1}^{(0)}(N,I) > 0, \quad n = 0, 1, 2, \ldots, \tag{1.13}$$

where the Toeplitz determinants $T_k^{(m)}(N,I)$ are defined by

$$T_0^{(m)}(N,I) := 1, \quad T_k^{(m)}(N,I) := \det(\mu_{m+j-n}^{(N,I)})_{j,n=0}^{k-1}, \quad \begin{matrix} k = 1, 2, 3, \ldots, \\ 1 \leq N, I \leq \infty. \end{matrix} \tag{1.14}$$

For $N = \infty$, $1 \leq I < \infty$, the bisequence $\{\mu_m^{(\infty,I)}\}_{m=-\infty}^{\infty}$ is positive $n_0(I)$-definite; that is

$$\mu_{-n}^{(\infty,I)} = \mu_n^{(\infty,I)}, \quad \Delta_n^{(\infty,I)} := T_{n+1}^{(0)}(\infty,I) > 0 \quad \text{for} \quad 0 \leq n \leq n_0(I) - 1, \tag{1.15a}$$

$$\Delta_{n_0(I)}^{(\infty,I)} := T_{n_0(I)+1}^{(0)}(\infty,I) = 0. \tag{1.15b}$$

For $1 \leq N, I \leq \infty$ we define $\langle \cdot, \cdot \rangle_{\psi_{N,I}}$ by

$$\langle f, g \rangle_{\psi_{N,I}} := \int_{-\pi}^{\pi} f(e^{i\theta}) \overline{g(e^{i\theta})} d\psi_{N,I}(\theta). \tag{1.16}$$

The monic Szegö polynomials $\rho_n(\psi_{N,I}; z)$ and reciprocal (reversed) polynomials $\rho_n^*(\psi_{N,I}; z) := z^n \overline{\rho_n(\psi_{n,I}; 1/\bar{z})} = z^n \rho_n(\psi_{N,I}; z^{-1})$ are defined by $\rho_0(\psi_{N,I}; z) = \rho_0^*(\psi_{N,I}; z) = 1$ and, for $1 \le n < n_0(I) + 1$ where $n_0(\infty) := \infty$, by

$$
\rho_n(\psi_{N,I}; z) := \frac{1}{\Delta_{n-1}^{(N,I)}}
\begin{vmatrix}
\mu_0^{(N,I)} & \mu_{-1}^{(N,I)} & \cdots & \mu_{-n}^{(N,I)} \\
\mu_1^{(N,I)} & \mu_0^{(N,I)} & \cdots & \mu_{-n+1}^{(N,I)} \\
\vdots & \vdots & & \vdots \\
\mu_{n-1}^{(N,I)} & \mu_{n-2}^{(N,I)} & \cdots & \mu_{-1}^{(N,I)} \\
1 & z & \cdots & z^n
\end{vmatrix},
$$

$$
\rho_n^*(\psi_{N,I}; z) := \frac{1}{\Delta_{n-1}^{(N,I)}}
\begin{vmatrix}
\mu_0^{(N,I)} & \mu_1^{(N,I)} & \cdots & \mu_n^{(N,I)} \\
\mu_{-1}^{(N,I)} & \mu_0^{(N,I)} & \cdots & \mu_{n-1}^{(N,I)} \\
\vdots & \vdots & & \vdots \\
\mu_{-n+1}^{(N,I)} & \mu_{-n+2}^{(N,I)} & \cdots & \mu_1^{(N,I)} \\
z^n & z^{n-1} & \cdots & 1
\end{vmatrix}.
$$

$$(1.17)$$

They satisfy orthogonality relations, for $1 \le n < n_0(I) + 1$,

$$
(\rho_n(\psi_{N,I}; z), z^m)_{\psi_{N,I}} = \begin{cases} 0, & 0 \le m \le n-1 \\ \Delta_n^{(N,I)} / \Delta_{n-1}^{(N,I)}, & m = n, \end{cases}
$$

$$(1.18a)$$

$$
\langle \rho_n^*(\psi_{N,I}; z), z^m \rangle_{\psi_{N,I}} = \begin{cases} \Delta_n^{(N,I)} / \Delta_{n-1}^{(N,I)}, & m = 0 \\ 0, & 1 \le m \le n. \end{cases}
$$

$$(1.18b)$$

These yield the recurrence relations

$$
\rho_n(\psi_{N,I}; z) = z \rho_{n-1}(\psi_{N,I}; z) + \delta_n^{(N,I)} \rho_{n-1}^*(\psi_{N,I}; z), \quad 1 \le n < n_0(I) + 1, \quad (1.19a)
$$

$$
\rho_n^*(\psi_{N,I}; z) = \delta_n^{(N,I)} z \rho_{n-1}(\psi_{N,I}; z) + \rho_{n-1}^*(\psi_{N,I}; z), \quad 1 \le n < n_0(I) + 1, \quad (1.19b)
$$

where the reflection coefficients $\delta_n^{(N,I)} := \rho_n(\psi_{N,I}; 0)$ satisfy

$$
\delta_n^{(N,I)} = \frac{(-1)^n T_n^{(-1)}(N, I)}{T_n^{(0)}(N, I)} = -\frac{\sum_{j=0}^{n-1} q_j^{(n-1,N,I)} \mu_{j+1-n}^{(N,I)}}{\sum_{j=0}^{n-1} q_j^{(n-1,N,I)}}, \quad 1 \le n < n_0(I) + 1, \quad (1.20)
$$

where $\sum_{j=0}^{n-1} q_j^{(n-1,N,I)} z^j := \rho_{n-1}(\psi_{N,I}; z)$. Levinson's algorithm utilizes (1.18) and (1.19) to compute successively the $\delta_n^{(N,I)}$ and $q_j^{(n-1,N,I)}$.

2   ORTHOGONAL POLYNOMIALS

In this section we assume that $2 \leq I \leq \infty$, $1 \leq N < \infty$ and let $\{\omega_j\}$ and $\{r_j\}$ denote fixed sequences of numbers satisfying (1.1c, d, e). We then consider signals $x_{N,I}$ of the form (1.1) where the α_j satisfy (1.1). For $n \geq 1$, $(z_j(n, \psi_{N,I}))$ denotes the zeros of the Szegö polynomial $\rho_n(\psi_{N,I}; z)$ associated with $x_{N,I}$, the zeros being arranged so that (1.8) holds. The main result of this section is the following:

THEOREM 2.1   *If $1 \leq K < I \leq \infty$, there exists a constant C such that*

$$|z_j(n_0(K), \psi_{N,I}) - e^{i\omega_j}| \leq C \left(\frac{1}{N} + \sum_{m=K+1}^{I} |\alpha_m| \right), \quad j \in J(K), \quad 1 \leq N < \infty. \quad (2.1)$$

Our proof of Theorem 2.1 makes use of the following four lemmas.

LEMMA 2.2   *For $1 \leq I < \infty$, $1 \leq N < \infty$, $1 \leq n \leq n_0(I)$, let $\hat{\rho}_n(\psi_{N,I}; z)$ be defined by*

$$\rho_n(\psi_{N,I}; z) = \rho_n(\psi_{\infty,I}; z) + \frac{\hat{\rho}_n(\psi_{N,I}; z)}{N}. \quad (2.2)$$

Then for each $R > 0$, $1 \leq I < \infty$ and $0 \leq k \leq n \leq n_0(I)$, there exists a constant $\hat{\rho}_n^{(k)}(I, R)$ and there exists a constant $\hat{\rho}_n^{(k)}(\infty, R)$ independent of I such that

$$|\hat{\rho}_n^{(k)}(\psi_{N,I}; z)| \leq \hat{\rho}_n^{(k)}(I, R) \leq \hat{\rho}_n^{(k)}(\infty, R)$$
$$\text{for} \quad |z| \leq R, \ 1 \leq N < \infty \quad \text{and} \quad 1 \leq I < \infty. \quad (2.3)$$

Proof: Let $q_j^{(n,N,I)}$ be defined by

$$\rho_n(\psi_{N,I}; z) = \sum_{j=0}^{n} q_j^{(n,N,I)} z^j, \quad 1 \leq N \leq \infty, \ 1 \leq I \leq \infty, \ 1 \leq n \leq n_0(I). \quad (2.4)$$

In [5, Lemma 4.9] it is shown that if $\hat{q}_j(n, N, I)$ is defined by

$$q_j^{(n,N,I)} = q_j^{(n,\infty,I)} + \frac{\hat{q}_j^{(n,N,I)}}{N} \quad (2.5a)$$

then there exist numbers $\hat{q}_j(n, I)$ dependent on I and $\hat{q}_j(n, \infty)$ independent of I such that

$$|\hat{q}_j^{(n,N,I)}| \leq \hat{q}_j(n, I) \leq \hat{q}_j(n, \infty), \quad \begin{array}{l} 0 \leq j \leq n, \ 1 \leq n \leq n_0(I), \\ 1 \leq N < \infty, \ 1 \leq I < \infty. \end{array} \quad (2.5b)$$

From this and (2.2) we obtain for all $|z| \leq R$, $0 \leq k \leq n \leq n_0(I)$, $1 \leq N < \infty$,

$1 \leq I < \infty,$

$$|\hat{\rho}_n^{(k)}(\psi_{N,I}; z)| = N|\rho_n^{(k)}(\psi_{N,I}; z) - \rho_n^{(k)}(\psi_{\infty,I}; z)|$$

$$= N \left| \sum_{j=k}^{n} (q_j^{(n,N,I)} - q_j^{(n,\infty,I)}) j(j-1) \cdots (j-k+1) z^{j-k} \right|$$

$$= N \left| \sum_{j=k}^{n} \frac{\hat{q}_j^{(n,N,I)}}{N} \frac{j!}{(j-k)!} z^{j-k} \right|$$

$$\leq \sum_{j=k}^{n} \hat{q}_j(n,I) \frac{j!}{(j-k)!} R^{j-k} =: \hat{\rho}_n^{(k)}(I,R)$$

$$\leq \sum_{j=k}^{n} \hat{q}_j(n,\infty) \frac{j!}{(j-k)!} R^{j-k} =: \hat{\rho}_n^{(k)}(\infty,R). \quad \square$$

LEMMA 2.3 *For $1 \leq N < \infty$, $2 \leq I \leq \infty$ and $1 \leq n < \infty$, let $\tilde{\rho}_n(\psi_{N,I}; z)$ be defined by*

$$\rho_n(\psi_{N,I}; z) = \rho_n(\psi_{N,I-1}; z) + |\alpha_I|\tilde{\rho}_n(\psi_{N,I}; z) \tag{2.6}$$

when $\alpha_I \neq 0$. Then for each $R > 0$, $2 \leq I < \infty$ and $0 \leq k \leq n$, there exists a constant $\tilde{\rho}_n^{(k)}(I,R)$ and there exists a constant $\tilde{\rho}_n^{(k)}(\infty,R)$ independent of I such that

$$|\tilde{\rho}_n^{(k)}(\psi_{N,I}; z)| \leq \tilde{\rho}_n^{(k)}(I,R) \leq \tilde{\rho}_n^{(k)}(\infty,R), \quad |z| \leq R, \ 1 \leq N < \infty, \ 2 \leq I < \infty. \tag{2.7}$$

Proof: In [5, Lemma 4.10] it is shown that if $\tilde{q}_j^{(n,N,I)}$ is defined by

$$q_j^{(n,N,I)} = q_j^{(n,N,I-1)} + |\alpha_I|\tilde{q}_j^{(n,N,I)}, \tag{2.8a}$$

then there exist numbers $\tilde{q}_j(n,I)$ dependent on I and $\tilde{q}_j(n,\infty)$ independent of I such that

$$|\tilde{q}_j^{(n,N,I)}| \leq \tilde{q}_j(n,I) \leq \tilde{q}_j(n,\infty), \quad 0 \leq j \leq n, \ 2 \leq I < \infty. \tag{2.8b}$$

From this and (2.6) we obtain for all $|z| \leq R$ and $0 \leq k \leq n < \infty$,

$$|\tilde{\rho}_n^{(k)}(\psi_{N,I}; z)| = \frac{1}{|\alpha_I|} \left| \rho_n^{(k)}(\psi_{N,I}; z) - \rho_n^{(k)}(\psi_{N,I-1}; z) \right|$$

$$= \frac{1}{|\alpha_I|} \left| \sum_{j=k}^{n} (q_j^{(n,N,I)} - q_j^{(n,N,I-1)}) \frac{j!}{(j-k)!} z^{j-k} \right|$$

$$\leq \sum_{j-k}^{n} \tilde{q}_j(n,I) \frac{j!}{(j-k)!} R^{j-k} =: \tilde{\rho}_m^{(k)}(I,R)$$

$$\leq \sum_{j=k}^{n} \tilde{q}_j(n,\infty) \frac{j!}{(j-k)!} R^{j-k} =: \tilde{\rho}_n^{(k)}(\infty,R). \quad \square$$

LEMMA 2.4   *For $|z| \leq R$, $1 \leq N < \infty$, $2 \leq I \leq \infty$, $1 \leq K < I$, $0 \leq k \leq n_0(K)$,*

$$|\rho_{n_0(K)}^{(k)}(\psi_{N,I}; z) - \rho_{n_0(K)}^{(k)}(\psi_{\infty,K}; z)| \leq \tilde{\rho}_{n_0(K)}^{(k)}(\infty, R)\alpha(K, I) + \frac{1}{N}\,\hat{\rho}_{n_0(K)}^{(k)}(\infty, R). \quad (2.10)$$

Proof: First we assume $2 \leq I < \infty$. By Lemmas 2.2 and 2.3 we obtain for $|z| \leq R$, $0 \leq k \leq n_0(K) < n_0(I)$, $1 \leq N < \infty$,

$$|\rho_{n_0(K)}^{(k)}(\psi_{N,I}; z) - \rho_{n_0(K)}^{(k)}(\psi_{\infty,K}; z)|$$

$$= \left| \sum_{j=K+1}^{I} (\rho_{n_0(K)}^{(k)}(\psi_{N,j}; z) - \rho_{n_0(K)}^{(k)}(\psi_{N,j-1}; z)) + \rho_{n_0(K)}^{(k)}(\psi_{N,K}; z) - \rho_{n_0(K)}^{(k)}(\psi_{\infty,K}; z) \right|$$

$$\leq \sum_{j=K+1}^{I} \left| \rho_{n_0(K)}^{(k)}(\psi_{N,j}; z) - \rho_{n_0(K)}^{(k)}(\psi_{N,j-1}; z) \right| + \left| \rho_{n_0(K)}^{(k)}(\psi_{N,K}; z) - \rho_{n_0(K)}^{(k)}(\psi_{\infty,K}; z) \right|$$

$$\leq \sum_{j=K+1}^{I} |\alpha_j| |\tilde{\rho}_{n_0(K)}^{(k)}(\psi_{N,j}; z)| + \frac{1}{N}|\hat{\rho}_{n_0(K)}^{(k)}(\psi_{N,K}; z)|$$

$$\leq \sum_{j=K+1}^{I} |\alpha_j| \tilde{\rho}_{n_0(K)}^{(k)}(I, R) + \frac{1}{N}\,\hat{\rho}_{n_0(K)}^{(k)}(I, R)$$

$$\leq \sum_{j=K+1}^{I} |\alpha_j| \tilde{\rho}_{n_0(K)}^{(k)}(\infty, R) + \frac{1}{N}\,\hat{\rho}_{n_0(K)}^{(k)}(\infty, R).$$

To extend this result to $I = \infty$, we let $I \to \infty$ in (2.10) and use the fact that $\lim_{I\to\infty} \rho_{n_0(K)}(\psi_{N,I}; z) = \rho_{n_0(K)}(\psi_{N,\infty}; z)$.   □

Let A be a subset of $(0, a)$ having zero as an accumulation point.

LEMMA 2.5   *Let $[P_{\alpha,N}(z) : \alpha \in A,\ N \in \mathbb{N}]$ be a family of monic polynomials in z of degree $n \geq 1$ converging, as $\alpha \to 0$ and $N \to \infty$, locally uniformly on $\mathbb{C}$ to a monic polynomial $P(z)$ of degree n. For $\alpha \in A$ and $N \in \mathbb{N}$, let $E(\alpha, N, z)$ be defined by*

$$P_{\alpha,N}(z) = P(z) + E(\alpha, N, z). \quad (2.11)$$

Suppose that for each $\alpha \in A$, $N \in \mathbb{N}$ and $0 < R < \infty$, there exist constants $\hat{E}(\alpha, N, R)$ and $\hat{D}(\alpha, N, R)$ such that

$$|E(\alpha, N, z)| \leq \hat{E}(\alpha, N, R) \quad and \quad |E'(\alpha, N, z)| \leq \hat{D}(\alpha, N, R) \quad for \quad |z| \leq R, \quad (2.12)$$

and

$$\lim_{\substack{N\to\infty \\ \alpha\to 0}} \hat{E}(\alpha, N, R) = \lim_{\substack{N\to\infty \\ \alpha\to 0}} \hat{D}((\alpha, N, R) = 0. \quad (2.13)$$

Then: (A) If $P(z)$ is written in the form

$$P(z) = \prod_{j=1}^{r}(z - z_j)^{\lambda_j}, \quad z_1, z_2, \ldots, z_r \text{ distinct}, \quad \sum_{j=1}^{r} \lambda_j = n, \quad (2.14)$$

then each $P_{\alpha,N}(z)$ can be written in the form

$$P_{\alpha,N}(z) = \prod_{j=1}^{r} \prod_{\nu=1}^{\lambda_j} (z - z_\nu(j,\alpha,N)), \qquad (2.15)$$

where

$$\lim_{\substack{N\to\infty \\ \alpha\to 0}} z_\nu(j,\alpha,N) = z_j \quad \text{for} \quad \nu = 1,2,\ldots,\lambda_j \quad \text{and} \quad j = 1,2,\ldots,r. \qquad (2.16)$$

(B) *There exist constants B, R, α_1 and N_1 such that, for all $0 < \alpha < \alpha_1$ and $N \geq N_1$,*

$$\left| \sum_{\nu=1}^{\lambda_j} (z_\nu(j,\alpha,N) - z_j) \right| \leq B[\widehat{D}(\alpha,N,R) + \widehat{E}(\alpha,N,R)]. \qquad (2.17)$$

Proof: (A): Let $\varepsilon > 0$ be chosen so that each closed disk $|z - z_j| \leq \varepsilon$, $j = 1,\ldots,r$, contains only one of the zeros $z_1,\ldots,z_r$ of $P(z)$. Let $R > 0$ be chosen so that the closed disk $|z| \leq R$ contains all of the circular paths $C_j : |z - z_j| = \varepsilon$, $j = 1,\ldots,r$, taken once in the positive direction. It follows that

$$\lambda_j = \frac{a}{2\pi i} \int_{C_j} \frac{P'(z)}{P(z)}\, dz, \qquad j = 1,2,\ldots,r. \qquad (2.18)$$

Since

$$\lim_{\substack{N\to\infty \\ \alpha\to 0}} \frac{1}{2\pi i} \int_{C_j} \frac{P'_{\alpha,N}(z)}{P_{\alpha,N}(z)}\, dz = \frac{1}{2\pi i} \int_{C_j} \frac{P'(z)}{P(z)}\, dz, \quad j = 1,2,\ldots,r \qquad (2.19)$$

and $\frac{1}{2\pi i} \int_{C_j} \frac{P'_{\alpha,N}(z)}{P_{\alpha,N}(z)}\, dz$ is the number of zeros of $P_{\alpha,N}(z)$ inside C_j, there exist α_0 and N_0 such that, for all $\alpha \geq \alpha_0$ and $N \geq N_0$, each set $\text{Int}(C_j)$ contains exactly λ_j zeros of $P_{\alpha,N}(z)$, $j = 1,\ldots,r$; we denote the zeros of $P_{\alpha,N}(z)$ in $\text{Int}(C_j)$ by $z_\nu(j,\alpha,N)$, $\nu = 1, z,\ldots,\lambda_j$. Since ε can be chosen arbitrarily small, this proves (A).

(B): It can be readily shown that

$$\frac{1}{2\pi i} \int_{C_j} \frac{zP'(z)}{P(z)}\, dz = \sum_{k=1}^{r} \frac{\lambda_k}{2\pi i} \int_{C_j} \frac{z}{z - z_k}\, dz = \lambda_j z_j, \quad j = 1,2,\ldots,r, \qquad (2.20)$$

and similarly

$$\frac{1}{2\pi i} \int_{C_j} \frac{zP'_{\alpha,N}(z)}{P_{\alpha,N}(z)}\, dz = \sum_{\nu=1}^{\lambda_j} z_\nu(j,\alpha,N), \quad j = 1,2,\ldots,r, \ \alpha \geq \alpha_0, \ N \geq N_0. \qquad (2.21)$$

It follows that, for $j = 1,2,\ldots,r$, $\alpha < \alpha_0$ and $N \geq N_0$,

$$\sum_{\nu=1}^{\lambda_j} z_\nu(j,\alpha,N) - \lambda_j z_j = \frac{1}{2\pi i} \int_{C_j} \left[\frac{P'_{\alpha,N}(z)}{P_{\alpha,N}(z)} - \frac{P'(z)}{P(z)} \right] dz$$

$$= \frac{1}{2\pi i} \int_{C_j} \frac{z[P(z)E'(\alpha,N,z) - P'(z)E(\alpha,N,z)]}{P(z)[P(z) + E(\alpha,N,z)]}\, dz.$$

Let τ and $\sigma > 0$ be chosen so that

$$\max_{|z|\leq R}|P(z)| \leq \tau, \quad \max_{|z|\leq R}|P'(z)| \leq \tau \quad \text{and} \quad \min_{\substack{z\in C_j \\ 1\leq j\leq r}}|P(z)| = \sigma > 0.$$

Moreover, let $0 < \alpha_1 < \alpha_0$ and $N_1 \geq n_0$ be chosen so that

$$\widehat{E}(\alpha, N, R) < \frac{\sigma}{2} \quad \text{for} \quad 0 < \alpha < \alpha_1 \quad \text{and} \quad N \geq n_1.$$

Then, for $j = 1, 2, \ldots, r$ and $0 < \alpha < \alpha_1$ and $N \geq N_1$,

$$\left|\sum_{\nu=1}^{\lambda_j} z_\nu(j, \alpha, N) - \lambda_j z_j\right| \leq \frac{2\pi\varepsilon}{2\pi} \max_{|z-z_j|=\varepsilon}\left|\frac{z[P(z)E'(\alpha, N, z) - P'(z)E(\alpha, N, z)]}{P(z)[P(z) + E(\alpha, N, z)]}\right|$$

$$\leq \frac{2\varepsilon\tau R}{\sigma^2}\left(\widehat{D}(\alpha, N, R) + \widehat{E}(\alpha, N, R)\right). \quad \square$$

Proof of Theorem 2.1: We apply Lemma 2.5 with

$$P(z) := \rho_{n_0(K)}(\psi_{\infty, K}; z) = (z-1)^L \prod_{j=1}^{K}(z - e^{i\omega_j})(z - e^{-i\omega_j})$$

and

$$P_{\alpha, N}(z) := \rho_{n_0(K)}(\psi_{N, I}; z),$$

$$\alpha := \alpha(K, I) := \sum_{j=K+1}^{I}|\alpha_j|,$$

$$E(\alpha, N, z) := \rho_{n_0(K)}(\psi_{N, I}; z) - \rho_{n_0(K)}(\psi_{\infty, K}; z).$$

By Lemma 2.4 we obtain, for $|z| \leq R$, $k = 0, 1$, $1 \leq N < \infty$,

$$|E^{(k)}(\alpha, N, z)| = |\rho_{n_0(K)}^{(k)}(\psi_{N, I}; z) - \rho_{n_0(K)}^{(k)}(\psi_{\infty, K}; z)|$$

$$\leq \tilde{\rho}_{n_0(K)}^{(k)}(\infty, R)\alpha(K, I) + \frac{1}{N}\,\hat{\rho}_{n_0(K)}^{(k)}(\infty, R)$$

$$=: \widehat{E}^{(k)}(\alpha, N, R), \quad k = 0, 1.$$

Clearly

$$\lim_{\substack{N\to\infty \\ \alpha\to 0}} \widehat{E}^{(k)}(\alpha, N, R) = 0, \qquad k = 0, 1.$$

Since the zeros $e^{i\omega_j}$, $j \in J(K)$, of $P(z)$ are all simple, in the application of Lemma 2.5 we have $\lambda_j = 1$ for $j = 1, 2, \ldots, r = n_0(K)$. Therefore Lemma 2.5 yields the existence of constants B, $R > 0$, α_1 and N_1 such that for $j \in J(K)$, $0 < \alpha < \alpha_1$, $N \geq N_1$,

$$|z_j(n_0(K), \psi_{N, I}) - e^{i\omega_j}| < B[\widehat{E}^{(0)}(\alpha, N, R) + \widehat{E}^{(1)}(\alpha, N, R)]$$

$$= B\left\{\left[\tilde{\rho}_{n_0(K)}^{(0)}(\infty, R)\alpha(K, I) + \frac{1}{N}\,\hat{\rho}_{n_0(K)}^{(0)}(\infty, R)\right]\right.$$

$$\left. + \left[\tilde{\rho}_{n_0(K)}^{(1)}(\infty, R)\alpha(K, I) + \frac{1}{N}\,\hat{\rho}_{n_0(K)}^{(1)}(\infty, R)\right]\right\}$$

$$\leq C\left[\alpha(K, I) + \frac{1}{N}\right]$$

where

$$C = 2B \left[\max_{k=0,1} (\tilde{\rho}^{(k)}_{n_0(K)}(\infty, R)) + \max_{k=0,1} (\tilde{\rho}^{(k)}_{n_0(K)}(\infty, R)) \right].$$

□

3 PARA-ORTHOGONAL POLYNOMIALS

THEOREM 3.1 *For every* $n \geq n_0(I)$, $N \geq 1$ *and* $|u| = 1$, *the zeros* $z_j(n, \psi_{N,I}; u)$ *of* $B_n(\psi_{N,I}; u; z)$ *can be ordered in such a manner that we have*

$$\lim_{N \to \infty} z_j(n, \psi_{N,I}; u) = e^{i\omega_j} \quad \text{for all} \quad j \in J(I). \tag{3.1}$$

Our proof of Theorem 3.1 makes use of the following result which we state and prove first.

THEOREM 3.2 *Let* $1 \leq I < \infty$ *be given and let* $\{N_k\}_{k=1}^{\infty}$ *be an arbitrary given subsequence of the sequence of natural numbers. Then: (A) There exists a subsequence* $\{N_{k_\nu}\}_{\nu=1}^{\infty}$ *and there exists a corresponding sequence of polynomials in* z $\{W_n(\{N_{k_\nu}\}, I; u; z)\}_{n=n_0(I)}^{\infty}$ *each of degree at most* $n - n_0(I)$, *such that for each* $n \geq n_0(I)$, $|u| = 1$,

$$\lim_{\nu \to \infty} B_n(\psi_{N_{k_\nu}, I}; u; z) = W_n(\{N_{k_\nu}\}, I; u; z) \rho_{n_0(I)}(\psi_{\infty, I}; z), \tag{3.2}$$

the convergence being locally uniform on $\mathbb{C}$.
    (B) *For each* $n \geq n_0(I)$, $\nu \geq 1$ *and* $|u| = 1$, *there exist zeros* $z_j(n, \psi_{N_{k_\nu}}, I; u)$ *of* $B_n(\psi_{N_{k_\nu}}, I; u; z)$ *such that*

$$\lim_{\nu \to \infty} z_j(n, \psi_{N_{k_\nu}}, I; u) = e^{i\omega_j} \quad \text{for all} \quad j \in J(I). \tag{3.3}$$

Proof: In [4, Lemma 3.3] it was shown that corresponding to each subsequence $\{N_k\}$ there exists a subsequence $\{N_{k_\nu}\}$ and an associated sequence of polynomials $\{U_n(\{N_{k_\nu}\}, I; z)\}_{n=n_0(I)}^{\infty}$, each of degree $n - n_0(I)$, such that

$$\lim_{\nu \to \infty} \rho_n(\psi_{N_{k_\nu}, I}; z) = U_n(\{N_{k_\nu}\}, I; z) \rho^*_{n_0(I)}(\psi_{\infty, I}; z), \quad n \geq n_0(I), \tag{3.4}$$

the convergence being locally uniform on $\mathbb{C}$. Therefore by (1.3) we obtain (3.2) where

$$W_n(\{N_{k_\nu}\}, I; u; z) := (-1)^L U_n(\{N_{k_\nu}\}, I; z) + u U^*_n(\{N_{k_\nu}\}, I; z) \tag{3.5a}$$

and

$$U^*_n(\{N_{k_\nu}\}, I; z) := z^{n - n_0(I)} U_n(\{N_{k_\nu}\}, I; z^{-1}). \tag{3.5b}$$

The local uniform convergence in (3.2) follows from the fact that the coefficients of individual powers of z in the polynomials on the left side of (3.2) are converging to

the corresponding coefficients on the right side (see, e.g., [4, Lemma 3.2]). (B) follows from an application of Hurwitz' theorem. $\square$

Proof of Theorem 3.1:   We assume that there exists an $n_1 \geq n_0(I)$, a subsequence $\{N_k\}$ of the natural number sequence, one of the critical points $e^{i\omega_\ell}$, $\ell \in J(I)$, a u_1 with $|u_1| = 1$, and an $\varepsilon > 0$ such that for every zero $z_j(n_1, \psi_{N_k,I}, u_1)$ of $B_{n_1}(\psi_{N_k,I}; u_1; z)$ we have

$$|z_j(n_1, \psi_{N_k,I}; u_1) - e^{i\omega_\ell}| \geq \varepsilon \quad \text{for all} \quad k = 1, 2, 3, \dots \ .$$

We then obtain an immediate contradiction from Theorem 3.2 (8). $\square$

THEOREM 3.3   If $1 \leq K < I \leq \infty$, $|u| = 1$, $(-1)^L u \neq -1$, then there exists a constant C such that

$$|z_j(n_0(K), \psi_{N,I}; u) - e^{i\omega_j}| \leq C \left(\frac{1}{N} + \sum_{m=K+1}^{I} |\alpha_m| \right), \quad j \in J(K), \ 1 \leq N < \infty. \quad (3.6)$$

Our proof makes use of the following lemmas.

LEMMA 3.4   For $1 \leq I < \infty$, $1 \leq N < \infty$, $1 \leq n \leq n_0(I)$ and $|u| = 1$, $(-1)^2 u \neq -1$, let $\widehat{B}_n(\psi_{N,I}; u; z)$ be defined by

$$B_n(\psi_{N,I}; u; z) = B_n(\psi_{\infty,I}; u; z) + \frac{1}{N} \ \widehat{B}_n(\psi_{N,I}; u; z). \quad (3.7)$$

Then for each $R > 0$, $1 \leq I < \infty$ and $0 \leq k \leq n \leq n_0(I)$, there exists a constant $\widehat{B}_n^{(k)}(I; u; R)$ and there exists a constant $\widehat{B}_n^{(k)}(\infty, u; R)$ independent of I such that

$$|\widehat{B}_n^{(k)}(\psi_{N,I}; u; z)| \leq \widehat{B}_n^{(k)}(I, u; R) \leq \widehat{B}_n^{(k)}(\infty, u; R)$$
$$\text{for} \quad |z| \leq R, \ 1 \leq N < \infty \quad \text{and} \quad 1 \leq I < \infty. \quad (3.8)$$

Proof: Following a line of proof similar to that used for Lemma 2.2, we obtain

$$|\widehat{B}_n^{(k)}(\psi_{N,I}; u; z)| = N|B_n^{(k)}(\psi_{N,I}; u; z) - B_n^{(k)}(\psi_{\infty,I}; u; z)|$$
$$\leq N|\rho_n^{(k)}(\psi_{N,I}; z) - \rho_n^{(k)}(\psi_{\infty,I}; z)| + N|u||\rho_n^{*(k)}(\psi_{N,I}; z) - \rho_n^{*(k)}(\psi_{\infty,I}; z)|$$
$$\leq \hat{\rho}_n^{(k)}(\infty, R) + |u|\hat{\rho}_n^{*(k)}(\infty, R) =: \widehat{B}_n^{(k)}(\infty; u; R),$$

where

$$\hat{\rho}_n^{*(k)}(I, R) := \sum_{j=k}^{n} \hat{q}_{n-j}(n, I) \frac{j!}{(j-k)!} R^{j-k}, \quad 1 \leq I \leq \infty. \quad \square \quad (3.9)$$

LEMMA 3.5   For $1 \leq N < \infty$, $2 \leq I < \infty$, $1 \leq n < \infty$ and $|u| = 1$, let $\widetilde{B}_n(\psi_{N,I}; u; z)$ be defined by

$$B_n(\psi_{N,I}; u; z) = B_n(\psi_{N,I}; u; z) + |\alpha_I|\widetilde{B}_n(\psi_{N,I}; u; z). \quad (3.10)$$

Then for each $R > 0$, $2 \leq I < \infty$, $0 \leq k \leq n$ and $|u| = 1$ with $(-1)^L u \neq -1$, there exists a constant $\widetilde{B}_n^{(k)}(I; u; R)$ and there exists a constant $\widetilde{B}_n^{(k)}(\infty; u; R)$ independent of I such that

$$|\widetilde{B}_n^{(k)}(\psi_{N,I}; u; z)| \leq \widetilde{B}_n^{(k)}(I; u; R) \leq \widetilde{B}_n^{(k)}(\infty; u; R) \tag{3.11}$$
$$\text{for } |z| \leq R, \ 1 \leq N < \infty \quad \text{and} \quad 2 \leq I < \infty.$$

A proof of Lemma 3.5 can be given that is analogous to that given for Lemma 2.3 (see also Lemma 3.4) and hence it is omitted.

LEMMA 3.6 *For $|z| \leq R$, $|u| = 1$ with $(-1)^L u \neq -1$, $1 \leq N < \infty$, $1 \leq K < I \leq \infty$ and $0 \leq k \leq n_0(K)$,*

$$|B_{n_0(K)}^{(k)}(\psi_{N,I}; u; z) - B_{n_0(K)}^{(k)}(\psi_{\infty,K}; u; z)|$$
$$\leq \widetilde{B}_{n_0(K)}^{(k)}(\infty; u; R)\alpha(K, I) + \frac{1}{N}\,\widehat{B}_{n_0(K)}^{(k)}(\infty; u; R). \tag{3.12}$$

A proof of Lemma 3.6 can be given that is analogous to that given for Lemma 2.4 by using Lemmas 3.4 and 3.5. It is omitted.

Proof of Theorem 3.3.   We apply Lemma 2.5 with

$$P(z) := B_{n_0(K)}(\psi_{\infty,K}; u; z) = (1 + (-1)^L u)\rho_{n_0(K)}(\psi_{\infty,K}; z)$$
$$= (1 + (-1)^L u)(z - 1)^L \prod_{j=1}^{K}(z - e^{i\omega_j})(z - e^{-i\omega_j}),$$

and

$$P_{\alpha,N}(z) := B_{n_0(K)}(\psi_{N,I}; u; z),$$
$$\alpha := \alpha(K, I) := \sum_{j=K+1}^{I} |\alpha_j|,$$
$$E(\alpha, N, z) := B_{n_0(K)}(\psi_{N,I}; u; z) - B_{n_0(K)}(\psi_{\infty,K}; u; z).$$

By Lemma 3.6 we obtain for $|z| \leq R$,

$$E^{(k)}(\alpha, N, z) = |B_{n_0(K)}^{(k)}(\psi_{N,I}; u; z) - B_{n_0(K)}^{(k)}(\psi_{\infty,K}; u; z)|$$
$$\leq \widetilde{B}_{n_0(K)}^{(k)}(\infty; u; R)\alpha(K, I) + \frac{1}{N}\widehat{B}_{n_0(K)}^{(k)}(\infty; u; R)$$
$$=: \widehat{E}^{(k)}(\alpha, N, R), \quad k = 0, 1.$$

Thus

$$\lim_{\substack{N \to \infty \\ \alpha \to 0}} \widehat{E}^{(k)}(\alpha, N, R) = 0, \qquad k = 0, 1.$$

Since the zeros $e^{i\omega_j}$, $j \in J(K)$, of $P(z)$ are all simple, an application of Lemma 2.5 with $\lambda_j = 1$ and $r = n_o(K)$ gives (3.6). $\square$

4   EXAMPLES

In this section results and methods discussed in previous sections are illustrated by means of three examples. For all three examples the input discrete time signals $x_{N,I}$ in (1.1a) are defined with $I = 4$. Values of α_j and ω_j used in these examples are given as follows:

Example 1

$j = 0$	1	2	3	4
$2\alpha_j = 0$	1	1	1	1
$\omega_j = 0$	$\pi/6$	$\pi/3$	$\pi/2$	$3\pi/4$

Example 2

$j = 0$	1	2	3	4
$2\alpha_j = 0$	1	1	1	10
$\omega_j = 0$	$\pi/6$	$\pi/3$	$\pi/2$	$3\pi/4$

Example 3

$j = 0$	1	2	3	4
$2\alpha_j = 0$	100	1	1	1
$\omega_j = 0$	$\pi/4$	$\pi/6$	$\pi/2$	$5\pi/6$

For each example we have computed the Szegö polynomials $\rho_k(\psi_{N,I}; z)$ and Szegö para-orthogonal polynomials $B_k(\psi_{N,I}; u; z)$ with $u = 1$ for several values of the degree k and input sample size N. We have also computed the zeros $z_j(k, \psi_{N,I})$ of $\rho_k(\psi_{N,I}; z)$ and $z_j(k, \psi_{N,I}; 1)$ of $B_k(\psi_{N,I}; 1; z)$.

Figures 1(a), 2(a) and 3(a) give graphs of $|z_j(k, N) - e^{i\omega_j}|$ versus N in a log-log scale where $z_j(k, N) := z_j(k, \psi_{N,I})$ for six representative values of k and for each of the 4 frequencies ω_j. In all cases we choose the zero $z_j(k, N)$ nearest to $e^{i\omega_j}$. For Example 1 the amplitudes α_j are all equal and the corresponding graphs in Figure 1(a) for different ω_j are all quite similar. For $\omega_4 = 3\pi/4$, $|z_j(8, N) - e^{i\omega_j}|$ is smaller for $j = 4$ than for other values of j; possibility due to the fact that $e^{i\omega_4}$ is somewhat isolated from the other critical points $e^{i\omega_j}$. In all of the graphs in Figure 1(a), $|z_j(k, N) - e^{i\omega_j}|$ appears to approach zero like $O\left(\frac{1}{N}\right)$ as expected from Theorem 2.1. In Example 2, the frequencies ω_j are the same as in Example 1, but the amplitude α_4 is 10 times the other amplitudes. Some consequences of this change can be seen in Figure 2(a). There $|z_j(k, N) - e^{i\omega_j}|$ tends to 0 as N increases more slowly for $k = 8$ than for other values of k except in the case $j = 4$ which has the largest amplitude. One also sees in

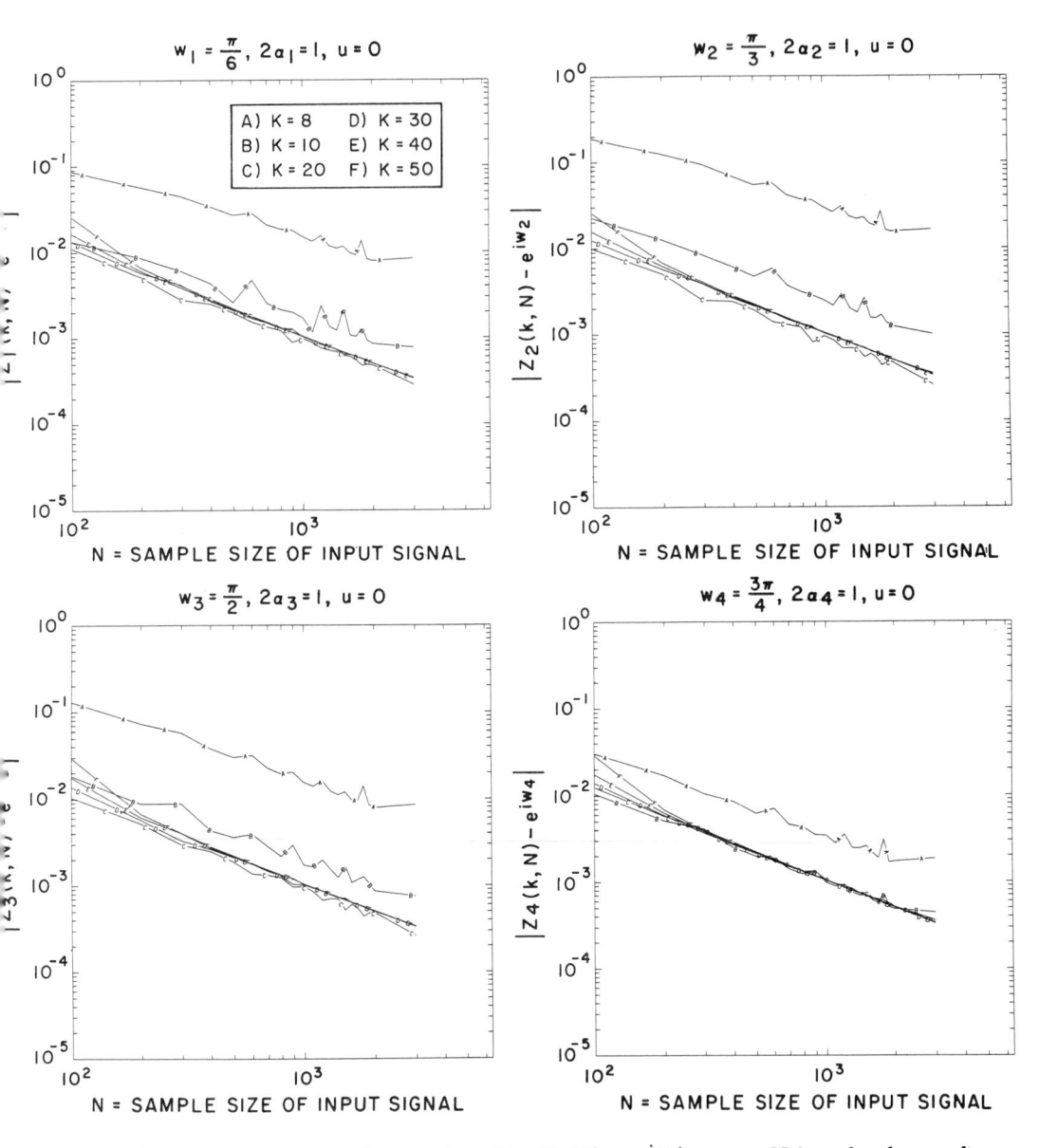

Figure 1(a) For Example 1 the graphs of $|z_j(k, N) - e^{i\omega_j}|$ versus N in a log-log scale, where $z_j(k, N) := z_j(k, \psi_{N,I})$ denotes a zero of the Szegö polynomial $\rho_k(\psi_{N,I}; z)$, $I = 4$.

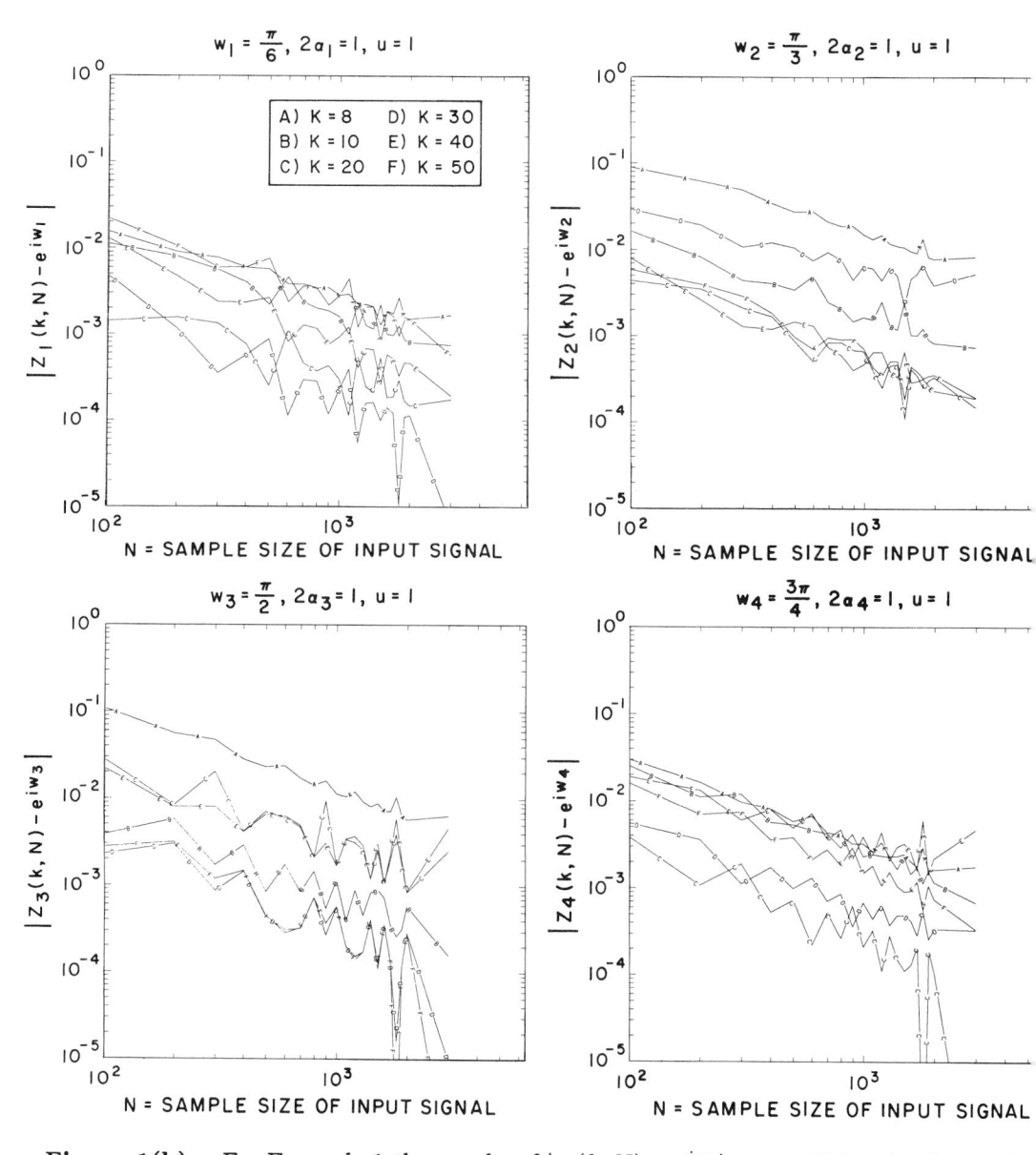

Figure 1(b)   For Example 1 the graphs of $|z_j(k, N) - e^{i\omega_j}|$ versus N in a log-log scale, where $z_j(k, N) := z_j(k, \psi_{N,I}; u)$ denotes a zero of the Szegö para-orthogonal polynomial $B_k(\psi_{N,I}; u; z)$, $u = 1$, $I = 4$.

ZEROS IN THE COMPLEX PLANE

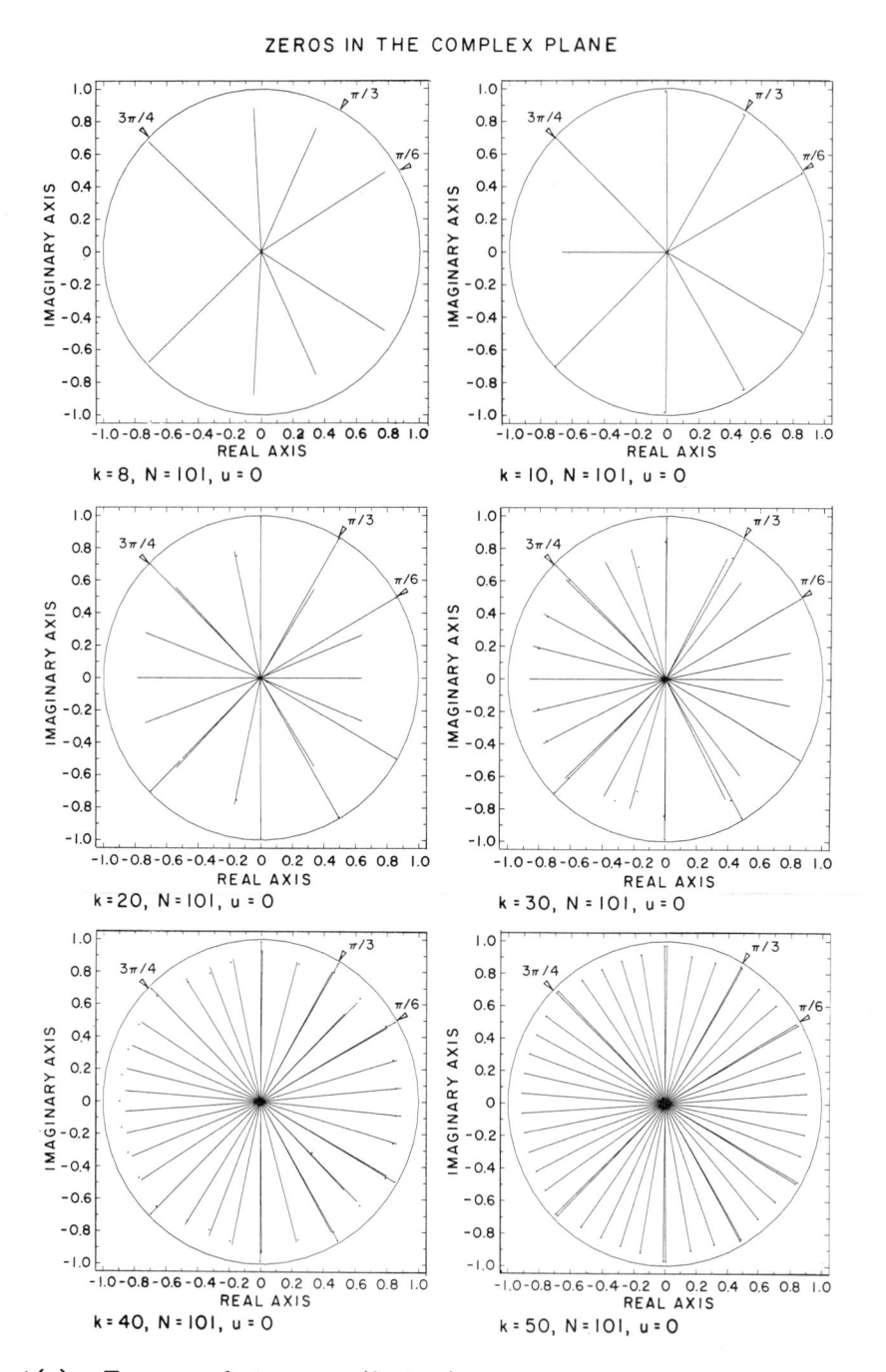

Figure 1(c) For example 1 zeros $z_j(k, \psi_{N,I})$ of Szegö polynomials $\rho_k(\psi_{N,I}; z)$ are shown as the endpoints of lines radiating from the origin; $N = 101$, $I = 4$.

ZEROS IN THE COMPLEX PLANE

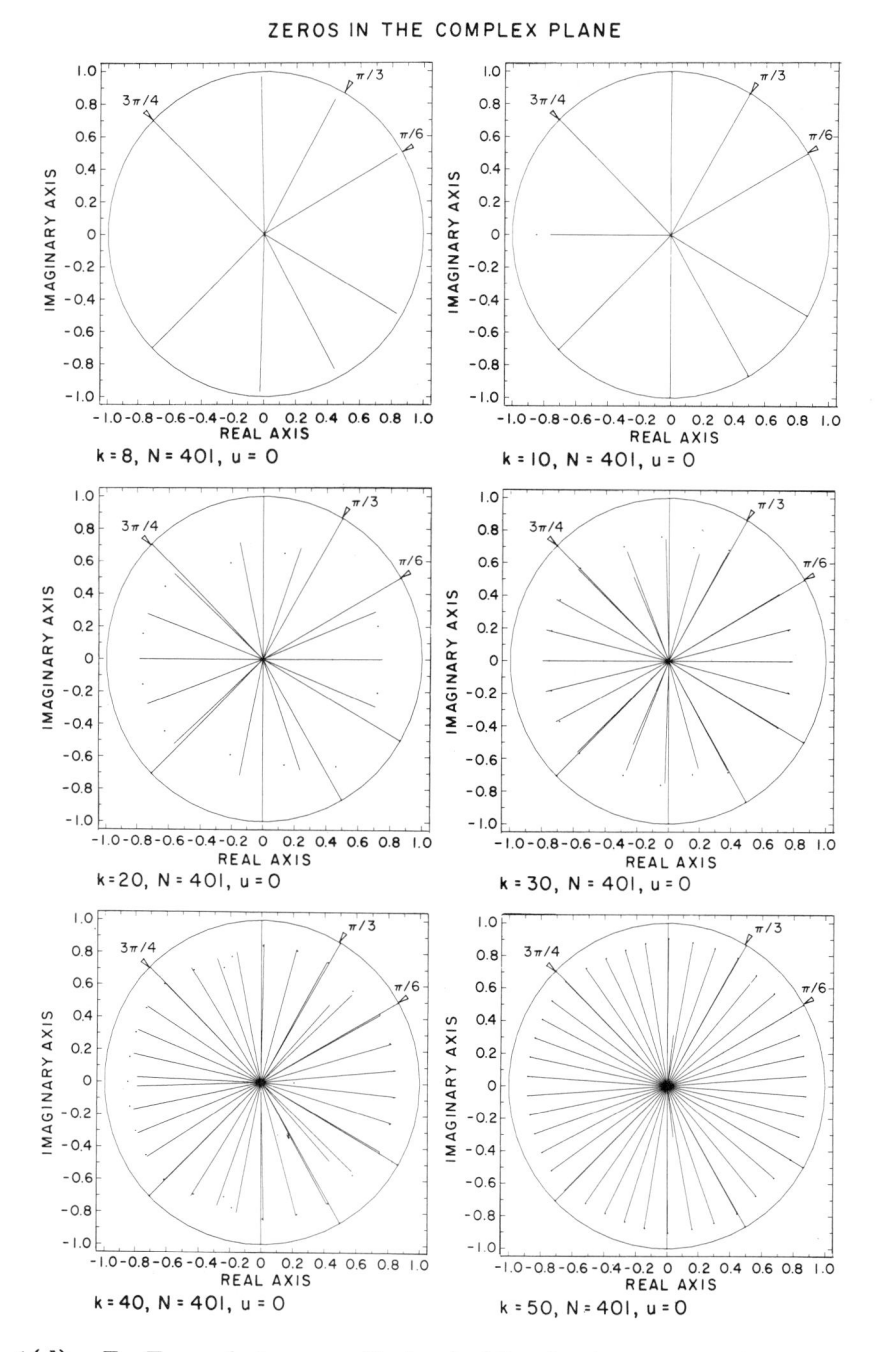

Figure 1(d)   For Example 1 zeros $z_j(k, \psi_{N,I})$ of Szegö polynomials $\rho_k(\psi_{N,I}; z)$ are shown as endpoints of lines radiating from the origin; $N = 401$, $I = 4$.

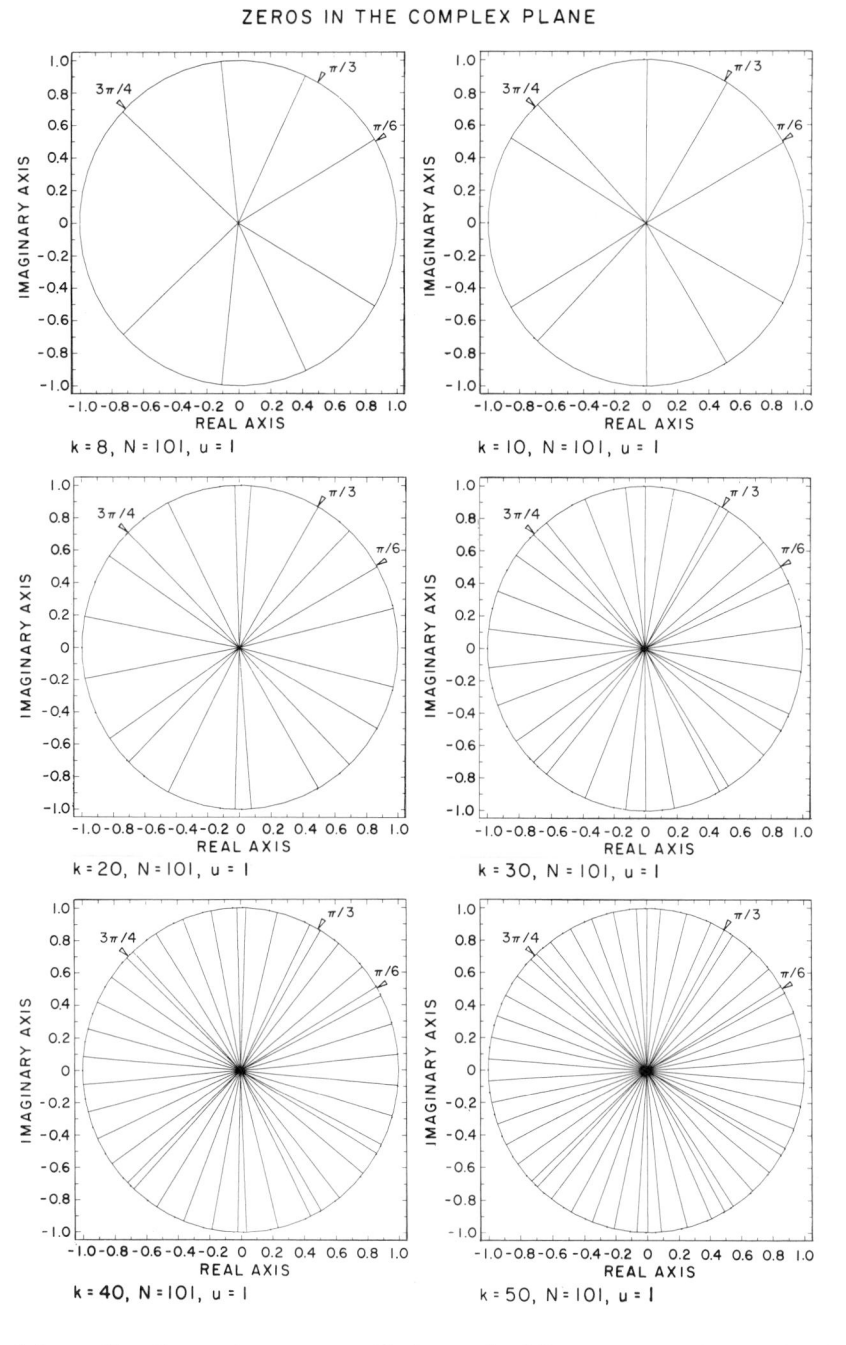

Figure 1(e)   For Example 1 zeros $z_j(k, \psi_{N,I}; u)$ of Szegö para-orthogonal polynomials $B_k(\psi_{N,I}; u; z)$ are shown as endpoints of lines radiating from the origin; $N = 101$,m $I = 4$, $u = 1$.

ZEROS IN THE COMPLEX PLANE

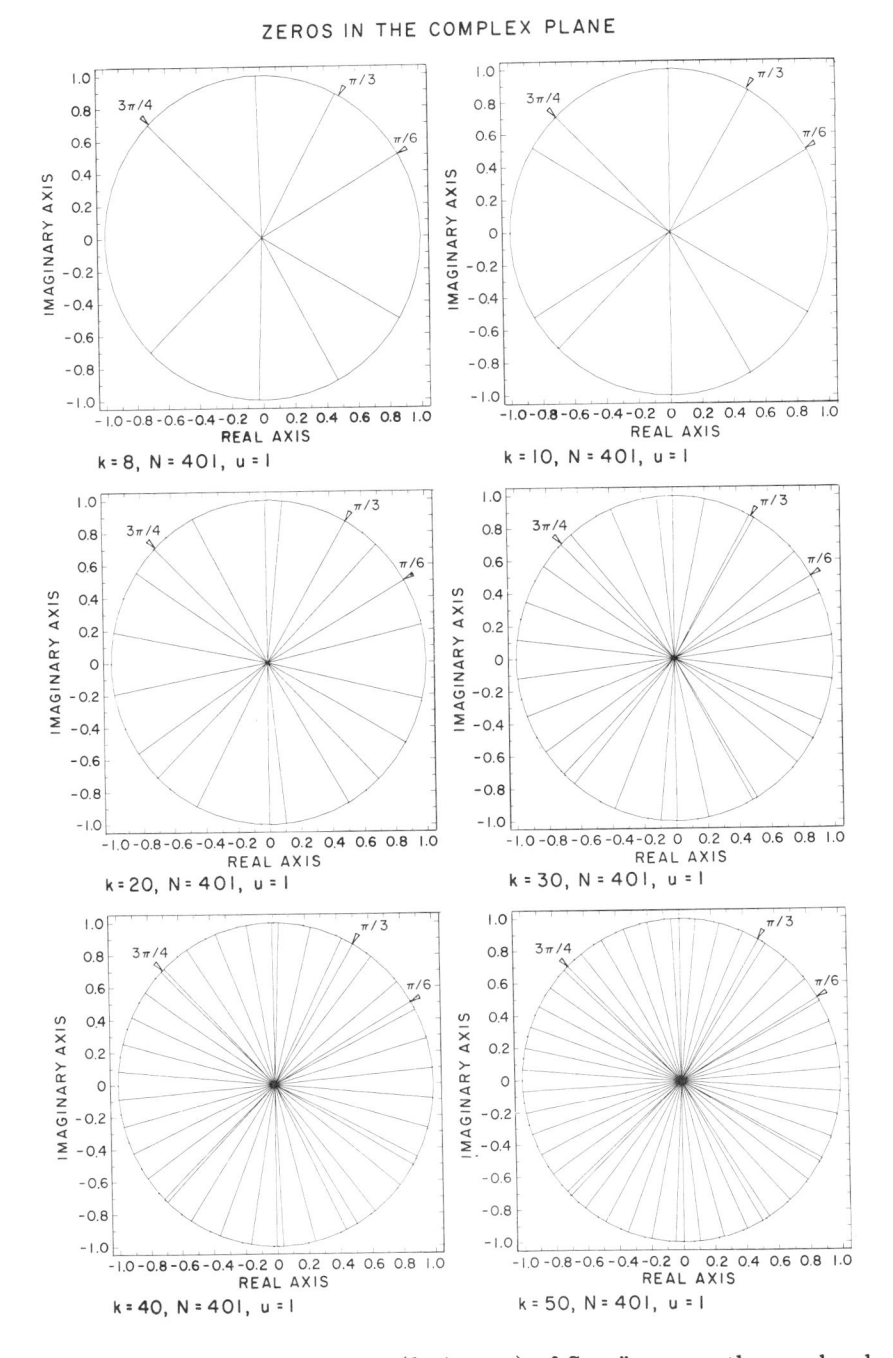

Figure 1(f)   For Example 1 zeros $z_j(k, \psi_{N,I}; u)$ of Szegö para-orthogonal polynomials $B_k(\psi_{N,I}; u; z)$ are shown as endpoints of lines radiating from the origin; $N = 401$, $II = 4$, $u = 1$.

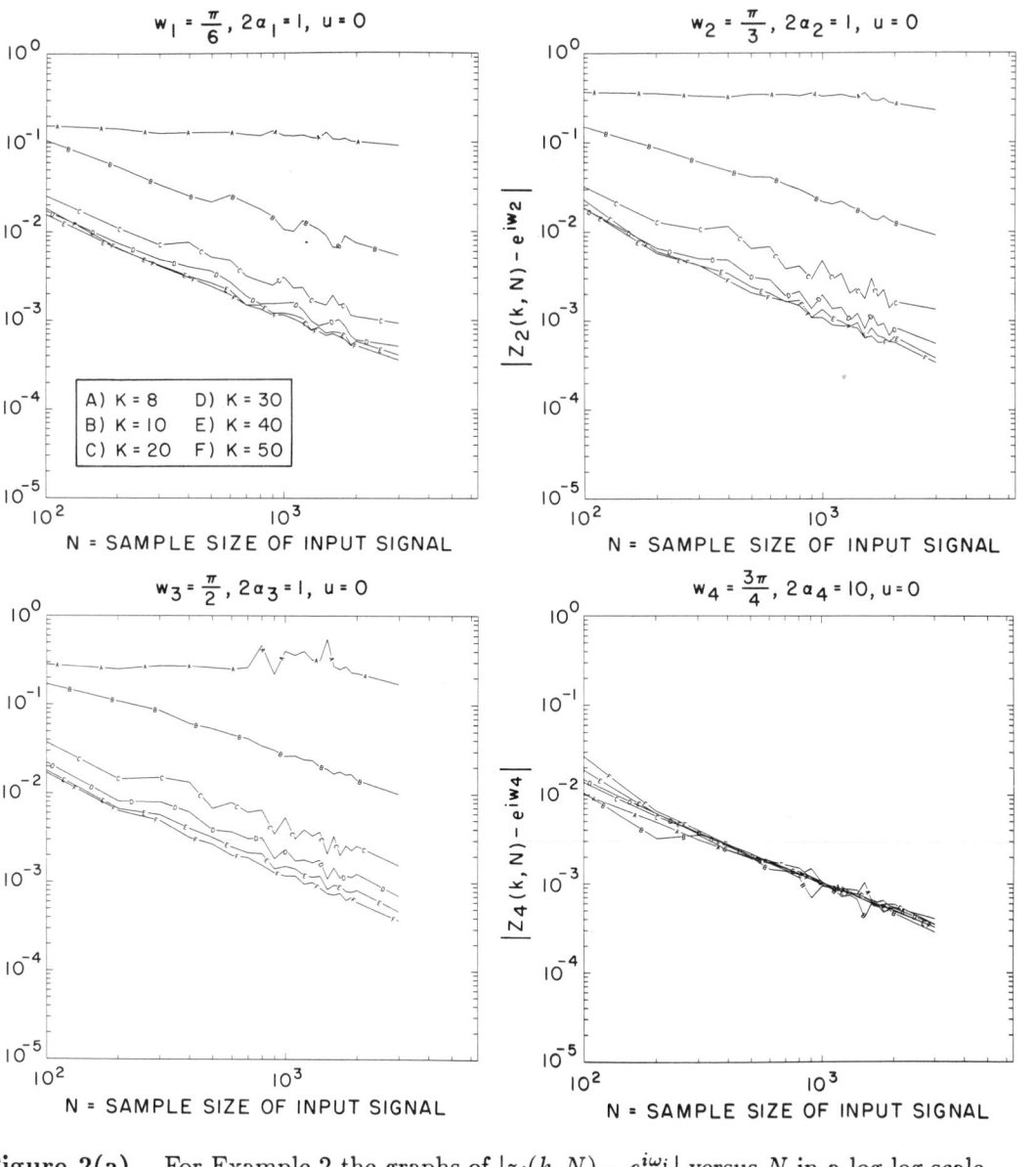

Figure 2(a) For Example 2 the graphs of $|z_j(k, N) - e^{i\omega_j}|$ versus N in a log-log scale, here $z_j(k, N) := z_j(k, \psi_{N,I})$ denotes a zero of the Szegö polynomial $\rho_k(\psi_{N,I}; z)$, $I = 4$.

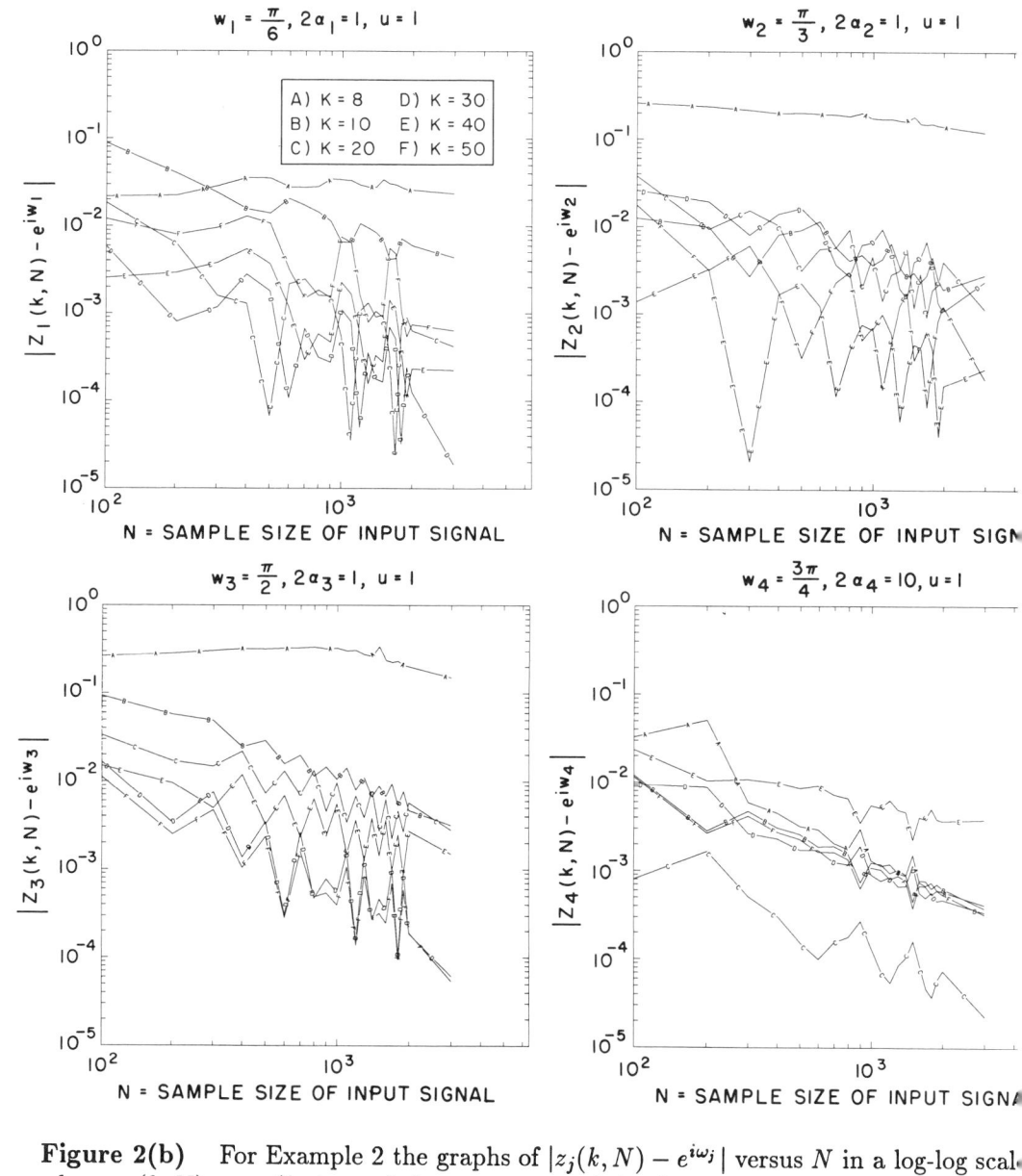

Figure 2(b)   For Example 2 the graphs of $|z_j(k, N) - e^{i\omega_j}|$ versus N in a log-log scale where $z_j(k, N) := z_j(k, \psi_{N,I}; u)$ denotes a zero of the Szegö para-orthogonal polynomial $B_k(\psi_{N,I}; u; z)$, $u = 1$, $I = 4$.

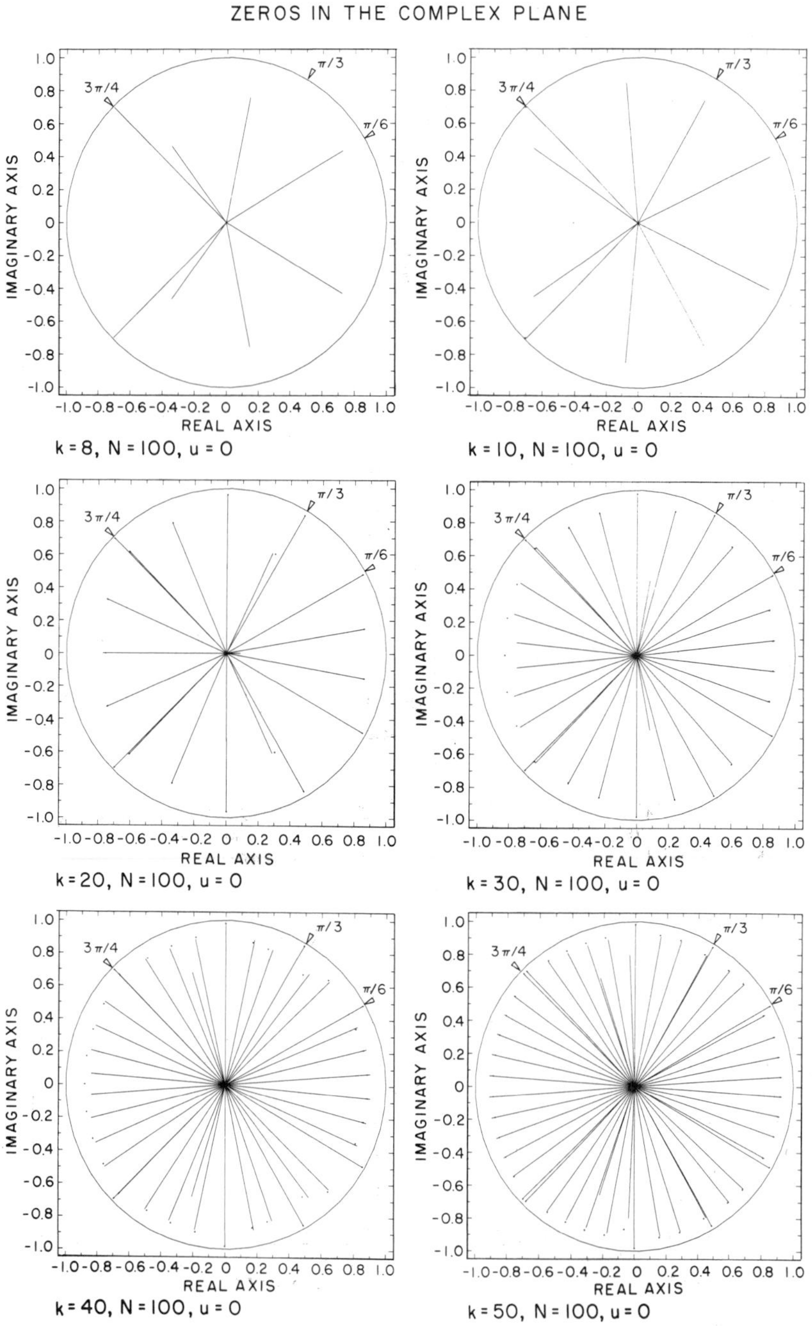

ZEROS IN THE COMPLEX PLANE

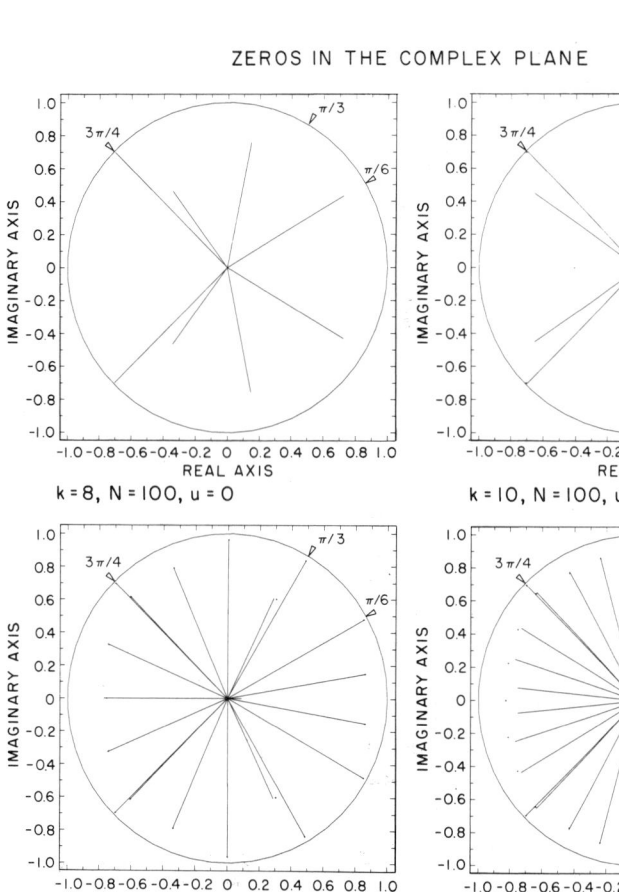

Figure 2(c)   For Example 2 zeros $z_j(k, \psi_{N,I})$ of Szegö polynomials $\rho_k(\psi_{N,I}; z)$ are shown as endpoints of lines radiating from the origin; $N = 100$, $I = 4$.

ZEROS IN THE COMPLEX PLANE

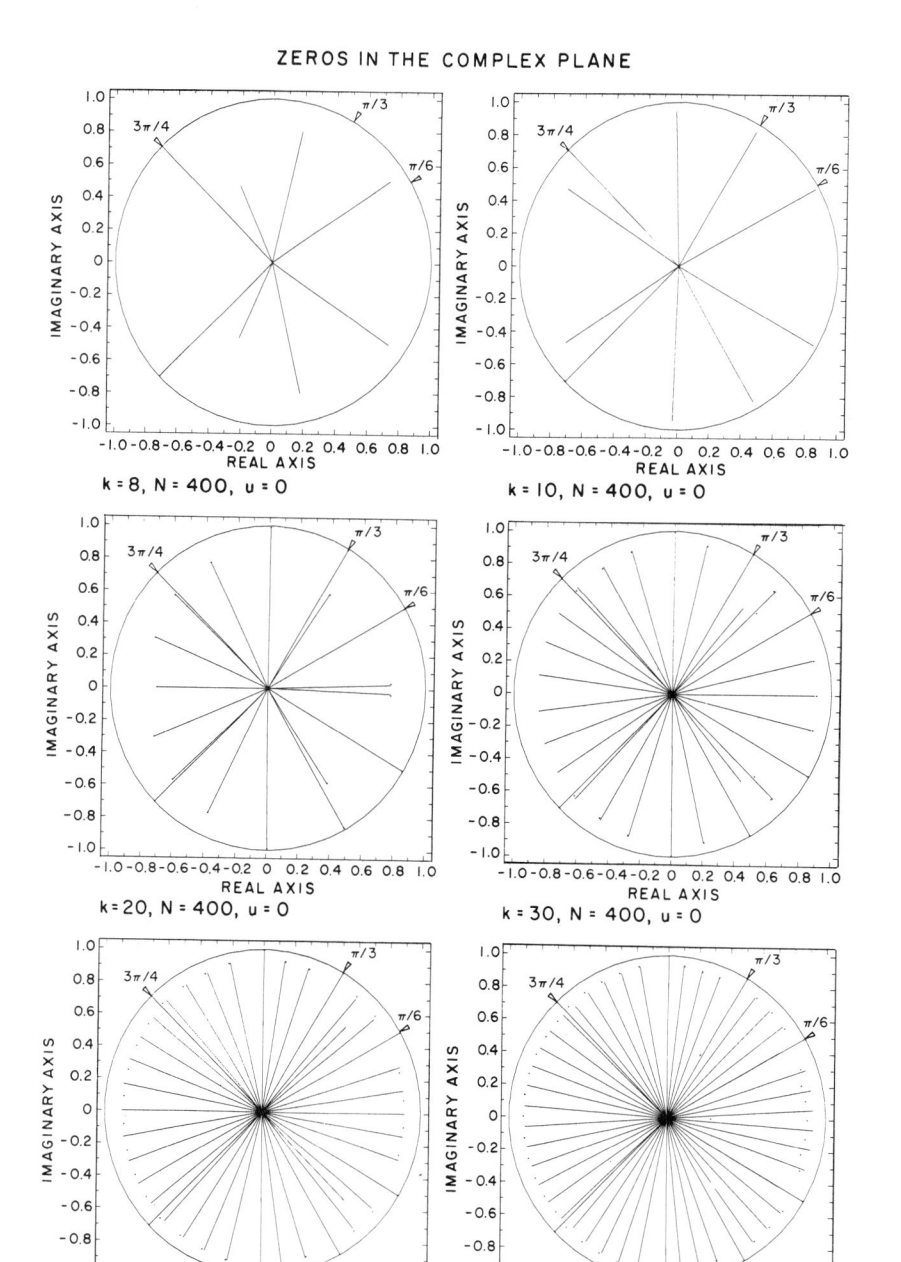

Figure 2(d)   For Example 2 zeros $z_j(k, \psi_{N,I})$ of Szegö polynomials $\rho_k(\psi_{N,I}; z)$ are shown as endpoints of lines radiating from the origin; $N = 400$, $I = 4$.

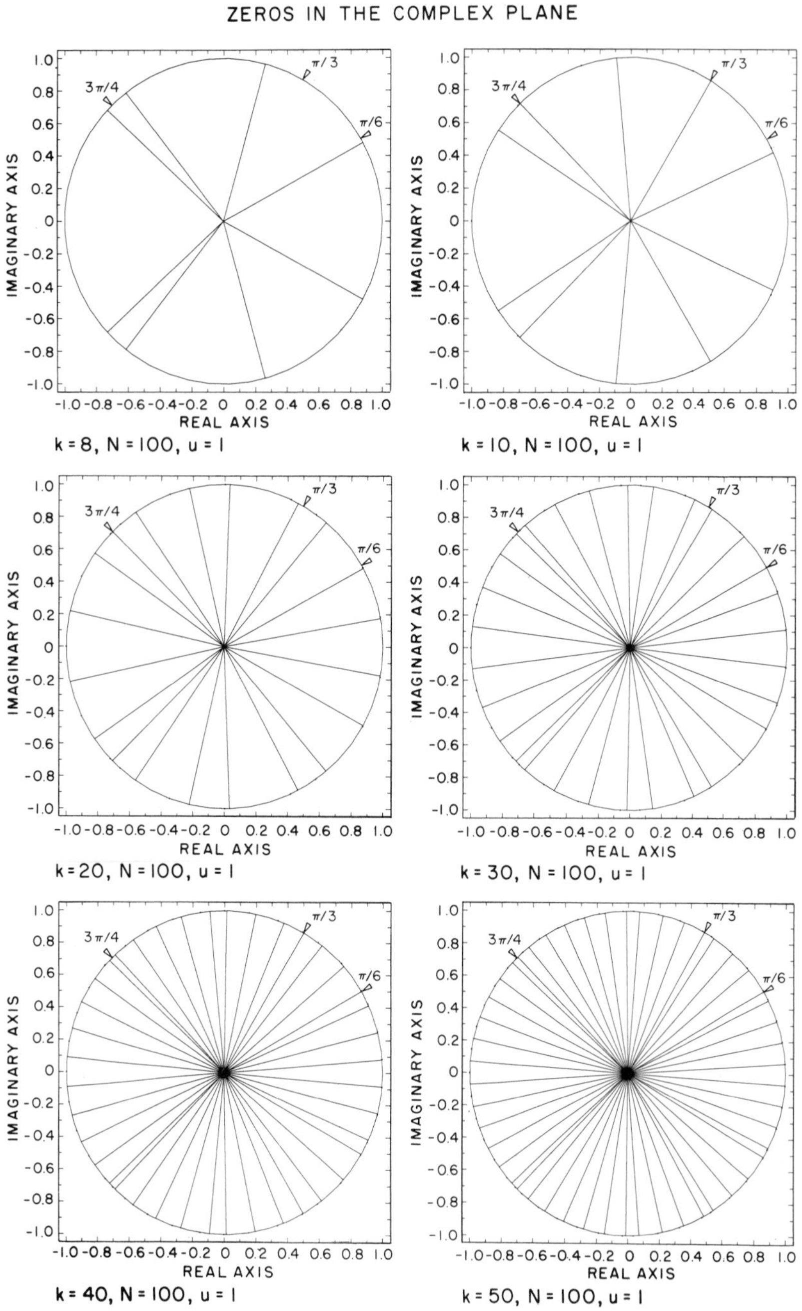

ZEROS IN THE COMPLEX PLANE

igure 2(e) For Example 2 zeros $z_j(k, \psi_{N,I}; u)$ of Szegö para-orthogonal polynomials $_k(\psi_{N,I}; u; z)$ are shown as endpoints of lines radiating from the origin; $N = 100$, $I = 4$, $= 1$.

ZEROS IN THE COMPLEX PLANE

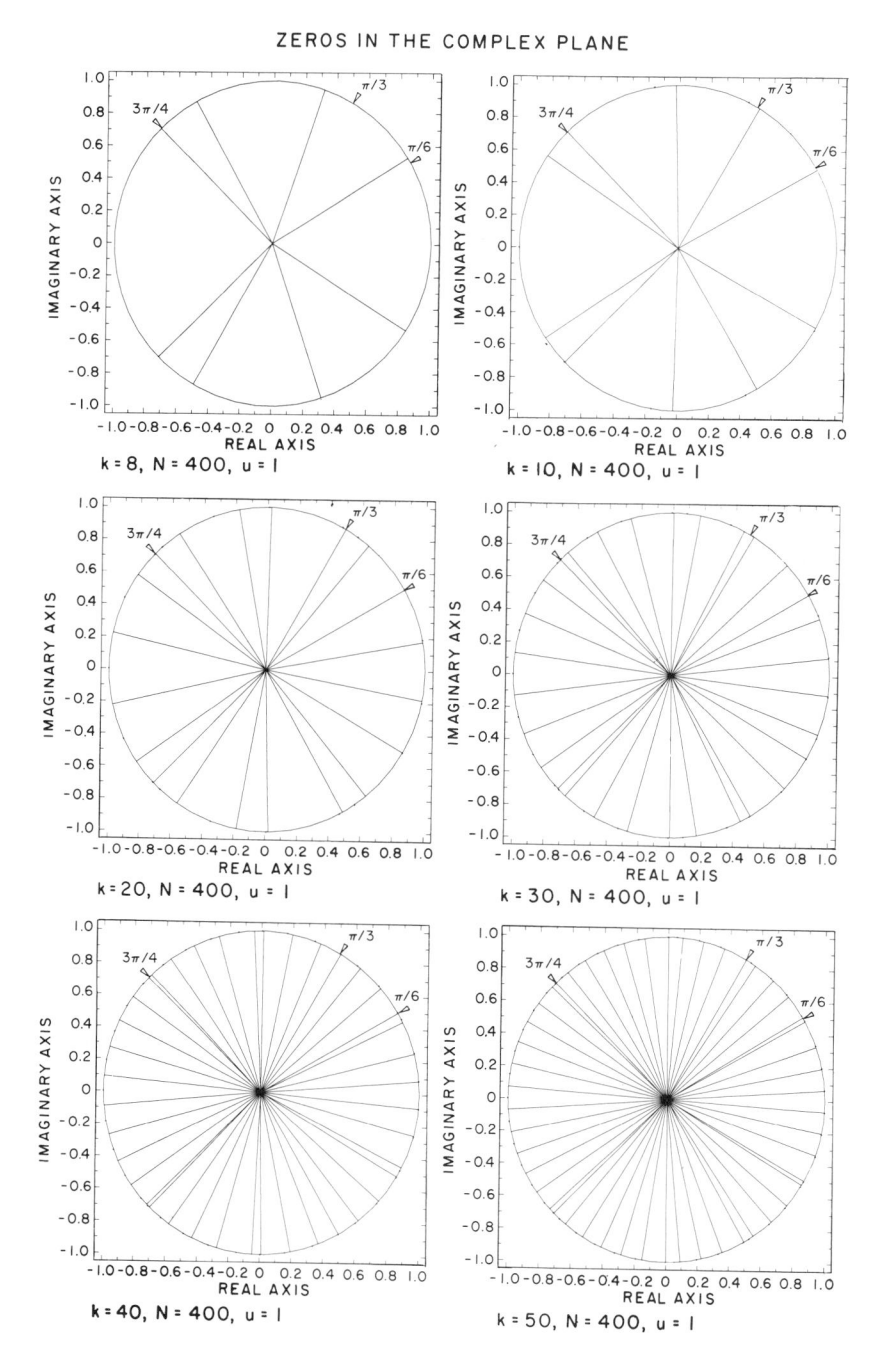

Figure 2(f)   For Example 2 zeros $z_j(k, \psi_{N,I}; u)$ of Szegö para-orthogonal polynomials $B_k(\psi_{N,I}; u; z)$ are shown as endpoints of lines radiating from the origin; $N = 400$, $I = 4$, $u = 1$.

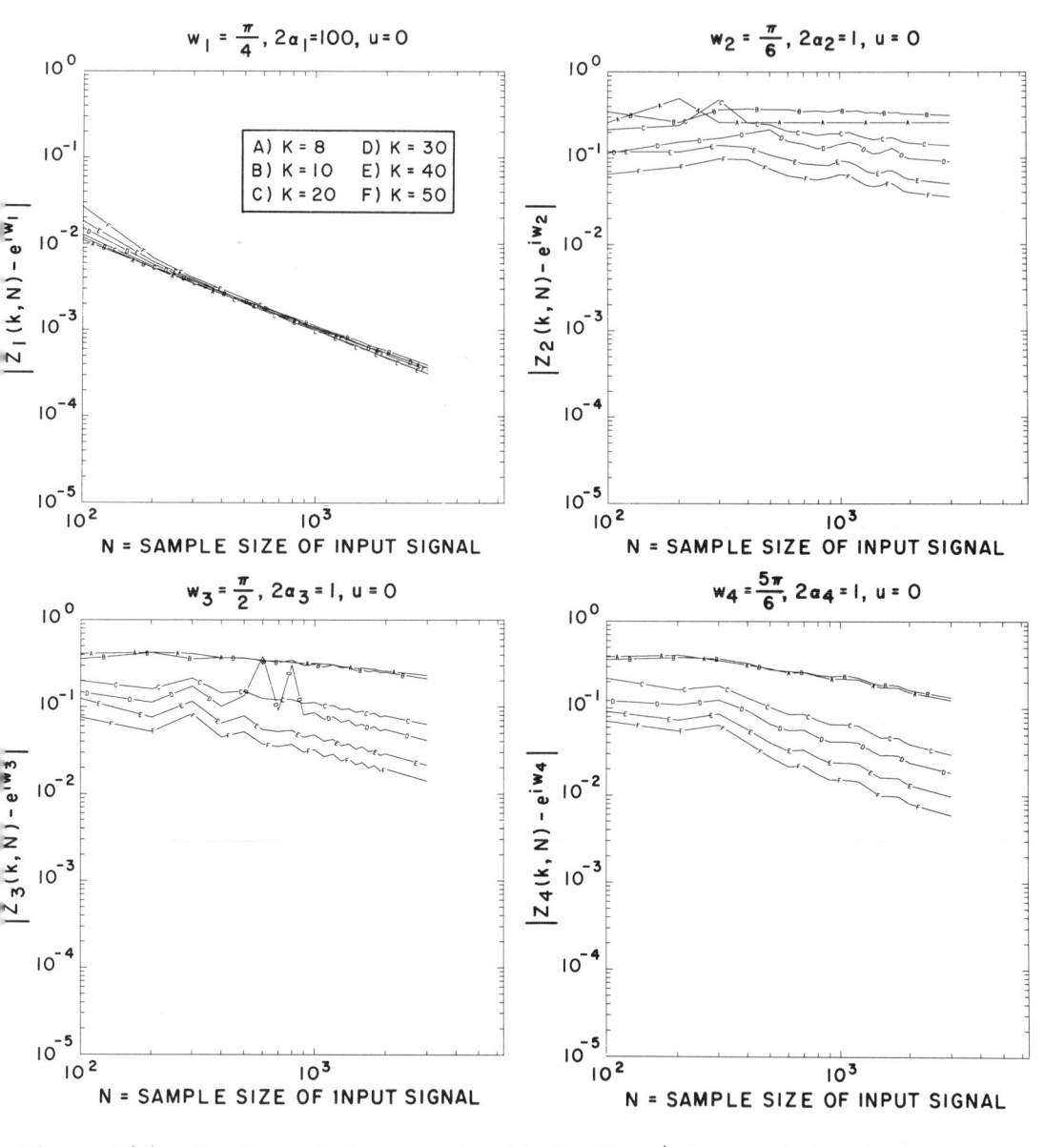

Figure 3(a)   For Example 3 the graphs of $|z_j(k, N) - e^{i\omega_j}|$ versus N in a log-log scale, where $z_j(k, N) := z_j(k, \psi_{N,I})$ denotes a zero of the Szegö polynomial $\rho_k(\psi_{N,I}; z)$, $I = 4$.

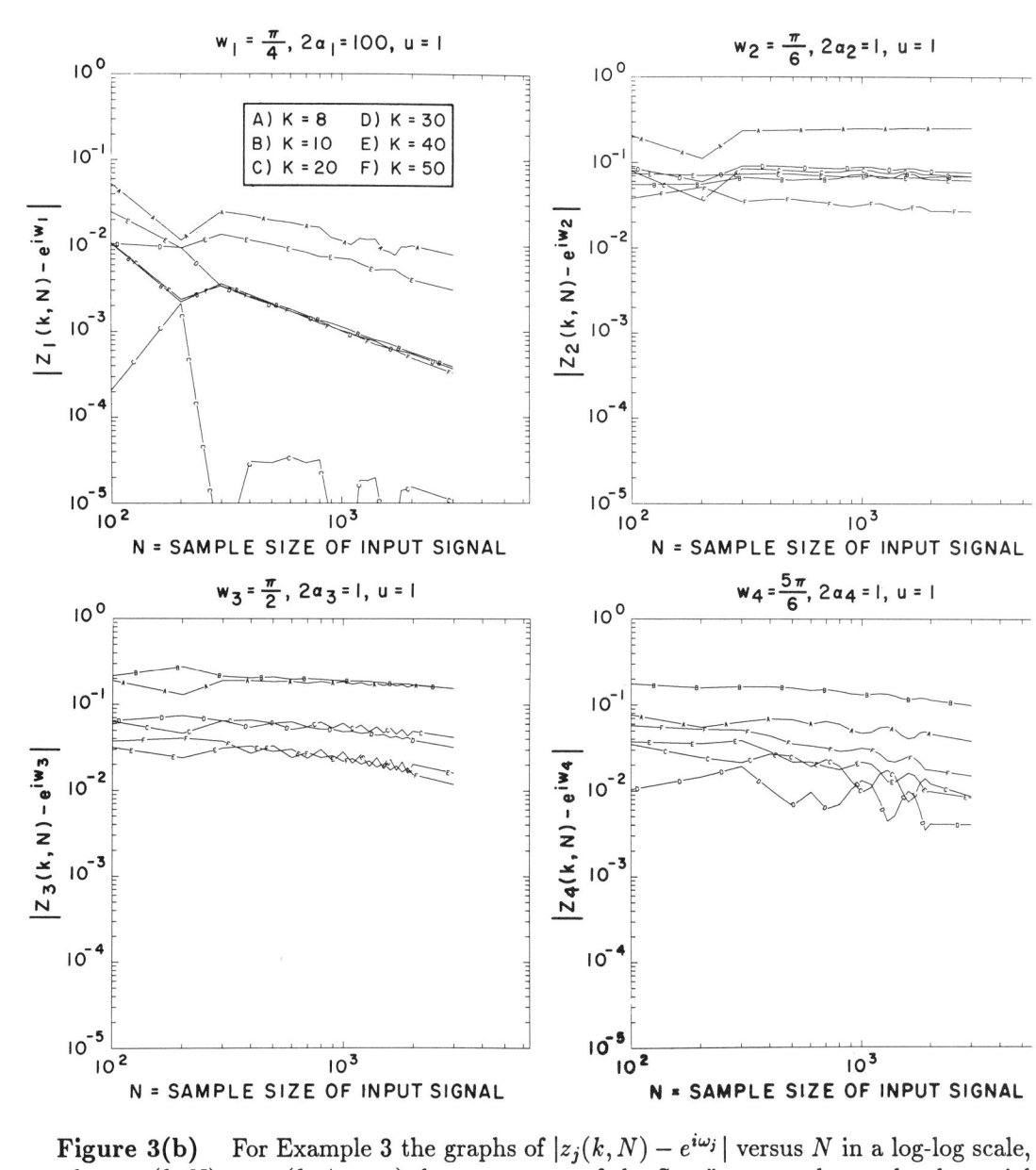

Figure 3(b)   For Example 3 the graphs of $|z_j(k, N) - e^{i\omega_j}|$ versus N in a log-log scale, where $z_j(k, N) := z_j(k, \psi_{N,I}; u)$ denotes a zero of the Szegö para-orthogonal polynomial $B_k(\psi_{N,I}; u; z)$, $u = 1$, $I = 4$.

ZEROS IN THE COMPLEX PLANE

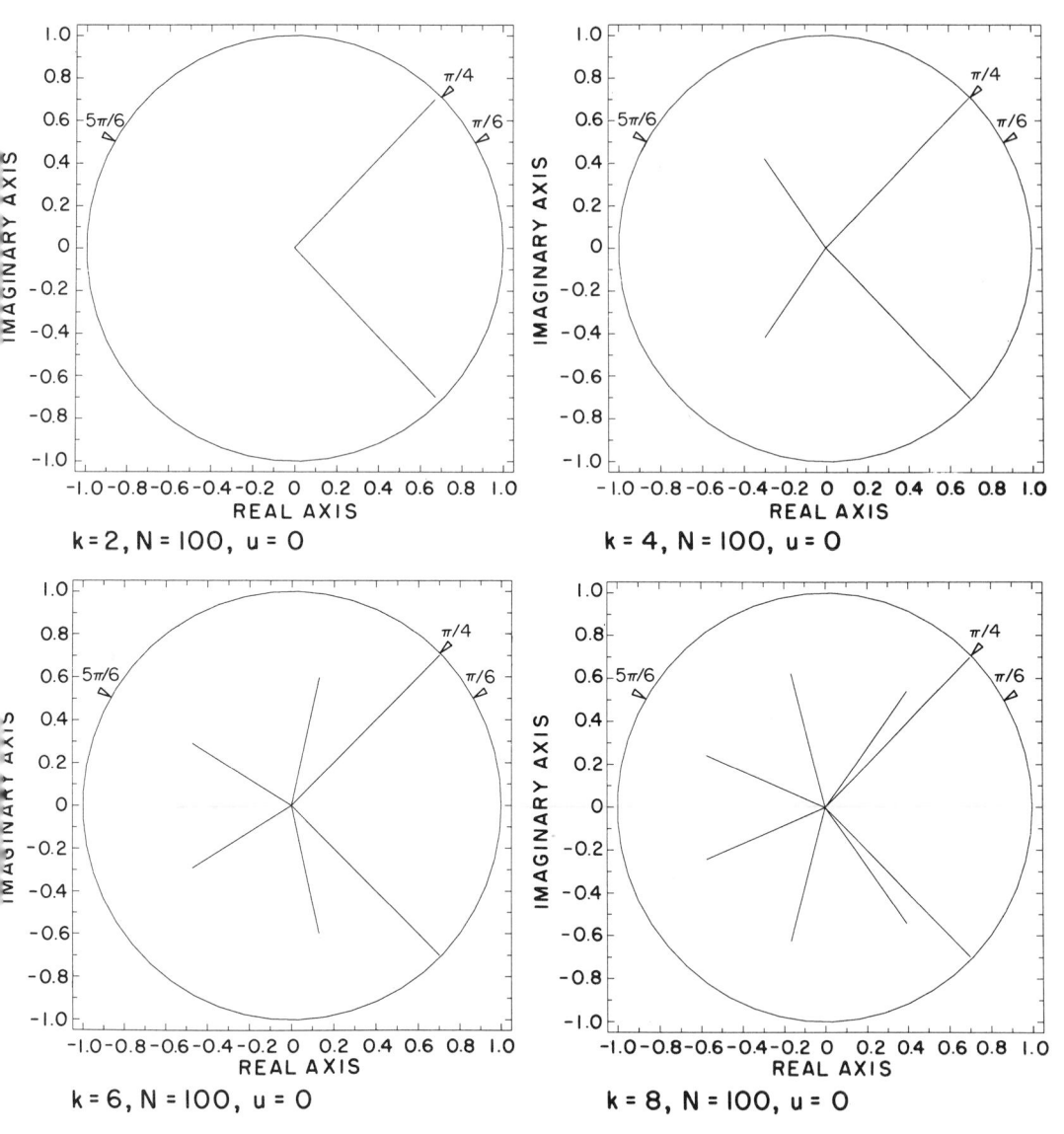

Figure 3(c)   For Example 3 zeros $z_j(k, \psi_{N,I})$ of Szegö polynomials $\rho_k(\psi_{N,I}; z)$ are shown as the endpoints of lines radiating from the origin; $N = 100$, $I = 4$.

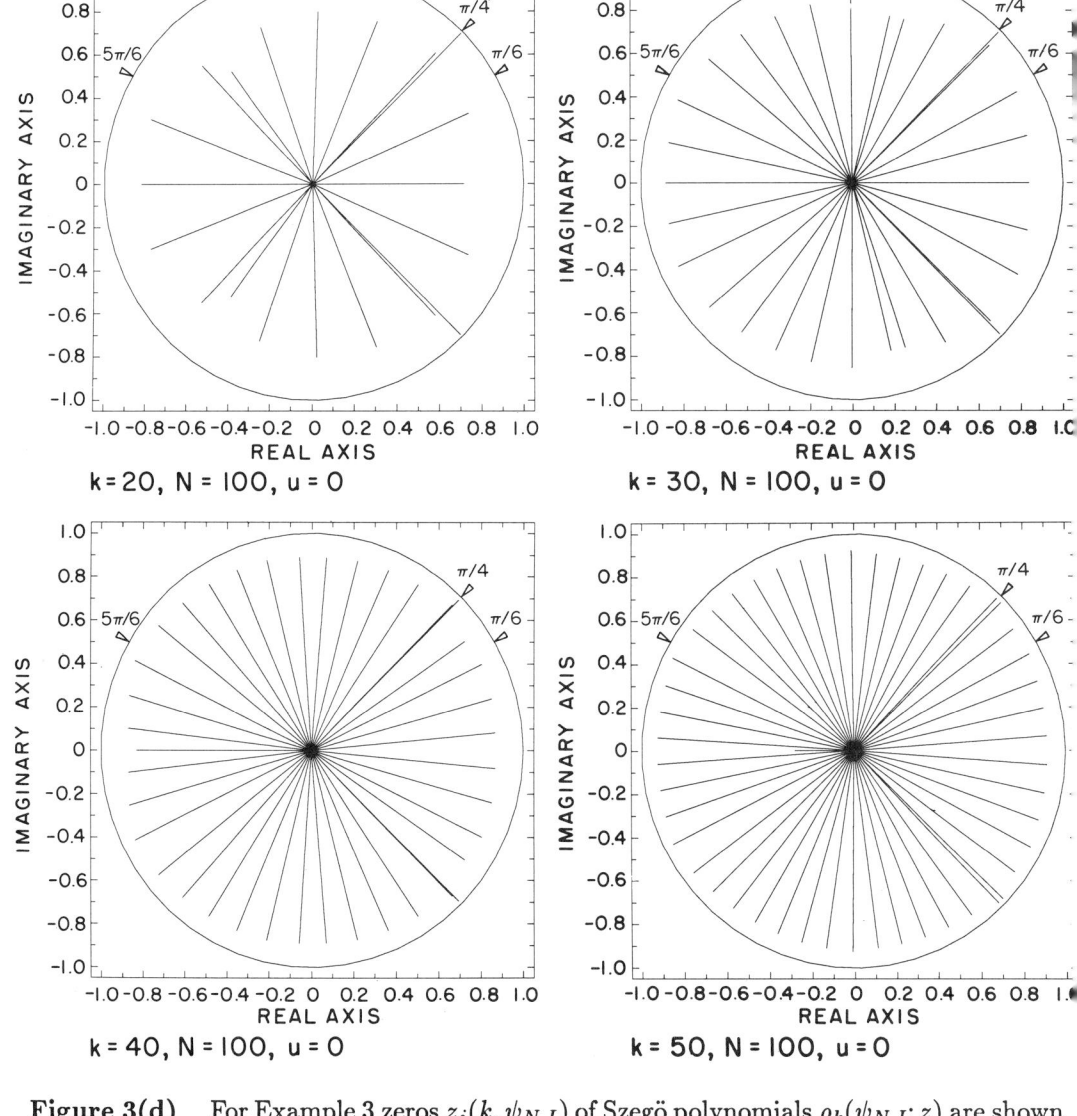

Figure 3(d)   For Example 3 zeros $z_j(k, \psi_{N,I})$ of Szegö polynomials $\rho_k(\psi_{N,I}; z)$ are shown as the endpoints of lines radiating from the origin; $N = 100$, $I = 4$.

ZEROS IN THE COMPLEX PLANE

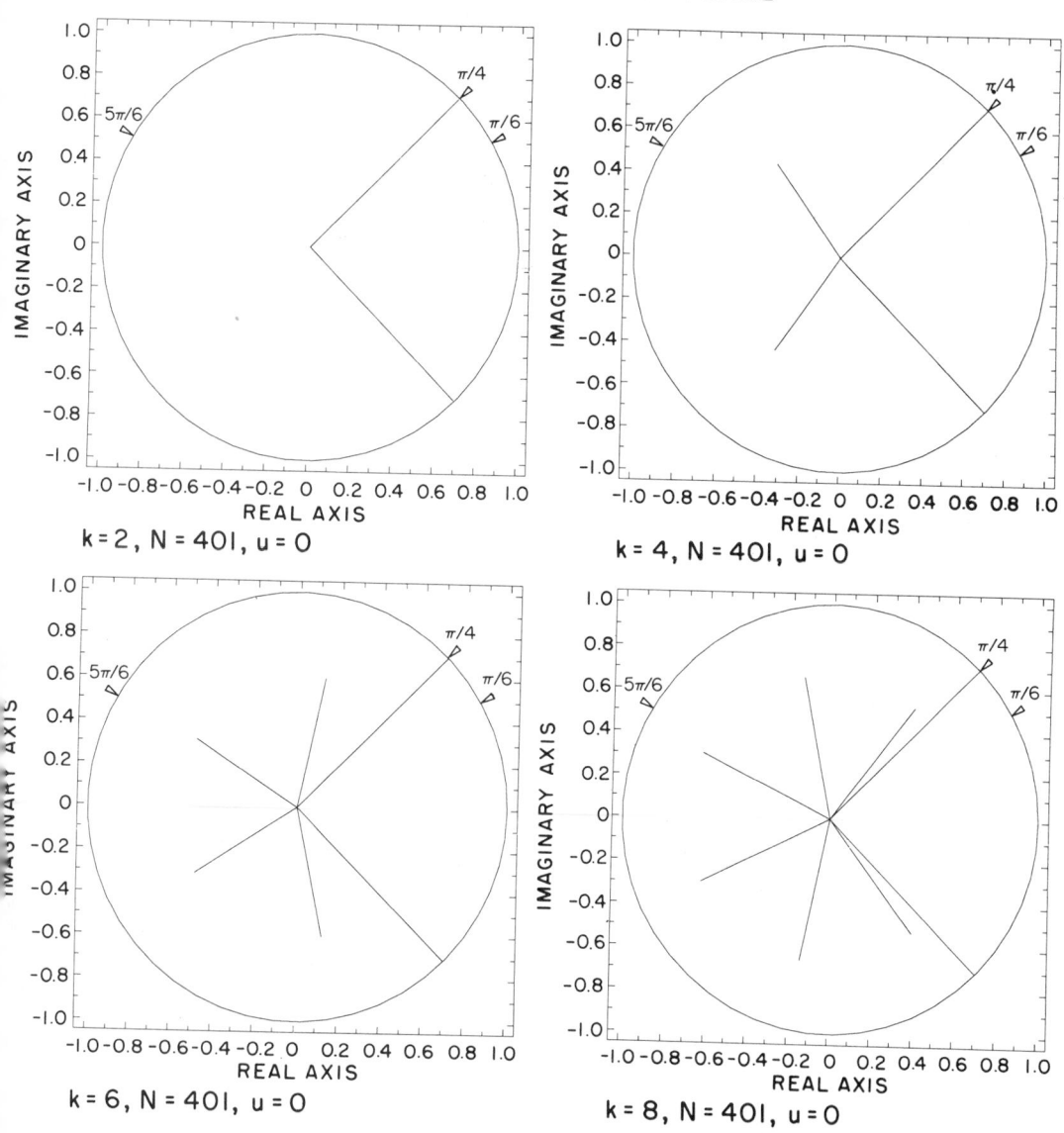

Figure 3(e) For Example 3 zeros $z_j(k, \psi_{N,I})$ of Szegö polynomials $\rho_k(\psi_{N,I}; z)$ are shown as the endpoints of lines radiating from the origin; $N = 401$, $I = 4$.

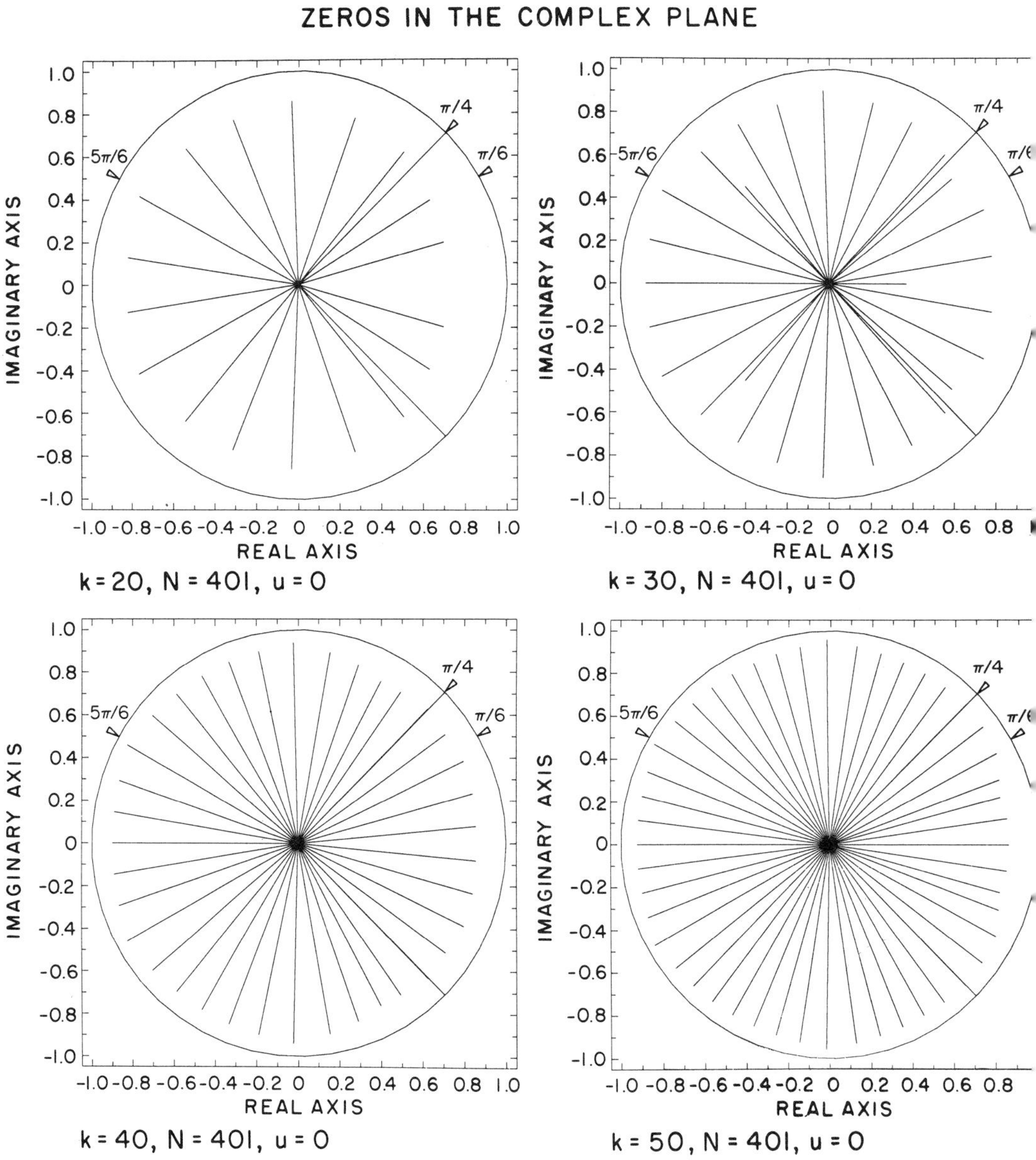

ZEROS IN THE COMPLEX PLANE

Figure 3(f)   For Example 3 zeros $z_j(k, \psi_{N,I})$ of Szegö polynomials $\rho_k(\psi_{N,I}; z)$ are shown as the endpoints of lines radiating from the origin; $N = 401$, $I = 4$.

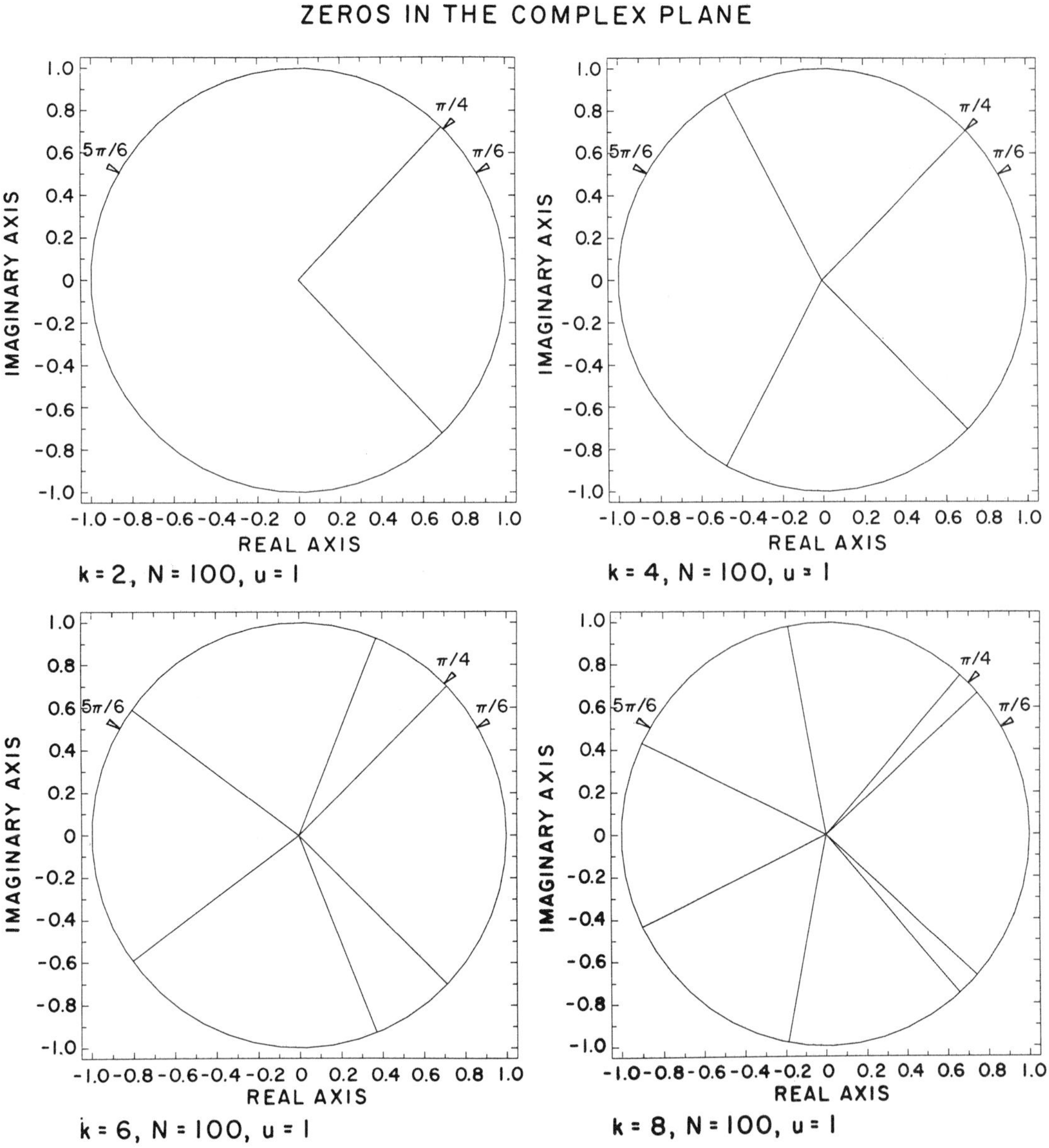

ZEROS IN THE COMPLEX PLANE

Figure 3(g) For Example 3 zeros $z_j(k, \psi_{N,I}; u)$ of Szegö para-orthogonal polynomials $B_k(\psi_{N,I}; u; z)$ are shown as endpoints of lines radiating from the origin; $N = 100$, $I = 4$, $u = 1$.

ZEROS IN THE COMPLEX PLANE

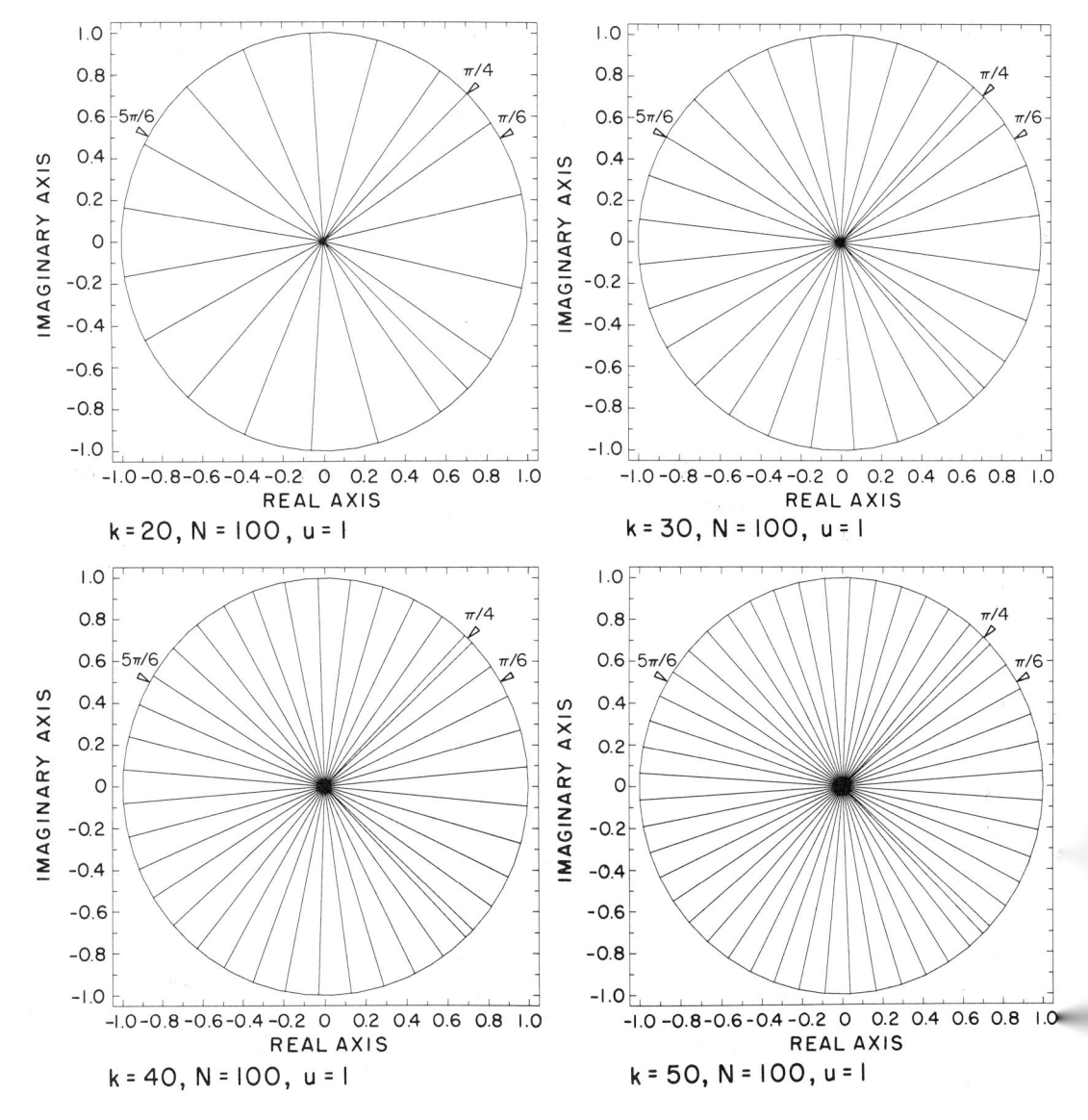

Figure 3(h)   For Example 3 zeros $z_j(k, \psi_{N,I}; u)$ of Szegö para-orthogonal polynomials $B_k(\psi_{n,i}; u; z)$ are shown as endpoints of lines radiating from the origin; $N = 100$, $I = 4$, $u = 1$.

ZEROS IN THE COMPLEX PLANE

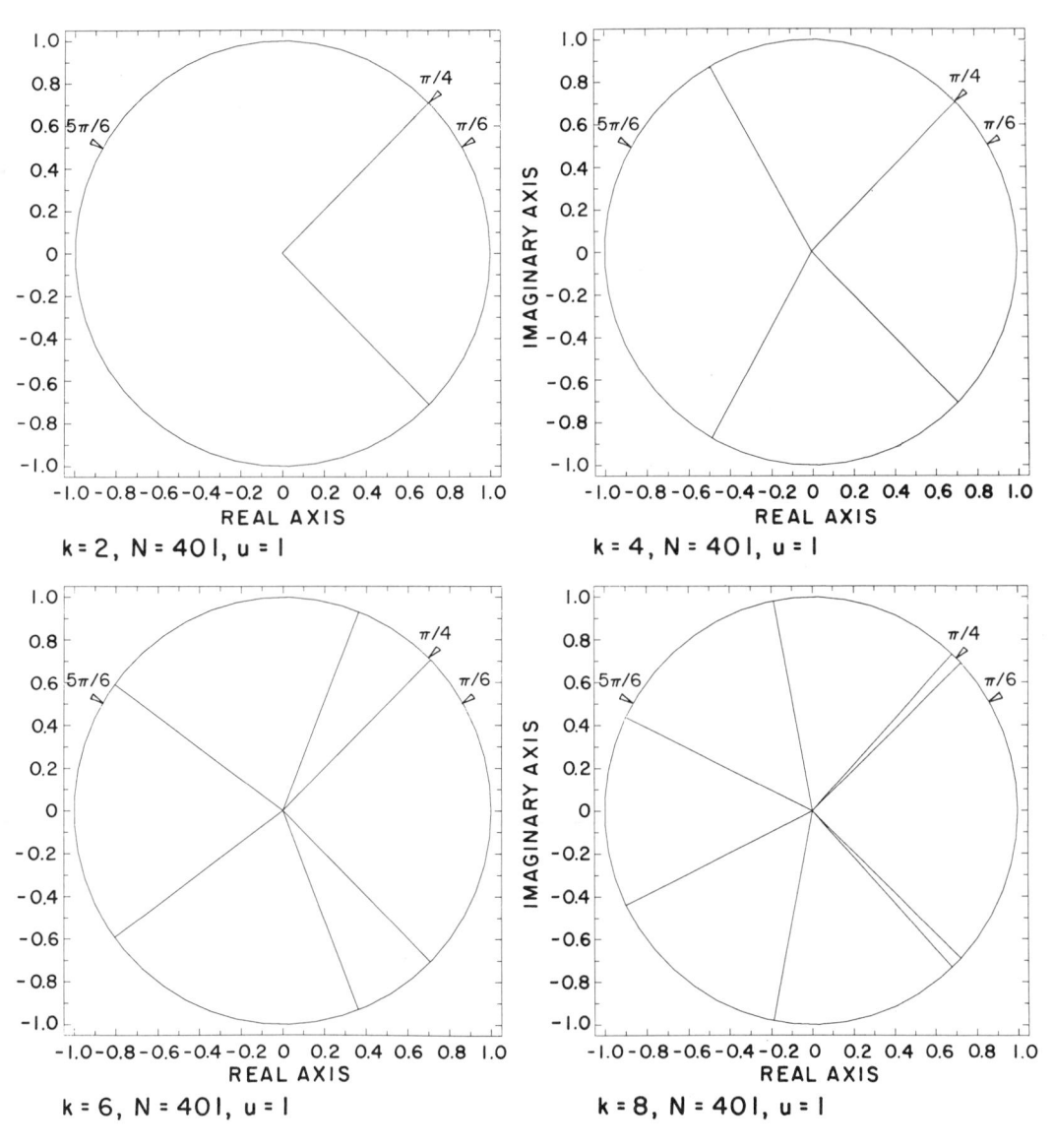

Figure 3(i) For Example 3 zeros $z_j(k, \psi_{N,I}; u)$ of Szegö para-orthogonal polynomials $\mathcal{B}_k(\psi_{N,I}; u; z)$ are shown as endpoints of lines radiating from the origin; $N = 401$, $I = 4$, $u = 1$.

ZEROS IN THE COMPLEX PLANE

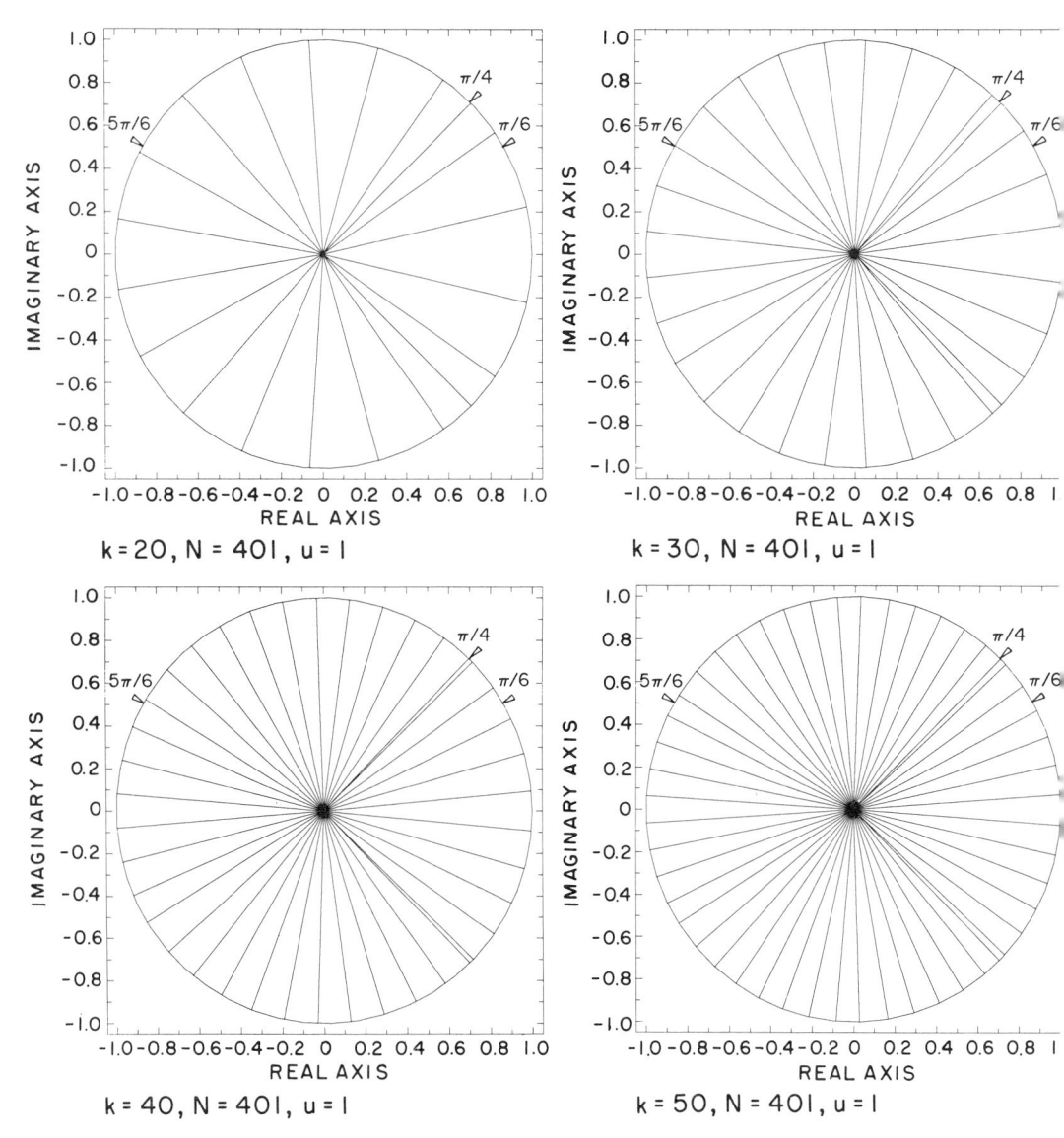

Figure 3(j) For Example 3 zeros $z_j(k, \psi_{N,I}; u)$ of Szegö para-orthogonal polynomials $B_k(\psi_{N,I}; u; z)$ are shown as endpoints of lines radiating from the origin; $N = 491$, $I = 4$, $u = 1$.

Figure 2(a) that for fixed $j \neq 4$, $|z_j(k, N) - e^{i\omega_j}|$ decreases as k increases and seems to approach a limiting situation at $k = 50$. In Example 3 the 4 frequencies are changed and the amplitude α_1 is 100 times greater than the other α_j. The graphs shown in Figure 3(a) have many features similar to those seen in the corresponding graphs of Figure 2(a). For the largest amplitude, $|z_j(k, N) - e^{i\omega_j}|$ tends to 0 as N increases at about the same rate in Figure 3(a) with $j = 1$ as in Figure 2(a) with $j = 4$. For other values of j, $|z_j(k, N) - e^{i\omega_j}|$ tends to zero more slowly (at least initially) in Figure 3(a) than in Figure 2(a). We believe this to be a consequence of the difference in the ratio of largest and smallest amplitudes in Example 3 (ratio = 100) and Example 2 (ratio = 10).

Some similar but much more complicated features can be seen in the graphs of $|z_j(k, N) - e^{i\omega_j}|$ versus N, where $z_j(k, N) := z_j(k, \psi_{N,I}; 1)$ are zeros of the Szegö paraorthogonal polynomials $B_k(\psi_{N,I}; 1; z)$, in Figures 1(b), 2(b) and 3(b) for Examples 1, 2, 3 respectively. For each graph we have used the zero $z_j(k, N)$ nearest to the critical point $e^{i\omega_j}$. In most cases $|z_j(k, N) - e^{i\omega_j}|$ decreases as N increases but the behavior as k increases is sporadic; at present we see no simple explanation for this.

Figure 1(c) shows the zeros $z_j(k, \psi_{N,I})$ as endpoints of lines radiating from the origin for $N = 101$ and several values of degree k (Example 1). The graphs illustrate the convergence of zeros $z_j(k, \psi_{N,I})$ to critical points as k increases. The effect of increasing the input sample size N to 401 can be seen in the corresponding graphs shown in Figure 1(d). Another interesting feature of these graphs is that the extraneous ("uninteresting") zeros have arguments $\arg z_j(k, \psi_{N,I})$ that are approximately uniformly distributed between $-\pi$ and π. The points in these graphs that are not endpoints of radial lines are the zeros $z_j(k - 1, \psi_{N,I})$ of the Szegö polynomial of degree $k - 1$. These empirical results indicate the possibility that there exists some kind of separation property of the zeros of Szegö polynomials. Similar properties can be seen in the graphs of zeros $z_j(k, \psi_{N,I}; 1)$ of $B_k(\psi_{N,I}; 1; z)$ shown in Figures 1(e) and (f). Here the zeros all lie on the unit circle and the separation of zeros $z_j(k, \psi_{N,I}; 1)$ by $z_j(k - 1, \psi_{N,I}; 1)$ is more apparent. Although some of the authors have proved separation in a few simple cases, a general separation property is at present an open problem to the best of our knowledge.

Graphs of zeros similar to Figures 1 (c,d,e,f) for Example 1 are given in Figures 2 (c,d,e,f) for Example 2 and in Figures 3 (c,d,e,f,g,h,i,j) for Example 3. In the illustrations for Example 3 we have included degrees $k = 2, 4, 6$ all less than $I = 4$ in order to show the convergence of zeros to $e^{i\pi/4}$ that is asserted in Theorems 2.1, 3.1 and 3.3. Some numerical tables corresponding to the illustrations in Example 3 have been given in [5].

Acknowledgements: The authors wish to thank Anne C. Jones for her able assistance in preparing the graphical illustrations in this paper.

REFERENCES

1. F. B. Hildebrand. *Introduction to Numerical Analysis*, McGraw-Hill Book Company, Inc., New York (1956).
2. William B. Jones, Olav Njåstad and E. B. Saff. Szegö polynomials associated with Wiener–Levinson filters, *J. of Comp. and Appl. Math 32*: (1990) 387–406.
3. William B. Jones, Olav Njåstad and W. J. Thron. Moment theory, orthogonal polynomials, quadrature, and continued fractions associated with the unit circle, *Bull. London Math. Soc. 21*: (1989) 113–152.
4. William B. Jones, Olav Njåstad, W. J. Thron and Haakon Waadeland. Szegö polynomials applied to frequency analysis, *J. Comp. and Appl. Math.*, (to appear).
5. William B. Jones, Olav Njåstad and Haakon Waadeland. Application of Szegö polynomials to frequency analysis, *SIAM Journal of Mathematical Analysis*, to appear.
6. R. Kumaresam, L. L. Scharf and A. K. Shaw. An algorithm for pole-zero modeling and spectral analysis, *IEEE Trans. ASSP 34, No. 3*: 637–640 (June 1986).
7. Norman Levinson. The Wiener RMS (root mean square) error criterion in filter design and prediction, *J. of Math. and Physics 225*: 2661–278 (1947).
8. K. Pan and E. B. Saff. Asymptotics for zeros of Szegö polynomials associated with trigonometric polynomial signals, *J. Approx. Theory 71*: (1992), 239–251.
9. A. K. Paul. Anharmonic frequency analysis, *Math. of Comp. 26, No. 118*: 437–447 (April 1972).
10. Norbert Wiener. *Extrapolation, Interpolation and Smoothing of Stationary Time Series*, published jointly by The Technology Press of the Massachusetts Institute of Technology, and John Wiley & Sons, Inc., New York (1949).

9

Continued Fractions and Iterated Function Systems

JOHAN KARLSSON Department of Mathematics, Chalmers Institute of Technology and University of Göteborg, S-412 96 Göteborg, Sweden

HANS WALLIN Department of Mathematics, University of Umeå, S-901 87 Umeå , Sweden

Dedicated to the memory of Arne Magnus

0. INTRODUCTION

The typical situation considered in this paper is the following. Let E be a finite set of complex numbers. We want to study continued fractions $K(a_n/1)$ and their modifications where each a_n is chosen at random from E. We are interested in convergence properties and a certain stability property of these continued fractions, and in describing the set of values of classes of convergent continued fractions. As tools we use generalized iteration of Möbius transformations and iterated function systems, and computer experiments. The main point of our paper is not to prove new technical results but to discuss the connection between continued fractions and generalized iteration.

We introduce a system $\{\mathcal{F}, V\}$ where $\mathcal{F}$ is a family of functions $f : V \to V$ and V a complete metric space. We shall study the (generalized) iteration

$$\widetilde{S}_n := s_n \circ \ldots \circ s_1$$

and the reversed iteration

$$S_n := s_1 \circ \ldots \circ s_n$$

where all $s_j \in \mathcal{F}$, and we are interested in the behaviour of $\widetilde{S}_n$ and S_n as n grows to infinity, in particular in the connection between $\widetilde{S}_n$ and S_n. We want to know if there is,

in some sense, an attractor for the process $z_0, z_1, z_2, \dots$, where $z_0 \in V$ is the initial value and

$$z_n := \widetilde{S}_n(z_0) = s_n(z_{n-1}), \quad n = 1, 2, \dots,$$

and if the process is stable in its dependence on the initial value z_0. We also want to discuss the set of possible values $\lim S_n(z)$, for a fixed $z \in V$, when s_j, $j = 1, 2, \dots$, are arbitrary functions in $\mathcal{F}$ and the limit exists.

Our main concern is the case when $V \subset \mathbb{C}$, the complex plane, and all $f \in \mathcal{F}$ are Möbius transformations of the form $f(z) = a/(1 + z)$, where $a \in \mathbb{C}$. The condition $f : V \to V$ then becomes $a/(1 + V) \subset V$, and $S_n(0)$ becomes the n:th *approximant* of a continued fraction [8] which is included in our general theory if $0 \in V$; the sequence $S_n(z)$, $n = 1, 2, \dots$, are *modified* approximants. In this case the set of possible values $\lim S_n(0)$ has been discussed in [14], [7], [15] and [16].

In general, the sequence of iterations $\widetilde{S}_n(z) = s_n \circ \dots \circ s_1(z)$ does not converge. In fact, if $\widetilde{S}_n(z)$ converges to a number $A \neq -1$ and $s_n(z) = a_n/(1 + z)$, $a_n \in \mathbb{C}$, then the relation $\widetilde{S}_n(z) = a_n/(1 + \widetilde{S}_{n-1}(z))$ shows that $a_n \to A(1 + A)$, i.e. the Möbius transformations $s_n(z)$ are determined by complex numbers a_n which converge. This is called the limit periodic case and was discussed in [10] together with many references. In this paper we investigate the case when $\{a_n\}$ does not converge.

Our main interest is when the functions $f \in \mathcal{F}$ are given by $f(z) = a/(1 + z)$ where $a \in E$ and $E \subset \mathbb{C}$ consists of a finite number of points; however, we also discuss the case when E is an infinite set and when the functions in $\mathcal{F}$ are not necessarily Möbius transformations. In the n:th step of the iteration $\widetilde{S}_n = s_n \circ \dots \circ s_1$ and the reversed iteration $S_n = s_1 \circ \dots \circ s_n$, the function s_n is allowed to be any of the functions in $\mathcal{F}$.

As a preparation, in Section 1 we discuss some computer experiments on attractors of processes $z_0, z_1, z_2, \dots$. In Section 2 we define a stability property of $\widetilde{S}_n$ which implies convergence of S_n. We also interpret the set of values $\lim S_n(z)$, $s_j \in \mathcal{F}$, as attractor of the system $\{\mathcal{F}, V\}$. In Section 3 we consider systems $\{\mathcal{F}, V\}$ which are hyperbolic in the sense that there is a constant λ less than 1 such that all $f \in \mathcal{F}$ are Lipschitz functions on V with a Lipschitz constant at most λ. For such systems $\widetilde{S}_n$ has the stability introduced in Section 2. In Section 4 we introduce the invariant (stationary) measure used in the theory of iterated function systems (see [6], [3] and [2]) and in discrete time Markov processes (see for instance [13], [11] and [4]; we are grateful to G. Högnäs for the references [13] and

[11]). We observe that it gives the distribution of values of classes of convergent continued fractions (Theorem 4; compare also [11]).

A natural tool in the investigation of attractors of processes $z_0, z_1, z_2, \ldots$ is the theory of Kleinian groups, in particular the notion of limit set. A preliminary study along these lines is given in [9].

Recent investigations somewhat related to ours are [1] and [12].

1. SOME PICTURES

The main thesis of [10] was that although the generalized iterates $\widetilde{S}_n = s_n \circ \ldots \circ s_1$ in general do not converge it was still possible in several cases to deduce the behaviour of the reversed iterates $S_n = s_1 \circ \ldots \circ s_n$ from that of $\widetilde{S}_n$. We shall discuss some pictures showing the behaviour for large n of the generalized iterates $\widetilde{S}_n$ when all the s_j are chosen at random from a finite family $\mathcal{F}$ of Möbius transformations of the form $a/(1+z)$, $a \in \mathbb{C}$. It is intuitively clear – and true in the case of Section 2 – that the behaviour of S_n is closely related to that of $\widetilde{S}_n$.

Let us start by recalling [10] what happens when $\mathcal{F}$ consists of *one* function s only. Then $\widetilde{S}_n = S_n$ and the only things that can happen are 1) $\widetilde{S}_n(z)$ diverges and they all lie on a circle which depends on the initial value z (Fig. 1) or 2) $\widetilde{S}_n(z)$ converges to a fixed point of s (Fig. 2). If $s(z) = a/(1+z)$ the former case corresponds to $a < -1/4$ and we say that s is of *divergence type*; for all other a we have the second case and say that s is of *convergence type*.

With Figure 1 and 2 as background let us now study what happens when we iterate with two or more Möbius transformations; in particular we want to see how the outcome depends on whether these are of convergence or divergence type. Hence, let $\mathcal{F}$ consist of more than one function; in the remaining pictures $\mathcal{F}$ consists of two or three functions of the form $a/(1+z)$. There are two different algorithms to generate pictures showing the behaviour of $\widetilde{S}_n$ for large n [2, §3.8]. In the first, the *deterministic algorithm*, we start from a subset L_0 of $\mathbb{C}$ and compute successively

$$L_n := \bigcup_{s \in \mathcal{F}} s(L_{n-1}), \quad n = 1, 2, \ldots .$$

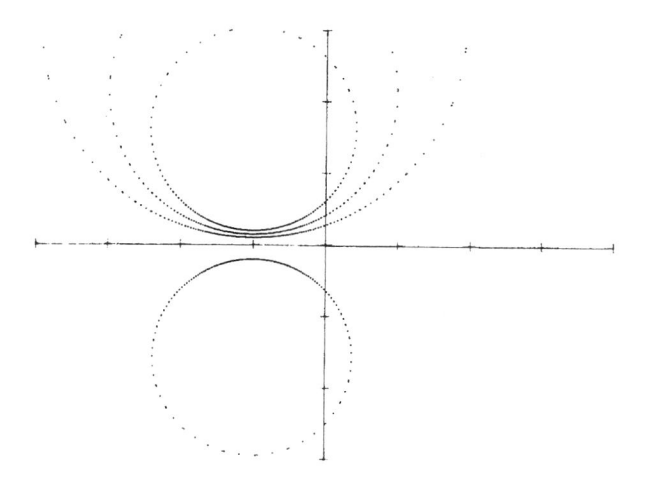

Fig. 1. Iteration of $s(z) = -0.4/(1 + z)$. The iterates lie on a circle which depends on the initial value.

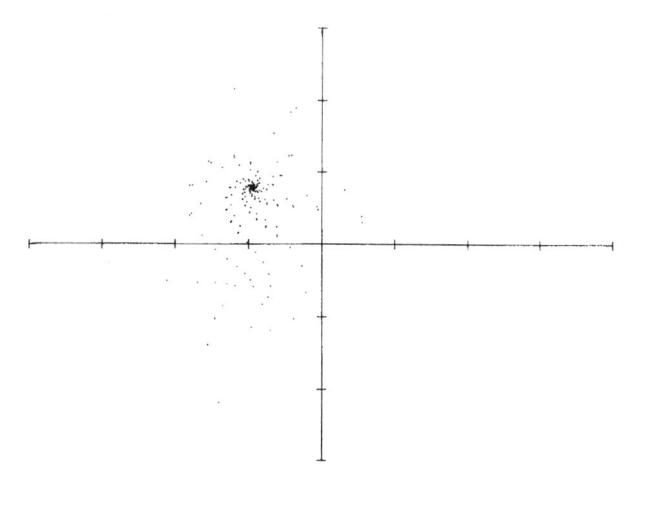

Fig. 2. Iteration of $s(z) = (-0.4 + 0.02i)/(1 + z)$. Convergence to a fixed point. The convergence is not along the spiral arms visible in the picture; the iterates jump between the arms. The arms rather show a strong stability; the iterates of various initial points tend to bunch together.

Under favourable circumstances L_n converges in some sense to an *attractor* L (see Section 2 - 4, in particular Definition 2 in Section 2, for more details); if we plot L_n for a "large" n we get a good picture of L. In the second, the *random iteration algorithm*, which is the one we have used to produce Figure 3 - 8, we start from an initial value $z_0 \in \mathbb{C}$ and compute successively the iterates $z_0, z_1, z_2, \dots$ where

$$z_n = \widetilde{S}_n(z_0) = s_n(z_{n-1}), \quad n = 1, 2, \dots .$$

In the n:th step s_n is chosen at random among the functions in $\mathcal{F}$. The first iterates are hidden (in our pictures 1000 iterates) and then a large number of them are plotted. It is expected that the points plotted, with probability 1, will fill out a picture which is a good approximation to the possible attractor L from the deterministic algorithm. In any case we expect our pictures to show an attractor loosely interpreted as a set which the iterates $\widetilde{S}_n(z_0)$ approach and stay in, and this attractor in our experiments seems to be independent of the initial value z_0.

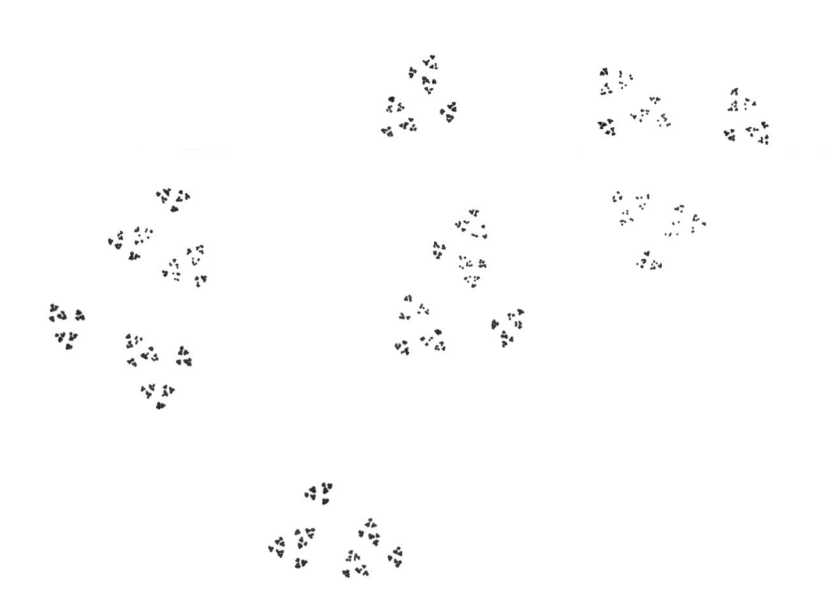

Fig. 3. This represents one of our first attempts on the lines of this paper. The picture is due to Jan Gelfgren in Umeå. It shows an attractor somewhat similar to a Sierpinski triangle. The picture is obtained from three Möbius transformations forming a hyperbolic system in the sense of Section 3 so the existence of an attractor L is assured by the theory if one takes initial values within the region where the system is hyperbolic. However, other

initial values, even very far from this region, produce the same attractor so we feel safe in conjecturing that this is what one gets regardless of where one begins.

The Möbius transformations involved are $(-0.25 + 0.5i)/(1 + z)$, $(0.5 + i)/(1 + z)$, and $(0.3025 + 0.15i)/(1 + z)$ and the area shown is $-0.25 < Rez < 1.08$, $-0.1 < Imz < 0.9$.

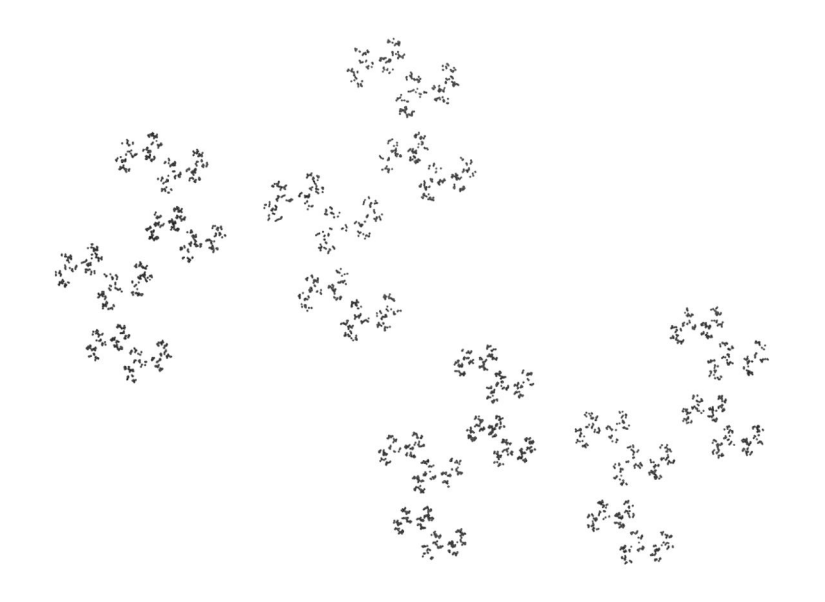

Fig. 4. Here two Möbius transformations of convergence type are used, $(-0.4+i)/(1+z)$ and $(-0.67+i)/(1+z)$. They do not form a hyperbolic system. The area shown is $-0.08 < Rez < 0.32$, $0.67 < Imz < 0.97$.

If we iterate by means of functions which are contractions the general theory (Section 3 and 2) guarantees the existence of an attractor L in the deterministic algorithm. Möbius transformations of convergence type will in general be contractions only close to the attractive fixed point. It is suggestive, however, that the dynamics of the individual transformations of convergence type are similar to those of contractions. Our experiments tend to bear this suggestion out, and we have never observed anything that indicates that the system $\mathcal{F}$ does not have an attractor. What is amazing is that even when the Möbius transformations are of divergence type we observe convergence to an attractor.

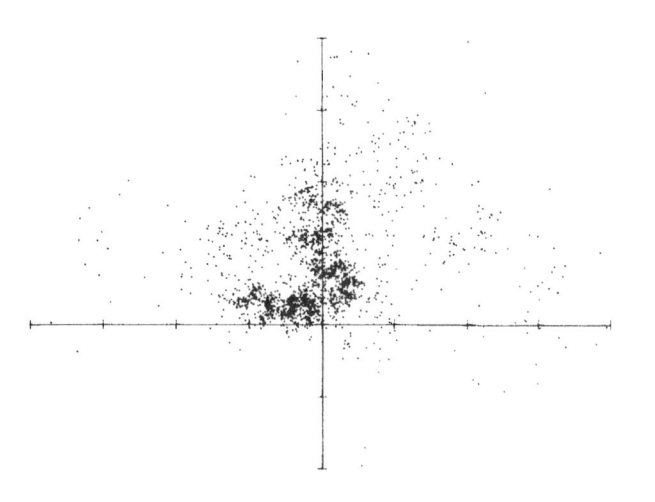

Fig. 5. Here we iterate $(-0.4 + i)/(1 + z)$ and $-0.67/(1 + z)$, of convergence and divergence type, respectively. Still we seem to get some kind of attractor. There are, however, occasional points in the outskirts of the picture which might indicate that on the average (probabilistically) we stay in an attractor but that there are rare sequences of $\widetilde{S}_n$ producing points more or less anywhere. The ticks on the axes are spaced one unit apart.

Fig. 6. This picture does not look very impressive but it is in fact our most interesting experiment. Two Möbius transformations, both of divergence type, are used, $-0.4/(1+z)$ and $-0.67/(1+z)$. The generalized iterates (for $n > 1000$) are all on the real axis (shown here from -2 to 2). We have observed convergence to a subset of the real axis for other sets of Möbius transformation of divergence type as well and conjecture that this is a general phenomena. That the attractor lies on the real axis, given that it exists, is perhaps not so surprising since Möbius transformations of the form $a/(1 + z)$ and divergence type have a common invariant circle (the real axis).If we knew that the attractor was independent of the initial value it would have to lie on the real axis since all iterates are real with real initial values.

Remark. It is important to note, in the case of Figure 6 that this is what the random iteration algorithm gives. The deterministic algorithm would produce a set of point dense in the plane as can be seen by considering periodic sequences $\{s_n\}$ (compare [10] Section 3). Such sequences, of course, occur with probability zero. The same phenomenon of the random algorithm producing smaller sets than the deterministic can be seen in [15].

2. STABILITY AND ATTRACTORS

2.1. General theory.

Let V be a complete metric space with metric d and $\mathcal{F}$ a family of functions $f : V \to V$. Typically, think of V as a closed subset of $\mathbb{C}$ with the usual metric. We refer to $\{\mathcal{F}, V\}$ as a generalized discrete dynamical system and define, for any given sequence $\{s_n\}$ $n = 1, 2, ...,$ in $\mathcal{F}$, the corresponding sequence of iterates $\widetilde{S}_n := s_n \circ ... \circ s_1$, and reversed iterates $S_n := s_1 \circ ... \circ s_n$, $n = 1, 2, ...$. We want to formalize the notion of stability meaning that the *orbit* (or $\{s_n\}$–orbit) $z_0, z_1, ...,$ with initial value $z_0 \in V$, defined by $z_n := s_n(z_{n-1}) = \widetilde{S}_n(z_0)$, $n = 1, 2, ...,$ is asymptotically stable in V, uniformly with respect to $\mathcal{F}$. We denote by $\widetilde{S}_n(V)$ the image of V under $\widetilde{S}_n$ and by $\operatorname{diam}\widetilde{S}_n(V)$ the diameter of $\widetilde{S}_n(V)$.

DEFINITION 1. Let $\{\rho_n\}$ be a sequence of positive numbers tending to zero. The system $\{\mathcal{F}, V\}$ is $\rho_n - stable$ if $\operatorname{diam}\widetilde{S}_n(V) \leq \rho_n$, for all n and all sequences $\{s_n\}$ in $\mathcal{F}$. The system $\{\mathcal{F}, V\}$ is (uniformly) *stable* if it is ρ_n–stable for some $\{\rho_n\}$.

Since in a ρ_n–stable system the same ρ_n is good enough for every choice of n functions from $\mathcal{F}$ we have the following property.

PROPOSITION 1. $\{\mathcal{F}, V\}$ *is* $\rho_n - stable$ *if and only if* $\operatorname{diam}S_n(V) \leq \rho_n$ *for all* n *and all sequences* $\{s_n\}$ *in* $\mathcal{F}$.

Stability of a system implies, and, in fact, is equivalent to, convergence of the reversed iterates in V in the following strong form.

THEOREM 1. *If a system* $\{\mathcal{F}, V\}$ *is* ρ_n*–stable then, for every sequence* $\{s_n\}$ *in* $\mathcal{F}$, *there exists an* $S_\infty \in V$ *such that the corresponding sequence* $\{S_n\}$ *converges to* S_∞ *in* V. *More precisely,*

$$d(S_n(z), S_\infty) \leq \rho_n, \quad for\ all \quad n \quad and\ all \quad z \in V. \tag{1}$$

Conversely, if for every $\{s_n\}$ *in* $\mathcal{F}$ *there is an* $S_\infty \in V$ *such that* (1) *holds, then* $\{\mathcal{F}, V\}$ *is* $2\rho_n$*-stable.*

Proof. The converse part follows immediately from the triangle inequality. To prove the direct part, assume that $\{\mathcal{F}, V\}$ is ρ_n–stable, and take $\{s_n\}$ in $\mathcal{F}$ and $z \in V$. For $k, l \geq n$ we observe that there are w_1 and w_2 in V such that $S_k(z) = s_1 \circ \ldots \circ s_k(z) = s_1 \circ \ldots \circ s_n(w_1) = S_n(w_1)$ and, analogously, $S_l(z) = S_n(w_2)$. This gives, by the stability $d(S_k(z), S_l(z)) = d(S_n(w_1), S_n(w_2)) \leq \mathrm{diam} S_n(V) \leq \rho_n$. Hence, $\{S_k(z)\}$ is a Cauchy sequence and since V is complete it has a limit $S_\infty(z)$ in V which satisfies $d(S_n(z), S_\infty(z)) \leq \rho_n$. Take $w \in V$, $w \neq z$. Then

$$d(S_\infty(w), S_\infty(z)) \leq d(S_\infty(w), S_n(w)) + d(S_n(w), S_n(z)) +$$
$$+ d(S_n(z), S_\infty(z)) \leq 3\rho_n,$$

i.e. $S_\infty(z)$ has a value S_∞ which is independent of $z \in V$, proving the theorem.

We shall now see that the set of values S_∞ which we obtain in Theorem 1, $\{S_\infty : \{s_n\}$ sequence in $\mathcal{F}\}$ can be interpreted as attractor in a very strong sense of the system $\{\mathcal{F}, V\}$. We first define the attractor in the sense we need it. For a given non-empty subset L_0 of V we define the sequence $\{L_n\}$ by

$$L_n := \bigcup_{s \in \mathcal{F}} s(L_{n-1}), \quad n = 1, 2, \ldots . \tag{2}$$

The *Hausdorff distance* between two sets A and B in V is the infimum of all $\varepsilon > 0$ such that A is in the ε–neighbourhood of B and B in the ε–neighbourhood of A [2].

DEFINITION 2. Let $\{\rho_n\}$ be a sequence of positive numbers tending to zero. The system $\{\mathcal{F}, V\}$ has a ρ_n–*attractor* L if L is closed, $L \subset V$ and, for every non-empty $L_0 \subset V$, the Hausdorff distance between L and the set L_n defined by (2) is at most ρ_n,

for all n. The system $\{\mathcal{F}, V\}$ has (a uniform) *attractor* L if L is ρ_n–attractor to $\{\mathcal{F}, V\}$ for some $\{\rho_n\}$.

We note that the attractor of a system, if it exists, is unique since we require it to be closed and the Hausdorff distance is a metric on the family of closed sets. If $\{\mathcal{F}, V\}$ has an attractor L then every orbit $\{\widetilde{S}_n(z_0)\}$, $z_0 \in V$, is attracted to L in the sense that $\widetilde{S}_n(z_0) \in L_n$, if L_0 contains z_0, and L_n tends to L; if we choose $L_0 = \{z_0\}$ we see that $L_n = \{\widetilde{S}_n(z_0) : s_j \in \mathcal{F}\}$ tends to L, i.e. is attracted by and fills out L.

THEOREM 2. *Suppose that the system* $\{\mathcal{F}, V\}$ *is* ρ_n*–stable. Put*

$$L := \{S_\infty : S_\infty = \lim s_1 \circ ... \circ s_n(z), z \in V, \{s_n\} \text{ sequence in } \mathcal{F}\}.$$

Then $\overline{L}$ *is* ρ_n*–attractor of* $\{\mathcal{F}, V\}$.

Proof. By Theorem 1 the definition of L makes sense and S_∞ is independent of z. If $S_\infty \in L$ and $\{s_n\}$ is a sequence in $\mathcal{F}$ such that $S_\infty = \lim S_n(z)$, $S_n = s_1 \circ ... \circ s_n$, $z \in V$, we conclude from Theorem 1 that $d\big(S_n(z), S_\infty\big) \leq \rho_n$, $z \in V$. Furthermore, for any $L_0 \subset V$ we see from (2) that

$$L_n = \{S_n(z) : s_j \in F, \quad \text{all } j, \quad z \in L_0\}.$$

These two facts imply

$$L \subset \{z \in V : d(z, L_n) \leq \rho_n\} \quad \text{and}$$

$$L_n \subset \{z \in V : d(z, L) \leq \rho_n\}.$$

These set inclusions mean that the Hausdorff distance from L_n to L is at most ρ_n, and the theorem is proved.

Theorem 2 means that for a stable system the orbits $\{\widetilde{S}_n(z_0)\}$ have an attractor which in a very strong sense is independent of z_0.

Remark. The set L in Theorem 2 need not be closed [5].

2.2 Stability of continued fractions.

We now give two examples of stability for systems of Möbius transformations. In each of these two systems $0 \in V$ and so, by Theorem 2, the set of possible values of the continued fractions is an attractor of the system.

EXAMPLE 1.   We consider the situation from the uniform para-
bola theorem [8, Theorem 4.40]. For $\alpha \in (-\pi/2, \pi/2)$ we introduce the parabola region
(see Figure 9 or [8, p. 99 and 76])

$$P_\alpha := \left\{ z \in \mathbb{C} : |z| - Re(ze^{-2i\alpha}) \leq \frac{1}{2} \cos^2 \alpha \right\},$$

and the half-plane (Figure 9 or [8 p.100 and 76])

$$V_\alpha := \left\{ z \in \mathbb{C} : Re(ze^{-i\alpha}) \geq -\frac{1}{2} \cos \alpha \right\}.$$

Now we fix $\alpha \in (-\pi/2, \pi/2)$ and a positive constant M and introduce

$$E := P_\alpha \cap \{ z \in \mathbb{C} : |z| \leq M \}.$$

Consider the system $\{\mathcal{F}, V\}$ where $V := V_\alpha$ with the usual metric in $\mathbb{C}$, and $f \in \mathcal{F}$ if
and only if $f(z) = a/(1+z)$ where $a \in E$. Then $f : V \to V$ for all $f \in \mathcal{F}$ [8,p. 100]
and the system $\{\mathcal{F}, V\}$ is ρ_n–stable with [8,p. 105]

$$\rho_n := \frac{2M}{(\cos \alpha) \prod_{m=1}^{n-1} \left(1 + \frac{\delta}{m}\right)}, \quad \text{where} \quad \delta = \frac{\cos^2 \alpha}{4M + 2\cos^2 \alpha}.$$

This example is illustrated in Figure 7 - 8 which are constructed as explained in Section 1.

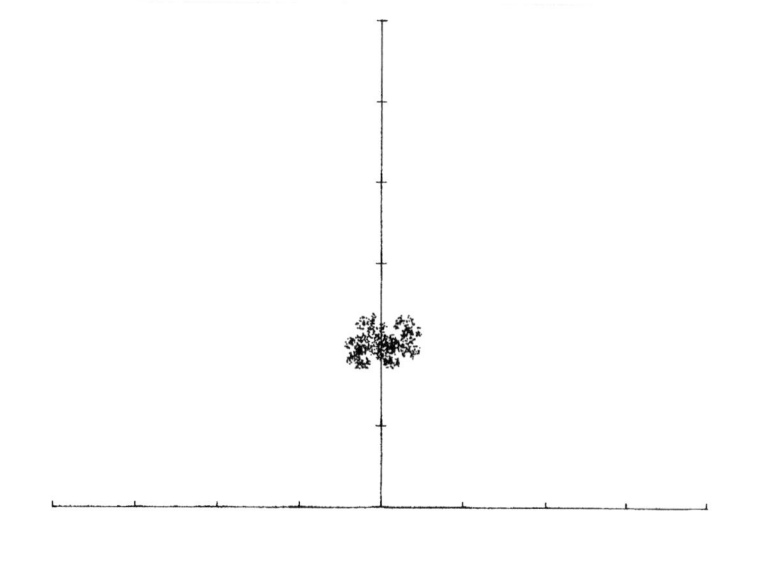

Fig. 7a.

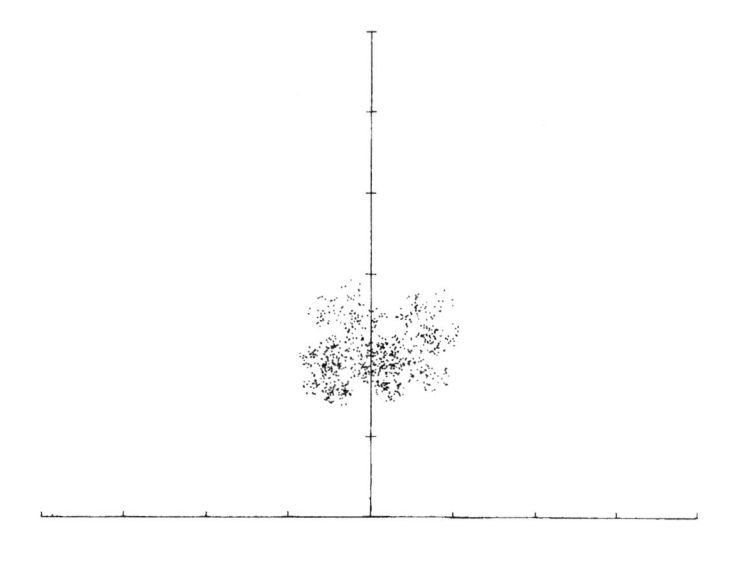

Fig. 7b.

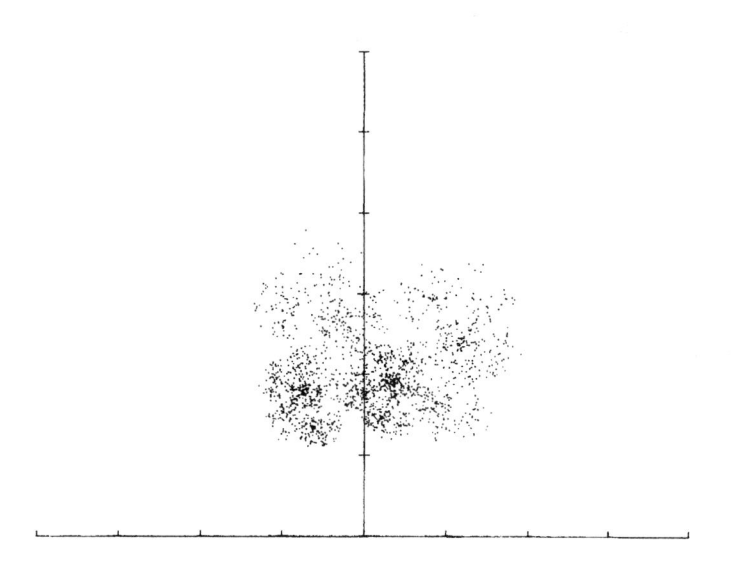

Fig. 7c.

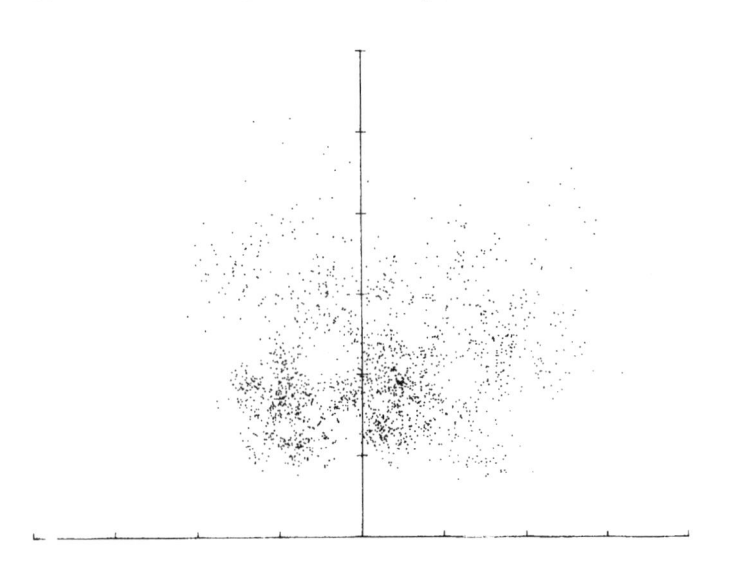

Fig. 7d.

Fig. 7a–d. To illustrate Example 1 where we have stability and hence the theory of Section 2 we take $\alpha = 3\pi/8$ so the axis of the parabola P_α has argument $3\pi/4$, and choose $a_{1,2} = -1 + i \pm t(1 + i)$ as the coefficients of two Möbius transformations. These a_k are symmetric with respect to the axis of the parabola and lie inside the parabola for $t < 0.326$. In Figure 7a-c, t=0.1, 0.2, and 0.3, respectively, while in 7d, t=0.4 so we have left the safety of Theorem 2. There does not seem to happen anything spectacular with the behaviour of $\widetilde{S}_n$ though. The tick spacing is 0.5.

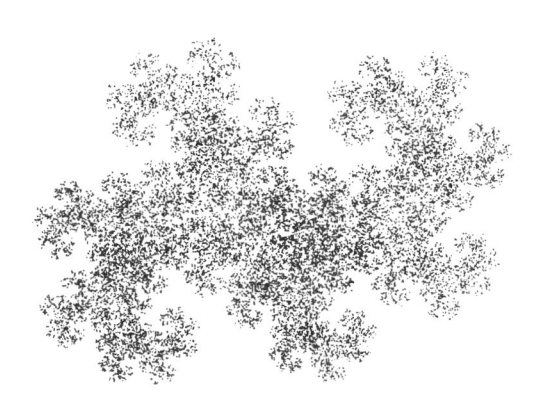

Fig. 8. Shows the attractor of Fig. 7a in more detail.

EXAMPLE 2. Here we treat the situation described by Theorem 4.45 in [8].

Case A.   The complex number a is given such that $|\arg a| < \pi$ and $|a| \geq 1/4$, and

$$E := \{z \in \mathbb{C} : |z - a| \leq \frac{1}{\sqrt{2}}(|a| + Rea)^{1/2}\}.$$

Introduce $\alpha := (\arg a)/2$ and P_α and V_α as in Example 1. Then $E \subset P_\alpha$ [9, p.112]. For $z \in E$ we also have $|z| \leq M$ where

$$M := |a| + \frac{1}{\sqrt{2}}(|a| + Rea)^{1/2}.$$

From the previous example we conclude that the system $\{\mathcal{F}, V\}$ is ρ_n–stable if $V := V_\alpha$, $f \in \mathcal{F}$ if and only if $f(z) = a/(1 + z)$, $a \in E$, and ρ_n is given by the same formula as in Example 1.

Case B. $a \in \mathbb{C}$ is given such that $|a| < 1/4$ and

$$E := \{z \in \mathbb{C} : |z - a| \leq |a + \frac{1}{4}|\}.$$

Let a', $a' \neq -1/4$, denote the point where the line passing through $-1/4$ and a meets the circle $|z| = 1/4$. Introduce $\alpha' := (\arg a')/2$. Then $\alpha' \in (-\pi/2, \pi/2)$ and $P_{\alpha'}$ and $V_{\alpha'}$

are defined as in Example 1. It follows [8, p. 112] that $E \subset P_{\alpha'}$. For $z \in E$ we also have $|z| \leq M$ where M is now given by

$$M := |a| + \left|a + \frac{1}{4}\right|.$$

From Example 1 we conclude that $\{\mathcal{F}, V\}$ is ρ_n–stable if $V := V_{\alpha'}$, $f \in \mathcal{F}$ if and only if $f(z) = a/(1 + z)$, $a \in E$, and ρ_n is given by the same formula as in Ex. 1.

3. HYPERBOLIC SYSTEMS

Let $\{\mathcal{F}, V\}$ be a system consisting of a complete metric space V with metric d and a family $\mathcal{F}$ of functions $f : V \to V$. We are interested in the case when the functions in $\mathcal{F}$ are contractions.

DEFINITION 3. Let λ be a number satisfying $0 \leq \lambda < 1$. $\{\mathcal{F}, V\}$ is a *hyperbolic system with Lipschitz constant* λ if, for every $f \in \mathcal{F}$,

$$d(f(x), f(y)) \leq \lambda\, d(x, y), \quad \text{for all} \quad x, y \in V.$$

For such a system we get

$$d(\widetilde{S}_n(x), \widetilde{S}_n(y)) \leq \lambda^n d(x, y), \quad \text{for all} \quad x, y \in V,$$

which gives the following theorem.

THEOREM 3. *If* $\{\mathcal{F}, V\}$ *is a hyperbolic system with Lipschitz constant* $\lambda < 1$, *and* V *is bounded, then* $\{\mathcal{F}, V\}$ *is* ρ_n–*stable with* $\rho_n = \lambda^n \cdot (\mathrm{diam} V)$.

EXAMPLE 3. Let $\alpha \in (-\pi/2, \pi/2)$ be fixed, and let P_α be the parabola region and V_α the half-plane in Example 1. Choose a finite set $E := \{a_1, ..., a_N\} \subset \mathbb{C}$, $N \geq 2$, where the elements a_j satisfy

$$\sqrt{|a_j|} < \frac{\cos \alpha}{2}, \quad 1 \leq j \leq N,$$

and introduce f_j by $f_j(z) = a_j/(1 + z)$, $1 \leq j \leq N$. The relevant geometric significance of $(\cos \alpha)/2$ is shown in Figure 9.

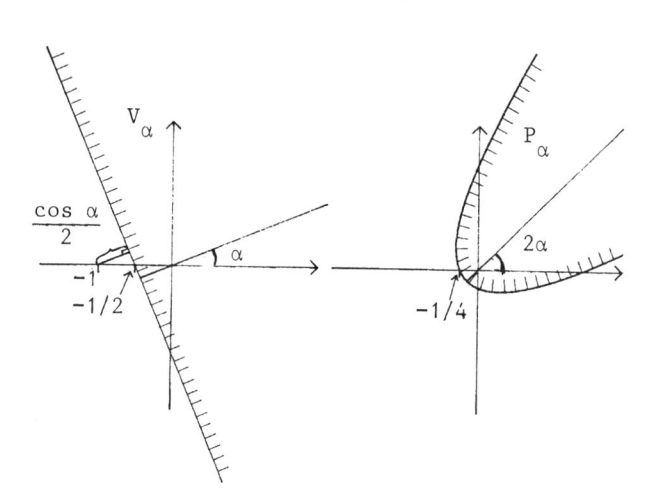

Fig. 9. The left part of the picture shows V_α and the right part P_α. For all α the boundary of P_α passes through $-1/4$ and the boundary of V_α through $-1/2$.

We conclude that, for $z \in V_\alpha$,

$$|1 + z| \geq \frac{\cos \alpha}{2} > \sqrt{|a_j|}.$$

Hence,

$$|f_j'(z)| \leq \frac{4|a_j|}{\cos^2 \alpha} < 1, \quad \text{for} \quad z \in V_\alpha, \quad \text{and}$$

$$\left| \frac{a_j}{1 + z} \right| < \sqrt{|a_j|} < \frac{\cos \alpha}{2}, \quad \text{for} \quad z \in V_\alpha.$$

It is also straightforward to check that $a_j \in P_\alpha$, for $1 \leq j \leq N$, and, consequently (see Example 1), that $f_j : V_\alpha \to V_\alpha$.

We now fix a constant $M \geq (\cos \alpha)/2$ and introduce $V := V_\alpha \cap \{z \in \mathbb{C} : |z| \leq M\}$ and $\mathcal{F} := \{f_1, ..., f_N\}$. Then the inequalities above quarantee that $f_j : V \to V$ and that $\{\mathcal{F}, V\}$ is a hyperbolic system with Lipschitz constant $\max\{4|a_j|/\cos^2 \alpha : 1 \leq j \leq N\}$.

4. INVARIANT MEASURES

4.1. Iterated function systems.

General references here are [6], [3] and [2]. Let $\{\mathcal{F}, V\}$ be a system where V is a compact metric space and $\mathcal{F}$ consists of finitely many, but at least 2, continuous functions $f_j : V \to V$, $1 \le j \le N$. Let p_j, $1 \le j \le N$, be positive constants with sum 1; you should think of p_j as the probability of choosing f_j in the next step of the iteration $\widetilde{S}_n$ and the reversed iteration S_n. Following [3] we refer to $\{\mathcal{F}, V\}$ as an iterated function system with associated probabilities $\{p_j\}$.

By a fixed point argument it is possible to prove [3, Theorem 1] that there exists a probability measure μ on V such that, if $f_j^{-1}(B)$ denotes the inverse image of B under f_j,

$$\mu(B) = \sum_{j=1}^{N} p_j \mu\big(f_j^{-1}(B)\big),$$

for all Borel subsets B of V. Such a measure μ is called an *invariant measure* for the iterated function system; if we would allow $N = 1$, it would be an invariant measure for a discrete dynamical system. It has the interpretation as a stationary probability measure for the discrete time Markov process $\{\mathcal{F}, \{p_j\}, V\}$ where the probability $P(z, B)$ of transfer from a point $z \in V$ to a Borel set $B \subset V$ is

$$P(z, B) = \sum_{j=1}^{N} p_j \delta_{f_j(z)}(B),$$

where $\delta_a(B)$ is 1 if $a \in B$ and 0 otherwise.

If the iterated function system is hyperbolic then [6, §3.1], [3, Theorem 3] there exists a unique compact set L which is *invariant* with respect to $\{\mathcal{F}, V\}$ in the sense that

$$L = \bigcup_{j=1}^{N} f_j(L).$$

The set L is the attractor of $\{\mathcal{F}, V\}$ in the sense of Definition 2 in Section 2. Furthermore, the invariant set L is the set of values $\{S_\infty : S_\infty = \lim s_1 \circ \ldots \circ s_n(z), z \in V, \{s_n\}$ sequence in $\mathcal{F}\}$, so that in particular, this set of values is closed (cf. Theorem 2). The invariant measure is then unique, has support L and is attractive (see [3], §2).

4.2. Now let $\{\mathcal{F}, V\}$, where $\mathcal{F} = \{f_1, ..., f_N\}$, be any iterated function system with probabilities p_j, $1 \leq j \leq N$. Assume that, for some $z_0 \in V$, the limit

$$S_\infty(z_0) := \lim s_1 \circ ... \circ s_n(z_0)$$

exists for any choice of sequence $\{s_n\}$ in $\mathcal{F}$. Furthermore, assume that, for each n, the probability that s_n is chosen as f_j is p_j. Let the measure ν be the probability distribution of the set of values

$$\{S_\infty(z_0) = \lim s_1 \circ ... \circ s_n(z_0) : \{s_n\} \text{ sequence in } \mathcal{F}\},$$

i.e. $\nu(B)$ is the probability that $S_\infty(z_0)$ belongs to B, for any Borel set $B \subset V$,

$$\nu(B) = \text{prob}\{S_\infty(z_0) \in B\}.$$

THEOREM 4.   *The measure ν is an invariant measure for the iterated function system $\{F, V\}$ with probabilities $\{p_j\}$.*

Proof.   If $S_\infty(z_0) := \lim s_1 \circ ... \circ s_n(z_0)$ we denote by $S_{1\infty}(z_0)$ the "first tail" (in continued fraction terminology)

$$S_{1\infty}(z_0) := \lim s_2 \circ ... \circ s_n(z_0).$$

Hence, $S_\infty(z_0) = s_1 \circ S_{1\infty}(z_0)$ and, consequently,

$$\text{prob}\{S_\infty(z_0) \in B\} = \sum_{j=1}^{N} p_j \cdot \text{prob}\{S_{1\infty}(z_0) \in f_j^{-1}(B)\}.$$

Since ν is also the probability distribution of the set of values of the first tails, this means that

$$\nu(B) = \sum_{j=1}^{N} p_j \cdot \nu\big(f_j^{-1}(B)\big),$$

i.e. ν is an invariant measure.

EXAMPLE 4.   The situation described in Example 1 - 3 falls under Theorem 4 if we in Example 1 and 2 restrict $\mathcal{F}$ to be a finite collection of Möbius transformations $f(z) = a/(1+z)$, $a \in E$, i.e. we restrict a to belong to a finite subset of E consisting of

at least two elements. As probabilities $\{p_j\}$ we choose any set of positive constants with sum 1. Notice that in Example 1 and 2 the set V which is chosen as V_α or $V_{a'}$ may be replaced by a suitable bounded subset of V_α or $V_{\alpha'}$ since $E/(1+V_\alpha)$ and $E/(1+V_{\alpha'})$ are bounded. The convergence of the sequence $S_n(z_0)$, $n = 1, 2, ...$, follows from Theorem 1.

In particular, we note that the situation described in [14, Ex. 10-12] falls under Theorem 4. A result closely related to Theorem 4 is given in [11] showing the great generality of this type of statement.

REFERENCES

1. B. Aebischer, The limiting behaviour of sequences of Möbius transformations, *Math. Z.* **205** (1990), 49-59.

2. M.F. Barnsley, "Fractals everywhere", Academic Press, New York, 1988.

3. M.F. Barnsley and S.G. Demko, Iterated function systems and the global construction of fractals, *Proc. Royal Soc. London* **A 399** (1985), 243-275.

4. M.F. Barnsley, S.G. Demko, J.H. Elton and J.S. Geronimo, Invariant measures for Markov processes arising from iterated function systems with place–dependent probabilities, *Ann. Inst. Henri Poincaré* **24** (1988), 367-394.

5. E. Gadde, Stable Iterated Function systems, University of Umeå ,Department of Mthematics, Doctoral Thesis No 4, 1992

6. J. Hutchinson, Fractals and self–similarity, *Indiana Univ. J. Math* **30** (1981), 713-747.

7. L. Jacobsen, W.J. Thron and H. Waadeland, Some observations on the distribution of values of contined fractions, *Numer. Math.* **55** (1989), 711-733.

8. W.B. Jones and W.J. Thron, "Continued fractions: Analytic theory and applications", Addison-Wesley Publ. Co., Reading, Mass., 1980.

9. J. Karlsson and H. Wallin, Kleinian groups, attractors and continued fractions, Preliminary manuscript, 1991.

10. J. Karlsson, H. Wallin and J. Gelfgren, Iteration of Möbius transforms and continued fractions, *Rocky Mountain J. Math* **21** (1991).

11. G. Letac, A contraction principle for certain Markov chains and its applications, *Contemp. Math* **50** (1986), 263-273.

12. L. Lorentzen, Compositions of contractions, *J. Comp. Appl. Math* **32** (1990), 1-10.

13. M. Rosenblatt, "Markov processes. Structure and asymptotic behaviour", Springer-Verlag, Berlin, 1971.

14. H. Waadeland, Where do all the values go? Playing with two-element continued fractions, *Rocky Mountain J. Math* **21** (1991).

15. H. Waadeland, The boundary version of Worpitzky's theorem: Some probabilistic remarks, Manuscript, 1990.

16. H. Waadeland, Boundary version of a twin region convergence theorem for continued fractions, Manuscript, 1990.

10

Strip Convergence Regions for Continued Fractions

L. J. LANGE   Department of Mathematics, University of Missouri, Columbia, Missouri 65211, USA

Dedicated to the memory of Arne Magnus

1   INTRODUCTION

In this work we shall be concerned with the convergence behavior of continued fractions of the form

$$1 + \mathop{\mathbf{K}}_{n=1}^{\infty} \frac{a_n}{1}, \quad a_n = -c_n^2. \tag{1.1}$$

We have found it convenient for geometric and algebraic reasons to formulate most of our results with conditions on the c_n rather than with conditions on the a_n directly. The fundamental goal that we had before us was to learn as much as we could about the following question: Under what boundary conditions will an infinite strip region C of constant width 1 at any height be a convergence region for continued fractions (c.f.'s) (1.1) in the sense that c_n in C guarantees the convergence of (1.1), provided a necessary growth condition on the c_n is met. Thron (1943) proved that the strips bounded by two parallel lines, one through each of the points $\pm 1/2$, are convergence regions in the above sense when he proved his parabola theorem which we give as Theorem 1.1 below. It is natural to ask how these straight strips might be bent and still be convergence regions. Thron (1944) made the first attempt to 'bend' these convergence strips and gave an interesting family of strip-like convergence regions. However, the strips in the latter pioneering effort do not meet our constant width requirement.

In Section 2 we derive necessary conditions for convergence regions for c.f.'s (1.1). In Section 3 we give a number of results for a family of strip regions which we call Worpitzky

strips because they contain the well known Worpitzky convergence disk for c.f.'s. Finally, in Section 4 we prove several results for a family of strips which we call transcendental strips because their boundaries are described in rectangular coordinates by transcendental functions.

Now let us introduce

$$t_n(z) = 1 + a_n/z \tag{1.2}$$

and

$$T_1(z) = t_1(z), \; T_n(z) = T_{n-1}(t_n(z)), \; n \ge 1. \tag{1.3}$$

Then, if we denote the numerator and denominator of the nth approximant of the c.f. (1.1) by A_n and B_n, respectively, we can write

$$\frac{A_n}{B_n} = 1 + \frac{a_1}{1} + \frac{a_2}{1} + \cdots + \frac{a_n}{1} = T_n(1) \tag{1.4}$$

and in general

$$T_n(z) = \frac{A_{n-1}(z-1) + A_n}{B_{n-1}(z-1) + B_n}. \tag{1.5}$$

The quantities A_n and B_n satisfy the well known recursion relations and determinant formula (see, e.g., Perron (1957, p.4, p.11))

$$\begin{bmatrix} A_0 \\ B_0 \end{bmatrix} = \begin{bmatrix} 1 \\ 1 \end{bmatrix}, \; \begin{bmatrix} A_1 \\ B_1 \end{bmatrix} = \begin{bmatrix} 1 + a_1 \\ 1 \end{bmatrix}, \; \begin{bmatrix} A_n \\ B_n \end{bmatrix} = \begin{bmatrix} A_{n-1} + a_n A_{n-2}, \; n \ge 2 \\ B_{n-1} + a_n B_{n-2}, \; n \ge 2 \end{bmatrix} \tag{1.6}$$

and

$$A_{n-1}B_n - A_n B_{n-1} = (-1)^n \prod_{k=1}^{n} a_k, \tag{1.7}$$

respectively. We shall now state the Parabola Theorem for continued fractions in a form suitable for our purposes.

THEOREM 1.1 (Parabola Theorem) *Let α be a fixed real number such that $-\pi/2 < \alpha < \pi/2$. If $c_n \in S_\alpha$, or equivalently, if $-c_n^2 \in P_\alpha$, $n = 1, 2, 3, \ldots$, where S_α is the strip region*

$$S_\alpha := \{ w \mid |\Re(we^{-i\alpha})| \le \frac{1}{2}\cos\alpha \}$$

and P_α is the parabolic region

$$P_\alpha := \{ w \mid |w| - \Re(we^{-2i\alpha}) \le \frac{1}{2}\cos^2\alpha \},$$

then the continued fraction

$$\mathop{K}_{n=1}^{\infty} \frac{-c_n^2}{1}$$

converges to a finite value iff the series

$$\sum_{n=1}^{\infty} \prod_{k=1}^{\infty} |c_k^2|^{(-1)^{n-k+1}}$$

diverges. The last series diverges iff at least one of the series

$$\sum_{n=1}^{\infty} \left| \frac{c_2 c_4 \cdots c_{2n}}{c_3 c_5 \cdots c_{2n+1}} \right|^2, \qquad \sum_{n=1}^{\infty} \left| \frac{c_3 c_5 \cdots c_{2n+1}}{c_4 c_6 \cdots c_{2n+2}} \right|^2$$

diverges.

This theorem was first proved by Thron (1943) by means of the Stieltjes-Vitali convergence extension theorem for analytic functions. He later gave (Thron 1963) an elementary proof of the Parabola Theorem using properties of linear fractional maps of the unit disk onto itself. In the late sixties Jones and Thron (1968) gave a very general extension of this theorem which they later called the Multiple Parabola Theorem in their book on continued fractions (Jones and Thron (1980)). For more on the Parabola Theorem and for a very useful list of related references we refer the reader to the recent book on continued fractions by Lorentzen and Waadeland (1992).

2 NECESSARY CONDITIONS FOR CONVERGENCE REGIONS

More than 30 years ago in Chapter I of his Ph.D. thesis, Lange (1960) studied periodic sequences of linear fractional transformations and periodic continued fractions and arrived at necessary conditions for convergence regions for continued fractions $1 + K(a_n/1)$. It was convenient in that work to set $a_n = c_n^2$ and express the necessary conditions derived for simple and twin convergence regions in terms of the c_n for continued fractions $1 + K(c_n^2/1)$. Though some of these results were employed in other published papers (see, e.g., Lange and Thron (1960)) many results of the Chapter which we feel are applicable to our present work have not been published elsewhere. We have had these results in the back of our mind througout our present study of strip convergence regions for continued fractions $1 + K(-c_n^2/1)$ and they were one of the motivating forces which led to the convergence theorems given in the next two sections. We shall give below some of the pertinent results for simple convergence regions that we derived in the 1960 work and omit the twin convergence region conclusions. The statements of the original results have been modified to fit the fact that we have here replaced the original condition $a_n = c_n^2$ by $a_n = -c_n^2$. We first provide some basic material to better understand the theorems. Let $A_n(B_n)$ be the nth numerator(denominator) of the continued fraction

$$1 + \operatorname*{K}_{n=1}^{\infty} \frac{a_n}{1}; \quad a_n = -c_n^2, \; n = 1, 2, \ldots . \tag{2.1}$$

and let

$$\Delta_k := \frac{A_{k-1} - B_{k-1} + B_k}{2 \prod_{m=1}^{k} c_m}. \tag{2.2}$$

By a *region* we shall mean an open connected set with all or part of its boundary. A region E is called a *simple convergence region* for continued fractions of the form (2.1) if any choice of the $a_n \in E$, $n = 1, 2, \ldots$ insures the convergence of the continued fraction. To distinguish the two notions and to save confusion, we shall call a region C a *simple c-convergence region* for c.f. (2.1) if it converges for every sequence $\{c_n\}$ of points $\in C$, $n \geq 1$. It is in this latter sense that we use C in Theorems 2.4-2.10 of this section. By the *width* $W(h)$ at height h of a region R we shall mean the number $W(h) = \sup\{ |\Re(w) - \Re(z)| \mid \Im(w) = \Im(z) = h, w, z \in R \}$. The

continued fraction (2.1) is said to be *periodic* of period k if $a_{kr+m} = a_m$ for all $r = 0, 1, 2, \ldots$ and $m = 1, 2, \ldots, k$. To avoid certain difficulties and to conform to the modern definition for the convergence of a continued fraction, we shall assume from now on that none of the a_n in the continued fraction (2.1) are zero. Finally, the nonsingular linear fractional transformation ($\ell.f.t.$)

$$s(z) = \frac{az + b}{cz + d}$$

with two distinct fixed points w_1, w_2 is said to be *elliptic* if $|\sigma| = 1, (\sigma \neq 1)$, where

$$\sigma = \begin{cases} \frac{a - cw_1}{a - cw_2} & \text{if } c \neq 0 \\ \frac{a}{d} & \text{if } c = 0. \end{cases} \tag{2.3}$$

We are now ready to present the theorems of this section.

THEOREM 2.1 *The nonsingular $\ell.f.t.$*

$$s(z) = \frac{az + b}{cz + d}$$

with two distinct fixed points w_1 and w_2 is elliptic iff δ is real and $0 \leq \delta < 1$, where

$$\delta = \frac{(a + d)^2}{4(ad - bc)}.$$

Proof: Set

$$\gamma = \frac{(1 + \sigma)^2}{4\sigma},$$

where σ is given by (2.3). If $0 \leq \gamma < 1$, then we can set $\gamma = \cos^2 \frac{\theta}{2}$, $0 < \theta < 2\pi$, and we obtain

$$\cos^2 \frac{\theta}{2} = \frac{(1 + \sigma)^2}{4\sigma} \Rightarrow$$

$$0 = (1 + \sigma)^2 - 4\sigma \cos^2 \frac{\theta}{2}$$

$$0 = \sigma^2 - 2\sigma \cos \theta + 1.$$

Solving the latter equation for σ we get

$$\sigma = \cos \theta \pm \sqrt{\cos^2 \theta - 1}$$

$$= \cos \theta \pm i \sin \theta = e^{\pm i\theta},$$

whence

$$|\sigma| = 1, \ \sigma \neq 1.$$

Conversely, if $\sigma = 1, (\sigma \neq 1)$, then σ can be expressed in the form $e^{i\theta}$ $(0 < \theta < 2\pi)$ and we have

$$\gamma = \frac{(1 + \sigma)^2}{4\sigma} = \frac{(1 + e^{i\theta})^2}{4e^{i\theta}} = \frac{(2e^{i\theta} \cos \frac{\theta}{2})^2}{4e^{i\theta}} = \cos^2 \frac{\theta}{2}.$$

Hence $|\sigma| = 1, (\sigma \neq 1)$ is equivalent to $0 \leq \gamma < 1$. Straightforward calculations will show that

$$\gamma = \frac{(1 + \sigma)^2}{4\sigma} = \frac{(a + d)^2}{4(ad - bc)}$$

and with this our proof is complete.

THEOREM 2.2 *A periodic c.f.* $1 + K(a_n/1)$ *of period* k *diverges if* $T_k(z)$ *is an elliptic transformation, where by formula (1.5),*

$$T_k(z) = \frac{A_{k-1}(z-1) + A_k}{B_{k-1}(z-1) + B_k}.$$

Proof: For a proof of this well known result we refer the reader to Lange (1960) and to the chapters on periodic continued fractions in the books by Jones and Thron (1980) and Lorentzen and Waadeland (1992).

THEOREM 2.3 *A periodic c.f.* $1 + K(-c_n^2/1)$ *of period* k *diverges if* $-1 < \Delta_k < 1$, *where* Δ_k *is give by formula (2.2) and, in particular,*

$$
\begin{aligned}
\Delta_1 &= \frac{1}{2c_1} \\
\Delta_2 &= \frac{1 - c_1^2 - c_2^2}{2c_1 c_2} \\
\Delta_3 &= \frac{1 - c_1^2 - c_2^2 - c_3^2}{2c_1 c_2 c_3} \\
\Delta_4 &= \frac{1 - c_1^2 - c_2^2 - c_3^2 - c_4^2 + c_1^2 c_3^2 + c_2^2 c_4^2}{2c_1 c_2 c_3 c_4}.
\end{aligned}
\tag{2.4}
$$

Proof: It follows from Theorem 2.1 that $T_k(z)$ given by (1.5) is elliptic if and only if $0 \leq \delta_k < 1$, where

$$\delta_k = \frac{(A_{k-1} - B_{k-1} + B_k)^2}{4 \prod_{m=1}^k c_m^2} = \Delta_k^2.$$

Hence the condition $-1 < \Delta_k < 1$ insures that $0 \leq \delta_k < 1$ so that $T_k(z)$ is elliptic. Thus the divergence of $1 + K(-c_n^2/1)$ follows from Theorem 2.2.

THEOREM 2.4 *If* C *is a simple c-convergence region for c.f.'s (2.1), then* $c \notin C$ *for all* c *for which* $\Im(c) = 0, |\Re(c)| > 1/2$.

Proof: Set $c_1 = c$ in the formula for Δ_1 in formulas (2.4) and apply Theorem 2.3.

THEOREM 2.5 *If* C *is a simple c-convergence region for c.f.'s (2.1) and if* $c \in C$, *then the points* $c \pm 1, -c \pm 1$ *cannot be interior points of* C.

Proof: The corresponding result of this type for c.f.'s $1 + K(c_n^2/1)$ was given by Thron (1943) in the important substantial paper in which he proved the parabola theorem among other results. To prove our theorem we set $c_1 = c$ (where c is fixed) and $c_2 = z$ in the formula for Δ_2 in (2.4) and obtain

$$\Delta_2(z) = \frac{1 - c^2 - z^2}{2cz}.$$

By Theorem (2.3) the periodic c.f. $1 + K(-c_n^2/1)$ with $c_{2n-1} = c$ and $c_{2n} = z$ diverges if z is a point on one of the two arcs defined by

$$\Im(\Delta(z)) = 0, \quad |\Re(\Delta(z))| < 1.$$

The endpoints of these two arcs are obtained by solving the equations

$$\Delta_2(z) = \pm 1$$

and they are $c \pm 1, -c \pm 1$. Clearly these points cannot be interior points of C and this completes our proof.

An interesting problem, still only partly solved , is to determine for all complex numbers c the largest r (depending on c) such that the closed disk $\overline{D}(c,r)$ with center c and radius r is a simple c-convergence region for continued fractions $1 + K(-c_n^2/1)$. By Theorem 2.4 it follows that for $\Im(c) = 0$, $|\Re(c)| > 1/2$ no region $\overline{D}(c,r)$ can exist. For all other values of c a simple c-convergence region $\overline{D}(c,r)$ exists for it is known (see Thron (1943, p.684)) that $r(c) \geq (1/2)\cos \arg c$. However, $r(c) \leq 1/2$ always, since, by Theorem 2.5, $c \pm 1/2$ cannot be interior points of $\overline{D}(c,r)$. We shall now prove a theorem which includes the above necessary restrictions on $r(c)$ and which, for certain values of c, gives a sharper upper bound for $r(c)$ than $1/2$.

THEOREM 2.6 *Let* $\overline{D}(c,r)$ *be the closed disk with center* c *and radius* $r(c)$. *If* $C :=$ $\overline{D}(c,r(c))$ *is a simple c-convergence region for c.f.'s (2.1), then*

$$r(c) \leq 1/2 \text{ if } \Re(c^2) \leq 1/4 \text{ and } r(c) < \sqrt{|\sin \gamma|}/2 \text{ if } \Re(c^2) > 1/4,$$

where $\gamma = arg(1/4 - c^2)$.

Proof: We have that the c.f.

$$1 - \frac{c_1^2}{1} - \frac{c_2^2}{1} - \frac{c_1^2}{1} - \frac{c_2^2}{1} - \cdots$$

of period 2 diverges if $-1 < \Delta_2 < 1$, where by formulas (2.4)

$$\Delta_2 = \frac{1 - c_1^2 - c_2^2}{2c_1 c_2}.$$

If we set $\Delta_2 = -\cos \phi$, then we have divergence if $0 < \phi < \pi$. Now set

$$c_1 = c + re^{i\theta}, \ c_2 = c - re^{i\theta},$$

where c is fixed. That is, we let c_1, c_2 be two points on a diameter of a circle with center c and radius r. Then $\Delta_2 = -\cos \phi$ implies

$$\cos \phi = \frac{(c + re^{i\theta})^2 + (c - re^{i\theta})^2 - 1}{2(c + r^{i\theta})(c - r^{i\theta})} \Rightarrow$$
$$0 = 2c^2(1 - \cos \phi) + 2r^2 e^{2i\theta}(1 + \cos \phi) - 1 \Rightarrow$$
$$0 = 4c^2 \sin^2 (\phi/2) + 4r^2 e^{2i\theta} \cos^2 (\phi/2) - 1.$$

We solve the last equation for r^2 and obtain

$$r^2 = \frac{1}{4} \left| \frac{1 - 4c^2 \sin^2 (\phi/2)}{\cos^2 (\phi/2)} \right|$$
$$= \frac{1}{4} \left| 1 + (1 - 4c^2) \tan^2 \frac{\phi}{2} \right|.$$

Now set

$$x = \tan^2 (\phi/2); \ 1 - 4c^2 = \rho e^{i\gamma}, \ 0 \leq \gamma \leq 2\pi.$$

Then

$$r = |1 + x\rho e^{i\gamma}|^{1/2}/2 \tag{2.5}$$

and the c.f. under consideration diverges for those values of r for which $x > 0$. A remaining problem is to find the minimum of r with respect to x for $x > 0$. It follows from equation (2.5) that

$$r^4 = (1 + 2x\rho\cos\gamma + \rho^2 x^2)/16;$$

so let us consider the function

$$g(x) = 1 + 2x\rho\cos\gamma + \rho^2 x^2.$$

If $\rho\cos\gamma = \Re(1 - 4c^2) \geq 0$, then all of the terms of $g(x)$ are nonnegative and g(x) decreases to 1 as x decreases to 0. Therefore, since $r^4 = g(x)/16$, a necessary condition in this case for $\overline{D}(c, r)$ to be a simple convergence region is that $r \leq 1/2$. If $\rho\cos\gamma < 0$, then $\rho > 0$ and differentiating $g(x)$ we obtain

$$g'(x) = 2\rho\cos\gamma + 2\rho^2 x.$$

Setting $g'(x) = 0$ and solving for x we get

$$x = -\frac{\cos\gamma}{\rho} > 0.$$

The function $g(x)$ (and hence $r = r(x)$) assumes its minimum for this value of x. After substituting the above value of x into equation (2.5) we obtain

$$r = |1 - \cos\gamma e^{i\gamma}|^{1/2}/2 = |e^{-i\gamma} - \cos\gamma|^{1/2}/2 = |\sin\gamma|^{1/2}/2.$$

It should be pointed out that the continued fraction still diverges for at least one point on the boundary of the circle with center c and radius r, where $r = |\sin\gamma|^{1/2}/2$, since $x = -(\cos\gamma)/\rho$ is within the range of values of x which guarantee divergence. Therefore, if $\rho\cos\gamma = \Re(1 - 4c^2) < 0$, a necessary condition for $\overline{D}(c, r)$ to be a simple convergence region for c.f.'s (2.1) is that $r < |\sin\gamma|^{1/2}$, it being understood that if $\gamma = \pi$ no such region exists (see Theorem 2.4). This completes the proof of our theorem.

THEOREM 2.7 *If C is a simple c-convergence region for continued fractions (2.1) and if $c \in C$, then the geometric means*

$$\pm\sqrt{(1/2)(1 + c)}, \quad \pm\sqrt{(1/2)(1 - c)}$$

cannot be interior points of C.

Proof: Let $c_1 = c_2 = z$ and $c_3 = c$ in the formula for Δ_3 in Theorem 2.3, where $c \in C$, and consider the function

$$\Delta_3(z) := \frac{1 - 2z^2 - c^2}{2z^2 c}.$$

The critical points in the statement of the theorem are the endpoints of the "bad" arcs determined by the conditions

$$\Im(\Delta_3(z)) = 0, \quad |\Re(\Delta_3(z))| < 1.$$

It can be shown that the arcs in question are portions of a lemniscate. By Theorem 2.3, if any of these critical points were an interior point of C and $c \in C$, there is a divergent c.f. $1 + K(-c_n^2/1)$ of period 3 with its elements $c_n \in C$. We omit the details of the proof.

THEOREM 2.8 *If C is a simple c-convergence region for c.f.'s (2.1) and if c and c' are any two elements of C, then the geometric means*

$$\pm\sqrt{(1+c)(1+c')}, \ \pm\sqrt{(1-c)(1-c')}, \ \pm\sqrt{(1+c)(1-c')}, \ \pm\sqrt{(1-c)(1+c')}$$

cannot be interior points of C.

Proof: To obtain this result we studied the convergence behavior of c.f.'s of period 4. We set $c_1 = c$, $c_3 = c'$; $c_2 = c_4 = z$ in Δ_4 of Theorem 2.3, where $c, c' \in C$, and obtained

$$\Delta_4(z) = \frac{1 - c^2 - (c')^2 - 2z^2 + c^2(c')^2 + z^4}{2cc'z^2}$$

The critical points in the theorem are the endpoints of the arcs determined by the divergence guaranteeing conditions

$$\Im(\Delta_4(z)) = 0, \ |\Re(\Delta_4(z))| < 1.$$

THEOREM 2.9 *If C is a simple c-convergence region for c.f.'s (2.1) and if c and c' are elements of C then the points*

$$1 - \frac{c^2}{1+c'}, \ 1 - \frac{c^2}{1-c'}, \ -1 + \frac{c^2}{1+c'}, \ -1 + \frac{c^2}{1-c'}$$

cannot be interior points of C.

Proof: The proof of this result is similar to the proof of the preceding result, the difference being that we take a different choice of the c_i in the formula for Δ_4. Here we set $c_1 = c_3 = c$, $c_2 = c'$, $c_4 = z$ in the formula. Again the critical points in this theorem are the endpoints of arcs which guarantee the divergence of the associated c.f. of period 4 when z is a point on one of these arcs.

THEOREM 2.10 *Let $c + 1/2$ and $c - 1/2$, where $\Im(c) \neq 0$, be two elements of a simple c-convergence region C for continued fractions (2.1). Let $K = K(ih, \sqrt{h^2 + 1/4})$, where $h = (|c^2| - 1/4))/(2\Im(c))$, be the circle with center ih and radius $\sqrt{h^2 + 1/4}$ which contains (and is therefore determined by) the three points $-1/2, c, 1/2$. In addition, let D_c denote the arc of K which contains c and which has the endpoints $\psi(c)$ and $-\psi(-c)$, where*

$$\psi(z) := \frac{1}{\sqrt{2}} \frac{z + \sqrt{2}/4}{z + \sqrt{2}/2}.$$

Then $z - 1/2$ and $z + 1/2$ cannot both be elements of C if z is an interior point of the arc D_c different from c.

Proof: To prove this theorem we again apply Theorem 2.3 to certain c.f.'s (2.1) of period 4. Here we set

$$c_1 = c - 1/2, \ c_2 = c + 1/2, \ c_3 = z - 1/2, \ c_4 = z + 1/2,$$

where c is fixed and $\Im(z) \neq 0$. If we make these substitutions for c_i, $i = 1, 2, 3, 4$, in the formula for Δ_4, then Δ_4 as a function of z, after some straightforward but tedious calculations, can be shown to be

$$\Delta_4(z) = 2 - \frac{1}{2} \frac{(z+1/2)(c-1/2)}{(z-1/2)(c+1/2)} - \frac{1}{2} \frac{(z-1/2)(c+1/2)}{(z+1/2)(c-1/2)}. \tag{2.6}$$

Now let $w = \phi(z)$, where $\phi(z)$ is the linear fractional transformation

$$\phi(z) = \frac{(z + 1/2)(c - 1/2)}{(z - 1/2)(c + 1/2)},$$

and set $\Delta_4 = x$, where $-1 < x < 1$. Then equation (2.6) becomes

$$x = 2 - \frac{w}{2} - \frac{1}{2w}, \quad \text{or equivalently,} \quad w = 2 - x \pm \sqrt{(2 - x)^2 - 1}.$$

It is easily seen that w ranges over the two open intervals $(3 - 2\sqrt{2}, 1)$, $(1, 3 + 2\sqrt{2})$ as x ranges over the interval $(-1, 1)$. Thus , in terms of w, it follows from Theorem 2.3 that the periodic c.f. under consideration diverges if $w \in I := (3 - 2\sqrt{2}, 3 + 2\sqrt{2})$, except possibly for $w = 1$. Our task now is to find the arc in the z-plane which maps onto the interval I under the transformation $\phi(z)$. Thus we consider the inverse transformation $z = \phi^{-1}(w)$, where

$$\phi^{-1}(w) = \frac{1}{2} \frac{(c + 1/2)w + (c - 1/2)}{(c + 1/2)w - (c - 1/2)}.$$

Under this transformation we see by inspection that the points $0, 1, \infty$ are mapped into the points $-1/2, c, 1/2$, respectively. Therefore, since $\phi^{-1}(w)$ is an $\ell.f.t.$, it follows that $\phi^{-1}(w)$ maps the real axis onto the circle determined by the three points $-1/2, c, 1/2$. If we let $K(c, r)$ denote the circle with center c and radius r, it is easily verified that this circle is the circle $K = K(h, \sqrt{h^2 + 1/4})$, where

$$h = \frac{|c|^2 - 1/4}{2\Im(c)}.$$

Thus the interval I must map onto an arc D_c with end points $\phi^{-1}(3 \pm 2\sqrt{2})$ of the circle K. To determine which of the two arcs with these end points is D_c we note that $1 \in I$ and $\phi^{-1}(1) = c$. Hence D_c must be the arc which contains the point c. Easy computations will show that

$$\phi^{-1}(3 + 2\sqrt{2}) = \frac{1}{\sqrt{2}} \left[\frac{c + \sqrt{2}/4}{c + \sqrt{2}/2} \right]; \quad \phi^{-1}(3 - 2\sqrt{2}) = -\frac{1}{\sqrt{2}} \left[\frac{c - \sqrt{2}/4}{c - \sqrt{2}/2} \right].$$

With this our proof is complete.

REMARKS 2.1 We would now like to make some comments about the role the preceding results have played in connection with our study of strip convergence regiojns for c.f.'s $1 + K(-c_n^2/1)$. We are interested in knowing what infinite strip regions C in the complex plane have the property that if $c_n \in C$ $\forall n$ and the c_n satisfy a certain necessary growth condition, then $1 + K(-c_n^2/1)$ converges. Such regions fall into the category of conditional c-convergence regions (see Definition 3.1). From Theorem 2.4 we have that no point on the real axis outside of the interval $[-1/2, 1/2]$ can be contained in such a strip. From Theorem 2.5 it is clear that a strip region C has width at most 1. Though Theorem 2.6 is a local neighborhood type of theorem, it does tell us that some strip regions must have width less than 1 at certain heights. Theorems 2.7, 2.8, 2.9, and 2.10 were all designed to help us gain information about bending conditions that convergence strips must satisfy. We have found Theorem 2.8 especially useful for deriving global bending conditions and Theorem 2.10 especially useful

for deriving local bending conditions. The implications of these two theorems had much to do with our discovery of the families of convergence regions that we give in the next two sections. For example, using Theorem 2.8, it can be shown that if $r = f(\theta)$, $\alpha - \pi/2 < \theta < \alpha + \pi/2$, where f is continuous, describes one of the boundaries of an infinite c-convergence strip of width 1 which is symmetric about the origin and which contains a line through the origin (not the x-axis) in its interior, then $f(\theta)$ must be logarithmically convex. Theorem 2.10 imposes restrictions on the slope at a point c on the center curve of a strip of constant width 1 relative to the slope at c of the circle that passes through the points $-1/2, c, 1/2$. Other conclusions in the spirit of those mentioned can be drawn from the results in this section, but for the sake of length of this paper and in order to meet certain deadlines we shall not continue our investigation in this area at this time.

3   WORPITZKY STRIPS

Let the angle α satisfy the condition $0 \leq \alpha < \pi/3$ and let $E(\alpha)$ be the strip region of width 1 defined by

$$E(\alpha) := C(\alpha) \cup -C(\alpha), \tag{3.1}$$

where $C(\alpha)$ is the closed region bounded by the following arcs described in rectangular coordinates

$$\left.\begin{array}{lll}
L_1(\alpha) & : & x\cos\alpha + y\sin\alpha = 1/2, \ -\infty < x \leq (1/2)\cos\alpha \\
C_1(\alpha) & : & y = \sqrt{1/4 - x^2}, \ (1/2)\cos\alpha \leq x \leq 1/2 \\
C_2(\alpha) & : & y = -\sqrt{1/4 - (x-1)^2}, \ 1/2 \leq x \leq 1 - (1/2)\cos\alpha \\
L_2(\alpha) & : & x\cos\alpha + y\sin\alpha = \cos\alpha - 1/2, \ 1 - (1/2)\cos\alpha \leq x < \infty \\
L_3(\alpha) & : & x\cos\alpha + y\sin\alpha = 0.
\end{array}\right\} \tag{3.2}$$

In polar coordinates

$$\left.\begin{array}{lll}
L_1(\alpha) & : & r = (1/2)\sec(\theta - \alpha), \ \alpha \leq \theta < \alpha + \pi/2 \\
C_1(\alpha) & : & r = 1/2, \ 0 \leq \theta \leq \alpha \\
C_2(\alpha) & : & r = \cos\theta - \sqrt{\cos^2\theta - 3/4}, \ -\pi/6 < \theta_0 \leq \theta \leq 0 \\
L_2(\alpha) & : & r = (\cos\alpha - 1/2)\sec(\theta - \alpha), \ \alpha - \pi/2 < \theta \leq \theta_0 \\
L_3(\alpha) & : & |\theta - \alpha| = \pi/2 \\
\theta_0 & : & = \arctan[(\sin\alpha)/(\cos\alpha - 2)].
\end{array}\right\} \tag{3.3}$$

Note that the region $E(\alpha)$ contains the Worpitzky disk $|z| \leq 1/2$. After restricting ϵ by $0 < \epsilon < \sin^2(\alpha/2)$, we define the modified strip $E(\alpha, \epsilon)$ by

$$E(\alpha, \epsilon) := C(\alpha, \epsilon) \cup -C(\alpha, \epsilon), \tag{3.4}$$

where $C(\alpha, \epsilon)$ is the closed region bounded by the following arcs described in rectangular coordinates

$$\left.\begin{array}{lll}
L_1(\alpha, \epsilon) & : & x\cos\alpha + y\sin\alpha = 1/2 - \epsilon, \ -\infty < x \leq (1/2 - \epsilon)\cos\alpha \\
C_1(\alpha, \epsilon) & : & y = \sqrt{(1/2 - \epsilon)^2 - x^2}, \ (1/2 - \epsilon)\cos\alpha \leq x \leq 1/2 - \epsilon \\
C_2(\alpha, \epsilon) & : & y = -\sqrt{(1/2 + \epsilon)^2 - (x-1)^2}, \ 1/2 - \epsilon \leq x \leq 1 - (1/2 + \epsilon)\cos\alpha \\
L_2(\alpha, \epsilon) & : & x\cos\alpha + y\sin\alpha = \cos\alpha - 1/2 - \epsilon, \ 1 - (1/2 + \epsilon)\cos\alpha < \infty \\
L_3(\alpha, \epsilon) & : & = L_3(\alpha).
\end{array}\right\} \tag{3.5}$$

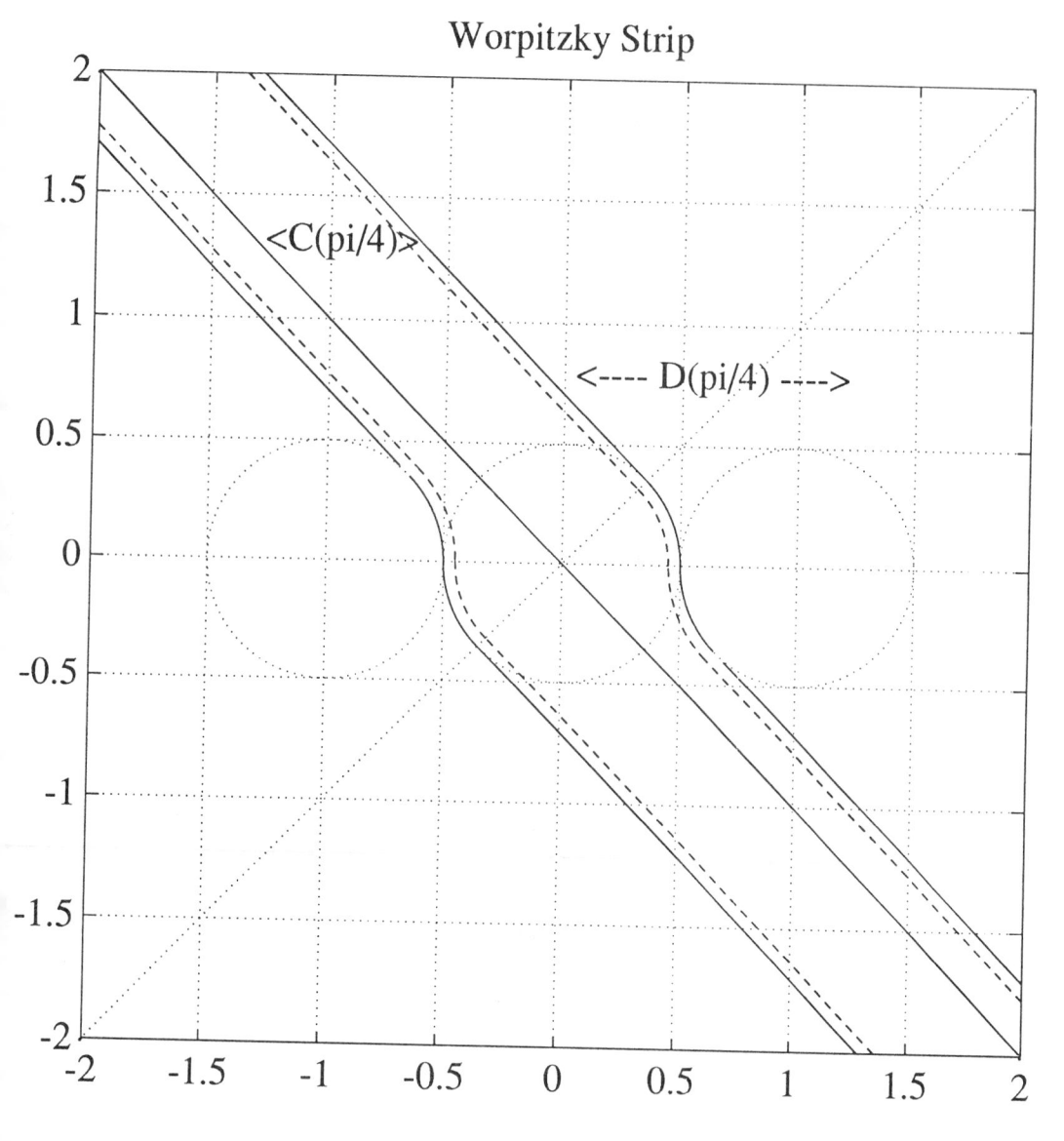

Worpitzky Strip

Figure 3.1 This is a graph of a bounded portion (solid line boundary arcs) of the Worpitzky strip region $E(\pi/4) = C(\pi/4) \cup -C(\pi/4)$. The closed region bounded by the solid line through the origin and the upper dashed line arc is $C(\pi/4, .05)$. The closed strip region bounded by the dashed line boundary arcs is the region $E(\pi/4, .05)$. If the c_n are elements of $E(\pi/4)$, then the approximants of $1 + K(-c_n^2/1)$ are in $D(\pi/4)$.

In polar coordinates

$$
\left.
\begin{array}{lll}
L_1(\alpha,\epsilon) & : & r = (1/2 - \epsilon)\sec(\theta - \alpha),\ \alpha \le \theta < \alpha + \pi/2 \\
C_1(\alpha,\epsilon) & : & r = 1/2 - \epsilon,\ 0 \le \theta < \alpha \\
C_2(\alpha,\epsilon) & : & r = \cos\theta - \sqrt{\cos^2\theta + (1/2+\epsilon)^2 - 1},\ -\pi/6 < \theta_1 \le 0 \\
L_2(\alpha,\epsilon) & : & r = (\cos\alpha - 1/2 - \epsilon)\sec(\theta - \alpha),\ \alpha - \pi/2 < \theta \le \theta_1 \\
\theta_1 & : & = \arctan[(1 + 2\epsilon)(\sin\alpha)/((1 + 2\epsilon)(\cos\alpha) - 2)] \\
L_3(\alpha,\epsilon) & : & |\theta - \alpha| = \pi/2.
\end{array}
\right\} \tag{3.6}
$$

Let $D(\alpha)$ denote closed unbounded region containing the point 1 with boundary $\partial D(\alpha)$ given by the following composition of arcs (see Figure)

$$
\partial D(\alpha) := L_1(\alpha) \cup C_1(\alpha) \cup C_2(\alpha) \cup L_2(\alpha). \tag{3.7}
$$

Finally, let $C^*(\alpha,\epsilon)$ denote the region that consists of all elements of $C(\alpha,\epsilon)$ that are in the open upper half plane $\Im(z) > 0$ plus all elements of $C(\alpha)$ that are in the closed lower half plane $\Im(z) \le 0$, that is,

$$
C^*(\alpha,\epsilon) := [C(\alpha,\epsilon) \cap \{z \mid \Im(z) > 0\}] \cup [C(\alpha) \cap \{z \mid \Im(z) \le 0\}]. \tag{3.8}
$$

We are now ready to give three useful lemmas.

LEMMA 3.1 *Let H be the closed half plane not containing the origin and bounded by the line L defined by*

$$
x\cos\omega + y\sin\omega = p;\ 0 \le \omega < 2\pi,\ p > 0
$$

in rectangular coordinates or by

$$
r(\theta) := p\sec(\theta - \omega)
$$

in polar coordinates. If all square roots are chosen to lie in the closed half plane bounded by the line $x\cos\omega + y\sin\omega = 0$ and containing the point $pe^{i\omega}$, then the geometric mean of any two points of L lies in H, i.e. if $z_1, z_2 \in L$, then $\sqrt{z_1 z_2} \in H$.

Proof: It is sufficient to show that

$$
\sqrt{r(\theta_1)r(\theta_2)} \ge r((\theta_1 + \theta_2)/2),\ \omega - \pi/2 < \theta_1, \theta_2 < \omega + \pi/2,
$$

and this inequality follows from the following sequence of inequalities:

$$
\begin{array}{rcl}
1 & \ge & \cos(\theta_1 - \theta_2) \Rightarrow \\
1 & \ge & \cos(\theta_1 - \omega)\cos(\theta_2 - \omega) + \sin(\theta_1 - \omega)\sin(\theta_2 - \omega) \Rightarrow \\
2\cos(\theta_1 - \omega)\cos(\theta_2 - \omega) & \le & 1 + \cos(\theta_1 - \omega)\cos(\theta_2 - \omega) - sin(\theta_1 - \omega)\sin(\theta_2 - \omega) \Rightarrow \\
\cos(\theta_1 - \omega)\cos(\theta_2 - \omega) & \le & \cos^2((\theta_1 + \theta_2)/2 - \omega) \Rightarrow \\
\sec(\theta_1 - \omega)\sec(\theta_2 - \omega) & \ge & \sec^2((\theta_1 + \theta_2)/2 - \omega).
\end{array}
$$

With this our proof is complete.

LEMMA 3.2 *If the c_n of the continued fraction $1 + K(-c_n^2/1)$ are contained in $C(\alpha)$, then its approximants f_n are contained in $D(\alpha)$.*

Proof: The proof consists in showing that

$$C(\alpha) = \bigcap_{z \in D(\alpha)} (z(1 - D(\alpha))^{1/2},$$

where the square root is chosen in the half plane containing the point $z = 1$ and bounded by line $L_3(\alpha)$. The facts are that D is starlike with respect to the origin and $1 - z \in \partial D$ whenever $z \in D$. Essentially one shows that the geometric mean of any two points in ∂D is in D under the latter restriction on the square root. If we set

$$r = f(\theta), \quad \alpha - \pi/2 < \theta < \alpha + \pi/2,$$

where $f(\theta)$ is determined by formulas (3.3), this amounts to showing that for each fixed θ

$$\sqrt{f(\theta - \delta)f(\theta + \delta)} \geq f(\theta) \tag{3.9}$$

for all $\delta > 0$ such that

$$\alpha - \pi/2 < \theta - \delta < \theta + \delta < \alpha + \pi/2.$$

To be convinced that (3.9) is satisfied, we suggest the following plan of attack. Let R_θ denote the ray $\arg z = \theta$ and consider the four cases in which R_θ intersects one-by-one each of the four arcs

$$L_1(\alpha), \; L_1(\alpha), \; L_1(\alpha), \; L_1(\alpha),$$

respectively. By making heavy use of Lemma 3.1, by noting there is equality in (3.9) for small δ when R_θ intersects $C_1(\alpha)$ at a point other than an endpoint, by employing the concavity property of C_α, and by using the starlikeness of ∂D with respect to the origin, it is a tedious but not difficult process to verify (3.9) under our restrictions on the parameters involved. We end our proof with these comments and omit the details of the proof.

Our next lemma is a technical lemma similar to Lemma 5.1 that appears in Thron (1943).

LEMMA 3.3 *Let two sequences $\{c_n\}$ and $\{a_n\}$ be given, where $|a_n| \leq a \in \mathcal{R} \; \forall n \in N$. Let $b \neq 0$ be a complex number. Further, let the sequences $\{b_n\}, \{\hat{b}_n\}$ be defined by*

$$b_0 = \hat{b}_0 = 1, \quad \frac{1}{b_n b_{n-1}} = -c_n^2, \quad \frac{1}{\hat{b}_n \hat{b}_{n-1}} = -(bc_n + a_n)^2, \; n \geq 1,$$

respectively. Then the two series $\sum |b_n|, \sum |\hat{b}_n|$ converge and diverge together.

Proof: It is clear that

$$\frac{\hat{b}_n \hat{b}_{n-1}}{b_n b_{n-1}} = \frac{c_n^2}{(bc_n + a_n)^2}.$$

Let $u_n = b_n/\hat{b}_n$. Then

$$u_n u_{n-1} = [b(1 + a_n/(bc_n))]^{-2}$$
$$u_{n+1} u_n = [b(1 + a_{n+1}/(bc_{n+1}))]^{-2}$$
$$\frac{u_{n+1}}{u_{n-1}} = \frac{[(1 + a_n/(bc_n))]^2}{[(1 + a_{n+1}/(bc_{n+1}))]^2} \implies$$
$$u_{2n+1} = u_1 \frac{\prod_{k=1}^n [(1 + a_{2k}/(bc_{2k}))]^2}{\prod_{k=1}^n [(1 + a_{2k+1}/(bc_{2k+1}))]^2} \implies$$
$$u_{2n} = u_0 \frac{[\prod_{k=1}^n (1 + a_{2k-1}/(bc_{2k-1}))]^2}{[\prod_{k=1}^n (1 + a_{2k}/(bc_{2k}))]^2}.$$

It is true that

$$\frac{1}{|c_n|} = \sqrt{|b_n||b_{n-1}|} \le \frac{|b_n| + |b_{n-1}|}{2}.$$

So if $\sum |b_n| < \infty$, then $\sum 1/|c_n| < \infty$. From the last two expressions above for u_{2n+1} and u_{2n} we have

$$|u_{2n+1}| \le |u_1| \frac{[\prod_{k=1}^n (1 + a/(|b||c_{2k}|))]^2}{[\prod_{k=1}^n (1 - a/(|b||c_{2k+1}|))]^2} \le M_1 < \infty$$

since the products

$$\prod_{k=1}^n (1 + a/(|b||c_{2k}|))], \quad \prod_{k=1}^n (1 - a/(|b||c_{2k+1}|))]$$

converge. Similarly, $|u_{2n}| \le M_2 < \infty$. Hence $|\hat{b}_n| \le M|b_{2n}|$, where $M = \max(M_1, M_2)$. Thus the convergence of $\sum |b_n| \implies$ the convergence of $\sum |\hat{b}_n|$. All we have to do is call $\hat{c}_n = bc_n + a_n$ so $c_n = \hat{c}_n/b - a_n/b$ and replace b by $1/b$ and a_n by $-a_n/b$ in our argument to get $|b_n| \le \rho|\hat{b}_n|$, etc. This completes our proof of this lemma. We now give a definition of what is meant by a conditional c-convergence region for continued fractions $1 + K(-c_n^2/1)$.

DEFINITION 3.1 *A subset Ω of the set of complex numbers $\mathcal{C}$ is called a* conditional c-convergence region *for continued fractions $1 + K(-c_n^2/1)$ if*

$$c_n \in \Omega, \ n = 1, 2, 3, \ldots, \quad and \quad \sum |b_n| = \infty$$

are sufficient for the convergence of $1 + K(-c_n^2/1)$. Here b_n is defined by

$$b_0 = 1, \ b_n b_{n-1} = -c_n^{-2}, \ n \ge 1.$$

The condition $\sum |b_n| = \infty$ is equivalent to the condition

$$\sum_{n=1}^{\infty} \prod_{k=1}^n |c_k^2(z)|^{(-1)^{n+k-1}} = \infty.$$

We are now ready to give our main theorem of this section.

THEOREM 3.1 *The region $C^*(\alpha, \epsilon)$ (or its conjugate $\overline{C}^*(\alpha, \epsilon)$) is a conditional c-convergence region for the continued fraction $K := 1 + K(-c_n^2/1)$.*

Proof: Let $c_n \in C^*(\alpha, \epsilon)$. Define analytic functions $c_n(z)$ by

$$c_n(z) = \begin{cases} c_n & \text{if } \Re(c_n e^{-i\alpha}) \le (1/2)\cos\alpha \\ c_n + e^{i\alpha}z & \text{otherwise.} \end{cases}$$

Let the domain G be defined by

$$G = \{ z \mid -\epsilon/\sqrt{2} < \Im(z) < \epsilon/\sqrt{2}, 1/2 - (3/2)\cos\alpha < \Re(z) < \epsilon/\sqrt{2} \},$$

where $0 < \epsilon$(fixed) $< \sin^2(\alpha/2)$. Then the approximants of $K(z) := 1 + K(-c_n^2(z)/1)$ are in $D(\alpha)$ if $z \in G$. In particular, if $z \in G$ and $\Re(z) = (1/2)\cos\alpha + \epsilon - 1/2$, then $c_n(z)$ satisfies $\Re(e^{-i\alpha}c_n(z)) \le (1/2)\cos\alpha$ so $K(z)$ converges conditionally for these values of z by Theorem 1.1. Hence, by the Stieltjes-Vitali Theorem, $K(z)$ converges uniformly on

every compact subset of G, in particular for $z = 0$. Therefore K converges conditionally if $c_n \in C^*(\alpha, \epsilon), n \geq 1$. The necessary condition that

$$\sum_{n=1}^{\infty} \prod_{k=1}^{n} |c_k^2(z)|^{(-1)^{n+k-1}}$$

diverge for these special values of $z \in G$ when the same series diverges with $c_k(z)$ replaced by c_k diverges follows from Lemma 3.3. Symmetry arguments can be used to show that $\overline{C}^*(\alpha, \epsilon)$ is also a conditional c-convergence region for K.

4 TRANSCENDENTAL STRIPS

Let d be a real number such that $0 < d \leq 1/2$ and let $E(d)$ be the strip region of width 1 defined by

$$E(d) := C(d) \cup -C(d), \tag{4.1}$$

where $C(d)$ is the closed region bounded by the y-axis and the arc $T(d)$ described in rectangular coordinates by

$$T(d) : \quad y = \tan\left[\pi(1/4 - x/2)/d\right], \ 0 < d \leq 1/2, \ |x - 1/2| < d. \tag{4.2}$$

Now let $0 < \epsilon < 1/2$ and let $C(d, \epsilon)$ denote the closed region bounded on the left by the line $x = 0$ and on the right by the arc $T(d, \epsilon)$ parameterized by $z(t) = x(t) + iy(t)$, where

$$T(d, \epsilon) : \quad x(t) = t + \epsilon f'(t)/\sqrt{1 + (f'(t))^2}, \ y(t) = f(t) - \epsilon/\sqrt{1 + (f'(t))^2}, \ |t - 1/2| < d \tag{4.3}$$

with $f(t) = \tan\left[\pi(1/4 - t/2)/d\right]$. That is, $C(d, \epsilon)$ is a proper subset of $C(d)$ with the same left boundary as $C(d)$, but whose right boundary is at a distance epsilon from the right boundary of $C(d)$. We let $D(d)$ denote the unbounded region which contains the point 1 and which has the arc $T(d)$ as its boundary. Finally, we define the region $C^*(d, \epsilon)$ by

$$C^*(d, \epsilon) := \{C(d) \cap \{z \mid \Im(z) \geq 0\}\} \cup \{C(d, \epsilon) \cap \{z \mid \Im(z) < 0\}\}. \tag{4.4}$$

That is $C^*(d, \epsilon)$ is the region that consists of all points of $C(d)$ that lie in the closed upper half plane plus all points of $C(d, \epsilon)$ that lie in the open lower half plane. The following sequence of lemmas are designed to aid us in establishing that, if the c_n in the continued fraction $1 + K(-c_n^2/1)$ are in $C(d)$, then its approximants are in $D(d)$.

LEMMA 4.1 *Let $r = f(\theta)$ be positive and continuous for $\theta \in I$, where*

$$I := \{\theta \mid \alpha - \pi/2 < \theta < \alpha + \pi/2\}.$$

Suppose also that $\log f(\theta)$ is convex on I. Then

$$\sqrt{f(\theta_1)f(\theta_2)} \geq f\left(\frac{\theta_1 + \theta_2}{2}\right) \ \forall \theta_1, \theta_2 \in I.$$

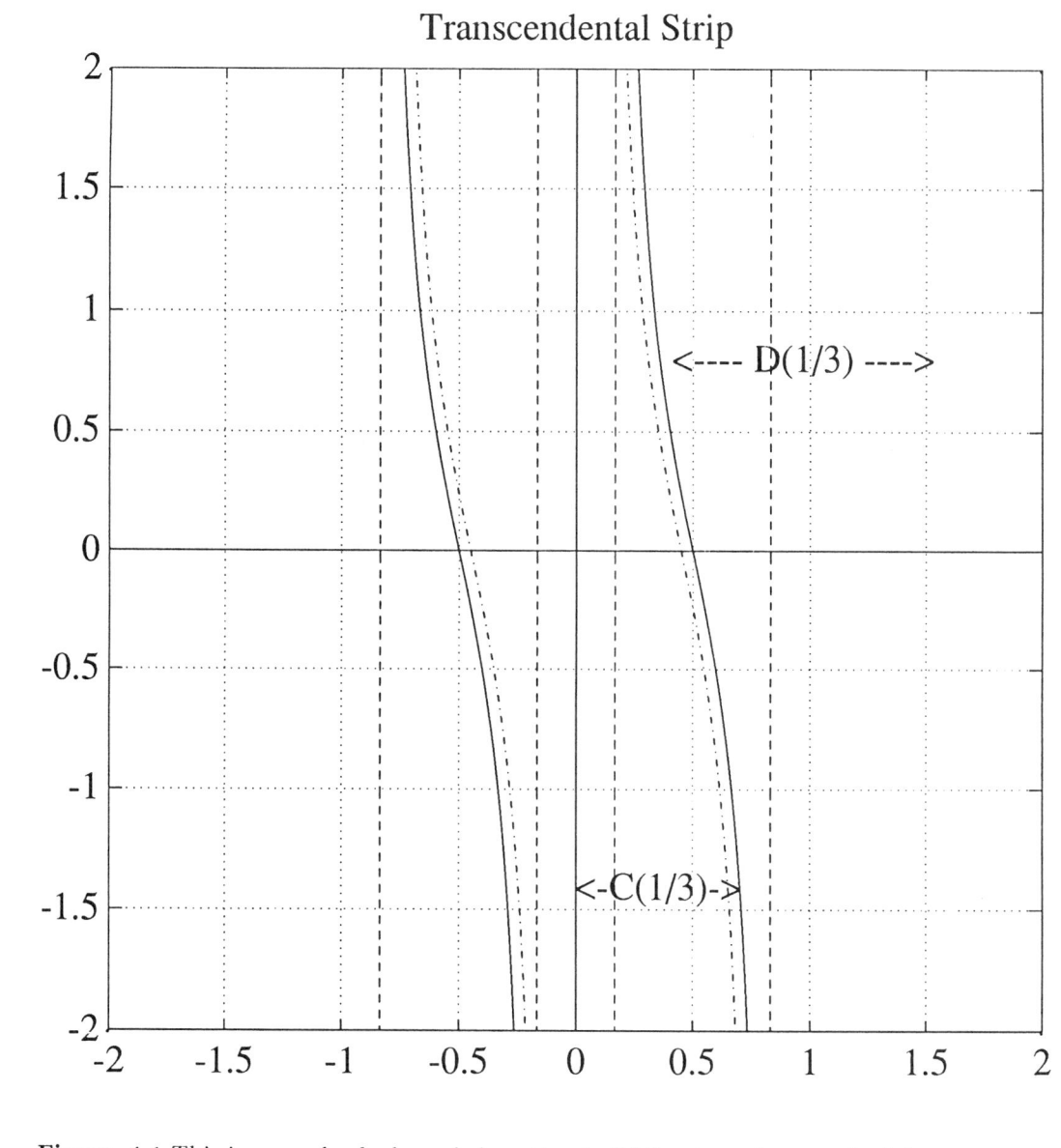

Figure 4.1 This is a graph of a bounded portion (solid line boundary arcs) of the transcendental strip region $E(1/3) = C(1/3) \cup -C(1/3)$. The closed region bounded by the solid line imaginary axis and the dot-dash arc on the right is $C(1/3, .05)$. The closed strip region bounded by the dot-dash boundary arcs is the region $E(1/3, .05)$. If the c_n are elements of $E(1/3)$, then the approximants of $1 + K(-c_n^2/1)$ are in $D(1/3)$.

Proof: Since f is logarithmically convex,

$$\log f\left(\frac{\theta_1 + \theta_2}{2}\right) \leq \frac{1}{2}\log[f(\theta_1)] + \frac{1}{2}\log[f(\theta_2)]$$

$$= \log[f(\theta_1)f(\theta_2)]^{1/2} \Rightarrow$$

$$f\left(\frac{\theta_1 + \theta_2}{2}\right) \leq \sqrt{f(\theta_1)f(\theta_2)},$$

so our lemma is true.

LEMMA 4.2 *Let* $r = f(\theta)$ *be positive and twice-differentiable for all* $\theta \in I$, *where* $I = (\alpha - \pi/2, \alpha + \pi/2)$. *Suppose also that* $\phi'(\theta) \geq 0$, *where* $\phi(\theta) := f'(\theta)/f(\theta)$. *Then*

$$\sqrt{f(\theta_1)f(\theta_2)} \geq f\left(\frac{\theta_1 + \theta_2}{2}\right) \forall \theta_1, \theta_2 \in I.$$

Proof: $f(\theta)$ is logarithmically convex and continuous so the result follows from Lemma 4.1.

LEMMA 4.3 *Let* $f(x)$ *be a twice-differentiable real function of* x *on a bounded open interval* $I = (a, b)$, *where* $a \geq 0$ *and let* $f(x) \to +\infty$ *at one endpoint of* I *and* $f(x) \to -\infty$ *at the other endpoint of* I *as* x *approaches these points from inside* I. *Let* $g(\theta) := \sqrt{x^2 + f^2(x)}$ *and* $\psi(\theta) := g'(\theta)/g(\theta)$, *where* θ *is the unique angle which satisfies the conditions*

$$\cos\theta = \frac{x}{\sqrt{x^2 + f^2(x)}}, \quad \sin\theta = \frac{f(x)}{\sqrt{x^2 + f^2(x)}}, \quad -\pi/2 < \theta < \pi/2.$$

Assume, also, that $\theta'(x) < 0 \ \forall \ x \in I$ *or* $\theta'(x) > 0 \ \forall \ x \in I$. *Then, if* $x \in I$ *and* $y = f(x)$, *we have that*

$$\frac{y''}{1 + (y')^2} \cdot \frac{x^2 + y^2}{xy' - y} \leq 1 \Rightarrow \psi'(\theta) \geq 0.$$

Proof: Let $\alpha(x) = \arctan f'(x)$. Then on the one hand

$$\frac{d\alpha}{d\theta} = 1 - \frac{\psi'(\theta)}{1 + \psi^2(\theta)},$$

and on the other hand

$$\frac{d\alpha}{d\theta} = \frac{d\alpha}{dx} \cdot \frac{dx}{d\theta} = \frac{y''}{1 + (y')^2} \cdot \frac{x^2 + y^2}{xy' - y}.$$

Note that the term $xy' - y$ in the last formula is not zero since

$$\frac{d\theta}{dx} = \frac{xy' - y}{x^2 + y^2} \neq 0$$

by our hypotheses. Thus the conclusion of our lemma follows.

LEMMA 4.4 *If*

$$y = f(x) = \tan[\pi(1/2 - x)/(2d)], \quad 0 < d \leq 1/2, \quad and \ |x - 1/2| < d,$$

then $y = f(x)$ *satisfies the convexity condition*

$$\frac{y''}{1 + (y')^2} \cdot \frac{x^2 + y^2}{xy' - y} \leq 1 \quad or \quad \frac{y''}{xy' - y} \leq \frac{1 + (y')^2}{x^2 + y^2}. \tag{4.5}$$

Proof:

$$y = \tan\left[\pi(1/2 - x)/(2d)\right] \implies \tag{4.6}$$

$$y' = -\frac{\pi}{2d}\sec^2\left[\pi(1/2 - x)/(2d)\right] = -\frac{\pi}{2d}(1 + y^2) \implies \tag{4.7}$$

$$y'' = \frac{\pi^2}{d^2}\sec^2\left[\pi(1/2 - x)/(2d)\right] \cdot \tan\left[\pi(1/2 - x)/(2d)\right] \tag{4.8}$$

$$= -\frac{\pi}{d}y'y = \frac{\pi^2}{2d^2}y(1 + y^2)$$

$$y(1/2) = 0, \ y'(1/2) = -\pi/(2d), \ y''(1/2) = 0 \tag{4.9}$$

Inequality (4.5) is equivalent to

$$\frac{y''}{xy' - y} \le \frac{1 + (y')^2}{x^2 + y^2} \tag{4.10}$$

Let

$$F(x) = xy' - y. \tag{4.11}$$

Then

$$F'(x) = xy'' \tag{4.12}$$

By (4.9) $F'(1/2) = 0$ and by (4.8) $F'(x) > 0$ if $x < 1/2$ and $F'(x) < 0$ if $x > 1/2$ so F has a maximum at $x = 1/2$. But $F(1/2) = -\pi/(4d) < 0$. Hence, with the aid of (4.8), the left side of (4.10) is ≤ 0 if $x \le 1/2$ so (4.10) is true for these values of x.

If $x > 1/2$, then $y < 0, y'' < 0$, so the left side of (4.10) is positive and (4.10) is equivalent (for $x > 1/2$) to

$$\frac{\pi^2 y(1 + y^2)}{2d^2(xy' - y)} \le \frac{1 + (y')^2}{x^2 + y^2} \tag{4.13}$$

which, with the aid of (4.6), (4.7), and (4.8), becomes

$$0 \le (y - xy')(1 + (y')^2) + \pi^2 y(1 + y^2)(x^2 + y^2)/(2d^2), \tag{4.14}$$

or equivalently,

$$0 \le [y + \pi x(1 + y^2)/(2d)](1 + (y')^2) + \pi^2 y(1 + y^2)(x^2 + y^2)/(2d^2). \tag{4.15}$$

Since $y < 0$), inequality (4.15) will certainly be true if

$$0 \le [y + \pi x(1 + y^2)/(2d)](1 + (y')^2) + \pi^2 y(1 + y^2)(1 + y^2)/(2d^2) \tag{4.16}$$

which is equivalent to

$$0 \le [y + \pi x(1 + y^2)/(2d)][1 + \pi^2(1 + y^2)^2/(4d^2)] + \pi^2 y(1 + y^2)^2/(2d^2) \tag{4.17}$$

which in turn is equivalent to

$$0 \le \left[y + \frac{\pi x}{2d}(1 + y)^2 - \frac{\pi xy}{d}\right]\left[1 + \frac{\pi^2}{4d^2}(1 + y^2)^2\right] + \frac{\pi^2}{2d^2}y(1 + y^2)^2. \tag{4.18}$$

Inequality (4.18) will be true if

$$0 \le y \left[1 - \frac{\pi x}{d}\right] \left[1 + \frac{\pi^2}{4d^2}(1+y^2)^2\right] + \frac{\pi^2}{2d^2}y(1+y^2)^2. \tag{4.19}$$

Since $y(1 - \pi x/d) > 0$, inequality (4.19) will certainly be true if

$$0 \le y \left[1 - \frac{\pi x}{d}\right] \left[\frac{\pi^2}{4d^2}(1+y^2)^2\right] + \frac{\pi^2}{2d^2}y(1+y^2)^2 \tag{4.20}$$

which is the same as

$$0 \le \frac{\pi^2}{4d^2}y(1+y^2)^2(3 - x\pi/d). \tag{4.21}$$

But $x > 1/2$ and $0 < d < 1/2$ so $x/d > 1$ and $3 - x\pi/d < 0$ which implies (4.21) is true since $y < 0$. Since the above steps are reversible for the $x > 1/2$ case, we have completed our proof. The statement of our next lemma contains one of the principal results we have been seeking.

LEMMA 4.5 *Let $D(d)$ be the unbounded closed region bounded on the left by the graph of*

$$y = f(x) = \tan\left[\pi(1/2 - x)/(2d)\right], \ 0 < d \le 1/2, \ |x - 1/2| < d.$$

If $c_n \in C(d), n = 1, 2, \cdots$, then the approximants of $1 + K(-c_n^2/1)$ lie in $D(d)$.

Proof: This lemma follows essentially from Lemmas 4.1-4.4, since $D(d)$ is starlike with respect to the origin and, if $z \in \partial D(d)$, then $1 - z \in \partial D(d)$. So

$$C(d) = \bigcap_{z \in D(d)} (z(1 - D(d))^{1/2},$$

where the square root in the right half plane is chosen.

Now let $0 < \epsilon < 1/2$ and define a new closed region $\hat{C}(d, \epsilon)$ in the right half plane, bounded on the left by the line $x = 0$ and on the right by the curve parameterized by $z(t) = x(t) + iy(t)$, where

$$x(t) = t + \epsilon f'(t)/\sqrt{1 + (f'(t))^2}, \ y(t) = f(t) - \epsilon/\sqrt{1 + (f'(t))^2}, \ |t - 1/2| < d$$

with $f(t) = \tan\left[\pi(1/2 - t/(2d)\right]$. That is, $\hat{C}(d, \epsilon)$ has the same left boundary as $C(d)$ but all points of its right boundary are at a distance $\ge \epsilon$ from the boundary of $C(d)$. We are now ready to define the region $C(d, \epsilon)$ by

$$C(d, \epsilon) := \{C(d) \cap \{z \mid \Im(z) \ge 0\}\} \cup \{\hat{C}(d, \epsilon) \cap \{z \mid \Im(z) < 0\}\}$$

THEOREM 4.1 *The region $C(d, \epsilon)$ (or its conjugate $\overline{C}(d, \epsilon)$) is a conditional c-convergence region for the continued fraction $1 + K(-c_n^2/1)$.*

Proof: Let

$$c_n(z) := \begin{cases} c_n & \text{if } 0 < \Re(c_n) \le 1/2 \\ c_n + z & \text{otherwise,} \end{cases}$$

where $c_n \in C(d, \epsilon)$. Define a domain G by

$$G := \{ z \mid d - 1 < \Re(z) < \epsilon/\sqrt{2} \} \cap \{ z \mid -\epsilon/\sqrt{2} < \Im(z) < \epsilon/\sqrt{2} \}.$$

Note that G contains the origin. If $z \in G$, then the approximants of $1 + K(-c_n^2(z)/1)$ are in $D(d)$. In particular, if $z \in G$ and $\Re(z) = \epsilon - d$, then the $c_n(z)$ satisfy $0 \le \Re(c_n(z)) \le 1/2$, so $1 + K(-c_n^2(z)/1)$ converges for these values of z under the growth assumption on the c_n by Lemma 3.3 and Theorem 1.1 with $\alpha = 0$. By the Stieltjes-Vitali Theorem $1 + K(-c_n^2(z)/1)$ converges for $z = 0$ and thus our proof is complete after we point out that symmetry arguments will show that $\overline{C}(d, \epsilon)$ is also a c-convergence for $1 + K(-c_n^2/1)$.

Our next result says that $D(d)$ of Lemma 4.5 and $D(\alpha)$ of Lemma 3.2 are the smallest value region regions corresponding to the element regions $C(d)$ and $C(\alpha)$, respectively. It parallels a similar result of Thron (1944, p.789).

THEOREM 4.2 *Let z be any point of $D(d)(D(\alpha))$. Then there exists a continued fraction $1 + K(-c_n^2(z)/1)$ whose elements $c_n(z)$ are in $C(d)(C(\alpha))$ and whose limit is z.*

Proof: We shall give a proof for the $D(d), C(d)$ pair only, since the proof for the $D(\alpha), C(\alpha)$ pair is similar. If $z \in \partial D(d)$ then so is $1 - z \in \partial D(d)$; but $z, 1 - z \in C(d)$, also. The continued fraction of period 2

$$1 - \frac{(1-z)^2}{1} \; - \; \frac{z^2}{1} \; - \; \frac{(1-z)^2}{1} \; - \cdots$$

is parabolic and converges to its common fixed point z. Let $z \in D(d)$ and choose z' on the boundary of $D(d)$. Then, since

$$C(d) = \bigcap_{z' \in D(d)} (z'(1 - D(d)))^{1/2},$$

there exists $c' \in C(d)$ such that $(c')^2 = z'(1 - z)$. Since there is a continued fraction $1 + K(-c_n^2/1)$ converging to z', we have $(c')^2/z' = 1 - z$ which implies $z = 1 - (c')^2/z'$ and therefore

$$z = 1 - \frac{(c')^2}{1} \; - \; \frac{c_1^2}{1} \; - \; \frac{c_2^2}{1} \; - \cdots .$$

Thus $D(d)$ is minimal.

REMARKS 4.1 It follows from Theorems 3.1 and 4.1 that the modified strips

$$E^*(\alpha, \epsilon) = C^*(\alpha, \epsilon) \cup -C^*(\alpha, \epsilon); \quad E^*(d, \epsilon) = C^*(d, \epsilon) \cup -C^*(d, \epsilon)$$

are conditional c-convergence regions for c.f.'s (2.1). We conjecture that the ϵ above can be removed or in other words that the Worpitzky strips $E(\alpha)$ and the transcendental strips $E(d)$ are actually conditional c-convergence regions for c.f.'s (2.1). If our conjecture is true, we would have the first examples of such regions of constant width 1 (they cannot be widened by Theorem 2.5) whose boundary arcs are not parallel lines through points $z = \pm 1/2$, respectively, as is the case with the strips S_α in Theorem 1.1. Because the

geometry becomes more involved (especially for the strips $E(d)$) when one tries to use the same method of proof that Thron (1943) used to prove Theorem 1.1, we have up-to-now not been successful in this approach to prove convergence. We have also not been able to prove the conjecture false through the study of periodic continued fractions. Theorem 1.1 appears all the more remarkable to us. We suspect that a new approach will be needed to settle the conjecture and we would be delighted with its resolution by anyone. In this regard the referee has informed us that some results in the doctoral thesis of Córdova (1992), which was recently accepted at the University of Würzburg, may have a bearing on some of our work here. Finally, we would like to point out that Theorems 3.1 and 4.1 also shed more light on the c-convergence neighborhood problem.

REFERENCES

1. Córdova Yévenes, A. (1992). A Convergence Theorem for Continued Fractions, Thesis, University of Würzburg, Am Hubland, 8700 Würzburg, Germany.

2. Jones, W. B. and Thron, W. J. (1968). Convergence of continued fractions, Can. J. Math. , 20: 1037-1055.

3. Jones, W. B. and Thron, W. J. (1980). Continued Fractions: Analytic Theory and Applications, Addison-Wesley, Reading, MA.

4. Lange, L. J. (1960). Divergence, Convergence, and Speed of Convergence of Continued Fractions $1 + K(a_n/1)$, Doctoral Thesis, University of Colorado, Boulder.

5. Lange, L. J. and Thron, W. J. (1960). A two-parameter family of best twin convergence regions for continued fractions, Math. Zeitschr., 73: 295-311.

6. Lorentzen, L. and Waadeland, H. (1992). Continued Fractions with Applications, North-Holland, Amsterdam.

7. Perron, O. (1957). Die Lehre von den Kettenbrüchen, 3rd ed. vol 2, Teubner, Stuttgart.

8. Thron, W. J. (1943). Two families of twin convergence regions for continued fractions, Duke Math. J., 10: 677-685.

9. Thron, W. J. (1944). A family of simple convergence regions for continued fractions, Duke Math. J., 11: 779-791.

10. Thron, W. J. (1963). Convergence of sequences of linear fractional transformations and of continued fractions, J. Indian Math. Soc., 27: 103-127.

11

Continued Fraction Representations for Functions Related to the Gamma Function

L. J. LANGE   Department of Mathematics, University of Missouri, Columbia, Missouri 65211, USA

Dedicated to the memory of Arne Magnus

1   INTRODUCTION

Our fundamental goal in this work is to present and justify a variety of representations for a multitude of complex functions related to the gamma function $\Gamma(z)$. For each function considered we give an explicit representation of it as a Laplace transform of another function, as the limit of a series or product of rational functions, and (most importantly to us) as the limit of one or more continued fractions. A functional equation that it satisfies is also associated with each function studied. With regard to the validity of the representations presented, we have primarily concerned ourselves with studying the convergence behavior and justifying the validity of the continued fraction (c.f.) expansions. We feel that another important contribution we have made in the continued fraction category is that for at least one continued fraction associated with each included special function we give an explicit formula for the modifying sequence $\{r_n\}$ used to determine a Bauer-Muir transform (see Theorem 3.4 for the definition) of the continued fraction. In light of the work of Jacobsen (1986) on the theory of general convergence, the sequences $\{r_n\}$ (which lead to a sequence of modified approximants for the given c.f.) could prove valuable in future studies on the speed of convergence of c.f. expansions. We use a technique to establish convergence which we shall refer to as the *Bauer-Muir transform method*. Simply put, this method amounts to the following: We first establish that a continued fraction K associated with a given function F converges on some subset S of the real line. Next we compute a transform $\hat{K}$ of K and use known results to establish that K and $\hat{K}$ converge to the same value $K(z)$ on S. But in our

cases, this implies that $K(z)$ satisfies a solvable functional relation. This leads to a series or product representation for $K(z)$ and to the conclusion that $K(z) = F(z)$ on S. Finally, it is established that these c.f. and non c.f. representations have analytic extensions to a domain D in the complex plane containing S, so that $K(z) = F(z)$ on D by the Identity Theorem for analytic functions.

Another objective that we had in mind when we wrote this paper was to make it reasonably self contained. In Sections 2 and 3 we have tried to provide the basic definitions, formulas, and theorems that we used to justify our material on our alphabet of functions in Section 4. Most of the continued fraction expansions that we give in Section 4 were derived from expansions (in most cases formal ones) given originally by Ramanujan (1927), Rogers (1907), and Stieltjes (1889, 1890, 1894). We were influenced greatly by the list of beautiful expansions given in the chapter entitled "Stieltjes Summability" in the Wall's 1948 book on continued fractions. For many years we wondered about what conditions on the parameters involved would make these expansions valid and about what more we could say about the limit functions. Our work on δ-fractions in the 1980's (see Lange (1983)) also led to this present study in the sense that it was our quest for δ-fraction expansions in $1/z$ for the Trigamma and Tetragamma functions, $\Psi'(z)$ and $\Psi''(z)$, respectively, that started us off on this endeavor. We believe that much of what we have given in Section 4 has not previously appeared in the literature and that our format for presenting the results there is unique. Finally, for a wealth of related information and results, we refer the reader to the papers or books by Berndt, Lamphere, and Wilson (1985), Berndt (1989), Jacobsen (1990), and Perron (1953).

2   SOME BASIC DEFINITIONS AND RESULTS

Let $A_n(B_n)$ denote the nth numerator(denominator) and let f_n denote the nth approximant of the continued fraction

$$b_0 + \frac{a_1}{b_1} + \frac{a_2}{b_2} + \frac{a_3}{b_3} + \cdots . \tag{2.1}$$

Set

$$h_n = B_n/B_{n-1}; \quad H_n = B_{n+2}/B_n. \tag{2.2}$$

Then

$$\begin{bmatrix} A_{-1} \\ B_{-1} \end{bmatrix} = \begin{bmatrix} 1 \\ 0 \end{bmatrix}, \begin{bmatrix} A_0 \\ B_0 \end{bmatrix} = \begin{bmatrix} b_0 \\ 1 \end{bmatrix}, \begin{bmatrix} A_n \\ B_n \end{bmatrix} = b_n \begin{bmatrix} A_{n-1} \\ B_{n-1} \end{bmatrix} + a_n \begin{bmatrix} A_{n-2} \\ B_{n-2} \end{bmatrix}, n \geq 1. \tag{2.3}$$

$$A_n B_{n-1} - A_{n-1} B_n = (-1)^{n-1} a_1 a_2 \cdots a_n \tag{2.4}$$

$$f_n = \frac{A_n}{B_n} = b_0 + \frac{a_1}{b_1} + \frac{a_2}{b_2} + \cdots + \frac{a_n}{b_n} \tag{2.5}$$

$$h_n = \frac{B_n}{B_{n-1}} = b_n + \frac{a_n}{b_{n-1}} + \frac{a_{n-1}}{b_{n-2}} + \cdots + \frac{a_2}{b_1} \tag{2.6}$$

$$\begin{aligned} H_0 &= b_1 b_2 + a_2 \\ H_1 &= b_1 b_3 + a_3 + \frac{a_2 b_3}{b_1} \\ H_n &= b_{n+1} b_{n+2} + a_{n+2} + \frac{a_{n+1} b_{n+2}}{b_{n+1}} - \frac{a_n a_{n+1} b_{n+2}}{b_n H_{n-2}}, n \geq 2. \end{aligned} \tag{2.7}$$

It follows easily from the above formulas that

$$|f_n - f_{n-1}| = \frac{|a_n||f_n - f_{n-2}|}{|H_{n-2} - a_n|}, \quad n \geq 2 \tag{2.8}$$

$$|f_n - f_{n+1}| = \frac{|H_n||f_n - f_{n+2}|}{|H_n - a_{n+2}|}, \quad n \geq 0 \tag{2.9}$$

and with the aid of these two formulas we obtain

$$|f_{2n} - f_{2n-1}| = \frac{|a_{2n}||f_{2n} - f_{2n-2}|}{|H_{2n-2} - a_{2n}|} \tag{2.10}$$

$$|f_{2n} - f_{2n-1}| = \frac{|H_{2n-1}||f_{2n+1} - f_{2n-1}|}{|H_{2n-1} - a_{2n+1}|} \tag{2.11}$$

$$|f_{2n+1} - f_{2n}| = \frac{|a_{2n+1}||f_{2n+1} - f_{2n-1}|}{|H_{2n-1} - a_{2n+1}|} \tag{2.12}$$

$$|f_{2n+1} - f_{2n}| = \frac{|H_{2n}||f_{2n+2} - f_{2n}|}{|H_{2n} - a_{2n+2}|}. \tag{2.13}$$

Thron (1989) has shown that the latter formulas and many others of a similar type can be derived using the invariance of the cross ratio under a linear fractional transformation. The *even part* of (2.1) has $f_{2n}(n = 0, 1, 2, \ldots)$ has its nth approximant and it is defined by

$$b_0 + \frac{a_1 b_2}{b_1 b_2 + a_2} \quad - \quad \frac{a_2 a_3 b_4}{(b_2 b_3 + a_3)b_4 + b_2 a_4} \quad - \quad \frac{a_4 a_5 b_2 b_6}{(b_4 b_5 + a_5)b_6 + b_4 a_6}$$

$$- \quad \frac{a_6 a_7 b_4 b_8}{(b_6 b_7 + a_7)b_8 + b_6 a_8} - \cdots \quad ; \ (b_{2n} \neq 0 \text{ for } n \geq 1). \tag{2.14}$$

The *odd part* of (2.1) has $f_{2n+1}(n = 0, 1, 2, \ldots)$ as its nth approximant and it is defined by

$$\frac{b_0 b_1 + a_1}{b_1} \quad - \quad \frac{a_1 a_2 b_3 / b_1}{(b_1 b_2 + a_2)b_3 + b_1 a_3} \quad - \quad \frac{a_3 a_4 b_1 b_5}{(b_3 b_4 + a_4)b_5 + b_3 a_5}$$

$$- \quad \frac{a_5 a_6 b_3 b_7}{(b_5 b_6 + a_6)b_7 + b_5 a_7} - \cdots \quad ; \ (b_{2n-1} \neq 0 \text{ for } n \geq 1). \tag{2.15}$$

Two continued fractions $b_0 + K(a_n/b_n)$ and $\hat{b}_0 + K(\hat{a}_n/\hat{b}_n)$ with nth approximants f_n and $\hat{f}_n$, respectively, are said to be *equivalent* if $f_n = \hat{f}_n$, $n = 0, 1, 2, \ldots$. We denote the equivalence of two continued fractions by writing

$$b_0 + K(a_n/b_n) \sim \hat{b}_0 + K(\hat{a}_n/\hat{b}_n).$$

The following theorem gives necessary and sufficient conditions for two continued fractions to be equivalent.

THEOREM 2.1 *Continued fractions $b_0 + K(a_n/b_n)$ and $\hat{b}_0 + K(\hat{a}_n/\hat{b}_n)$ are equivalent iff there exists a sequence of non-zero constants $\{\rho_n\}$ with $\rho_0 = 1$ such that*

$$\begin{bmatrix} \hat{a}_n \\ \hat{b}_n \end{bmatrix} = \begin{bmatrix} \rho_n \rho_{n-1} a_n, & n = 1, 2, 3, \ldots \\ \rho_n b_n & n = 0, 1, 2, \ldots \end{bmatrix}. \tag{2.16}$$

The mapping from $b_0 + K(n/b_n)$ to $\hat{b}_0 + K(\hat{a}_n/\hat{b}_n)$ defined by (2.16) is called an *equivalence transformation*. For a proof of this result see Jones and Thron (1980, p.31) or Perron (1957, p.5).

A continued fraction $\hat{b}_0 + K(\hat{a}_n/\hat{b}_n)$ with sequence of approximants $\{\hat{f}_n\}$ is said to be an *extension* (*contraction*) of the continued fraction $b_0 + K(a_n/b_n)$ with sequence of approximants $\{f_n\}$ if $\{f_n\}$ ($\{\hat{f}_n\}$) is a subsequence of $\{\hat{f}_n\}$ ($\{f_n\}$). We shall use the following result often in this paper.

THEOREM 2.2 *If the section*

$$\frac{a_k}{b_k} + \frac{a_{k+1}}{b_{k+1}}$$

in the continued fraction $b_0 + K(a_n/b_n)$ is replaced by

$$\frac{a_k}{b_k - \rho} + \frac{\rho}{1} - \frac{a_{k+1}/\rho}{b_{k+1} + a_{k+1}/\rho}; \ \rho \neq 0 \in \mathcal{C}$$

while the other partial numerators and denominators are left invariant and if $f_n = A_n/B_n$ and $\hat{f}_n$ denote the nth approximant of the given and modified continued fraction, respectively, then

$$\hat{f}_n = \begin{cases} f_n, & 0 \leq n \leq k-1 \\ \frac{A_k - \rho A_{k-1}}{B_k - \rho B_{k-1}}, & n = k \\ f_{n-1}, & n \geq k+1. \end{cases}$$

If, instead, the section

$$b_0 + \frac{a_1}{b_1}$$

is replaced by

$$b_0 - \rho + \frac{\rho}{1} - \frac{a_1/\rho}{b_1 + a_1/\rho},$$

then

$$\hat{f}_n = \begin{cases} \frac{A_0 - \rho}{B_0} = f_0 - \rho, & n = 0 \\ f_{n-1}, & n \geq 1. \end{cases}$$

See Jones and Thron(1980, p.430) or Perron (1957, p.16) for theorems essentially equivalent to the one above along with their proofs.

3   SOME USEFUL THEOREMS

THEOREM 3.1 (Seidel-Stern-Broman-Pringsheim) *The continued fraction*

$$b_0 + \mathop{\mathrm{K}}_{n=1}^{\infty} \frac{a_n}{b_n}; \ a_n > 0, \ n \geq 1; \ b_n \geq 0, \ n \geq 0 \tag{3.1}$$

converges iff the series

$$\sum_{n=1}^{\infty} b_n \prod_{k=1}^{n} a_k^{(-1)^{n-k+1}}$$

diverges. In particular, (3.1) converges if the series

$$\sum_{n=1}^{\infty} \sqrt{\frac{b_n b_{n+1}}{a_{n+1}}}$$

diverges.

For more on these classical results see Satz 2.10 and Satz 2.11 in Perron (1957).

THEOREM 3.2 (Alternating Continued Fractions) *A continued fraction of the form*

$$\mathop{K}_{n=1}^{\infty} \frac{(-1)^{n+\delta} a_n}{1}; \ a_n > 0, \ \delta \ (fixed) = 0 \ or \ 1 \tag{3.2}$$

is called an alternating continued fraction. The alternating c.f. (3.2), where the a_n satisfy the additional conditions

$$1 + (-1)^{\delta} a_{1+\delta} \geq 0, \ 1 + a_{2n+\delta} - a_{2n+1+\delta} \geq 0, \ n \geq 1 \tag{3.3}$$

converges iff its odd part (even part) converges when $\delta = 0$ ($\delta = 1$). If, in addition to (3.3), the series

$$\sum_{n=1}^{\infty} \sqrt{\frac{(1 + a_{2n+\delta} - a_{2n+1+\delta})(1 + a_{2n+2+\delta} - a_{2n+3+\delta})}{a_{2n+1+\delta} \, a_{2n+2+\delta}}} \tag{3.4}$$

diverges, then (3.2) converges.

Proof: We shall sketch a proof of this theorem for the $\delta = 1$ case. The proof for the $\delta = 0$ case is similar. The even part of (3.2) is

$$\frac{a_1}{1 - a_2} + \frac{a_2 a_3}{1 + a_3 - a_4} + \frac{a_4 a_5}{1 + a_5 - a_6} + \cdots . \tag{3.5}$$

Let $A_n(B_n)$ denote the nth numerator(denominator) of (3.2). Then

$$B_2/B_0 = 1 - a_2; \ B_{2n+2}/B_{2n} = 1 + a_{2n+1} - a_{2n+2} + a_{2n} a_{2n+1}/(B_{2n}/B_{2n-2}).$$

So, under our assumptions (3.3), we have that $B_{2n}/B_{2n-2} \geq 0$ for all $n \geq 1$. From the recurrence relations for the A_n, B_n we obtain, after letting $f_n = A_n/B_n$,

$$\begin{aligned} |f_{2n} - f_{2n+1}| &= \frac{|B_{2n+2}/B_{2n}||f_{2n} - f_{2n+2}|}{|B_{2n+2}/B_{2n} + a_{2n+2}|} \\ &\leq |f_{2n} - f_{2n+2}| \end{aligned}$$

since $B_{2n+2}/B_{2n} \geq 0$ and $a_{2n+2} > 0$. Hence (3.2) converges if (3.5) converges. By Theorem (3.1) above, (3.5) converges under the conditions (3.3) if the series (3.4) diverges. This completes our proof. Perron (1911; 1957, p.52) has given a number of other results concerning alternating continued fractions.

THEOREM 3.3 (Stieltjes) *Let*

$$\frac{a_1 z}{1} + \frac{a_2 z}{1} + \frac{a_3 z}{1} + \cdots \tag{3.6}$$

be an S-fraction, so that

$$a_n > 0, n = 1, 2, 3, \ldots.$$

(A) Then the even and odd parts of (3.6) converge to functions analytic in

$$R = \{\, z : |\arg z| < \pi \,\}$$

uniformly on every compact subset of R.
(B) The S-fraction (3.6) converges to a function analytic in R iff

$$\sum_{n=1}^{\infty} \prod_{k=1}^{n} |a_k|^{(-1)^{n-k+1}}$$

diverges.
(C) If (3.6) converges at a single point in R, then it converges $\forall z \in R$ to an analytic function
(D) The S-fraction (3.6) converges to a function analytic in R if $\exists\, M > 0 \ni |a_n| \leq M, n = 1, 2, 3, \ldots$.

For a proof of this theorem see Jones and Thron (1980, p.136).

THEOREM 3.4 (The Transformation of Bauer and Muir) *Let A_n and B_n denote the nth numerator and denominator, respectively, of the finite continued fraction*

$$b_0 + \frac{a_1}{b_1} + \frac{a_2}{b_2} + \cdots + \frac{a_n}{b_n} + \cdots,$$

where $a_n \neq 0, n = 1, 2, \ldots$. Furthermore, let $\{r_n\}[n \geq 0]$ be any sequence of complex numbers such that

$$\varphi_n = a_n - r_{n-1}(b_n + r_n) \neq 0 \qquad (n = 1, 2, 3, \ldots).$$

Then the continued fraction

$$b_0 + r_0 + \frac{\varphi_1}{b_1 + r_1} + \frac{\frac{a_1 \varphi_2}{\varphi_1}}{b_2 + r_2 - r_0 \frac{\varphi_2}{\varphi_1}} + \cdots + \frac{\frac{a_{n-1} \varphi_n}{\varphi_{n-1}}}{b_n + r_n - r_{n-2} \frac{\varphi_n}{\varphi_{n-1}}} + \cdots$$

has the sequence of numerators(denominators) $\{\hat{A}_n\}(\{\hat{B}_n\})$ defined by

$$\begin{bmatrix} \hat{A}_{-1} \\ \hat{B}_{-1} \end{bmatrix} = \begin{bmatrix} 1 \\ 0 \end{bmatrix}; \quad \begin{bmatrix} \hat{A}_n \\ \hat{B}_n \end{bmatrix} = \begin{bmatrix} A_n + r_n A_{n-1} \\ B_n + r_n B_{n-1} \end{bmatrix}, \quad n \geq 0.$$

We say that the second continued fraction is a Bauer-Muir transform of the first continued fraction.

For a proof of this result see Lorentzen and Waadeland (1992, p.76). Bauer (1872) and Muir (1877) derived a transform for, or modification of, the approximants for terminating continued fractions which led to the above extension for infinite continued fractions. For a statement of their result and a definition of the Bauer-Muir transform for terminating c.f.s see Perron (1957, Satz 1.12). Especially beautiful formulas can be obtained if, with the given a_n, b_n, the r_n can be chosen in a simple way such that the φ_n are independent of n.

THEOREM 3.5 (Perron) *a) If both continued fractions*

$$b_0 + \frac{a_1}{b_1} + \frac{a_2}{b_2} + \frac{a_3}{b_3} + \cdots$$

$$b_0 + r_0 + \frac{\varphi_1}{b_1 + r_1} + \frac{a_1 \frac{\varphi_2}{\varphi_1}}{b_2 + r_2 - r_0 \frac{\varphi_2}{\varphi_1}} + \frac{a_2 \frac{\varphi_3}{\varphi_2}}{b_3 + r_3 - r_1 \frac{\varphi_3}{\varphi_2}} + \cdots ,$$

where $\varphi_\nu = a_\nu - r_{\nu-1}(b_\nu + r_\nu)$, have positive elements and if both converge, then they have the same value. b) If the first continued fraction has positive elements and it converges and if from a certain ν on, $r_\nu \geq 0$, then the second continued fraction also converges and it has the same value as the first.

Proof of a): If $\frac{A_\nu}{B_\nu}$ is the νth approximant of the first continued fraction, then by Theorem 4

$$\frac{A_\nu + r_\nu A_{\nu-1}}{B_\nu + r_\nu B_{\nu-1}} \tag{3.7}$$

is the νth approximant of the second continued fraction. By our hypothesis, both of the following limits exist

$$\lim_{n \to \infty} \frac{A_\nu}{B_\nu} = \xi_0; \quad \lim_{n \to \infty} \frac{A_\nu + r_\nu A_{\nu-1}}{B_\nu + r_\nu B_{\nu-1}} = \eta_0.$$

We have only to show that they are equal. If $r_\nu \geq 0$ infinitely often, then the fraction (3.7) lies infinitely often between $\frac{A_\nu}{B_\nu}$ and $\frac{A_{\nu-1}}{B_{\nu-1}}$ and therefore comes arbitrarily near ξ_0. Therefore $\xi_0 = \eta_0$. There still remains the case where $r_\nu < 0$ from a certain ν on. Now, using the fact that $A_\nu = b_\nu A_{\nu-1} + a_\nu A_{\nu-2}$, we obtain

$$A_\nu + r_\nu A_{\nu-1} - \frac{a_\nu}{r_{\nu-1}}(A_{\nu-1} + r_{\nu-1}A_{\nu-2}) = (b_\nu + r_\nu - \frac{a_\nu}{r_{\nu-1}})A_{\nu-1} = \frac{-\varphi_\nu}{r_{\nu-1}}A_{\nu-1}.$$

Using the corresponding equation for the B_ν and division we obtain

$$\frac{A_\nu + r_\nu A_{\nu-1} - \frac{a_\nu}{r_{\nu-1}}(A_{\nu-1} + r_{\nu-1}A_{\nu-2})}{B_\nu + r_\nu B_{\nu-1} - \frac{a_\nu}{r_{\nu-1}}(B_{\nu-1} + r_{\nu-1}B_{\nu-2})} = \frac{A_{\nu-1}}{B_{\nu-1}}.$$

Since however the second continued fraction has positive elements-therefore positive denominators-and converges and since the $\frac{a_\nu}{r_{\nu-1}}$ are always negative, one can see from this equation that $\frac{A_{\nu-1}}{B_{\nu-1}}$ lies between two successive approximants of the second continued fraction. Therefore the limits ξ_0 and η_0 are again the same.

Proof of b): The first continued fraction has positive elements so the B_ν are also positive; furthermore by hypothesis $r_\nu \geq 0$ for ν large enough. Therefore the fraction (3.7) lies between $\frac{A_\nu}{B_\nu}$ and $\frac{A_{\nu-1}}{B_{\nu-1}}$. So, if $\{\frac{A_\nu}{B_\nu}\}$ has a limit, then the sequence whose νth term is (3.7) must have the same limit.

THEOREM 3.6 (Jacobsen) *Let*

$$b_0 + \frac{a_1}{b_1} + \frac{a_2}{b_2} + \frac{a_3}{b_3} + \cdots , \tag{3.8}$$

where $a_n, b_n \in \mathcal{C}$, $a_n \neq 0$, converge to a value $f \in \hat{\mathcal{C}} = \mathcal{C} \cup \{\infty\}$. Let $\{g_n\}$ satisfy

$$g_n \in \hat{\mathcal{C}}, \quad g_0 \neq f - b_0, \quad g_{n-1} = a_n/(b_n + g_n), \quad n = 1, 2, 3, \ldots . \tag{3.9}$$

Finally, let $\{\varphi_n\}$ and $\{w_n\}$ satisfy

$$\varphi_n = a_n - w_{n-1}(b_n + w_n) \neq 0, \quad w_{n-1} \in \mathcal{C}, \quad n = 1, 2, 3, \ldots. \quad (3.10)$$

If $\lim \inf d(w_n, g_n) > 0$, where $d(\cdot, \cdot)$ denotes the chordal metric on the Riemann sphere $\hat{\mathcal{C}}$ then the Bauer-Muir transform

$$b_0 + w_0 + \cfrac{\varphi_1}{b_1 + w_1} + \cfrac{a_1 \frac{\varphi_2}{\varphi_1}}{b_2 + w_2 - w_0 \frac{\varphi_2}{\varphi_1}} + \cfrac{a_2 \frac{\varphi_3}{\varphi_2}}{b_3 + w_3 - w_1 \frac{\varphi_3}{\varphi_2}} + \cdots \quad (3.11)$$

and (3.8) converge to the common limit f.

For a proof of the above theorem in an equivalent form, see Jacobsen (1989, pp.497-498).

4 AN ALPHABET OF SPECIAL FUNCTION REPRESENTATIONS

Because there are a large number of formulas and equations in this section and because the order in which the material is presented is considerably different from the order in which the work was orginally done, we have found it convenient to use a mixed numbering system consisting of a capital letter from the English alphabet followed by an arabic numeral. The title of this section also reflects the fact that we consider 26 cases; thus a case corresponding to each letter of the alphabet.

FUNCTION A (The Trigamma Function Ψ') :

$$
\begin{aligned}
\Psi'(z) &= \frac{d^2}{dz^2} \log \Gamma(z) = \frac{d}{dz}\left(\frac{\Gamma'(z)}{\Gamma(z)}\right) = \frac{d}{dz}\Psi(z) \\
&= \int_0^\infty \frac{ue^{-zu}}{1 - e^{-u}}\, du, \quad \Re(z) > 0 & (A.1) \\
&= \int_0^\infty \frac{2u}{\sinh u} e^{(1-2z)u}\, du, \quad \Re(z) > 0 & (A.2) \\
&= \Psi'(z+1) + \frac{1}{z^2}, \quad \text{Recurrence formula.} & (A.3) \\
&= \pi^2 \csc^2 \pi z - \Psi'(1-z), \quad \text{Reflection formula.} & (A.4) \\
&= \sum_{k=0}^\infty \frac{1}{(z+k)^2}, \quad z \neq 0, -1, -2, \ldots, \quad \text{Series expansion.} & (A.5) \\
&\asymp \frac{1}{z} + \frac{1}{2z^2} + \sum_{k=1}^\infty \frac{B_{2k}}{z^{2k+1}}, \quad z \to \infty, \ |\arg(z)| < \pi, \quad \text{Asymptotic formula.} & (A.6) \\
&\asymp \frac{1}{z} + \frac{1}{2z^2} + \frac{1}{6z^3} - \frac{1}{30z^5} + \frac{1}{42z^7} - \frac{1}{30z^9} + \cdots. \\
&= \frac{1}{z-\frac{1}{2}} + \frac{\alpha_1}{z-\frac{1}{2}} + \frac{\alpha_2}{z-\frac{1}{2}} + \cdots; \ \alpha_n = \frac{n^4}{4(2n-1)(2n+1)}, \ \Re(z) > 1/2 & (A.7) \\
&= \frac{2}{1(2z-1)} + \frac{1^4}{3(2z-1)} + \frac{2^4}{5(2z-1)} + \frac{3^4}{7(2z-1)} + \cdots, \ \Re(z) > 1/2 & (A.8)
\end{aligned}
$$

$$= 2\left[b_0(z) + \mathop{\mathbf{K}}_{n=1}^{\infty} \frac{a_n}{b_n(z)}\right]^{-1}, \quad \Re(z) > 1/2, \; where \tag{A.9}$$

$$a_n = n^4; \; b_n(z) = (2n+1)(2z-1)$$

$$= \frac{1}{z} + \frac{1}{2z^2}\left[1 + \frac{1}{3z} + \frac{\frac{1\cdot 2^2\cdot 3}{4}}{5z} + \frac{\frac{2\cdot 3^2\cdot 4}{4}}{7z} + \cdots\right], \quad \Re(z) > 0 \tag{A.10}$$

$$= \frac{1}{z} + \frac{1}{2z^2} + \frac{1}{2z^2}\left[d_0(z) + \mathop{\mathbf{K}}_{n=1}^{\infty} \frac{c_n}{d_n(z)}\right]^{-1}, \quad \Re(z) > 0, \; where \tag{A.11}$$

$$c_n = n(n+1)^2(n+2)/4; \; d_n(z) = (2n+3)z.$$

$$= \frac{\sigma_1 w}{1} + \frac{\sigma_2 w}{1} + \frac{\sigma_3 w}{1} + \cdots, \quad w = 1/z, \; \Re(z) > 1/2, \; where \tag{A.12}$$

$$\sigma_1 = 1, \; \sigma_{2n} = \frac{-n^2}{2(2n-1)}, \; \sigma_{2n+1} = \frac{n^2}{2(2n+1)}, \; or, \; equivalently,$$

$$\sigma_1 = 1, \; \sigma_n = \frac{[n/2]^2}{1 + (-1)^{n-1}(2n-1)}, \; n \geq 2, \; [x] := \; greatest \; integer \; function.$$

Proof: The c.f. expansion (A.7) is given without proof in Wall (1948,(94.13)) and Stieltjes (1890, p.387). The identical c.f.'s (A.8) and (A.9) were obtained from (A.7) by an equivalence transformation. We define the Bauer-Muir sequences for the c.f. $b_0(z)+K(a_n/b_n(z))$ in (A.9) by

$$\left.\begin{array}{rcl} r_n(z) &=& n^2 + 2(1-z)n + (2z^2 - 2z + 1), \; n \geq 0 \\ \varphi_n(z) &=& a_n(z) - r_{n-1}(b_n(z) + r_n) = -4z^2, \; n \geq 1. \end{array}\right\} \tag{A.13}$$

Now let z be real and $z > 1/2$. Then by Theorem 3.1 the c.f.'s (A.7)-(A.9) converge, since the series $\sum 1/\sqrt{a_n}$ diverges. We set $F(z)$ equal to the limit of (A.9), apply a B-M transformation to $b_0 + K$ in the denominator of (A.9) with $\{r_n(z)\}$ given by (A.13) and arrive at

$$\begin{aligned} F(z) &= \frac{2}{b_0(z) + r_0(z)} - \frac{4z^4}{b_1(z) + r_1(z) + K_{n=1}^{\infty}\frac{a_n}{b_n(z+1)}}, \quad z > 1/2 \\ &= \frac{2}{2z^2} - \frac{4z^4}{2z^2 + 2z + 1 + K_{n=1}^{\infty}\frac{a_n}{b_n(z+1)}} \\ &= \frac{2}{2z^2} - \frac{4z^4}{2z^2 + 2/F(z+1)} \\ &= 1/z^2 + F(z+1). \end{aligned} \tag{A.14}$$

From the functional equation (A.14) it is easily derived that

$$F(z) - F(z+n+1) = \sum_{k=0}^{n} \frac{1}{(z+k)^2}. \tag{A.15}$$

From the c.f. (A.9) we have immediately that

$$0 \leq F(z+n+1) \leq 1/(z+n+1/2)$$

so that $F(z+n+1) \to 0$ as $n \to \infty$. Thus it follows from (A.15) and the partial fraction series (A.5) for $\Psi'(z)$ that

$$F(z) = \sum_{k=0}^{\infty} \frac{1}{(z+k)^2} = \Psi'(z), \; z > 1/2.$$

We would now like to show that the c.f.'s (A.7)-(A.9) converge to $\Psi'(z)$ if z is complex and $\Re(z) > 1/2$. By an equivalence transformation, the c.f. (A.7) can be put into the form

$$\frac{1/(z-1/2)}{1} + \frac{\alpha_1/(z-1/2)^2}{1} + \frac{\alpha_2/(z-1/2)^2}{1} + \cdots \tag{A.16}$$

if $\Re(z) > 1/2$. Hence the c.f.

$$\frac{1/(z-1/2)^2}{1} + \frac{\alpha_1/(z-1/2)^2}{1} + \frac{\alpha_2/(z-1/2)^2}{1} + \cdots \tag{A.17}$$

is an S-fraction in the variable $w = 1/(z-1/2)^2$. By Theorem 3.3, the c.f. (A.7) converges to an analytic function if $\operatorname{Re} z > 1/2$ since the series $\sum 1/\sqrt{\alpha_n}$ diverges. The function

$$\frac{1}{z - \frac{1}{2}} \sum_{k=0}^{\infty} \frac{1}{(z+k)^2} \tag{A.18}$$

is also analytic if $\Re(z) > 1/2$. It follows from what we have proved above that the functions represented by (A.17) and (A.18) are equal if z is real and $z > 1/2$. By the Identity Theorem for analytic functions these two functions are the same if $\Re(z) > 1/2$. But the function to which the series converges is $\Psi'(z)/(z-1/2)$. It follows that (A.7)-(A.9) are valid if $\Re(z) > 1/2$. We now proceed to verify the validity of expansions (A.10) and (A.11). Rogers (1907, p.84) derived the formal correspondence

$$\sum_{k=1}^{\infty} B_{2k} z^{2k-2} = \frac{1}{6} + \frac{1 \cdot 2^2 \cdot 3z^2}{10} + \frac{2 \cdot 3^2 \cdot 4z^2}{14} + \frac{3 \cdot 4^2 \cdot 5z^2}{18} + \cdots$$

$$= \frac{1}{2 \cdot 3} + \frac{1 \cdot 2^2 \cdot 3z^2}{2 \cdot 5} + \frac{2 \cdot 3^2 \cdot 4z^2}{2 \cdot 7} + \frac{3 \cdot 4^2 \cdot 5z^2}{2 \cdot 9} + \cdots . \tag{A.19}$$

By an equivalence transformation we get

$$\sum_{k=1}^{\infty} B_{2k} z^{2k-2} = \frac{\frac{1}{2}}{3} + \frac{1 \cdot 2^2 \cdot 3\frac{z^2}{4}}{5} + \frac{2 \cdot 3^2 \cdot 4\frac{z^2}{4}}{7} + \frac{3 \cdot 4^2 \cdot 5\frac{z^2}{4}}{9} + \cdots . \tag{A.20}$$

The B_{2k} are the Bernoulli numbers. The series on the left is everywhere divergent. The known asymptotic series for Ψ' is given by (A.6). So using (A.6) and (A.20) we formally arrive at

$$\Psi'(z) = \frac{1}{z} + \frac{1}{2z^2} + \frac{2}{z} \left[\frac{\frac{z^{-2}}{4}}{3} + \frac{1 \cdot 2^2 \cdot 3\frac{z^{-2}}{4}}{5} + \frac{2 \cdot 3^2 \cdot 4\frac{z^{-2}}{4}}{7} + \cdots \right]. \tag{A.21}$$

Again by an equivalence transformation, (A.21) can be put into the form (A.11). The task we choose now is to prove that the continued fractions (A.11) and (A.21) actually converge to $\Psi'(z)$ if z is real and positive. So let us assume that z is real and positive, Then (A.21), and hence (A.11), converges by applying Theorem 3.1 for continued fractions with positive elements. Let the B-M sequences for the c.f. $d_0(z) + K(c_n/d_n(z))$ be defined by

$$\begin{aligned} r_n(z) &= n^2/2 + n(3-2z)/2 + z^2 - z + 1, \ n \geq 0 \\ \phi_n(z) &= c_n - r_{n-1}(z)(d_n(z) + r_n(z)) = -z^2(z+1)^2. \end{aligned} \tag{A.22}$$

Let $F(z)$ denote the limit function of c.f.'s (A.11) and (A.21). Since $r_n(z)$ is eventually positive when $z > 0$, we can by Theorem 3.4 apply a Bauer-Muir transformation to $d_0 + K(c_n/d_n)$ in (A.11) using (A.22) and obtain

$$F(z) = \frac{1}{z} + \frac{1}{2z^2} + \frac{z^{-2}/2}{3z + r_0(z)} - \frac{z^2(z+1)^2}{z^2 + 3(z+1)} + \frac{a_1}{5(z+1)} + \frac{a_2}{7(z+1)} + \cdots$$

$$= \frac{1}{z} + \frac{z^{-2}/2}{(z+1)^2} - \frac{z^2(z+1)^2}{z^2 + g(z+1)}.$$

But

$$1/g(z) = 2z^2 F(z) - 2z - 1,$$

so

$$g(z+1) = 1/(2(z+1)^2 F(z+1) - 2z - 3).$$

Hence

$$F(z) = \frac{1}{z} + \frac{1}{2z^2} + \frac{z^{-2}/2}{(z+1)^2} - \frac{z^2(z+1)^2}{z^2} + \frac{1}{2(z+1)^2 F(z+1) - 2z - 3}$$

$$= \frac{1}{z} + \frac{1}{2z^2} + F(z+1) - \frac{2z^3 + 3z^2 - 1}{2z^2(z+1)^2} = \frac{1}{z^2} + F(z+1), \quad z > 0.$$

Thus it follows that

$$F(z) - F(z+n+1) = \sum_{k=0}^{n} \frac{1}{(z+k)^2}.$$

Since $F(z+n+1) \to 0$ as $n \to \infty$, we have that

$$F(z) = \sum_{k=0}^{\infty} \frac{1}{(z+k)^2} = \Psi'(z), \quad z > 0.$$

We are now ready to investigate the convergence behavior of continued fractions (A.11) and (A.21) for complex values of z. If we let $w = z^{-2}/4$, the continued fraction enclosed in brackets in formula (A.21) can be written as

$$\frac{w}{3} + \frac{(1 \cdot 2^2 \cdot 3)w}{5} + \frac{(2 \cdot 3^2 \cdot 4)w}{7} + \cdots \tag{A.23}$$

which is equivalent to the c.f.

$$\frac{\frac{w}{3}}{1} + \frac{\frac{(1 \cdot 2^2 \cdot 3)w}{3 \cdot 5}}{1} + \frac{\frac{(2 \cdot 3^2 \cdot 4)w}{5 \cdot 7}}{1} + \cdots \;. \tag{A.24}$$

The c.f. (A.24) is an S-fraction in w that converges for positive real values of w and therefore converges to an analytic function of w in the cut plane $R = w : |\arg w| < \pi$. It follows that the c.f. (A.21) converges to an analytic function of z for all z such that $\Re(z) > 0$. Since the c.f. (A.21) and

$$\Psi'(z) = \sum_{k=0}^{\infty} \frac{1}{(z+k)^2}$$

are each analytic in $\Re(z) > 0$ and since they are equal for positive real z, it follows from the Identity Theorem that (A.11) and the following equivalent expansions are valid when $\Re(z) > 0$.

$$
\begin{aligned}
\Psi'(z) &= \frac{1}{z} + \frac{1}{2z^2} + \frac{2}{z}\left[\frac{w}{3} + \frac{(1\cdot2^2\cdot3)w}{5} + \frac{(2\cdot3^2\cdot4)w}{7} + \cdots\right], \quad w = \frac{1}{4z^2} \\
&= \frac{1}{z} + \frac{1}{2z^2} + \frac{2}{z}\left[\frac{\frac{w}{1\cdot3}}{1} + \frac{\frac{(1\cdot2^2\cdot3)w}{3\cdot5}}{1} + \frac{\frac{(2\cdot3^2\cdot4)w}{5\cdot7}}{1} + \cdots\right] \\
&= \frac{1}{z} + \frac{1}{2z^2} + \frac{1}{z^2}\left[\frac{1}{2\cdot3z} + \frac{1\cdot2^2\cdot3}{2\cdot5z} + \frac{2\cdot3^2\cdot4}{2\cdot7z} + \cdots\right].
\end{aligned}
$$
(A.25)

These c.f.s seem to converge quite rapidly for large positive real z in particular. We now proceed to verify expansion (A.12). The even part of (A.12) is

$$
\frac{\sigma_1 w}{1 + \sigma_2 w} - \frac{\sigma_2\sigma_3 w^2}{1 + (\sigma_3 + \sigma_4)w} - \frac{\sigma_4\sigma_5 w^2}{1 + (\sigma_5 + \sigma_6)w} - \cdots,
$$
(A.26)

which becomes

$$
\frac{w}{1 - \frac{w}{2}} + \frac{\frac{1^2 w^2}{4\cdot1\cdot3}}{1 - \frac{w}{2}} + \frac{\frac{2^2 w^2}{4\cdot3\cdot5}}{1 - \frac{w}{2}} + \frac{\frac{3^2 w^2}{4\cdot5\cdot7}}{1 - \frac{w}{2}} + \cdots.
$$
(A.27)

A substitution of $w = 1/z$ in (A.27) leads to the equivalent c.f.

$$
F(z) = \frac{2}{2\cdot1(z - \frac{1}{2})} + \frac{1^4}{2\cdot3(z - \frac{1}{2})} + \frac{2^4}{2\cdot5(z - \frac{1}{2})} + \cdots.
$$
(A.28)

By applying an equivalence transformation to the previously validated expansion (A.7), $F(z) = \Psi'(z)$ if $\Re(z) > 1/2$. Let $P_n(Q_n)$ denote the nth numerator(denominator) of (A.27), and let $\rho_n = \sigma_{2n} w$. Then, if f_{2n-1} is the $(2n-1)$st approximate of (A.12), we have that

$$
f_{2n-1} = \frac{P_n - \rho_n P_{n-1}}{Q_n - \rho_n Q_{n-1}},
$$

so

$$
|f_{2n-1} - P_n/Q_n| = \frac{|\sigma_{2n}||P_n/Q_n - P_{n-1}/Q_{n-1}|}{|zQ_n/Q_{n-1} + |\sigma_{2n}||}.
$$

Clearly $f_{2n-1} \to \Psi'(z) = \lim P_n/Q_n$ if the latter limit exists and $\Re(zQ_n/Q_{n-1}) \geq 0$. We now give an inductive proof showing that $\Re(zQ_n/Q_{n-1}) > 0$ if $\Re(z) > 1/2$. Since

$$
Q_1/Q_0 = 1 - w/2 = 1 - 1/(2z)
$$

and

$$
\Re(z(1 - 1/(2z)) = \Re(z - 1/2),
$$

it follows that $\Re(zQ_1/Q_0) > 0$ if $\Re(z) > 1/2$. Now assume $\Re(zQ_n/Q_{n-1}) > 0$ and $\Re(z) > 1/2$ for some n. Then

$$
\begin{aligned}
\Re(zQ_{n+1}/Q_n) &= \Re\left[z\left(1 - \frac{w}{2} - \frac{\sigma_{2n}\sigma_{2n+1}w^2}{Q_n/Q_{n-1}}\right)\right] \\
&= \Re\left(z - \frac{1}{2} + \frac{|\sigma_{2n}||\sigma_{2n+1}|}{zQ_n/Q_{n-1}}\right) > 0,
\end{aligned}
$$

since $\Re(z) > 1/2$ and $\Re(zQ_n/Q_{n-1}) > 0$ by hypothesis. Hence, by induction, $\Re(zQ_n/Q_{n-1}) > 0$ for every positive integer n when $\Re(z) > 1/2$. Thus we conclude that the expansion (A.12) is valid if $\Re(z) > 1/2$.

FUNCTION B (The Tetragamma Function Ψ'') :

$$\Psi''(z) \ = \ \frac{d^2}{dz^2}\Psi(z) = \frac{d^3}{dz^3}\log\Gamma(z) \tag{B.1}$$

$$= \ -\int_0^\infty \frac{u^2 e^{-zu}}{1 - e^{-u}}\,du, \ \Re(z) > 0 \tag{B.2}$$

$$= \ -\int_0^\infty \frac{4u^2}{\sinh u}e^{-(2z-1)u}\,du, \ \Re(z) > 0 \tag{B.3}$$

$$= \ -\frac{2}{z^3} + \Psi''(z+1), \quad Recurrence\ formula \tag{B.4}$$

$$= \ -\pi\frac{d^2}{dz^2}\cot\pi z + \Psi''(1-z), \quad Reflection\ formula \tag{B.5}$$

$$= \ -\sum_{k=0}^\infty \frac{2}{(z+k)^3}, \ z \neq 0, -1, -2, \ldots, \ Series\ expansion \tag{B.6}$$

$$\asymp \ -\frac{1}{z^2} - \frac{1}{z^3} - \sum_{k=1}^\infty \frac{(2k+1)B_{2k}}{z^{2k+2}} \ (z \to \infty, \ |\arg(z)| < \pi), \ Asymptotic\ formula \tag{B.7}$$

$$\asymp \ -\frac{1}{z^2} - \frac{1}{z^3} - \frac{1}{2z^4} + \frac{1}{6z^6} - \frac{1}{6z^8} + \frac{3}{10z^{10}} - \frac{5}{6z^{12}} + \cdots.$$

$$= \ -w + \underset{n=1}{\overset{\infty}{\mathrm{K}}}\frac{d_n w}{1}, \ w = 1/z, \ \Re(z) > 1, \ where \tag{B.8}$$

$$d_1 = 1; \ d_{4n-1} = -1 - d_{4n-2} = \frac{1+n^2}{1-2n}; \ d_{4n} = -d_{4n+1} = \frac{n^3}{2(1+n^2)}$$

$$= \ -\underset{n=1}{\overset{\infty}{\mathrm{K}}}\frac{c_n\sigma}{n}, \ \sigma = \frac{1}{z(z-1)}, \ \Re(z) > 1/2, \ z \notin (1/2, 1], \ where \tag{B.9}$$

$$c_1 = 1; \ c_{2n} = c_{2n+1} = n^4$$

$$= \ -\frac{1}{z}\left[b_0(z) + \underset{n=1}{\overset{\infty}{\mathrm{K}}}\frac{a_n(z)}{b_n(z)}\right]^{-1}, \ \Re(z) > 1/2, \ z \notin (1/2, 1], \ where \tag{B.10}$$

$$a_{2n-1}(z) = a_{2n}(z) = n^4(z-1)/z; \ b_n(z) = (n+1)(z-1)$$

$$= \ -z^{-2} - z^{-3} - \frac{z^{-3}/2}{z} + \frac{p_1}{z} + \frac{q_1}{z} + \frac{p_2}{z} + \frac{q_2}{z} + \cdots, \ where \tag{B.11}$$

$$\Re(z) > 0; \ p_n = \frac{n^2(n+1)}{4n+2}; \ q_n = \frac{n(n+1)^2}{4n+2}$$

$$= \ -z^{-2} - z^{-3} - z^{-3}\left[\hat{b}_0(z) + \underset{n=1}{\overset{\infty}{\mathrm{K}}}\frac{\hat{a}_n}{\hat{b}_n(z)}\right]^{-1}, \ \Re(z) > 0, \ where \tag{B.12}$$

$$\hat{a}_{2n-1} = (n+1)n^3; \ \hat{a}_{2n} = n(n+1)^3; \ \hat{b}_n(z) = (n+2)z.$$

Proof: The odd part of c.f. (B.8) is given by

$$\Psi''(z) \ = \ -\frac{w^2}{1-w} - \frac{d_3 d_4 w^2}{1} - \frac{d_5 d_6 w^2}{1-w} - \frac{d_7 d_8 w^2}{1} - \frac{d_9 d_{10} w^2}{1-w} - \cdots$$

$$= \frac{-1/z}{z-1} + \frac{1^4}{2z} + \frac{1^4}{3(z-1)} + \frac{2^4}{4z} + \frac{2^4}{5(z-1)} + \frac{3^4}{6z} + \cdots$$

and these continued fractions are equivalent to c.f. (B.10). Now let

$$\begin{aligned}
r_{2n}(z) &= (n+1)(n-z+1) + z^2/2, \ n \geq 0 \\
r_{2n-1}(z) &= [(z-1)/z]r_{2n-2}(z), \ n \geq 1 \\
\varphi_n(z) &= z^3(1-z)/4.
\end{aligned} \tag{B.13}$$

It is easily verified that

$$\begin{aligned}
b_{2n}(z) + r_{2n}(z) - r_{2n-2}(z) &= b_{2n-1}(z+1) = 2nz \\
b_{2n+1}(z) + r_{2n+1}(z) - r_{2n-1}(z) &= (2n+1)(z-1/z) = (1-1/z^2)b_{2n}(z+1) \quad \text{(B.14)} \\
a_n(z) - r_{n-1}(z)(b_n(z) + r_n(z)) &= \varphi_n(z).
\end{aligned}$$

The continued fraction (B.10) clearly converges by Theorem (3.1) if z is real and $z > 1$. We set $F(z)$ equal to the limit of this continued fraction and we apply a B-M transformation to the c.f. $b_0 + K(a_n/b_n)$ in (B.10) to obtain (for $z > 1$)

$$F(z) = \frac{-1/z}{b_0 + r_0} + \frac{\varphi_1}{b_1 + r_1} + \frac{\frac{a_1\varphi_2}{\varphi_1}}{b_2 + r_2 - r_0\frac{\varphi_2}{\varphi_1}} + \frac{\frac{a_2\varphi_3}{\varphi_2}}{b_3 + r_3 - r_1\frac{\varphi_2}{\varphi_2}} + \cdots . \tag{B.15}$$

All terms in continued fraction (B.15) are evaluated at z. With the aid of formulas (B.13) and (B.14), straightforward calculations will show that (B.15) implies

$$F(z) = F(z+1) - 2/z^3, \ z > 1. \tag{B.16}$$

From the functional equation (B.16) and the fact that $F(z) \to 0$ as $z \to \infty$, we obtain

$$F(z) = \Psi''(z), \ z > 1.$$

Thus the odd part of the c.f. (B.8) converges to $\Psi''(z)$ on $(1,\infty)$. The continued fraction (B.10) can also be written in the form

$$-\frac{d_1d_2w^2}{1 + (d_2+d_3)w} - \frac{d_3d_4w^2}{1 + (d_4+d_5)w} - \frac{d_5d_6w^2}{1 + (d_6+d_7)w} - \cdots . \tag{B.17}$$

Let P_n/Q_n be the nth approximant of (B.17), where

$$P_0 = 0, \ P_1 = -d_1d_2w^2; \ Q_0 = 1, \ Q_1 = 1 + (d_2+d_3)w$$

and P_n and Q_n satisfy the difference equation

$$X_n = (1 + (d_{2n}+d_{2n+1})w)X_{n-1} - d_{2n-1}d_{2n}w^2 X_{n-2}.$$

If we set $\rho_n = d_{2n+1}w$, $n \geq 0$, then the approximants of (B.8) are given by

$$\frac{P_0 - \rho_0}{Q_0}, \frac{P_0}{Q_0}, \frac{P_1 - \rho_1 P_0}{Q_1 - \rho_1 Q_0}, \frac{P_1}{Q_1}, \frac{P_2 - \rho_2 P_1}{Q_2 - \rho_2 Q_1}, \ldots .$$

If we denote the nth approximant of (B.8) by f_n, then

$$f_{2n+1} = \frac{P_n}{Q_n}; \quad f_{2n} = \frac{P_n - \rho_n P_{n-1}}{Q_n - \rho_n Q_{n-1}}, \quad n \geq 0.$$

Hence,

$$|f_{2n} - f_{2n+1}| = \left| \frac{P_n - \rho_n P_{n-1}}{Q_n - \rho_n Q_{n-1}} - \frac{P_n}{Q_n} \right|$$

$$= \frac{|d_{2n+1}| \left| \frac{P_n}{Q_n} - \frac{P_{n-1}}{Q_{n-1}} \right|}{\left| z \frac{Q_n}{Q_{n-1}} + |d_{2n+1}| \right|}.$$

It is easy to show that $\mathrm{Re}\,(zQ_n/Q_{n-1}) > 0$ if $\mathrm{Re}\,z > 1$, so that

$$|f_{2n} - f_{2n+1}| \leq |f_{2n+1} - f_{2n-1}|$$

under this restriction on z. Hence $f_{2n} \to L$ if $f_{2n+1} \to L$ as $n \to \infty$.
Continued fraction (B.10) is equivalent to continued fraction (B.9) which we repeat here

$$-\frac{\sigma}{1} + \frac{1^4 \sigma}{2} + \frac{1^4 \sigma}{3} + \frac{2^4 \sigma}{4} + \frac{2^4 \sigma}{5} + \frac{3^4 \sigma}{6} + \cdots,$$

where

$$\sigma = \frac{1}{z(z-1)}.$$

The question now is: For what values of z is $\Im(\sigma) = 0$ and $\Re(\sigma) < 0$? Set $\sigma = -t$, where $t > 0$. Then

$$z^2 - z = -1/t$$

and the zeros of this equation are

$$z = \frac{1 \pm \sqrt{1 - 4/t}}{2}.$$

When $t \geq 4$ we have $0 \leq z \leq 1$. When $0 < t < 4$,

$$z = \frac{1}{2} \pm i \frac{\sqrt{4/t - 1}}{2}.$$

Thus it follows that

$$\{ z : \Im(z) = 0, \Re(z) < 0 \} \cap \{ z : \Re(z) > 1/2 \setminus (1/2, 1] \} = \emptyset.$$

Thus, since c.f. (B.9) is equal to -1 times an S-fraction, for all z in the cut halfplane $\{ z : \Re(z) > 1/2 \setminus (1/2, 1] \}$, we have c.f.'s (B.9) and (B.10) converge to an analytic function $H(z)$ on this cut halfplane. Since $\Psi''(z)$ is also analytic in this region and we have shown that $\Psi''(z) = H(z)$ for all $z \in (1, \infty)$, it follows from the Identity Theorem that c.f.'s (B.9) and (B.10) represent $\Psi''(z)$ in the region. Since we showed above that the odd and even approximants of the continued fraction (B.8) converge to the same limit when $\Re(z) > 1$, it follows also that (B.8) is valid in this halfplane.

Wall(1948,(94.14)) credits Stieltjes with the formal c.f. expansion (B.11) (The expansions given by Wall (1948, p.373) and Shenton and Bowman (1971, formula 7(b)) both contain misprints). Continued fraction (B.11) is equivalent to

$$\Psi''(z) = -z^{-2} - z^{-3} - \frac{z^{-3}/2}{z} + \frac{\alpha_1}{\beta_1 z} + \frac{\alpha_2}{\beta_2 z} + \frac{\alpha_3}{\beta_3 z} + \cdots, \qquad (\text{B.18})$$

where

$$\alpha_{2n-1} = n^2(n+1); \ \alpha_{2n} = n(n+1)^2; \ \beta_{2n-1} = 4n+2; \ \beta_{2n} = 1.$$

Let $\{\rho_n\}_{n\geq 0}$ be an arbitrary sequence with no zero terms. Then the following c.f. is equivalent to (B.18).

$$\Psi''(z) = -z^{-2} - z^{-3} - \frac{\rho_0 z^{-3}/2}{\rho_0 z} + \frac{\alpha_1 \rho_0 \rho_1}{\beta_1 \rho_1 z} + \cdots + \frac{\alpha_n \rho_{n-1} \rho_n}{\beta_n \rho_n z} + \cdots. \qquad (\text{B.19})$$

Choose

$$\rho_{2n-1} = 1/2; \ \rho_{2n} = 2n + 2.$$

With this choice for $\{\rho_n\}$, the c.f. (B.19), whose limit we now denote by $F(z)$, becomes

$$F(z) = -z^{-2} - z^{-3} - z^{-3}\left[\hat{b}_0 + \underset{n=1}{\overset{\infty}{\mathbf{K}}} \frac{\hat{a}_n}{\hat{b}_n(z)}\right]^{-1}, \qquad (\text{B.20})$$

where

$$\hat{a}_{2n-1} = (n+1)n^3; \ \hat{a}_{2n} = n(n+1)^3; \ \hat{b}_n(z) = (n+2)z.$$

and which is the same as the c.f. (B.12). We now apply a B-M transformation to the c.f. in brackets in (B.20) for $z > 0$ using the Bauer-Muir sequences defined by

$$\begin{aligned}
r_{2n}(z) &= n(n+2-z) + \frac{z^3 - z^2 + z + 1}{2z + 1} \\
r_{2n+1}(z) &= n(n+3-z) + \frac{z^3 - 2z^2 + 3z + 2}{2z + 1} \qquad (\text{B.21}) \\
\varphi_n(z) &= -\frac{z^3(z+1)^3}{(2z+1)^2}.
\end{aligned}$$

Then from (B.20) and (B.21) we obtain (assuming $z > 0$)

$$F(z) = -z^{-2} - z^{-3} - \frac{z^{-3}}{\hat{b}_0(z) + r_0(z)} + \frac{\varphi(z)}{\hat{b}_1(z) + r_1(z)} + \frac{\hat{a}_1}{\hat{b}_1(z+1)} + \frac{\hat{a}_2}{\hat{b}_2(z+1)} + \cdots$$

from which it follows that

$$F(z + 1) = 2/z^3 + F(z).$$

Hence

$$\sum_{k=0}^{n}[F(z+k+1) - F(z+k)] = F(z+n+1) - F(z) = \sum_{k=0}^{n} \frac{2}{(z+k)^3}.$$

Since $F(z+n+1) \to 0$ as $n \to \infty$ when $z > 0$, it follows that

$$F(z) = -\sum_{k=0}^{\infty} \frac{2}{(z+k)^3} = \Psi''(z)$$

if $z > 0$. Since $F(z)$ and $\Psi''(z)$ are analytic in the halfplane $\Re(z) > 0$, it is true that $F(z) = \Psi''(z)$ for all z in this halfplane. Finally, we would like to point out that, in view of recurrence formula (B.4), each of the continued fractions (B.8), (B.9), and (B.10) can be used to evaluate $\Psi''(z)$ on the halfplane $\Re(z) > 0$.

FUNCTION C (The Small Beta Function) :

$$\beta(z) = \sum_{n=0}^{\infty} \frac{(-1)^n}{n+z} \tag{C.1}$$

$$= \int_0^{\infty} \frac{e^{-(2z-1)u}}{\cosh u} \, du, \ \Re(z) > 0 \tag{C.2}$$

$$= \frac{1}{2}\left[\Psi\left(\frac{z+1}{2}\right) - \Psi\left(\frac{z}{2}\right)\right], \ \Re(z) > 0 \tag{C.3}$$

$$= 1/z - \beta(z+1) \tag{C.4}$$

$$= \left[2z - 1 + \mathop{K}_{n=1}^{\infty} \frac{n^2}{2z-1}\right]^{-1}, \ \Re(z) > \frac{1}{2} \tag{C.5}$$

$$= \mathop{K}_{n=1}^{\infty} \frac{d_n z^{-1}}{1}, \ d_1 = \frac{1}{2}, d_n = \frac{(-1)^{n-1}}{2}\left\lceil\frac{n}{2}\right\rceil, \ n \geq 2, \ Re(z) > \frac{1}{2}, \tag{C.6}$$

where $\lceil\cdot\rceil$ denotes the greatest integer function.

Proof: Let $H(z)$ be the limit function of the continued fraction (C.5), and at this point let us assume z is real and $z > 1/2$. We apply a B-M transformation to the c.f. in brackets in (C.5) using

$$r_n(z) = n + 1 - z; \ \varphi_n(z) = z^2 \tag{C.7}$$

and obtain

$$H(z) = \frac{1}{z} + \frac{z^2}{z+1} + \frac{1^2}{2z+1} + \frac{2^2}{2z+1} + \frac{3^2}{2z+1} + \cdots . \tag{C.8}$$

From (C.8) we get

$$H(z) = 1/z - H(z+1).$$

Hence

$$H(z+n+1) + H(z+n) = \frac{1}{z+n}$$

and

$$\sum_{k=0}^{\infty} (-1)^k [H(z+k+1) + H(z+k)] = \sum_{k=0}^{\infty} \frac{(-1)^k}{z+k}. \tag{C.9}$$

Let F_n denote the nth partial sum of the series (C.9). Then

$$F_{2n} = \sum_{k=0}^{2n-1} [H(z+k+1) + H(z+k)]$$

$$= H(z) - H(z+2n) \to H(z) \text{ as } n \to \infty.$$

and

$$F_{2n+1} = F_{2n} + H(z+2n+1) + H(z+2n)$$

$$= H(z) + H(z+2n+1) \to H(z) \text{ as } n \to \infty.$$

Hence,

$$H(z) = \beta(z) = \sum_{k=0}^{\infty} \frac{(-1)^k}{z+k}, \quad \Re(z) > 1/2.$$

Since $H(z)$ and $\beta(z)$ are both analytic in the halfplane $\Re(z) > 1/2$, it follows that $H(z) = \beta(z)$ for all z in this halfplane. By extending (C.5) we obtain

$$\beta(z) = \frac{1}{2z} - \frac{1}{1} + \frac{1}{2z} - \frac{2}{1} + \frac{2}{2z} - \frac{3}{1} + \frac{3}{2z} - \cdots \tag{C.10}$$

which is equivalent to c.f. (C.6) repeated below in a continued form

$$\beta(z) = \frac{\frac{1}{2}z^{-1}}{1} - \frac{\frac{1}{2}z^{-1}}{1} + \frac{\frac{1}{2}z^{-1}}{1} - \frac{\frac{2}{2}z^{-1}}{1} + \frac{\frac{2}{2}z^{-1}}{1} - \frac{\frac{3}{2}z^{-1}}{1} + \frac{\frac{3}{2}z^{-1}}{1} - \cdots .$$

The continued fraction (C.6) is the regular δ-fraction in the variable $1/z$ for $\beta(z)$. The c.f. (C.5) is the even part of the c.f. (C.10). If $A_n(B_n)$ denotes the nth numerator(denominator) of (C.10), then

$$\frac{B_{2n}}{B_{2n-2}} = 2z - 1 + \frac{(n-1)^2}{2z-1} + \frac{(n-2)^2}{2z-1} + \cdots + \frac{1^2}{2z-1}. \tag{C.11}$$

Now assume $\Re(z) > 1/2$. Then, using (C.11),

$$
\begin{aligned}
|f_{2n} - f_{2n-1}| &= \frac{|a_{2n}||f_{2n-2} - f_{2n}|}{|B_{2n}/B_{2n-2} - a_{2n}|}, \quad f_n = A_n/B_n \\
&= \frac{n|f_{2n-2} - f_{2n}|}{|B_{2n}/B_{2n-2} + n|} \le |f_{2n-2} - f_{2n}|,
\end{aligned}
$$

since $\Re(B_{2n}/B_{2n-2} > 0$ when $\Re(z) > 1/2$. Since $f_n \to \beta(z), \Re(z) > 1/2$, it follows that (C.10) and (C.6) converge to $\beta(z)$ when $\Re(z) > 1/2$. $\square$

FUNCTION D (The Derivative of the Small Beta Function) :

$$-\beta'(z) = \sum_{n=0}^{\infty} \frac{(-1)^n}{(z+n)^2} \tag{D.1}$$

$$= \int_0^{\infty} \frac{2u}{\cosh u} e^{-(2z-1)u}\, du, \quad \Re(z) > 0 \tag{D.2}$$

$$= \frac{1}{z^2} + \beta'(z+1) \tag{D.3}$$

$$= \frac{1}{4}\left[\Psi'\left(\frac{z}{2}\right) - \Psi'\left(\frac{z+1}{2}\right)\right], \quad \Re(z) > 0 \tag{D.4}$$

$$= \frac{\hat{z}/2}{1} + \frac{1^2\hat{z}}{1} + \frac{1^2\hat{z}}{1} + \frac{2^2\hat{z}}{1} + \frac{2^2\hat{z}}{1} + \cdots, \quad where \tag{D.5}$$

$$\hat{z} = \frac{1}{z(z-1)}; \ z \in D = \{z \mid \Re(z) > 1/2\} \setminus (0,1]$$

$$= -w + \mathop{K}_{n=1}^{\infty} \frac{d_n w}{1}, \quad w = 1/z, \ z \in (3/2, \infty], \ where \tag{D.6}$$

$$d_1 = 1; \ d_{4n+2} = -d_{4n-1} = n - 1/2; \ d_{4n} = -d_{4n+1} = n^2/(n-1/2).$$

Proof: From Perron (1957, p.30) for $\alpha > 1$

$$\int_0^\infty \frac{t}{\cosh t} e^{-\alpha t}\, dt = \frac{1}{\alpha^2 - 1} + \frac{4 \cdot 1^1}{1} + \frac{4 \cdot 1^2}{\alpha^2 - 1} + \frac{4 \cdot 2^2}{1} + \frac{4 \cdot 2^2}{\alpha^2 - 1} + \cdots . \tag{D.7}$$

Let $z = (\alpha + 1)/2$. Then for $z > 1$ it follows from (D.7) that

$$\int_0^\infty \frac{2t}{\cosh t} e^{-(2z-1)t}\, dt = \frac{z^{-1}/2}{z - 1} + \frac{1^2}{z} + \frac{1^2}{z - 1} + \frac{2^2}{z} + \frac{2^2}{z - 1} + \cdots . \tag{D.8}$$

The continued fraction in (D.8) is easily seen to be equivalent to each of the two following continued fractions whose limit function we denote by $F(z)$.

$$F(z) = \frac{\hat{z}/2}{1} + \frac{1^2 \hat{z}}{1} + \frac{1^2 \hat{z}}{1} + \frac{2^2 \hat{z}}{1} + \frac{2^2 \hat{z}}{1} + \cdots , \quad \hat{z} = \frac{1}{z(z - 1)} \tag{D.9}$$

$$= \frac{1}{2z} \left[b_0(z) + \mathop{\mathrm{K}}_{n=1}^\infty \frac{a_n(z)}{b_n(z)} \right]^{-1}, \quad \text{where} \tag{D.10}$$

$$a_{2n-1}(z) = a_{2n}(z) = n^2(z - 1)/z; \ b_n(z) = z - 1.$$

After applying a B-M transformation to the continued fraction $b_0 + K(a_n/b_n)$ in (D.10), with the associated sequences determined by

$$r_{2n}(z) = n + 1 - \frac{z}{2}; \ r_{2n-1}(z) = \frac{(z - 1)(2n - z)}{2z}; \ \varphi_n(z) = \frac{z(z - 1)}{2}$$

and using Theorems (3.1) and (3.5) we arrive at the functional relation

$$F(z) = 1/z^2 - F(z + 1). \tag{D.11}$$

Since $F(z + n) \to 0$ as $n \to \infty$, it follows from (D.11) that

$$F(z) = -\beta'(z) = \sum_{n=0}^\infty \frac{(-1)^n}{(z + n)^2}, \quad z > 1. \tag{D.12}$$

We observe that c.f. (D.9) is an S-fraction in the variable $\hat{z}$ so it converges to an analytic function for all z in a domain for which $|\arg \hat{z}| < \pi$. Straightforward calculations will show that this condition is satisfied in the cut halfplane $D = \{z \mid \Re(z) > 1/2\} \setminus (0, 1]$. Hence $F(z) = -\beta'(z)$ in this region also.

The task before us now is to justify expansion (D.6). We start by extending c.f. (D.8) to

$$\frac{\frac{1}{2z}}{z - 1 - \rho_1} + \frac{\rho_1}{1} - \frac{\frac{1^2}{\rho_1}}{z + \frac{1^2}{\rho_1} - \rho_2} + \frac{\rho_2}{1} - \frac{\frac{1^2}{\rho_2}}{z - 1 + \frac{1^2}{\rho_2} - \rho_3} + \cdots , \tag{D.13}$$

and choose

$$\rho_{2n} = -n; \ \rho_{2n+1} = -(n + 1)$$

With this choice of $\{\rho_n\}$, (D.13) becomes

$$\frac{w^2/2}{1} - \frac{w}{1} + \frac{w}{1} - \frac{w}{1} + \frac{w}{1} - \frac{2w}{1} + \frac{2w}{1} - \frac{2w}{1} + \frac{2w}{1} - \cdots . \tag{D.14}$$

We extend (D.14) by "inserting" terms at the beginning and arrive at

$$- w + \frac{w}{1} - \frac{w/2}{w/2 + K},$$ (D.15)

where K denotes the c.f. in (D.14) that divides the first partial numerator $w^2/2$. By an equivalence transformation (D.15) can be written as

$$- w + \frac{w}{1} - \frac{w/2}{1 + w/2} + \frac{w}{b_1} + \frac{w}{b_2} + \frac{w}{b_3} + \cdots,$$ (D.16)

where

$$b_{4n} = 1, \; b_{4n-2} = -1, \; b_{4n-3} = -1/n, \; b_{4n-1} = 1/n.$$

Using theorems from Perron (1957, pp.139-141), the c.f. (D.16) corresponds to

$$- w + \frac{w}{1} - \frac{w/2}{1} + \frac{w}{h_1} + \frac{w}{h_2} + \frac{w}{h_3} + \cdots,$$ (D.17)

where

$$h_{4n-3} = -\frac{2}{2n-1}, \; h_{4n-1} = \frac{2}{2n-1}, \; h_{4n-2} = -\frac{(2n-1)^2}{4n^2}, \; h_{4n} = 1.$$

Finally, after applying an appropriate equivalence transformation to c.f. (D.17), we obtain our desired continued fraction

$$- w + \mathop{\mathbf{K}}_{n=1}^{\infty} \frac{d_n w}{1}, \; w = 1/z, \text{ where}$$ (D.18)

$$d_1 = 1; \; d_{4n+2} = -d_{4n-1} = n - 1/2; \; d_{4n} = -d_{4n+1} = n^2/(n - 1/2)$$

and which is the same as (D.6). It turns out that the odd part of (D.18) is

$$\frac{w^2/2}{1 - w} + \frac{1^2 w^2}{1} + \frac{1^2 w^2}{1 - w} + \frac{2^2 w^2}{1} + \frac{2^2 w^2}{1 - w} + \cdots$$ (D.19)

which in turn is equivalent to both (D.9) and (D.10). The even part of $K(d_n w/)$ in (D.18) is given by

$$\frac{d_1 w}{1 + d_2 w} - \frac{d_2 d_3 w^2}{1 + (d_3 + d_4)w} - \frac{d_4 d_5 w^2}{1 + (d_5 + d_6)w} - \cdots.$$ (D.20)

Since the elements of c.f. (D.20) are positive if $z > 3/2$, the c.f. converges under this restriction on z by Theorem 3.1. Since $K(d_n w/1)$ is an alternating c.f. when $z > 3/2$, it now follows from Theorem 3.2 that it (and therefore also (D.18)) converges if $z > 3/2$. Since we have shown that the odd part of (D.18) converges to $-\beta'(z)$ when $\Re(z) > 1/2, z \ni (0,1]$ we finally have that (D.18) and hence (D.6) converges to $-\beta'[(z)$ for $z \in (3/2, \infty)$. We have not yet investigated the convergence behavior of (D.6) elsewhere in the complex plane.

FUNCTION E (Wall(1948, 93.11)) :

$$F(t, z) = \int_0^\infty \frac{1-t}{e^{(1-t)u} - t} e^{-zu}\, du, \ t > 0 \tag{E.1}$$

$$= \sum_{k=1}^\infty \frac{(1-t)^{k-1}}{z + k(1-t)}, \ 0 < t < 1 \tag{E.2}$$

$$= \sum_{k=1}^\infty \frac{(t-1)t^{-k}}{z + (k-1)(t-1)}, \ t > 1 \tag{E.3}$$

$$= \frac{1}{z} + \frac{1}{1} + \frac{t}{z} + \frac{2}{1} + \frac{2t}{z} + \frac{3}{1} + \frac{3t}{z} + \cdots, \ |\arg z| < \pi \tag{E.4}$$

$$= \left[z + 1 + \mathop{K}_{n=1}^\infty \frac{-n^2 t}{z + 1 + n(t+1)} \right]^{-1}, \ |\arg z| < \pi \tag{E.5}$$

$$F(1, t) = \int_0^\infty \frac{e^{-u}}{z + u}\, du, \ t = 1 \ \text{(Wall, 1948, p.355)} \tag{E.6}$$

$$= \frac{1}{z} + \frac{1}{1} + \frac{1}{z} + \frac{2}{1} + \frac{2}{z} + \frac{3}{1} + \frac{3}{z} + \cdots, \ |\arg z| < \pi \tag{E.7}$$

$$= \frac{1}{z+1} - \frac{1^2}{z+3} - \frac{2^2}{z+5} - \frac{3^2}{z+7} - \cdots, \ |\arg z| < \pi \tag{E.8}$$

$$= -e^z E_i(-z) \ \text{(Wall,1948, 92.16)} \tag{E.9}$$

$$= -e^z \left[\gamma + Log\, z + \sum_{n=1}^\infty \frac{(-z)^n}{n \cdot n!} \right] \tag{E.10}$$

$$= z^{-1/2} e^{z/2} W_{-1/2,0}(z) \ \text{(Oberhettinger and Badii,1970, p.419)}, \tag{E.11}$$

where

$$E_i(-z) = -\int_z^\infty \frac{e^{-u}}{u}\, du.$$

Proof: The continued fractions (E.5) and (E.8) are the even parts of the continued fractions (E.4) and (E.7), respectively. Now let $H(t, z)$ denote the limit function of the continued fraction (E.4). Then by an equivalence transformation

$$H(t, z) = \left[b_0 + \mathop{K}_{n=1}^\infty \frac{a_n}{b_n} \right]^{-1}, \tag{E.12}$$

where

$$a_{2n-1} = nt, \ a_{2n} = nt^2; \ b_{2n-1} = t, \ b_{2n} = z$$

We first prove that $H(t, z) = F(t, z)$ if $t \neq 1$ and $z > 0$. To accomplish this we use the Bauer-Muir transform of $b_0 + K(a_n/b_n)$ in (E.12) with

$$r_{2n} = \frac{2}{t-1}; \ r_{2n-1} = \frac{n + |1 - t|}{z} + \frac{t}{|1 - t|}; \ \varphi_n = \frac{zt^2}{|1 - t|(1 - t)}. \tag{E.13}$$

If $t > 1$, the Bauer-Muir transform determined by (E.13) along with Theorem (3.5) give

$$H(t, z + t - 1) - tH(t, z) = (1 - t)/z$$

which leads to

$$H(t,z) = \sum_{k=1}^{\infty} \frac{(t-1)t^{-k}}{z+(k-1)(t-1)} = F(t,z), \; z > 0, \; t > 1. \tag{E.14}$$

Similarly, if $0 < t < 1$, we obtain

$$H(t,z) - tH(t, z+1-t) = \frac{1-t}{z+1-t},$$

whence

$$H(t,z) = \sum_{k=1}^{\infty} \frac{(1-t)t^{k-1}}{z+k(1-t)} = F(t,z), \; z > 0, \; 0 < t < 1. \tag{E.15}$$

Since the sums in (E.14) and (E.15) and the limit functions of the continued fractions (E.5) and (E.6) are analytic if $|\arg z| < \pi$, it follows that $F(t,z) = H(t,z)$ when $0 < t < 1$ or $t > 1$ and $|\arg z| < \pi$. When $t = 1$, the convergence of (E.7) and (E.8) to

$$\int_0^\infty \frac{e^{-u}}{z+u}\,du$$

for all z satisfying $|\arg z| < \pi$ follows from formula (92.6) in Wall (1948). Stieltjes moment theory was used to derive the latter result.

FUNCTION F (Wall(1948, 94.1)) :

$$
\begin{aligned}
F_m(a,z) &= \int_0^\infty \frac{e^{-zu}}{(\cosh u + a \sinh u)^m}\,du, \; m > 0, \; a \geq 0, \; \Re(z) > -m \tag{F.1}\\[2mm]
&= \frac{2^{m-1}}{(1+a)^m} \int_0^\infty \frac{e^{-\frac{z+m}{2}u}}{(1+\frac{1-a}{1+a}e^{-u})^m}\,du \tag{F.2}\\[2mm]
&= \frac{2^m}{(1+a)^m} \sum_{n=0}^\infty (-1)^n \frac{(m)_n}{n!} \frac{\left(\frac{1-a}{1+a}\right)^n}{z+m+2n} \tag{F.3}\\[2mm]
&= \frac{2}{(1+a)(z+m)} - \frac{(1-a)(z+2-m)}{(1+a)(z+m)} F_m(a, z+2) \tag{F.4}\\[2mm]
&= \left[z + ma + \underset{n=1}{\overset{\infty}{\mathrm{K}}} \frac{n(m+n-1)(1-a^2)}{z+(m+2n)a} \right]^{-1}, \; m > 0, \; valid\ if \tag{F.5}\\
&\qquad (a)\ 0 \leq a < 1, \; z > -ma, \; or\\
&\qquad (b)\ \sqrt{m/(m+1)} \leq a < 1, \; \Re(z) > (2-m)a, \; or\\
&\qquad (c)\ a > 1, \; \Re(z) > -m\\[2mm]
&= \left[b_0(z) + \underset{n=1}{\overset{\infty}{\mathrm{K}}} \frac{a_n}{b_n(z)} \right]^{-1}, \; m > 0, \; a > 1, \; \Re(z) > -m, \; where \tag{F.6}\\
&\qquad a_{2n} = n(a+1); \; a_{2n-1} = (m+n-1)(a-1); \; b_{2n}(z) = z+m, \; b_{2n-1}(z) = 1
\end{aligned}
$$

$$F_m(0,z) = \frac{1}{z} + \frac{m}{z} + \frac{2(m+1)}{z} + \frac{3(m+1)}{z} + \cdots, \; m > 0, \; \Re(z) > 0 \tag{F.7}$$

$$F_m(1,z) = \frac{1}{z+m}, \; m > 0, \; \Re(z) > 0. \tag{F.8}$$

Proof: Wall (1948, p.366) has verified the special $a = 0$ case, that is the validity of expansion (F.7), using Stieltjes moment theory and the theory of J-fractions. Termwise integration of a series expansion of the integrand of (F.2) leads to the series expansion (F.3). The functional relationship (F.4) is easily verified using (F.3). We shall now sketch proofs using Bauer-Muir transforms justifying expansions (F.5) and (F.6) for the indicated ranges of the parameters involved. The B-M sequences for the continued fraction in brackets in (F.5) are

$$
\begin{aligned}
r_{2n}(z) &= 2(1-a)n + (1-a)(m-z)/2 \\
r_{2n+1}(z) &= 2(1-a)n + (1-a)(m+2-z)/2 \\
\varphi_n(z) &= (1-a^2)(z+2-m)(z+m)/4
\end{aligned}
\right\}
\qquad (F.9)
$$

when $0 \leq a < 1$. When $a > 1$ the B-M sequences for the c.f. in brackets in (F.6) are

$$
\begin{aligned}
r_{2n}(z) &= (a-1)(z+m)/2 \\
r_{2n-1}(z) &= [4n + (a-1)(z+2-m)]/[2(z+m)] \\
\varphi_n(z) &= (1-a^2)(z+2-m)/4
\end{aligned}
\right\}.
\qquad (F.10)
$$

Now let us assume that $0 \leq a < 1$, z is real, and $z > -ma$. Then the elements of the continued fraction (F.5) are positive real numbers and under these assumptions it converges by Theorem 3.1. We let $H(z)$ denote the limit function of (F.5), calculate the Bauer-Muir transform of the c.f. in brackets in (F.5) using (F.9), employ Theorem 3.5, and arrive at the functional relation

$$
H(z) = \frac{2}{(1+a)(z+m)} - \frac{(1-a)(z+2-m)}{(1+a)(z+m)} H(z+2)
$$

which is the same as (F.4) with $F_m(z)$ replaced by $H(z)$. We point out that the sequence $\{r_n\}$ in (F.9) is eventually positive under our assumptions so we can make use of part b) of Theorem 3.5. Solving the functional equation above, using the fact that $H(z+n+2) \to 0$ as $n \to \infty$, we find that $H(z)$ is equal to the sum (F.3) if $z > -ma$. Hence $H(z) = F_m(a, z)$ if $0 \leq a < 1$ and $z > -ma$. It follows that (F.5) is valid if $a = 0$ and $\Re(z) > 0$, since both the series (F.3) and the c.f. (F.7) converge to analytic functions in the open right half plane. The reason (F.7) converges to an analytic function for $\Re(z) > 0$ is that the c.f. equal to z^{-1} times the c.f. (F.7) is equivalent to an S-fraction in the variable z^{-2}.

Our goal now is to shed further light on the convergence behavior of c.f. (F.5). Let

$$
\sqrt{m/(m+1)} \leq a < 1, \quad \Re(z) > (2-m)a, \quad m > 0.
\qquad (F.11)
$$

Then $1/2$ times the c.f. (F.5) can be put in the form

$$
\underset{n=0}{\overset{\infty}{\mathrm{K}}} \frac{\alpha_n(z)}{1},
\qquad (F.12)
$$

where

$$
\alpha_0(z) = \frac{1}{2(z+ma)}, \quad \alpha_n(z) = \frac{n(n+m-1)(1-a^2)}{[z+(m+2n-2)a][z+(m+2n)a]}, \quad n \geq 1.
$$

Let $z = x + iy$. Then clearly

$$
|\alpha_0(z)| \leq 1/[2(x+ma)] \leq 1/4
$$

and for $n \geq 1$

$$|\alpha_n(z)| \leq \frac{n(n+m-1)(1-a^2)}{[x+(m+2n-2)a][x+(m+2n)a]}$$

$$\leq \frac{(n+m-1)(1-a^2)}{4na^2} \leq \frac{m(1-a^2)}{4a^2} \leq \frac{1}{4}.$$

Since $|\alpha_n(z)| \leq 1/4$ under the conditions (F.11), it follows fro, Worpitzky's Theorem (Theorem 10.1, Wall(1948, p.42)) the c.f. (F.12) converges uniformly under these conditions on z and the constants a and m. Futhermore, the approximants of (F.12) lie in the closed unit disk $\{z : |z| \leq 1/2\}$. Thus the approximants of (F.12) are analytic functions of z if $\Re(z) > (2-m)a$, and therefore (F.12) converges to an analytic function on this half plane. Since the c.f. (F.5) is a constant multiple of the C.f. (F.12), it also converges to an analytic function under conditions (F.11). We proved above that c.f. (F.5) converges to the function (F.3) when $z > -ma$, so it follows from the Identity Theorem that (F.5) converges to $F_m(a, z)$ under conditions (F.11). That (F.5) is valid under conditions (c) follows from the fact the c.f. (F.5) is the even part of c.f. (F.6).

It remains to justify the validity of expansion (F.6). Toward this end we let $G(z)$ denote the limit function of (F.6), calculate the Bauer-Muir transform of the c.f. in brackets in (F.6) using (F.10), employ Theorem 3.5, and arrive at the functional equation

$$G(z) = \frac{2}{(1+a)(z+m)} - \frac{(1-a)(z+2-m)}{(1+a)(z+m)}G(z+2)$$

which is the same as (F.4) with $F_m(z)$ replaced by $G(z)$. Theorem 3.1 can be used to establish that c.f. (F.6) converges for real values of z greater than $-m$. With the aid of the above functional equation it is not difficult to confirm that the limit in this case is the sum (F.3). Since expansion (F.6) is equivalent to an S-fraction in the variable $(z+m)^{-1}$, it follows from Theorem 3.3 that the c.f. converges to an analytic function in the half plane $\Re(z) > -m$. Thus, since the series (F.3) converges to an analytic function on the same half plane, expansion (F.6) is valid as asserted.

FUNCTION G (Wall(1948, 94.2)) :

$$F(z) = \int_0^\infty \tanh u\, e^{-zu}\, du = \int_0^\infty \frac{e^{-zu}}{z\cosh^2 u}\, du, \ \Re(z) > 0 \tag{G.1}$$

$$= \frac{1}{2}\left[\Psi\left(\frac{z+2}{4}\right) - \Psi\left(\frac{z}{4}\right)\right] - \frac{1}{z}, \ \Re(z) > 0 \tag{G.2}$$

$$= \sum_{k=0}^\infty \frac{2(-1)^k}{(z+2k+1)^2-1}, \ \Re(z) > 0 \tag{G.3}$$

$$= \frac{1}{z^2} + \frac{1\cdot 2}{1} + \frac{2\cdot 3}{z^2} + \frac{3\cdot 4}{1} + \frac{4\cdot 5}{z^2} + \cdots, \ \Re(z) > 0 \tag{G.4}$$

$$= \frac{1}{z}\left[z + \underset{n=1}{\overset{\infty}{\mathrm{K}}}\frac{n(n+1)}{z}\right]^{-1}, \ \Re(z) > 0. \tag{G.5}$$

$$= w - \frac{w}{1} + \frac{w}{1} - \frac{w}{1} + \frac{2w}{1} - \frac{2w}{1} + \frac{3w}{1} - + \cdots, \ w = \frac{1}{z}, \ \Re(z) > 0 \tag{G.6}$$

$$= -w + \frac{w}{1} - \frac{w}{1} + \frac{w}{1} - \frac{2w}{1} + \frac{2w}{1} - \frac{3w}{1} + - \cdots, \ w = \frac{1}{z}, \ \Re(z) > 1. \tag{G.7}$$

Proof: It is easy to verify that c.f. (G.5) is equivalent to c.f. (G.4) when $\Re(z) > 0$. The validity of expansions (G.4) and (G.5) follow from our arguments for Function G, since

$$zF(z) = F_m(0, z) = \int_0^\infty \frac{e^{-zu}}{\cosh^2 u} \, du.$$

However, for the sake of completeness we shall include a brief argument here. Let

$$r_n(z) = n + 1 - z/2, \ n \geq 0; \varphi_n(z) = (1 + z/2)/z/2. \tag{G.8}$$

Now let us assume that z is a positive real naumber. Then (G.5) has positive elements and it converges by Theorem 3.1. We apply a B-M transformation to the continued fraction in brackets in (G.5) with $r_n(z)$ and $\varphi_n(z)$ given by (G.8) and obtain

$$
\begin{aligned}
F(z) &= \frac{\frac{1}{z}}{1 + \frac{z}{2}} + \frac{\frac{z}{2}(1 + \frac{z}{2})}{2 + \frac{z}{2}} + \frac{1 \cdot 2}{z + 2} + \frac{2 \cdot 3}{z + 2} + \cdots \\
&= \frac{1/z}{1 + z/2} + \frac{z(1 + z/2)/2}{-z/2} + \frac{1}{(z + 2)F(z + 2)} \\
&= -F(z + 2) + \frac{2}{(z + 1)^2 - 1}, \ z > 0. \tag{G.9}
\end{aligned}
$$

From (G.11) we get

$$\sum_{k=0}^{n} (-1)^k [F(z + 2k) + F(z + 2k + 2)] = F(z) + (-1)^n F(z + 2n + 2)$$

Hence

$$F(z) = \sum_{k=0}^{\infty} \frac{2(-1)^k}{(z + 2k + 1)^2 - 1}, \ z > 0, \tag{G.10}$$

since $F(z + 2n + 2) \to 0$ as $n \to \infty$. But

$$
\begin{aligned}
\sum_{k=0}^{\infty} \frac{2(-1)^k}{(z + 2k + 1)^2 - 1} &= \frac{1}{2}\left[\Psi\left(\frac{z + 2}{4}\right) - \Psi\left(\frac{z}{2}\right)\right] - \frac{1}{z} \\
&= \int_0^\infty \tanh u \, e^{-zu} \, du \tag{G.11}
\end{aligned}
$$

when $\Re(z) > 0$. The c.f. (G.5) is equivalent to the following S-fraction in w^2, where $w = 1/z$

$$\frac{w^2}{1} + \frac{1 \cdot 2w^2}{1} + \frac{2 \cdot 3w^2}{1} + \frac{3 \cdot 4w^2}{1} + \cdots \tag{G.12}$$

and which converges therefore by Theorem 3.3 to an analytic function on the half plane $\Re(z) > 0$. This function must agree there with the limit of the series in (G.11) above. Thus expansions (G.4) and (G.5) are valid for $\Re(z) > 0$.

We shall extend (G.12) in two different ways. We first obtain the extension

$$\hat{F}(z) = w - \frac{w}{1} + \frac{w}{1} - \frac{w}{1} + \frac{2w}{1} - \frac{2w}{1} + \frac{3w}{1} - \frac{3w}{1} + \cdots \tag{G.13}$$

The odd part of (G.13) is (G.12). Let $P_n(Q_n)$ denote the nth numerator(denominator) of (G.13). Then

$$
\left| \frac{P_{2n}}{Q_{2n}} - \frac{P_{2n+1}}{Q_{2n+1}} \right| = \frac{|a_{2n+1}| \left| \frac{P_{2n+1}}{Q_{2n+1}} - \frac{P_{2n-1}}{Q_{2n-1}} \right|}{\left| \frac{Q_{2n+1}}{Q_{2n-1}} - a_{2n+1} \right|}, \; n \geq 1
$$

$$
= \frac{|nw| \left| \frac{P_{2n+1}}{Q_{2n+1}} - \frac{P_{2n-1}}{Q_{2n-1}} \right|}{\left| \frac{Q_{2n+1}}{Q_{2n-1}} + nw \right|}
$$

$$
= \frac{n \left| \frac{P_{2n+1}}{Q_{2n+1}} - \frac{P_{2n-1}}{Q_{2n-1}} \right|}{\left| \frac{1}{w} \frac{Q_{2n+1}}{Q_{2n-1}} + n \right|}.
$$

It is easy to verify by induction that

$$
\Re \left(\frac{1}{w} \frac{Q_{2n+1}}{Q_{2n-1}} \right) = \Re \left(z \frac{Q_{2n+1}}{Q_{2n-1}} \right) > 0
$$

if $\Re(z) > 0$. It follows that $\hat{F}(z) = F(z)$ if $\Re(z) > 0$.

Now let us extend (G.12) to the following regular δ-fraction in w

$$
F^*(z) = -w + \frac{w}{1} - \frac{w}{1} + \frac{w}{1} - \frac{2w}{1} + \frac{2w}{1} - + \cdots \tag{G.14}
$$

The odd part of (G.14) is (G.12), so the odd part converges to $F(z)$ if $\Re(z) > 0$. The even part of (G.14) is

$$
-w + \frac{w}{1-w} + \frac{w^2}{1-w} + \frac{2^2 w^2}{1-w} + \frac{3^2 w^2}{1-w} + \cdots \tag{G.15}
$$

which is equivalent to

$$
-w + \frac{\hat{w}}{1} + \frac{\hat{w}^2}{1} + \frac{2^2 \hat{w}^2}{1} + \frac{3^2 w^2}{1} + \cdots , \tag{G.16}
$$

where $\hat{w} = w/(1 - w)$. Since $w = 1/z$, (G.16) is equal to

$$
\tilde{F}(z) = -\frac{1}{z} + \frac{\hat{z}}{1} + \frac{1^2 \hat{z}^2}{1} + \frac{2^2 \hat{z}^2}{1} + \frac{3^2 \hat{z}^2}{1} , \tag{G.17}
$$

where $\hat{z} = 1/(z - 1)$. The continued fraction (G.17) converges to an analytic function if $\Re(z) > 1$. But with the aid of expansion (F.7) with $m = 1$ for Function G, c.f. (G.17) converges to

$$
\tilde{F}(z) = -\frac{1}{z} + \int_0^\infty \frac{e^{-(z-1)u}}{\cosh u} \, du, \; \Re(z) > 1. \tag{G.18}
$$

The expression on the right of (G.18) is

$$
\hat{F}(z) = -\frac{1}{z} + \frac{1}{2} \left[\Psi \left(\frac{z+2}{4} \right) - \Psi \left(\frac{z}{4} \right) \right]
$$

$$
= \int_0^\infty \tanh u \, e^{-zu} \, du. \tag{G.19}
$$

Hence

$$
\hat{F}(z) = F^*(z) = \tilde{F}(z) = F(z), \; \Re(z) > 1
$$

and our argument is complete.

FUNCTION H (Wall(1948, 94.3)) :

$$F(z) = \int_0^\infty \frac{c\sinh au \sinh bu}{\sinh cu} e^{-zu}\, du, \quad c = a+b,\ a>0,\ b>0,\ \Re(z)>0 \tag{H.1}$$

$$= F(z+2c) + \frac{4abc(z+c)}{z(z+2a)(z+2b)(z+2c)} \tag{H.2}$$

$$= \sum_{k=0}^\infty \frac{4abc(z+c+2kc)}{[(z+c+2kc)^2 - (a+b)^2][(z+c+2kc)^2 - (a-b)^2]} \tag{H.3}$$

$$= \frac{1}{4}\left[\Psi\left(\frac{z+2b}{2c}\right) - \Psi\left(\frac{z}{2c}\right) - \Psi\left(\frac{z+2c}{2c}\right) + \Psi\left(\frac{z+2a}{2c}\right)\right] \tag{H.4}$$

$$= \frac{ab}{z}\left[z + \mathop{\mathrm{K}}_{n=1}^{\infty}\frac{a_n}{(n+1)z}\right]^{-1}, \quad \Re(z)>0,\ \text{where} \tag{H.5}$$

$$a_{2n-1} = 4n^2(nc-a)(nc-b); \quad a_{2n} = 4n^2(nc+a)(nc+b).$$

$$= \frac{ab}{z^2+\beta_1} - \frac{\alpha_1}{z^2+\beta_2} - \frac{\alpha_2}{z^2+\beta_3} - \cdots, \quad \Re(z)>0,\ \text{where} \tag{H.6}$$

$$\beta_n = (2n^2-2n+1)c^2 - a^2 - b^2$$

$$\alpha_n = 4n^2(n^2c^2-a^2)(n^2c^2-b^2)/(4n^2-1)$$

$$= -z^{-1} + \frac{z^{-1}}{1} - \frac{\sigma_1 z^{-1}}{1} + \frac{\sigma_1 z^{-1}}{1} - \frac{\sigma_2 z^{-1}}{1} + \frac{\sigma_2 z^{-1}}{1} - \cdots, \quad \text{where} \tag{H.7}$$

$$c = a+b = 2, z > 1+(1-a)^2;\ \sigma_{2n} = 2n,\ \sigma_{2n+1} = \frac{(nc+a)(nc+b)}{2n+1}$$

$$= z^{-1} - \frac{z^{-1}}{1} + \frac{\sigma_1 z^{-1}}{1} - \frac{\sigma_1 z^{-1}}{1} + \frac{\sigma_2 z^{-1}}{1} - \frac{\sigma_2 z^{-1}}{1} + \cdots, \quad \text{where} \tag{H.8}$$

$$\sigma_{2n} = 2n,\ \sigma_{2n+1} = (nc+a)(nc+b)/(2n+1);\ \Re(z)>0.$$

Proof: Wall (1948, p.370) lists the expansion (H.6) for $F(z)$ without proof and without any restrictions on the parameters such as the ones we have imposed in (H.1). Expansion (H.6) was originally derived by Stieltjes (1890, p.383), formally without an investigation of its validity. We shall not attempt here to investigate the validity of this expansion in full generality. The fact that the expansion (H.6) is valid under the conditions $c = a+b, a > 0, b > 0$ when $\Re(z) > 0$ follows from the validity of (H.5) since c.f. (H.6) is the even part of c.f. (H.5). In order to verify that c.f. (H.5) converges when $\mathrm{Re}\, z > 0$ to $F(z)$ as given by (H.1) and (H.3), we use a B-M transformation of $z + K(a_n/[(n+1)z])$ with the modifying sequences determined by

$$r_{2n}(z) = 2n^2 c + n(2c-z) - z + \frac{(z+2a)(z+2b)(z+2c)}{4c(z+c)}$$

$$r_{2n+1}(z) = 2n^2 c + n(4c-z) + \frac{zr_0(z) + 4c^2}{z+2c} \tag{H.9}$$

$$\varphi_n(z) = [z+r_0(z)][2c - z - r_1(z)].$$

With the aid of this transform we are led to the relation (H.2) when $z > 0$ and to the fact that c.f. (H.5) converges to the sum (H.3) on the positive real axis. Expansion (H.5) is equivalent to

$$\frac{abz^{-2}}{1} + \frac{\hat a_1 z^{-2}}{1} + \frac{\hat a_2 z^{-2}}{1} + \frac{\hat a_3 z^{-2}}{1} + \cdots,$$

where $\hat{a}_n = a_n/[n(n+1)]$ and which is an S-fraction in the variable z^{-2} that converges to an analytic function in the half plane $\Re(z) > 0$ by Theorem 3.3. Since the series (H.3) also converges to an analytic function on this half plane and this function agrees with $F(z)$ on the positive real axis, the validity of (H.5) follows.

Our goal now is to obtain the regular δ-fraction expansion for $F(z)$ with respect to the variable z^{-1}. As a first step toward this goal we observe that c.f. (H.5) is equivalent to

$$F(z) = \frac{ab}{z^2} + \frac{a_1}{2} + \frac{a_2}{3z^2} + \frac{a_3}{4} + \frac{a_4}{5z^2} + \cdots$$

and this c.f. is equivalent to

$$F(z) = \frac{ab}{z^2} + \frac{c_1}{1} + \frac{c_2}{3z^2} + \frac{c_3}{1} + \frac{c_4}{5z^2} + \cdots , \tag{H.10}$$

where

$$c_{2n-1} = 2n(nc - a)(nc - b); \quad c_{2n} = 2n(nc + a)(nc + b).$$

Using the fact that $c = a + b$, it is easy to see that

$$\prod_{k=1}^{n} \frac{c_{2k-1}}{c_{2k}} = \frac{ab}{(nc + a)(nc + b)}.$$

With the aid of this result we are able to deduce that c.f. (H.10) is equivalent to

$$F(z) = \frac{z^{-2}}{d_1} + \frac{z^{-2}}{d_2} + \frac{z^{-2}}{d_3} + \cdots , \tag{H.11}$$

where

$$d_{2n} = \frac{1}{2n}; \quad d_{2n+1} = \frac{2n+1}{(nc + a)(nc + b)}.$$

We extend c.f. (H.11) to obtain

$$F(z) = -z^{-1} + \frac{z^{-1}}{1} - \frac{z^{-1}}{d_1} + \frac{z^{-1}}{1} - \frac{z^{-1}}{d_2} + \frac{z^{-1}}{1} - \frac{z^{-1}}{d_3} + \cdots . \tag{H.12}$$

The continued fraction (H.12) is equivalent to

$$F(z) = -z^{-1} + \frac{z^{-1}}{1} - \frac{\sigma_1 z^{-1}}{1} + \frac{\sigma_1 z^{-1}}{1} - \frac{\sigma_2 z^{-1}}{1} + \frac{\sigma_2 z^{-1}}{1} - \cdots , \tag{H.13}$$

where $\sigma_n = 1/d_n$ and which is our desired δ-fraction. More explicitly,

$$\sigma_{2n} = 2n; \quad \sigma_{2n+1} = \frac{(nc + a)(nc + b)}{2n + 1}.$$

It is interesting to note that if $a = b$, then

$$\sigma_{2n+1} = a^2(2n + 1).$$

It turns out that the odd part of c.f. (H.13) is c.f. (H.11). Now let us take another look at c.f. (H.13). Its even part is given by

$$-w + \frac{w}{1 - \sigma_1 w} + \frac{\sigma_1^2 w^2}{1 + (\sigma_1 - \sigma_2)w} + \frac{\sigma_2^2 w^2}{1 + (\sigma_2 - \sigma_3)w} + \cdots , \tag{H.14}$$

where $w = 1/z$. Here we set

$$c = a + b = 2.$$

Then

$$\sigma_{2n} = 2n; \quad \sigma_{2n+1} = 2n + 1 - \frac{(1-a)^2}{2n+1}$$

$$\sigma_{2n+1} - \sigma_{2n+2} = -1 - \frac{(1-a)^2}{2n+1}$$

$$\sigma_{2n} - \sigma_{2n+1} = -1 + \frac{(1-a)^2}{2n+1}.$$

We claim c.f. (H.14) converges by Theorem 3.1 when $z > 1 + (1-a)^2$. If f_n is the nth approximant of (H.13) and H_n is given by (2.2), then from formula (2.13)

$$|f_{2n+1} - f_{2n}| = \frac{|H_{2n}||f_{2n+2} - f_{2n}|}{|H_{2n} + \sigma_{n+1}w|}$$

$$\leq |f_{2n+2} - f_{2n}|, z > 1 + (1-a)^2.$$

Since $\{f_{2n}\}$ converges when $z > 1 + (1-a)^2$, it follows that $\{f_{2n+1}\}$ also converges and to the same limit. Thus the c.f.'s (H.13) and (H.7), under the restriction $a + b = 2$, converge to $F(z)$ when $z > 1 + (1-a)^2$. We feel that expansion (H.7) is valid for other values of z, but so far we have not seen how to extend our convergence set significantly for this continued fraction.

We shall now go about the task of establishing the validity of expansion (H.8). The continued fraction (H.11) can be extended to

$$F(z) = z^{-1} - \frac{z^{-1}}{1} + \frac{z^{-1}}{d_1} - \frac{z^{-1}}{1} + \frac{z^{-1}}{d_2} - \frac{z^{-1}}{1} + \frac{z^{-1}}{d_3} - \frac{z^{-1}}{1} + \cdots$$

which is equivalent to

$$F(z) = z^{-1} - \frac{z^{-1}}{1} + \frac{\sigma_1 z^{-1}}{1} - \frac{\sigma_1 z^{-1}}{1} + \frac{\sigma_2 z^{-1}}{1} - \frac{\sigma_2 z^{-1}}{1} + \cdots . \tag{H.15}$$

C.f. (H.15) which is the same as c.f. (H.8) is not a regular δ-fraction in z^{-1}, but it is the negative of such a continued fraction. From formula (2.12) we have that

$$|f_{2n+1} - f_{2n}| = \frac{|\gamma_{2n+1}||f_{2n+1} - f_{2n-1}|}{|H_{2n-1} - \gamma_{2n+1}|}, \tag{H.16}$$

where H_n is given by (2.7), f_n is the nth apparoximant of (H.15), and γ_n is the nth partial numerator of (H.15). It is not difficult to derive from (H.16) that

$$|f_{2n+1} - f_{2n}| = \frac{|\sigma_n||f_{2n+1} - f_{2n-1}|}{|z H_{2n-1} + \sigma_n|}, \tag{H.17}$$

where we define $\sigma_0 = 1$. It is easily established by induction that $\Re(z H_{2n-1}) > 0$ when $\Re(z) > 0$, so it follows from (H.17) that

$$|f_{2n+1} - f_{2n}| \leq |f_{2n+1} - f_{2n-1}|.$$

Since the odd part of (H.15) is (H.11) and c.f. (H.11) converges to $F(z)$ when $\Re(z) > 0$, we have finally established the validity of (H.8).

FUNCTION I (Wall(1948, 94.4)) :

$$F(z) = \int_0^\infty \frac{b \sinh au}{\sinh bu} e^{-zu}\, du, \ b > a > 0, \ \Re(z) > a - b \tag{I.1}$$

$$= \frac{2ab}{(z+b)^2 - a^2} + F(z + 2b) \tag{I.2}$$

$$= \sum_{k=0}^\infty \frac{2ab}{[z + (2k+1)b]^2 - a^2} \tag{I.3}$$

$$= \frac{1}{2}\left[\Psi\left(\frac{z+a+b}{2b}\right) - \Psi\left(\frac{z-a+b}{2b}\right) \right] \tag{I.4}$$

$$= a\left[z + \mathop{\text{K}}_{n=1}^\infty \frac{n^2(n^2b^2 - a^2)}{(2n+1)z} \right]^{-1}, \ b > a > 0, \ \Re(z) > 0. \tag{I.5}$$

Proof: We define the modifying sequences for a B-M transform of the c.f. in the brackets in (I.5) by

$$r_n(z) = bn^2 + (b + (-1)^n z)n + \frac{z^2 + b^2 - a^2}{2b} \tag{I.6}$$

$$\varphi_n(z) = -\left[\frac{(z+b)^2 - a^2}{2b} \right]^2. \tag{I.7}$$

By Theorem 3.1 the continued fraction (I.5) converges to a function $G(z)$ for all positive real values of z. If (I.5) is modified by replacing the c.f. in brackets by its transform using (I.6) and (I.7), then the limit of this new c.f. is also $G(z)$ for $z > 0$ by Theorem 3.5. But this forces the relation

$$G(z) = \frac{2ab}{(z+b)^2 - a^2} + G(z + 2b)$$

which is easily solved to give

$$G(z) = F(z) = \sum_{k=0}^\infty \frac{2ab}{[z + (2k+1)b]^2 - a^2}, \ z > 0.$$

C.f. (I.5) multiplied by the factor $1/z$ is equivalent to the c.f.

$$\mathop{\text{K}}_{n=0}^\infty \frac{a_n z^{-2}}{1}, \ \text{where}$$

$$a_0 = a, \ a_n = n^2(n^2b^2 - a^2)/(4n^2 - 1), \ n \geq 1.$$

This c.f. is an S-fraction in z^{-2} and it converges to an analytic function in the half plane $\Re(z) > 0$ by Theorem 3.3. Since the series (I.3) also converges to an analytic function on this half plane, it now follows that expansion (I.5) is valid under the conditions asserted.

FUNCTION J (Wall(1948, 94.5)) :

$$H(a,b,z) = \int_0^\infty F(a,b,(a+b+1)/2; -\sinh^2 u)e^{-zu}\, du \tag{J.1}$$

$$= \frac{1}{z} + \frac{a_1}{z} + \frac{a_2}{z} + \frac{a_3}{z} + \cdots, \quad where\ a > 0, b > 0, a+b > 1 \tag{J.2}$$

$$a_{n+1} = \frac{4(n+a)(n+b)(n+1)(n+a+b-1)}{(2n+a+b-1)(2n+a+b+1)}, \quad \Re(z) > 0$$

$$= \frac{c_0}{b_0(z)} + \frac{c_1}{b_1(z)} + \frac{c_3}{b_3(z)} + \cdots, \quad \Re(z) > 0,\ where \tag{J.3}$$

$$c_0 = (a+b-1)/2; \quad c_n = (a+n-1)(b+n-1)n(a+b+n-2),\ n \geq 1$$

$$b_n(z) = [n + (a+b-1)/2]z,\ n \geq 0$$

$$= \frac{(z+4-2a)(z+4-2b)}{(z+2a)(z+2b)}H(a,b,z+4) + \frac{4(a+b-1)}{(z+2a)(z+2b)} \tag{J.4}$$

$$= \frac{a+b-1}{4} \frac{\Gamma\left(\frac{z+2a}{4}\right)\Gamma\left(\frac{z+2b}{4}\right)}{\Gamma\left(\frac{z+4-2a}{4}\right)\Gamma\left(\frac{z+4-2b}{4}\right)} \sum_{k=0}^\infty \frac{\Gamma\left(\frac{z+4-2a}{4}+k\right)\Gamma\left(\frac{z+4-2b}{4}+k\right)}{\Gamma\left(\frac{z+4+2a}{4}+k\right)\Gamma\left(\frac{z+4+2b}{4}+k\right)} \tag{J.5}$$

$$H\left(\frac{m}{2}, 1+\frac{m}{2}, z\right) = \int_0^\infty \frac{e^{-zu}}{\cosh^m u}\, du \tag{J.6}$$

$$H\left(1+\frac{a}{2}, 1-\frac{a}{2}, z\right) = \frac{1}{a}\int_0^\infty 2\frac{\sinh(au)}{\sinh(2u)}e^{-zu}\, du, \quad a \neq 0 \tag{J.7}$$

$$= \frac{1}{2a}\left[\Psi\left(\frac{z+2+a}{4}\right) - \Psi\left(\frac{z+2-a}{4}\right)\right] \tag{J.8}$$

$$H(1,1,z) = \frac{1}{4}\Psi'\left(\frac{z}{4}+\frac{1}{2}\right) \tag{J.9}$$

$$H(3/2,3/2,z) = -2z + 8\left[\frac{\Gamma\left(\frac{z+3}{4}\right)}{\Gamma\left(\frac{z+1}{4}\right)}\right]^2 \tag{J.10}$$

$$H(a,3-a,z) = \frac{z}{2(a-1)(a-2)} - \frac{2}{(a-1)(a-2)}\frac{\Gamma\left(\frac{z+2a}{4}\right)\Gamma\left(\frac{z+6-2a}{4}\right)}{\Gamma\left(\frac{z+2a-2}{4}\right)\Gamma\left(\frac{z+4-2a}{4}\right)}, a \neq 1,2 \tag{J.11}$$

$$H(2,2;z) = -\frac{3z}{8} + \frac{3z^2}{64}\left[\Psi'\left(\frac{z}{4}\right) + \Psi'\left(\frac{z+4}{4}\right)\right] \tag{J.12}$$

$$H(1,3;z) = \frac{3z}{16} - \frac{3}{64}(z^2-4)\Psi'\left(\frac{z+2}{4}\right) \tag{J.13}$$

$$H(1,2;z) = \int_0^\infty \frac{e^{-zu}}{\cosh^2 u}\, du \tag{J.14}$$

$$= \frac{z}{4}\left[\Psi\left(\frac{z+2}{4}\right) - \Psi\left(\frac{z}{4}\right) - \Psi\left(\frac{z+4}{4}\right) + \Psi\left(\frac{z+2}{4}\right)\right]. \tag{J.15}$$

Proof: Hopefully the above formulas are strong evidence that the Laplace transform (J.1) provides a rich source of special functions upon varying the parameters a and b. The continued fraction (J.2) was derived by Stieltjes (1890, p.389) and is listed in Wall (1948, p.370). Neither Stieltjes nor Wall gave any indication of the properties of the transform or of the convergence nature of the associated continued fraction. We arrived at our formulas after much calculation and a considerable amount of study. We do not claim to know the full story about $H(a,b,z)$ and the continued fractions (J.2) and (J.3). Below we give various

Bauer-Muir modifying sequences that we have derived to aid in justifying our assertion through the Bauer-Muir transform technique and we give some brief arguments supporting our claims.

C.f. (J.3) is easily obtained from c.f. (J.2) by an equivalence relation. The modifying sequences for the continued fraction

$$b_0(z) + \overset{\infty}{\underset{n=1}{\mathrm{K}}} \frac{c_n}{b_n(z)} \tag{J.16}$$

dividing c_0 in (J.3) are

$$
\begin{aligned}
r_{2n+1}(z) &= 4n^2 + [2(a+b+1)-z]n + r_1(z) \\
r_{2n}(z) &= 4n^2 + [2(a+b-1)-z]n + r_0(z) \\
\varphi_n(z) &= [z(a+b-1)/2 + r_0(z)][2(a+b-1)-z-r_1(z)],
\end{aligned}
\tag{J.17}
$$

where

$$
\begin{aligned}
r_0(z) &= [z^2 - 2(a+b-2)z + 4ab]/8 \\
r_1(z) &= [z^2 - 2(a+b)z + 4ab + 8(a+b)]/8.
\end{aligned}
\tag{J.18}
$$

It is convenient to have the next two formulas, also.

$$
\begin{aligned}
b_0(z) + r_0(z) &= (z+2a)(z+2b)/8 \\
z - 2(a+b-1) + r_1(z) &= (z+4-2a)(z+4-2b)/8.
\end{aligned}
\tag{J.19}
$$

With the aid of Theorem 3.5 and the Bauer-Muir transform of (J.16) using (J.17) we were able to discover the functional equation (J.4). It follows from (J.4) that

$$
\begin{aligned}
\frac{H(2,2;z)}{z^2} &= \frac{H(2,2;z+4)}{(z+4)^2} + \frac{12}{z^2(z+4)^2} \\
&= \frac{H(2,2;z+4)}{(z+4)^2} + \frac{3}{8}\left[\frac{1}{z+4} - \frac{1}{z}\right] + \frac{3}{4}\left[\frac{1}{z^2} + \frac{1}{(z+4)^2}\right].
\end{aligned}
\tag{J.20}
$$

Solving (J.20) leads to (J.12). If $b = 4 - a$ and $a \neq 1, 2, 3$, then letting

$$H(z) = H(a, 4-a; z)/[(z-4+2a)(z+4-2a)] \tag{J.21}$$

it follows that

$$
\begin{aligned}
H(z) &= H(z+4) + \frac{12}{(z-4+2a)(z+4-2a)(z+2a)(z+8-2a)} \tag{J.22} \\
&= H(z+4) + \frac{3}{16(2-a)}\left[\frac{1}{z-4+2a} - \frac{1}{z+4-2a}\right] \tag{J.23} \\
&\quad - \frac{3}{16(3-a)}\left[\frac{1}{z-4+2a} - \frac{1}{z+8-2a}\right] \\
&\quad - \frac{3}{16(1-a)}\left[\frac{1}{z+2a} - \frac{1}{z+4-2a}\right] \\
&\quad + \frac{3}{16(2-a)}\left[\frac{1}{z+2a} - \frac{1}{z+8-2a}\right] \\
&= \frac{3}{64(1-a)(2-a)}\left[\Psi\left(\frac{z+2a}{4}\right) - \Psi\left(\frac{z+4-2a}{4}\right)\right] \\
&\quad + \frac{3}{64(2-a)(3-a)}\left[\Psi\left(\frac{z+8-2a}{4}\right) - \Psi\left(\frac{z-4+2a}{4}\right)\right]. \tag{J.24}
\end{aligned}
$$

It can be shown by letting $a \to 1$ in (J.34) that

$$
\begin{aligned}
\frac{H(1,3;z)}{z^2 - 4} &= \frac{3}{128} \left[\Psi\left(\frac{z+6}{4}\right) - \Psi\left(\frac{z-2}{4}\right) \right] - \frac{3}{64} \Psi'\left(\frac{z+2}{4}\right) \\
&= \frac{3z}{16(z^2-4)} - \frac{3}{64} \Psi'\left(\frac{z+2}{4}\right).
\end{aligned}
$$
(J.25)

By computing the limit as $a \to 2$ in (J.24) we arrive at

$$
\begin{aligned}
\frac{H(2,2;z)}{z^2} &= \frac{3}{32} \left[\Psi\left(\frac{z}{4}\right) - \Psi\left(\frac{z+4}{4}\right) \right] + \frac{3}{64} \left[\Psi'\left(\frac{z}{4}\right) + \Psi'\left(\frac{z+4}{4}\right) \right] \\
&= -\frac{3}{8z} + \frac{3}{64} \left[\Psi'\left(\frac{z}{4}\right) + \Psi'\left(\frac{z+4}{4}\right) \right]
\end{aligned}
$$
(J.26)

from which (J.12) can be derived. From (J.4) we have

$$
\begin{aligned}
\frac{H(1,3;z)}{z^2 - 2} &= \frac{H(1,3;z+4)}{(z+2)(z+6)} + \frac{12}{(z-2)(z+2)^2(z+6)} \\
&= \frac{H(1,3;z+4)}{z+6} + \frac{3}{32}\left(\frac{1}{z-2} - \frac{1}{z+6}\right) - \frac{3}{4}\frac{1}{(z+2)^2}.
\end{aligned}
$$
(J.27)

Formula (J.13) can be derived again by solving the functional equation (J.27). Similarly by solving the functional equation

$$
\frac{H(1,2;z)}{z} = \frac{H(1,2;z+4)}{z+4} + \frac{1}{z} - \frac{2}{z+2} + \frac{1}{z+4}
$$
(J.28)

we obtain formula (J.15).

We now set about the task of justifying the representation (J.5). From (J.4) we have (after letting $F(z) = H(a,b;z)$)

$$
F(z+4) = \frac{(z+2a)(z+2b)z}{(z+4-2a)(z+4-2b)} F(z) - \frac{4(a+b-1)}{(z+4-2a)(z+4-2b)}.
$$
(J.29)

It follows from the functional equation (J.29) that

$$
F(z+4k+4) = \frac{(z+4k+2a)(z+4k+2b)z}{(z+4k+4-2a)(z+4k+4-2b)} F(z) - \frac{4(a+b-1)}{(z+4k+4-2a)(z+4k+4-2b)}.
$$
(J.30)

Equation (J.30), for a fixed z, is equivalent to the difference equation

$$
a_{k+1} = b_{k+1} a_k - c_{k+1}
$$
(J.31)

where a_k, b_k. and c_k have the obvious definitions. To solve this difference equation, we let $a_k = d_k \prod_{m=0}^{k} b_m$. Then

$$
d_{k+1} - d_k = -c_{k+1} / \prod_{m=0}^{k+1} b_m.
$$
(J.32)

Equation (J.32) leads to

$$
d_{n+1} = d_0 - \sum_{k=0}^{n} c_{k+1} / \prod_{m=0}^{k+1} b_m
$$
(J.33)

from which it follows that

$$a_{n+1}/\prod_{m=1}^{n+1} b_m = a_0 - \sum_{k=0}^{n} c_{k+1}/\prod_{m=1}^{k+1} b_m. \tag{J.34}$$

If we have not made a mistake in our estimates, the left side of equation (J.34) $\to 0$ as $n \to \infty$ so that we have

$$F(z) = \sum_{k=0}^{\infty} c_{k+1}/\prod_{m=1}^{k+1} b_m \tag{J.35}$$

$$= H(z) \sum_{k=0}^{\infty} \frac{\Gamma\left(\frac{z+4-2a}{4} + k\right)\Gamma\left(\frac{z+4-2b}{4} + k\right)}{\Gamma\left(\frac{z+4+2a}{4} + k\right)\Gamma\left(\frac{z+4+2b}{4} + k\right)}, \tag{J.36}$$

where

$$H(z) = \frac{a+b-1}{4} \frac{\Gamma\left(\frac{z+2a}{4}\right)\Gamma\left(\frac{z+2b}{4}\right)}{\Gamma\left(\frac{z+4-2a}{4}\right)\Gamma\left(\frac{z+4-2b}{4}\right)}. \tag{J.37}$$

Formula (J.5) now follows from (J.36) and (J.37).

To justify the convergence of c.f.'s (J.2) and (J.3) to some of the functions listed under Function J we used Bauer-Muir transforms of the continued fraction

$$z + \underset{n=1}{\overset{\infty}{\mathbf{K}}} \frac{a_n}{z} \tag{J.38}$$

which is the reciprocal of c.f. (J.2). Formulas for the sequences used to determine these transforms are given below. For (J.10) we chose

$$r_n(z) = n + 1 - z/2; \; \varphi_n(z) = (z+1)^2/4, \tag{J.39}$$

for (J.6) we chose

$$r_n(z) = n + \frac{m-z}{2}; \; \varphi_n(z) = (z+2-m)(z+m)/4, \tag{J.40}$$

and for (J.7) we chose

$$r_n(z) = 2n^2 + (2 + (-1)^n z)n + \frac{z^4 + 4 - a^2}{4}; \; \varphi_n(z) = -\left[\frac{(z+2)^2 - a^2}{4}\right]^2. \tag{J.41}$$

To determine the Bauer-Muir transform of c.f. (J.38) for the remaining (J.11) case we set

$$r_n(z) = n + 1 - z/2; \; \varphi_n(z) = (2-a)(a-1) + \frac{z}{2}\left(1 + \frac{z}{2}\right). \tag{J.42}$$

Also for this case we set

$$G(z) = H(a, 3 - a, z) - \frac{z}{2(a-1)(a-2)} \tag{J.43}$$

and made use of the fact that

$$G(z)G(z+2) = \frac{(z-2+2a)(z+4-2a)}{[2(a-1)(a-2)]^2} \tag{J.44}$$

as an aid in arriving at the limit function.

FUNCTION K (Wall(1948, 94.6)) :

$$F(z) = \int_0^\infty \frac{u}{\sinh u} e^{-zu} \, du, \ \Re(z) > -1 \tag{K.1}$$

$$= \frac{2}{(z+1)^2} + F(z+2) \tag{K.2}$$

$$= \sum_{k=0}^\infty \frac{2}{(z+1+2k)^2} \tag{K.3}$$

$$= \frac{1}{2} \Psi' \left(\frac{z+1}{2} \right) \tag{K.4}$$

$$= \left[z + \underset{n=1}{\overset{\infty}{\mathrm{K}}} \frac{n^4}{(2n+1)z} \right]^{-1}, \ \Re(z) > 0. \tag{K.5}$$

Proof: We use the modifying sequences determined by

$$r_n(z) = n^2 + (1-z)n + \frac{1+z^2}{2}; \ \ \varphi_n(z) = -\frac{(z+1)^4}{4}. \tag{K.6}$$

to compute a Bauer-Muir transform of

$$z + \underset{n=1}{\overset{\infty}{\mathrm{K}}} \frac{n^4}{(2n+1)z} \tag{K.7}$$

which is the reciprocal of the c.f. (K.5). We also let $G(z)$ denote the limit function of c.f. (K.5) which converges, in particular, for all positive real values of z by Theorem 3.1. With the aid of the Bauer-Muir transform of (K.7) using (K.6) and Theorem 3.5 we find that

$$G(z) = \frac{2}{(z+1)^2} + G(z+2) \tag{K.8}$$

which is the same as the functional realtion (K.2). We solve (K.8) for $z > 0$ making use of the fact that $G(z) \to 0$ as $z \to \infty$ and arrive at

$$G(z) = \sum_{k=0}^\infty \frac{2}{(z+1+2k)^2}.$$

Thus in view of (K.3) we have that $G(z) = F(z)$ for $z > 0$. After dividing c.f. (K.5) by z and applying an equivalence relation we obtain

$$\frac{G(z)}{z} = \underset{n=0}{\overset{\infty}{\mathrm{K}}} \frac{c_n z^{-2}}{1}, \tag{K.9}$$

where

$$c_0 = 1, c_n = n^4/(4n^2 - 1), n \geq 1.$$

C.f. (K.9) is an S-fraction in the variable z^{-2} so it follows from Theorem 3.3 that G(z) is analytic in the half plane $\Re(z) > 0$. Since $F(z)$ is also analytic on this half plane we have that $F(z) = G(z)$ there and we are done.

FUNCTION L (Wall(1948, 94.7)) :

$$F(z) = \int_0^\infty \frac{u}{\cosh u} e^{-zu}\, du, \; \Re(z) > -1 \tag{L.1}$$

$$= \frac{2}{(z+1)^2} - F(z+2) \tag{L.2}$$

$$= \sum_{k=0}^\infty \frac{2(-1)^n}{(z+1+2n)^2} \tag{L.3}$$

$$= \frac{1}{8}\left[\Psi'\left(\frac{z+1}{4}\right) - \Psi'\left(\frac{z+3}{4}\right)\right] \tag{L.4}$$

$$= \frac{1}{z^2-1} + \frac{4\cdot 1^2}{1} + \frac{4\cdot 1^2}{z^2-1} + \frac{4\cdot 2^2}{1} + \frac{4\cdot 2^2}{z^2-1} + \cdots \tag{L.5}$$

$$= \left[b_0(z) + \mathop{\mathrm{K}}_{n=1}^\infty \frac{a_n}{b_n(z)}\right]^{-1}, \; \Re(z) > -1, \; where \tag{L.6}$$

$$a_{2n-1} = a_{2n} = 4n^2; \; b_{2n-1}(z) = 1; \; b_{2n}(z) = z^2 - 1.$$

Proof: The proof is very similar to the proof for Function K so we will not go into detail. The key ingredient is to compute the Bauer-Muir transform of the continued fraction

$$b_0(z) + \mathop{\mathrm{K}}_{n=1}^\infty \frac{a_n}{b_n(z)} \tag{L.7}$$

using modifying sequences defined by

$$r_{2n-1}(z) = \frac{2n}{z+1} - \frac{1}{2}; \; r_{2n}(z) = (z+1)\left[2n + \frac{3-z}{2}\right]; \; \varphi_n(z) = \frac{(z+1)^2}{4}. \tag{L.8}$$

FUNCTION M (Wall(1948, 94.8)) :

$$F(z) = \exp\int_0^\infty \frac{1}{u}\left(1 - \frac{\cosh 2au}{\cosh 2u}\right)e^{-zu}\, du \tag{M.1}$$

$$= \frac{8}{z} \cdot \frac{\Gamma(\frac{z+6-2a}{8})\Gamma(\frac{z+6+2a}{8})}{\Gamma(\frac{z+2-2a}{8})\Gamma(\frac{z+2+2a}{8})} \tag{M.2}$$

$$= 1 + \frac{2(1-a^2)}{z^2} + \frac{3^2-a^2}{1} + \frac{5^2-a^2}{z^2} + \cdots, \; 0 \le a^2 < 1, \; \Re(z) > 0. \tag{M.3}$$

Proof: Let $H(z)$ denote the limit function of c.f. (M.2) and define $K(z)$ by

$$K(z) = 1 + \frac{1^2-a^2}{z^2} + \frac{3^2-a^2}{1} + \frac{5^2-a^2}{z^2} + \frac{7^2-a^2}{1} + \cdots \tag{}$$

$$= b_0(z) + \mathop{\mathrm{K}}_{n=1}^\infty \frac{a_n}{b_n(z)}, \tag{M.4}$$

where

$$a_n = (2n-1)^2 - a^2, \; n \ge 1; \; b_{2n-1}(z) = z^2, \; n \ge 1; \; b_{2n}(z) = 1, \; n \ge 0.$$

Then K and H are related by

$$K(z) = [1 + H(z)]/2. \tag{M.5}$$

By applying a B-M transformation to the c.f. (M.4) with

$$r_n(z) = 2n/z - 1/2, \; n \geq 0; \varphi_n(z) = (1/z + 1/2)^2 - a^2/z^2, \; n \geq 1 \tag{M.6}$$

and using (M.5) we obtain (if $z > 0$) that

$$H(z)H(z + 4) = \frac{(z + 2 - 2a)(z + 2 + 2a)}{z(z + 4)}. \tag{M.7}$$

Hence

$$\frac{H(z)}{H(z + 8)} = \frac{z + 8}{z} \cdot \frac{(z + 2 - 2a)(z + 2 + 2a)}{(z + 6 - 2a)(z + 6 + 2a)}. \tag{M.8}$$

From (M.8) we get

$$\prod_{k=0}^{n} \frac{H(z + 8k)}{H(z + 8k + 8)} = \frac{H(z)}{H(z + 8n + 8)} \tag{M.9}$$

$$= \frac{\Gamma(\frac{z}{8})\Gamma(\frac{z+6-2a}{8})\Gamma(\frac{z+6+2a}{8})}{\Gamma(\frac{z+8}{8})\Gamma(\frac{z+2-2a}{8})\Gamma(\frac{z+2+2a}{8})} \tag{M.10}$$

$$\cdot \frac{\Gamma(\frac{z+8+8n+8}{8})\Gamma(\frac{z+2-2a+8n+8}{8})\Gamma(\frac{z+2+2a+8n+8}{8})}{\Gamma(\frac{z+8n+8}{8})\Gamma(\frac{z+6-2a+8n+8}{8})\Gamma(\frac{z+6+2a+8n+8}{8})}.$$

By letting $n \to \infty$ in (M.9) and (M.10), we arrive at

$$H(z) = \frac{8}{z} \cdot \frac{\Gamma(\frac{z+6-2a}{8})\Gamma(\frac{z+6+2a}{8})}{\Gamma(\frac{z+2-2a}{8})\Gamma(\frac{z+2+2a}{8})}. \tag{M.11}$$

Hence $H(z) = F(z)$ if $z > 0$. Since H and F are analytic in the plane $\Re(z) > 0$, it follows that $H(z) = F(z)$ in this region.

FUNCTION N (Wall(1948, 94.9)) :

$$F(z) = \tanh\left(\int_0^\infty \frac{\sinh 2au}{2u \cosh u} e^{-zu} \, du\right), \; a > 0, \; \Re(z) > 2a - 1 \tag{N.1}$$

$$= \frac{G(z) - 1}{G(z) + 1}, \; where \; G(z) = \frac{\Gamma(\frac{z+3+2a}{4})\Gamma(\frac{z+1-2a}{4})}{\Gamma(\frac{z+1+2a}{4})\Gamma(\frac{z+3-2a}{4})} \tag{N.2}$$

$$= a\left[z + \mathop{K}_{n=1}^\infty \frac{n^2 - a^2}{z}\right]^{-1}, \; 0 < a < 1, \; \Re(z) > max\{0, 2a - 1\}. \tag{N.3}$$

Proof: We choose the B-M pair $\{r_n\}, \{\varphi_n\}$ for the c.f. $z + K((n^2 - a^2)/z)$ as follows:

$$r_n(z) = n + (1 - z)/2; \; \varphi_n(z) = -a^2 + (z + 1)^2/4. \tag{N.4}$$

After computing the B-M transform using (n.4) and setting $H(z)$ equal to the limit of (N.3), we obtain

$$H(z) = \frac{2a - (z + 1)H(z + 2)}{z + 1 - 2aH(z + 2)}, \; z > max\{0, 2a - 1\}. \tag{N.5}$$

Now set

$$R(z) = \frac{1 + H(z)}{1 - H(z)}. \tag{N.6}$$

Then, with the aid of (N.5), we obtain

$$R(z)R(z+2) = \frac{z+1+2a}{z+1-2a}. \tag{N.7}$$

From (N.7) we get

$$\frac{R(z)}{R(z+4)} = \frac{(z+1+2a)(z+3-2a)}{(z+1-2a)(z+3+2a)}. \tag{N.8}$$

Employing (N.8) we derive

$$\prod_{k=0}^{n} \frac{R(z+4k)}{R(z+4k+4)} = \prod_{k=0}^{n} \frac{(z+1+2a+4k)(z+3-2a+4k)}{(z+1-2a+4k)(z+3+2a+4k)} \tag{N.9}$$

which leads to

$$\frac{R(z)}{R(z+4n+4)} = \frac{\Gamma(\frac{z+1-2a}{4})\Gamma(\frac{z+3+2a}{4})\Gamma(\frac{z+1+2a+4n+4}{4})\Gamma(\frac{z+3-2a+4n+4}{4})}{\Gamma(\frac{z+1+2a}{4})\Gamma(\frac{z+3-2a}{4})\Gamma(\frac{z+1-2a+4n+4}{4})\Gamma(\frac{z+3+2a+4n+4}{4})}. \tag{N.10}$$

We take the limit of both sides of (N.10) as $n \to \infty$ and obtain

$$R(z) = \frac{\Gamma(\frac{z+1-2a}{4})\Gamma(\frac{z+3+2a}{4})}{\Gamma(\frac{z+1+2a}{4})\Gamma(\frac{z+3-2a}{4})}. \tag{N.11}$$

From (N.6) we get

$$H(z) = \frac{R(z)-1}{R(z)+1}. \tag{N.12}$$

Since $R(z) = G(z)$, it follows that expansion (N.3) is valid if $z > \max\{0, 2a-1\}$. Since H and F are analytic if $\Re(z) > \max\{0, 2a-1\}$, it follows that (N.3) is valid on this half plane.

FUNCTION O (Wall(1948, 94.10)) :

$$F(z) = \tanh \int_0^\infty \frac{\sinh au}{u \cosh u} e^{-zu}\, du, \; a > 0, \; \Re(z) > a - 1 \tag{O.1}$$

$$= \frac{G^2(z)-1}{G^2(z)+1}, \; where \; G(z) = \frac{\Gamma(\frac{z+3+a}{4})\Gamma(\frac{z+1-a}{4})}{\Gamma(\frac{z+1+a}{4})\Gamma(\frac{z+3-a}{4})} \tag{O.2}$$

$$= a\left[z + \mathop{\mathrm{K}}_{n=1}^{\infty} \frac{n^2 - a^2(1+(-1)^n)/2}{z}\right]^{-1}, \; 0 < a < 2, \; \Re(z) > \max\{0, a-1\} \tag{O.3}$$

Proof: Let $H(z)$ denote the limit function of the c.f. (O.3). Furthermore, define

$$r_n(z) = n + \frac{1-z}{2} + \frac{(-1)^n}{2}\frac{a^2}{z+1}; \; \varphi_n(z) = \frac{1}{4}\left[z+1+\frac{(-1)^n a^2}{z+1}\right]^2. \tag{O.4}$$

We apply a B-M transformation to the c.f. in brackets in (O.3) with $r_n(z)$ and $\varphi_n(z)$ given in (O.4) to obtain

$$H(z) = \frac{2a(z+1) - [(z+1)^2 + a^2]H(z+2)}{(z+1)^2 + a^2 - 2a(z+1)H(z+2)}. \tag{O.5}$$

Let

$$R(z) = \frac{1 + H(z)}{1 - H(z)}. \tag{O.6}$$

Then

$$R(z)R(z+2) = \left(\frac{z+1+a}{z+1-a}\right)^2, \tag{O.7}$$

whence

$$\frac{R(z)}{R(z+4)} = \left[\frac{(z+1+a)(z+3-a)}{(z+1-a)(z+3+a)}\right]^2. \tag{O.8}$$

From (O.8) we get

$$\prod_{k=0}^{n} \frac{R(z+4k)}{R(z+4k+4)} = \left[\prod_{k=0}^{n} \frac{(z+1+a+4k)(z+3-a+4k)}{(z+1-a+4k)(z+3+a+4k)}\right]^2. \tag{O.9}$$

Finally, it follows from (O.9) and properties of $\Gamma(z)$ that

$$\frac{R(z)}{R(z+4n+4)} = \left[\frac{\Gamma(\frac{z+1-a}{4})\Gamma(\frac{z+3+a}{4})\Gamma(\frac{z+1+a+4n+4}{4})\Gamma(\frac{z+3-a+4n+4}{4})}{\Gamma(\frac{z+1+a}{4})\Gamma(\frac{z+3-a}{4})\Gamma(\frac{z+1-a+4n+4}{4})\Gamma(\frac{z+3+a+4n+4}{4})}\right]^2. \tag{O.10}$$

Now let $n \to \infty$ on both sides of (O.10) to obtain

$$R(z) = \left[\frac{\Gamma(\frac{z+1-a}{4})\Gamma(\frac{z+3+a}{4})}{\Gamma(\frac{z+1+a}{4})\Gamma(\frac{z+3-a}{4})}\right]^2. \tag{O.11}$$

Thus $R(z) = G(z)$ if $z > \max\{0, a-1\}$, so $F(z) = H(z)$ for these values of z. Since F and H are both analytic on the half plane $\Re(z) > \max\{0, a-1\}$, it follows that $F(z) = H(z)$ there.

FUNCTION P :

$$F(z) = \int_0^\infty \frac{u^2 e^{-zu}}{\sinh u}\, du, \quad \Re(z) > -1 \tag{P.1}$$

$$= \sum_{k=0}^{\infty} \frac{4}{(z+2k+1)^3} \tag{P.2}$$

$$= \frac{4}{(z+1)^3} + F(z+2) \tag{P.3}$$

$$= -\frac{1}{4}\Psi''\left(\frac{z+1}{2}\right) \tag{P.4}$$

$$= \frac{1}{z^2-1} + \frac{2\cdot 1^3}{1} + \frac{2\cdot 1^3}{3(z^2-1)} + \frac{2\cdot 2^3}{1} + \frac{2\cdot 2^3}{5(z^2-1)} + \cdots, \tag{P.5}$$
$$\Re(z) > 0, z \notin [0,1]$$

$$= \frac{1}{z+1}\left[b_0(z) + \mathop{\mathrm{K}}_{n=1}^{\infty} \frac{a_n(z)}{b_n(z)}\right]^{-1}, \quad \Re(z) > 0, z \notin [0,1], \quad where \tag{P.6}$$

$$a_{2n-1}(z) = a_{2n}(z) = 4n^4(z-1)/(z+1); \quad b_n(z) = (n+1)(z-1).$$

Proof: Ramanujan (1927, p.XXVIII) lists the expansion (P.5) without proof for the special case $z = \sqrt{3}$. Perron (1953) claims that Preece gave the same expansion with $\sqrt{3}$ replaced by a variable, and in this same paper Perron gives a proof for the validity of expansion (P.5) for real $z > 1$. Our approach to investigating the convergence behavior of c.f. (P.5) is considerably different than Perron's. We begin by deriving c.f. (P.6) from c.f. (P.5) by an equivalence relation. This new expansion may not be as elegant as expansion (P.5) but we found it lends itself better to our Bauer-Muir technique. It is not difficult to see, using Theorem 3.1, that c.f.'s (P.5) and (P.6) converge if $z > 1$. By computing the Bauer-Muir transform of the c.f. $b_0 + K(a_n/b_n)$ in brackets in (P.6) and Theorem 3.5 with $\{r_n\}$ and $\{\varphi_n\}$ defined by

$$
\left.
\begin{array}{rcl}
r_{2n}(z) &=& 2n^2 + n(3 - z) + 1 - z + (z + 1)^2/4 \\
r_{2n-1}(z) &=& r_{2n-2}(z)(z - 1)/(z + 1) \\
\varphi_n(z) &=& -(z + 1)(z - 1)/4.
\end{array}
\right\}
\tag{P.7}
$$

we are able to show that the limit function of c.f. (P.6) satisfies the functional equation (P.3) when $z > 1$. With the aid of this relation and the fact that $F(z) \to 0$ when $z \to \infty$ it is not difficult to show that c.f. (P.6) converges to the same value at z as the series (P.2) if $z > 1$. To see that c.f. (P.5) (and therefore (P.6)) converges to an analytic function on the region $\Re(z) > 0, z \notin [0,1]$ we convert the continued fraction into an S-fraction in the variable $w = 1/(z^2 - 1)$ and employ Theorem 3.3. The validity of each of the expansions (P.5) and (P.6) now follows from the Identity Theorem for analytic functions.

FUNCTION Q :

$$
\begin{align}
F(z) &= \int_0^\infty 2(-u + \coth u)e^{-zu}\, du, \quad \Re(z) > -2 \tag{Q.1} \\
&= \sum_{k=0}^\infty \frac{1}{(1 + z/2 + k)^2} \tag{Q.2} \\
&= \Psi'(1 + z/2) \tag{Q.3} \\
&= \frac{2}{z + 1} + \frac{1^4}{3(z + 1)} + \frac{2^4}{5(z + 1)} + \frac{3^4}{7(z + 1)} + \cdots, \quad \Re(z) > -1 \tag{Q.4} \\
&= \frac{d_1 z^{-1}}{1} + \frac{d_2 z^{-1}}{1} + \frac{d_3 z^{-1}}{1} + \cdots, \quad z > 0, \; where \tag{Q.5}
\end{align}
$$

$$
d_1 = 2; \quad d_{2n+1} = \frac{-n^2}{2n + 1}; \quad d_{2n+2} = \frac{(n + 1)^2}{2n + 1}, \quad n \geq 0.
$$

Proof: The validity of (Q.4) follows from (A.8) in Function A. One can use the Bauer-Muir technique to verify (Q.4) directly by choosing

$$
\left.
\begin{array}{rcl}
r_{2n}(z) &=& 4n^2 - 2nz + z^2/2 + z + 1 \\
r_{2n+1}(z) &=& 4n^2 + 2n(2 - z) + z^2/2 + 2 \\
\phi_n(z) &=& -(z + 2)^4/4
\end{array}
\right\}
\tag{Q.6}
$$

and using (Q.6) to determine the Bauer-Muir transfom of the c.f.

$$
z + 1 + \mathop{\mathrm{K}}_{n=1}^\infty \frac{n^4}{(2n + 1)(z + 1)}
\tag{Q.7}
$$

whose reciprocal times 2 is the same as (Q.4). The even part of (Q.5) is (Q.4). The odd part of c.f. (Q.5) is equal to $2/z$ minus a continued fraction with positive real elements when

$z > -1$, and it converges for these values of z by Theorem 3.1. By using formula (2.12) and math induction to establish that $\Re(zH_{2n-1} > 0$ when $z > 0$ we obtain that the even and odd parts of (Q.5) converge to the same value on the positive real axis. We could have established the convergence of c.f. (Q.5) also by observing that the negative of this continued fraction is an alternating continued fraction and then applying Theorem 3.2. to the negative of c.f. (Q.5). We have not yet investigated what the convergence behavior of (Q.5) is like when z is nonreal.

FUNCTION R (Frame (1979)) :

$$F(z) = \int_0^\infty (-1 + u \coth u) e^{-zu} \, du, \quad \Re(z) > 0 \tag{R.1}$$

$$= \frac{4}{[z(z+2)]^2} + F(z+2) \tag{R.2}$$

$$= \sum_{k=0}^\infty \frac{4}{[(z+2k)(z+2k+2)]^2} \tag{R.3}$$

$$= \frac{1}{2}\Psi'\left(\frac{z}{2}\right) - \frac{1}{z^2} - \frac{1}{z} \tag{R.4}$$

$$= \frac{z^{-3}}{\frac{1}{1}+\frac{1}{2}} + \frac{z^{-2}}{\frac{1}{2}+\frac{1}{3}} + \frac{z^{-2}}{\frac{1}{3}+\frac{1}{4}} + \frac{z^{-2}}{\frac{1}{4}+\frac{1}{5}} + \cdots, \quad \Re(z) > 0 \tag{R.5}$$

$$= \frac{2z^{-2}}{3z} + \frac{1 \cdot 2^2 \cdot 3}{5z} + \frac{2 \cdot 3^2 \cdot 4}{7z} + \frac{3 \cdot 4^2 \cdot 5}{9z} + \cdots, \quad \Re(z) > 0 \tag{R.6}$$

$$= \frac{2}{z^2}\left[b_0(z) + \mathop{\mathbf{K}}_{n=1}^{\infty} \frac{a_n}{b_n(z)}\right]^{-1}, \quad \Re(z) > 0, \; where \tag{R.7}$$

$$a_n = n(n+1)^2(n+2); \quad b_n(z) = (2n+3)z.$$

Proof: The expansion (R.5) was derived by Frame (1979, p.816) for positive real values of z. We obtained c.f. (R.6) from (R.5) by an equivalence transformation. C.f. (R.7) is just c.f. (R.6) in closed form. In view of the functional relation (R.2) and the series representation (R.3) for $F(z)$, the Bauer-Muir method can be readily applied to establish the validity of (R.7) (and therefore of (R.5) and (R.6)) once an appropriate Bauer-Muir transform is chosen. We have found that the transform of the c.f. $b_0(z) + K(a_n/b_n(z))$ with the modifying sequences defined by

$$r_n(z) = n^2 + n(3-z) + \frac{z^2 - 2z + 4}{2}; \quad \varphi_n(z) = -\frac{z^2(z+2)^2}{4}. \tag{R.8}$$

serves our purposes.

The continued fractions above can be extended to the c.f.

$$-z^{-1} + \frac{z^{-1}}{1 - z^{-1}} + \frac{z^{-1}}{1} + \frac{d_1 z^{-1}}{1} - \frac{d_1 z^{-1}}{1} + \frac{d_2 z^{-1}}{1} - \frac{d_2 z^{-1}}{1} + \frac{d_3 z^{-1}}{1} - \cdots, \tag{R.9}$$

where, if $\mathcal{H}_0 = 0$ and $\mathcal{H}_n = 1 + \frac{1}{2} + \cdots + \frac{1}{n}$, $n \geq 1$, then

$$d_{2n} = \frac{\mathcal{H}_{2n+1}\mathcal{H}_{2n-1}}{\mathcal{H}_{2n+1} - \mathcal{H}_{2n-1}}; \quad d_{2n+1} = \frac{1}{(\mathcal{H}_{2n+2} - \mathcal{H}_{2n})\mathcal{H}_{2n+1}^2}.$$

It remains to investigate the convergence behavior of the continued fraction (R.9). It is a regular δ-fraction in the variable $1/z$.

FUNCTION S (Rogers (1907)) :

$$F(z) = \exp \int_0^\infty \frac{\tanh u}{u} e^{-zu}\, du, \quad \Re(z) > 0 \tag{S.1}$$

$$= \frac{z}{4}\left[\frac{\Gamma(z/4)}{\Gamma(z/4 + 1/2)}\right]^2 \tag{S.2}$$

$$= \frac{z}{z-1} + \frac{1^2}{2(z-1)} + \frac{3^2}{2(z-1)} + \frac{5^2}{2(z-1)} + \cdots, \quad \Re(z) > 1. \tag{S.3}$$

Proof: Rogers (1907, p.85) formally derived a continued fraction for the Laplace transform (S.1) equivalent to the continued fraction (S.3) for real values of z. We let $H(z)$ denote the limit of (S.3) and we claim that expansion (S.3) is actually valid for $\Re(z) > 1$. Continued fraction (S.3) is equivalent to the continued fraction

$$H(z) = \frac{\frac{z}{z-1}}{1} + \frac{2 \cdot \frac{1^2}{4(z-1)^2}}{1} + \frac{\frac{3^2}{4(z-1)^2}}{1} + \frac{\frac{5^2}{4(z-1)^2}}{1} + \cdots . \tag{S.4}$$

But we recognize that c.f. (S.4) is very simply related to another continued fraction when $z - 1$ is replaced by z. In fact,

$$H(z) = \frac{z}{(z-1)G(z(z-1))}, \tag{S.5}$$

where

$$G(z) = \exp \int_0^\infty \frac{1}{u}(1 - \operatorname{sech} 2u)e^{-zu}\, du \tag{S.6}$$

$$= 1 + \frac{2(\frac{1}{z})^2}{1} + \frac{(\frac{3}{z})^2}{1} + \frac{(\frac{5}{z})^2}{1} + \frac{(\frac{7}{z})^2}{1} + \cdots \tag{S.7}$$

$$= 1 + \frac{2}{z^2} + \frac{3^2}{1} + \frac{5^2}{z^2} + \frac{7^2}{1} + \cdots . \tag{S.8}$$

The function $G(z)$ is just the special case $a = 0$ of Function M that we dealt with earlier. Since expansions (S.7) and (S.8) are valid for $G(z)$ when $\Re(z) > 0$, it now follows that (S.3) is valid on the half plane $\Re(z) > 1$.

FUNCTION T (Rogers (1907)) :

$$F(z) = \int_0^\infty \frac{\cosh au}{\cosh bu} e^{-zu}\, du, \quad a \in \mathcal{R},\ b > 0,\ \Re(z) > |a| - b \tag{T.1}$$

$$= \frac{2(z+b)}{(z+b)^2 - a^2} - F(z + 2b) \tag{T.2}$$

$$= \sum_{k=0}^\infty \frac{2(-1)^k[z + (2k+1)b]}{[z + (2k+1)b]^2 - a^2} \tag{T.3}$$

$$= \frac{1}{4b}\left[\Psi\left(\frac{z - a + 3b}{4b}\right) - \Psi\left(\frac{z - a + b}{4b}\right) + \Psi\left(\frac{z + a + 3b}{4b}\right) - \Psi\left(\frac{z + a + b}{4b}\right)\right] \tag{T.4}$$

$$= \frac{1}{z} + \frac{1^2b^2 - a^2}{z} + \frac{2^2b^2}{z} + \frac{3^2b^2 - a^2}{z} + \frac{4^2b^2}{z} + \frac{5^2b^2 - a^2}{z} + \cdots \tag{T.5}$$

$$= \left[z + \mathop{\mathbf{K}}_{n=1}^\infty \frac{a_n}{z}\right]^{-1}, \quad b > |a| \geq 0,\ \Re(z) > 0\ \text{where} \tag{T.6}$$

$$a_{2n-1} = (2n-1)^2 b^2 - a^2; \quad a_{2n} = 4n^2 b^2.$$

Proof: Expansion (T.5) is simply expansion (T.6) in a more displayed form. Rogers (1907, p.394) listed a continued fraction expansion without proof for the above Laplace transform (T.1) for the case $b = 1$. We derived our continued fraction (T.6) from the one he gave by an equivalence transformation. We define the B-M transform sequences $\{r_n\}, \{\varphi_n\}$ for the c.f. $z + K(a_n/z)$ by

$$r_n(z) = nb + \frac{b^2 + (-1)^{n-1}a^2 - z^2}{2(z+b)}; \quad \varphi_n(z) = \left[\frac{(z+b)^2 - a^2}{2(z+b)}\right]^2. \tag{T.7}$$

Note that $r_n(z)$ is eventually positive when z is positive, since $b > 0$. This allows us to employ Theorem 3.5 which leads us to the conclusion that the limit function of (T.6) satisfies (T.2) when $z > 0$. From this result we can easily deduce that (T.6) converges to the sum (T.3) on the positive real axis. Since the analyticity of the functions defined by the expansion (T.6) and the series (T.3) is easily established, we have that (T.6) is valid under the stated conditions.

FUNCTION U (Rogers (1907)) :

$$F(z) = \int_0^\infty \frac{\sinh au}{\cosh u} e^{-zu}\, du, \; a \in \mathcal{R}, \; \Re(z) > |a| - 1 \tag{U.1}$$

$$= -F(z+2) + \frac{2a}{(z+1)^2 - a^2} \tag{U.2}$$

$$= \sum_{k=0}^\infty \frac{2a(-1)^k}{(z+2k+1)^2 - a^2} \tag{U.3}$$

$$= \frac{1}{4}\left[\Psi\left(\frac{z-a+3}{4}\right) - \Psi\left(\frac{z+a+3}{4}\right) + \Psi\left(\frac{z+a+1}{4}\right) - \Psi\left(\frac{z-a+1}{4}\right)\right] \tag{U.4}$$

$$= \frac{a}{z^2-1} + \frac{2^2 - a^2}{1} + \frac{2^2}{z^2-1} + \frac{4^2 - a^2}{1} + \frac{4^2}{z^2-1} + \cdots, \; a \in \mathcal{R}, |a| < 2 \tag{U.5}$$

$$= \frac{a}{z-1}\left[b_0(z) + \mathop{K}_{n=1}^\infty \frac{a_n(z)}{b_n(z)}\right]^{-1}, \; a \in \mathcal{R}, |a| < 2, \Re(z) > 0, z \notin [0,1], \; where \tag{U.6}$$

$$\phi(z) = (z-1)/(z+1); \; b_n(z) = z-1, \; n \geq 0 \; and$$
$$a_{2n-1}(z) = (4n^2 - a^2)\phi(z); \; a_{2n}(z) = 4n^2\phi(z), \; n \geq 1.$$

Proof: Rogers (1907, p.394) listed without proof a c.f. expansion for the Laplace transform (U.1) equivalent to both expansions (U.5) and (U.6) which are equivalent to each other. If $z \neq \pm 1$, c.f.'s (U.5) and (U.6) are each equivalent to

$$H(z) = \frac{\frac{a}{z^2-1}}{1} + \frac{\frac{2^2-a^2}{z^2-1}}{1} + \frac{\frac{2^2}{z^2-1}}{1} + \frac{\frac{4^2-a^2}{z^2-1}}{1} + \frac{\frac{4^2}{z^2-1}}{1} + \cdots. \tag{U.7}$$

By Theorem 3.3, c.f. (U.7) converges to an analytic function on the region $\Re(z) > 0, z \notin [0,1]$ and, in particular, it converges for all real $z > 1$. Now let the sequences $\{r_n\}, \{\varphi_n\}$ be given by

$$r_{2n}(z) = 2n + 1 + \frac{1 - z^2 - a^2}{2(z+1)}, \; n \geq 0$$

$$r_{2n-1}(z) = \left[2n + \frac{a^2 - (z+1)^2}{2(z+1)}\right]\left[\frac{z-1}{z+1}\right], \quad n \geq 1 \tag{U.8}$$

$$\varphi_n(z) = \left[\frac{(z+1)^2 - a^2}{2(z+1)}\right]\left[\frac{z-1}{z+1}\right].$$

We use (U.8) for the B-M transformation of $b_0(z) + K(a_n(z)/b_n(z))$ in (U.6) with $z > 1$ and obtain the functional relation

$$H(z) = -H(z+2) + \frac{2a}{(z+1)^2 - a^2}. \tag{U.9}$$

We solve this functional equation and arrive at

$$H(z) = \sum_{k=0}^{\infty} \frac{2a(-1)^k}{(z + 2k + 1)^2 - a^2}. \tag{U.10}$$

Hence $H(z) = F(z)$ if $z > 1$. Let

$$D = \{\, z \mid \mathrm{Re}\, z > 0, z \notin [0,1]\,\}.$$

Then since both c.f. (U.7) and series (U.3) converge to analytic functions on D, it follows that $H(z) = F(z)$ on D.

FUNCTION V :

$$F(z) = \int_0^\infty \frac{e^{-zu}}{\cosh u}\, du, \quad \Re(z) > -1 \tag{V.1}$$

$$= \sum_{k=0}^{\infty} \frac{2(-1)^k}{z + 2k + 1} \tag{V.2}$$

$$= \frac{2}{z+1} - F(z+2) \tag{V.3}$$

$$= \frac{1}{2}\left[\Psi\left(\frac{z+3}{4}\right) - \Psi\left(\frac{z+1}{4}\right)\right] \tag{V.4}$$

$$= \left[z + \overset{\infty}{\underset{n=1}{\mathrm{K}}}\frac{n^2}{z}\right]^{-1}, \quad \Re(z) > 0. \tag{V.5}$$

Proof: Although the Laplace transform (V.1) is a special case of each of the transforms (T.1) in Function T, (J.6) in Function J, and (F.1) in Function F, we include it here for the elegance of its c.f. representation (V.5) and for the simplicity of the other representations of the function it represents. The sequences $\{r_n\}$ and $\{\varphi_n\}$ defined by

$$\left.\begin{array}{rcl} r_n(z) &=& n - (z-1)/2, \ n \geq 0 \\ \varphi_n(z) &=& (z+1)^2/4, \ n \geq 1. \end{array}\right\} \tag{V.6}$$

can be used to compute the Bauer-Muir transform of the c.f. in brackets in (V.5), and then with the aid of this transform, the Bauer-Muir method and Theorem 3.3 can be employed in the manner we have illustrated often to complete the verification of (V.5). After we had derived the sequences (V.6), we learned that Perron (1953, p.32) had already derived the sequence $\{r_n\}$ and proved the validity of (V.5) for positive real values of z.

The following four functions $F_i(z, b), i = 1, 2, \ldots$ and there their corresponding continued fraction expansions are listed by Wall (1948) without proof in the chapter on Stieltjes summability in his book on continued fractions. With the aid of the table of Laplace transforms by Oberhettinger and Badii (1973), we were able to supply the corresponding Laplace transform representation for each of the functions $F_i(z, b)$. Once we arrived at these transforms we recognized these functions as special cases of functions that we have considered earlier. We shall be more specific about this last assertion ahead.

FUNCTION W (Wall (1948, 94.11)) :

$$F_1(z, b) = \frac{1}{2}[\Psi(z + b) - \Psi(z + 1 - b)] \tag{W.1}$$

$$= \frac{2b - 1}{2}\left[z + \underset{n=1}{\overset{\infty}{\mathbf{K}}}\frac{a_n}{z}\right]^{-1}, \quad a_n = \frac{n^2(n + 1 - 2b)(n - 1 + 2b)}{4(2n - 1)(2n + 1)}, \quad \text{where} \tag{W.2}$$

$$0 < b < 1, \Re(z) > 0$$

$$= \int_0^\infty \frac{\sinh[(b - 1/2)u]}{2\sinh(u/2)} e^{-zu}\, du, \ 0 < b < 1, \ \Re(z) > b - 1. \tag{W.3}$$

Proof: We point out that, in view of the transform (W.3), this case is essentially a special case of Function I so that expansion (W.2) is valid under the given conditions.

FUNCTION X (Wall (1948, 94.12)) :

$$F_2(z, b) = \frac{1}{2}[\Psi(z + b) + \Psi(z + 1 - b) - \Psi(z) - \Psi(z + 1)] \tag{X.1}$$

$$= \frac{b(1 - b)}{2}\left[b_0(z) + \underset{n=1}{\overset{\infty}{\mathbf{K}}}\frac{a_n}{b_n(z)}\right]^{-1}, \quad \Re(z) > 0, \ b_{2n-1}(z) = 1, \ b_{2n}(z) = z^2, \tag{X.2}$$

$$a_{2n-1} = \frac{n(n - b)(n - 1 + b)}{2(2n - 1)}, \quad a_{2n} = \frac{n(n + b)(n + 1 - b)}{2(2n + 1)}$$

$$= \int_0^\infty \frac{\sinh(bu/2)\sinh[(1 - b)u/2]}{\sinh(u/2)} e^{-zu}\, du, \ 0 < b < 1, \ \Re(z) > 0. \tag{X.3}$$

Proof: Function X is a special case of Function H. After an appropriate identification and pairing of the constant parameters involved, it is easily seen that c.f. (X.2) is equivalent to c.f. (H.5) when $\Re(z) > 0$. Thus expansion (X.2) is valid under the asserted conditions.

FUNCTION Y (Wall (1948, 94.15)) :

$$F_3(z, b) = \frac{1}{2}\left[\Psi\left(\frac{z + 1 + b}{2}\right) + \Psi\left(\frac{z + 2 - b}{2}\right) - \Psi\left(\frac{z + b}{2}\right) - \Psi\left(\frac{z + 1 - b}{2}\right)\right] \tag{Y.1}$$

$$= \left[z + \underset{n=1}{\overset{\infty}{\mathbf{K}}}\frac{a_n}{z}\right]^{-1}, \quad 0 < b < 1, \ \Re(z) > 0 \tag{Y.2}$$

$$a_{2n} = n^2, \ a_{2n+1} = (n + b)(n + 1 - b)$$

$$= \int_0^\infty \frac{\cosh[(b - 1/2)u]}{\cosh(u/2)} e^{-zu}\, du, \ \Re(z) > |b - 1/2| - 1/2. \tag{Y.3}$$

Proof: Function Y and the expansion (Y.8) are clearly special cases of Function T and its associated expansion (T.6).

FUNCTION Z (Wall (1948, 94.16)) :

$$F_4(z,b) = \frac{1}{2}\left[\Psi\left(\frac{z+b}{2}\right) + \Psi\left(\frac{z+2-b}{2}\right) - \Psi\left(\frac{z+1-b}{2}\right) - \Psi\left(\frac{z+1+b}{2}\right)\right] \quad (Z.1)$$

$$= \frac{2b-1}{2}\left[z^2 + d_0 + \mathop{\text{K}}_{n=1}^{\infty}\frac{a_n}{z^2+d_n}\right]^{-1}, \; b \in \mathcal{R}, \; |b-1/2| < 1, \quad (Z.2)$$

$$a_n = -n^2[n^2 - (b-1/2)^2]; \; d_n = b - b^2 + 2(n+1/2)^2, \; \Re(z) > 0, z \notin [0,1/2]$$

$$= \int_0^\infty \frac{\sinh[(b-1/2)u]}{\cosh(u/2)}\, e^{-zu}\, du, \; b \in \mathcal{R}, \; \Re(z) > |z-1/2| - 1/2 \quad (Z.3)$$

$$= 2\int_0^\infty \frac{\sinh[(2b-1)u]}{\cosh u}e^{-2zu}\, du, \; b \in \mathcal{R}, \; \Re(z) > |z-1/2| - 1/2. \quad (Z.4)$$

Proof: The Laplace transform (Z.4) tells us that Function Z is equal to 2 times the value of Function U above with the parameter a in (U.1) set equal to $2b - 1$ and with z replaced by $2z$. C.f. (Z.2) is the even part of 2 times c.f. (U.5) with $a = 2b - 1$ and z replaced by $2z$. When the conditions for the validity of expansion (U.6) (which are the conditions for (U.5) also) are are modified to fit the changes in constant and variable we have indicated, it follows that expansion (Z.2) is a valid as asserted.

REFERENCES

1. Abramowitz, M. and Stegun, I. A. (1970). Handbook of Mathematical Functions Dover Publications, Inc., New York.

2. Bauer, G. (1872). Von einem Kettenbruch von Euler und einem Theorem von Wallis, Abh. Bayr. Akad. Wiss., Munchen, 11: 99-116.

3. Berndt, B. C., Lamphere, R. L., and Wilson, R. M. (1985). Chapter 12 of Ramanujan's notebook: continued fractions, Rocky Mtn. J. Math., 15: 235-310.

4. Berndt, B. C. (1989). Ramanujan's Notebooks, Part II, Springer-Verlag, New York.

5. Frame, J. S. (1979). The Hankel power sum matrix inverse and the Bernoulli continued fraction, Math. Comp., 33: 815-816.

6. Jacobsen, L. (1986). General convergence of continued fractions, Trans. Amer. Math. Soc., 294: 477-485.

7. Jacobsen, L. (1990). On the Bauer-Muir transformation for continued fractions and its applications, J. Math. Annal. Appl., 152: 496-514.

8. Jones, W. B. and Thron, W. J. (1980). Continued Fractions: Analytic Theory and Applications, Addison-Wesley, Reading, MA.

9. Lange, L. J. (1983). δ-fraction expansions of analytic functions, SIAM J. Math. Anal., 14: 323-368.

10. Lorentzen, L. and Waadeland, H. (1992). Continued Fractions with Applications, North-Holland, Amsterdam.

11. Muir, T. (1877). A theorem in continuants, Phil. Mag., (5) , 3: 137-138.

12. Perron, O. (1911). Einige Konvergenz- und Divergenzkriterien für alternierende Kettenbrüche, <u>Sb. Münch.</u>: 205-216.

13. Perron, O. (1953). Über die Preece'schen Kettenbrüche, <u>Sb. Münch.</u>: 21-56.

14. Perron, O. (1957). <u>Die Lehre von den Kettenbrüchen, 3rd ed. vol 2</u>, Teubner, Stuttgart.

15. Ramanujan, S. (1927). <u>Collected Papers</u>, Cambridge.

16. Rogers, L. J. (1907). On the representation of certain asymptotic series as convergent continued fractions, Proc. Lond. Math. Soc., (2), <u>4</u>: 72-89.

17. Shenton, L. R. and Bowman, K. O. (1971). Continued fractions for the psi function and its derivatives, <u>SIAM J. Appl. Math.</u>, <u>20</u>: 547-554.

18. Stieltjes, T. J. (1889). Sur la réduction en fraction continue d'une série précédent suivant les pouissances descendants d'une variable, <u>Ann. Fac. Sci. Toulouse</u>, <u>3</u>: 1-17; <u>Oeuvres</u>, <u>2</u>: 184-200.

19. Stieltjes, T. J. (1890). Sur quelques intégrales définies et leur développment en fractions continues, <u>Quart. Jour. of Math.</u>, <u>24</u>: 370-382; <u>Oeuvres</u>, <u>2</u>: 378-394.

20. Stieltjes, T. J. (1894). Recherches sur les fractions continues, <u>Ann. Fac, Sci. Toulouse</u>, <u>8</u>: 1-122; <u>9</u>: 1-47; <u>Oeuvres</u>, <u>2</u>: 402-566.

21. Thron, W. J. (1989). Continued fraction identities derived from the crossratio under ℓ.f.t., <u>Analytic Theory of Continued Fractions III</u> (L. Jacobsen, ed.), Lecture Notes in Math., <u>1406</u>, Springer, Berlin, pp. 124-134.

22. Wall, H. S. (1948). <u>Analytic Theory of Continued Fractions</u>, Van Nostrand, New York.

12

A Convergence Property for Sequences of Linear Fractional Transformations

LISA LORENTZEN Division of Mathematical Sciences, University of Trondheim, NTH, N–7034 Trondheim, Norway

Dedicated to the memory of ARNE MAGNUS

1. INTRODUCTION AND MAIN RESULTS

Let $\{t_n\}$ be a sequence of (non–singular) linear fractional transformations such that

$$t_n(U) \subseteq U \quad \text{for} \quad n = 1, 2, 3, \ldots, \qquad \text{where} \quad U := \{w \in \mathbf{C} : |w| < 1\}, \tag{1.1}$$

and let

$$T_n := t_1 \circ t_2 \circ \cdots \circ t_n \qquad \text{for} \quad n = 1, 2, 3, \ldots \tag{1.2}$$

be generated from $\{t_n\}$ by composition. Then also $T_n(U) \subseteq U$. In fact

$$\Delta_{n+1} := T_{n+1}(\overline{U}) = T_n(t_{n+1}(\overline{U})) \subseteq T_n(\overline{U}) = \Delta_n, \tag{1.3}$$

so that $\{\Delta_n\}$ is a sequence of non–empty closed, nested sets in $\overline{U}$, and thus the limiting set

$$\Delta := \lim_{n \to \infty} \Delta_n = \cap_{n=1}^{\infty} \Delta_n \subseteq \overline{U} \tag{1.4}$$

exists and is non–empty. Here, $\overline{A}$ denotes the closure of a set $A \subseteq \mathbf{C}$. This should not be confused with the notation $\bar{a}$ for the complex conjugate of a number $a \in \mathbf{C}$. We also write $\hat{\mathbf{C}} := \mathbf{C} \cup \{\infty\}$. The purpose of this paper is to establish sufficient conditions for

$$\lim_{n \to \infty} T_n(w) = T(w) \equiv c \in \overline{U} \qquad \text{for all} \quad w \in U. \tag{1.5}$$

Actually we shall find sufficient conditions for either of the two statements

---

The author has changed her name from Lisa Jacobsen.

(S1.1) $\{T_n(w)\}$ converges uniformly in $\overline{U}$ to a constant $c \in \overline{U}$.

(S1.2) $\{T_n(w)\}$ converges locally uniformly in $\hat{C} \setminus \partial U$ to a constant $c \in \overline{U}$.

(S1.1) evidently holds if and only if Δ is a one–point set, $\Delta = \{c\}$. This is called the limit point case. If Δ is larger, then Δ is a circular disk, the limit circle case, and we shall see that (S1.2) follows under our conditions.

Of course, (1.5), (S1.1) and (S1.2) are asymptotic properties. Hence it suffices that (1.1) holds from some n on, if we allow convergence to values $\notin \overline{U}$.

Throughout this paper, $\{t_n\}$ and $\{T_n\}$ shall always be as defined above. In addition we shall use the notation

$r_n,\ R_n$:	radii of $t_n(U),\ T_n(U)$	
$c_n,\ C_n$:	centers of $t_n(U),\ T_n(U)$	
$\zeta_n,\ Z_n$:	poles of $t_n(w),\ T_n(w)$	
$q_n,\ Q_n$:	$q_n := 1/\overline{\zeta_n},\ Q_n := 1/\overline{Z_n}$	
$k_n,\ K_n$:	$k_n := t_n(\infty),\ K_n := T_n(\infty)$	
$\tilde{r}_n,\ \tilde{R}_n$:	radii of $t_n^{-1}(U),\ T_n^{-1}(U)$ if $\infty \notin t_n^{-1}(\overline{U}), T_n^{-1}(\overline{U})$,	
	$\tilde{r}_n, \tilde{R}_n := \infty$ otherwise	
$\tilde{c}_n,\ \tilde{C}_n$:	centers of $t_n^{-1}(U),\ T_n^{-1}(U)$ if $\infty \notin t_n^{-1}(\overline{U}), T_n^{-1}(\overline{U})$	
$R,\ C$:	$R := \lim R_n,\ C := \lim C_n$	

The limits R and C exist since they are the radius and center of Δ in the limit circle case, and $R = 0$, $C = c$ in the limit point case. The poles ζ_n or Z_n may well be at infinity, in which case $q_n = 0$ or $Q_n = 0$. Expressions involving q_n and/or Q_n (ζ_n and/or Z_n) are then to be interpreted in the natural way. (We are working with linear fractional transformations which are univalent on the Riemann sphere $\hat{C}$.)

The following suggestive theorem from 1965 is due to Hillam and Thron, although ideas in their paper can be traced back to [11]:

THEOREM A1.1, [1], [5, Lemma 4.38, p.95], [10, Lemma 13, p.120].   Let $k \in U$ be a fixed number, and let

$$t_n(\infty) = k \qquad \text{for all } n. \tag{1.6}$$

Then (1.5) holds.

(Hillam and Thron actually worked with the closed unit disk, and had $|k| < 1$, but of course, this makes no difference since $t_n(\overline{U}) \subseteq \overline{U}$ if and only if $t_n(U) \subseteq U$ for non–singular linear fractional transformations.) The idea of their proof can be described as follows: If $R_n \to 0$, then there is no problem. The limit point case occurs, and (S1.1) follows trivially. Hence we just have to consider the limit circle case (the only alternative to the limit point case). Then $R_n \to R > 0$. They then proved that condition (1.6) is sufficient to ensure that (1.5) still holds.

Their idea can also be used in our general setting. For the limit circle case we get:

LEMMA 1.1

A. If $R > 0$ and either $\limsup r_n < 1$ or $\liminf \tilde{r}_n > 1$, then $\sum(1 - |Q_n|) < \infty$.

B. If $R > 0$, $\sum(1 - |Q_n|) < \infty$, and there exists a sequence $\{w_n\}$ of numbers from $\hat{C}$ such that

$$\liminf \big| |w_n| - 1 \big| > 0 \quad \text{and} \quad \liminf \big| |t_n(w_n)| - 1 \big| > 0, \tag{1.7}$$

then (S1.2) holds.

Lemma 1.1 is equivalent to two lemmas from 1970 by Jones and Thron, [4, Lemma 4.1, Lemma 4.2]. This will be proved in Section 2. Those two lemmas were designed to prove convergence of continued fractions with twin value sets, and they are therefore quite involved. We shall therefore offer a separate proof of Lemma 1.1 in Section 3, where most of the results in this section are proved.

The essence of Lemma 1.1 is the following main result:

THEOREM 1.2 If (1.7) holds for a sequence $\{w_n\}$ of numbers from $\hat{C}$, and either $\limsup r_n < 1$ or $\liminf \tilde{r}_n > 1$, then (S1.1) or (S1.2) holds.

REMARKS R1.1.

1. In the limit point case we naturally have that $\lim T_n(w) = C$ for all $w \in \overline{U}$. In the limit circle case it follows from the proof of Lemma 1.1 that $\lim T_n(w) = \lim T_n(0) = C - Re^{i\theta}$ for all $w \notin \partial U$, where $\theta := \lim \arg(C_n - T_n(0)) = \lim \arg(C_n - T_n(\infty))$ under the conditions of Theorem 1.2.

2. We shall see later (Observation 1.3C) that if for instance $\liminf |\zeta_n| > 1$, then

$$\limsup r_n < 1 \qquad \Leftrightarrow \qquad \liminf \tilde{r}_n > 1.$$

However, this equivalence does not hold in general. For instance,

$$t_n(w) := \frac{n + 1/2}{n} - \frac{1}{2} \cdot \frac{\left(\frac{n+1}{n}\right)^2 - 1}{\frac{n+1}{n} - w} \quad \text{for } n = 1, 2, 3, \ldots$$

satisfy (1.1) with radii $r_n = 1/2$, whereas $\tilde{r}_n \to 1$ as $n \to \infty$. Similarly,

$$t_n(w) := \frac{1}{n^2} - \frac{n-1}{n} \cdot \frac{w - n/(n+1)}{1 - wn/(n+1)} \quad \text{for } n = 2, 3, 4, \ldots$$

satisfy (1.1) with radii $r_n = 1 - 1/n \to 1$, whereas $t_n^{-1}(U)$ is a halfplane, such that $\tilde{r}_n = \infty$ for all n.

3. Since $2r_n < 2 - |k|$ for all n in Theorem A1.1, it follows that $\limsup r_n \le 1 - |k|/2 < 1$ under those conditions. In fact we also have $\tilde{r}_n = \infty$ when (1.6) holds, so that $\liminf \tilde{r}_n > 1$. The choice $w_n := \infty$ for all n in Theorem 1.2 therefore gives Theorem A1.1. Hence Theorem 1.2 represents a generalization of Theorem A1.1.

Condition (1.7) is not always so convenient. Can it be formulated in another way? And how do the radii r_n and $\tilde{r}_n$ relate? The following observation gives some answers:

OBSERVATION 1.3

A.
$$r_n|\zeta_n| - (1 - r_n) \le |k_n| \le r_n|\zeta_n| + (1 - r_n) \le |\zeta_n| \quad \text{for all } n \qquad (1.8)$$

and

$$|k_n| \le \tilde{r}_n|k_n| - (\tilde{r}_n - 1) \le |\zeta_n| \le \tilde{r}_n|k_n| + (\tilde{r}_n - 1) \quad \text{if } |k_n| > 1. \qquad (1.9)$$

Moreover, $|k_n| = |\zeta_n|$ *if and only if* $\zeta_n = \infty$ *or* $r_n = 1$. *Finally,* $r_n = 1$ *if and only if* $\tilde{r}_n = 1$.

B. *If* $|k_n| > 1$, *then*

$$\tilde{r}_n = \begin{cases} r_n \dfrac{|\zeta_n|^2 - 1}{|k_n|^2 - 1} & \text{if } \zeta_n \ne \infty, \\ 1/r_n & \text{if } \zeta_n = \infty. \end{cases} \qquad (1.10)$$

C. *Let* $\{n_j\}$ *be the set of indices where* $|k_n| > 1$. *If* $\{n_j\}$ *is a finite set, or if* $\liminf_{j \to \infty} |\zeta_{n_j}| > 1$, *then*

$$\limsup r_n < 1 \qquad \Rightarrow \qquad \liminf \tilde{r}_n > 1. \qquad (1.11)$$

Let $\varepsilon > 0$, *and let* $\{n_j\}$ *be the set of indices where* $|k_n| \ge 1 - \varepsilon$. *If* $\{n_j\}$ *is a finite set, or if* $\liminf_{j \to \infty} |\zeta_{n_j}| > 1$, *then*

$$\liminf \tilde{r}_n > 1 \qquad \Rightarrow \qquad \limsup r_n < 1. \qquad (1.12)$$

D. *If there exist positive constants* ε_1, $\varepsilon_2 > 0$ *such that*

$$|\zeta_n| < 1 + \varepsilon_1 \qquad \Rightarrow \qquad ||k_n| - 1| \ge \varepsilon_2 \qquad \text{for all } n, \qquad (1.13)$$

then there exists a sequence $\{w_n\} \subseteq \hat{\mathbf{C}}$ *such that (1.7) holds.*

E. *If there exists a subsequence* $\{n_m\}$ *of* $\mathbf{N}$ *such that*

$$\lim_{m \to \infty} |\zeta_{n_m}| = 1 \quad \text{and} \quad \lim_{m \to \infty} |k_{n_m}| = 1, \qquad (1.14)$$

then there exists no sequence $\{w_n\} \subseteq \hat{\mathbf{C}}$ *such that (1.7) holds.*

The limit point case (S1.1) has several advantages. It is therefore of interest to identify situations where this occurs.

THEOREM 1.4

A. *If* $\sum(1 - (r_n + |c_n|)) = \infty$, *then (S1.1) holds.*

B. *If* $\sum \text{dist}(\partial U, t_n^{-1}(\partial U)) = \infty$, *where* $\text{dist}(J_1, J_2)$ *is the euclidean distance between two curves* J_1 *and* J_2 *in* $\hat{\mathbf{C}}$, *then (S1.1) holds.*

C. *If*

$$\lim_{n \to \infty} r_n \frac{|\zeta_n| + 1}{|\zeta_n| - 1} \prod_{j=1}^{n-1} \kappa_j = 0, \quad \text{where } \kappa_j := r_j \frac{|\zeta_j|^2 - 1}{(|\zeta_j - c_{j+1}| - r_{j+1})^2} \le r_j \frac{|\zeta_j| + 1}{|\zeta_j| - 1}, \qquad (1.15)$$

then (S1.1) holds. This is in particular so if $\prod_{j=1}^{\infty} r_j(|\zeta_j| + 1)/(|\zeta_j| - 1) = 0$.

REMARKS R1.2

1. We always have $r_n + |c_n| \le 1$ by (1.1). Moreover, $r_n + |c_n| = 1$ if and only if the boundary of $t_n(U)$ meets the boundary of U. Hence the condition in Theorem 1.4A just says that $\sum \text{dist}(\partial U, t_n(\partial U)) = \infty$, similarly to the condition in Theorem 1.4B. These two conditions are independent of $\{\zeta_n\}$, and they hold even in some cases where $|c_n| \to 1$, or $r_n \to 1$ and $\tilde{r}_n \to 1$.

2. In Theorem 1.4C we also allow $r_n + |c_n| = 1$ for all n. The condition (1.15) relates r_n to the location of the pole ζ_n. If in particular all these poles are at infinity; i.e.

$$t_n(w) = c_n + p_n w \qquad \text{for} \quad n = 1, 2, 3, \ldots, \tag{1.16}$$

then $\prod \kappa_n = \prod r_n = 0$ if and only if $\sum(1 - r_n) = \infty$. Hence, in this case the limit point case holds if $\sum(1 - r_n) = \infty$. (See also Example 4.1.)

3. We have adopted the convention that a linear fractional transformation is non-singular by definition. If we allow t_n to be singular for an $n \in \mathbf{N}$, then $\text{diam}(\Delta_n) = 0$ for this n, and thus $\{T_m(w)\}$ converges uniformly in $\overline{U}$ to $c := T_n(0) \in \overline{U}$ as $m \to \infty$.

THEOREM 1.5 *If $k_n = k$ and $\zeta_n = \zeta$ are independent of n and $|\zeta| \neq |k|$, then (S1.1) holds.*

This result was essentially proved in [9, Thm. 2.3]. We shall return to its proof in Section 5.

REMARKS R1.3

1. If $k = \infty$, and thus also $\zeta = \infty$ in Theorem 1.5, then all t_n have the form (1.16), and (S1.1) holds by virtue of Theorem 1.4C if $\sum(1 - r_n) = \infty$. (See Remark R1.2.2.)

2. Our condition that $|k| \neq |\zeta|$ is not so very strong. (See Observation 1.3A.) Actually, if $r_n < 1$ for one index n, then $r_n < 1$ for all n by Observation 1.3A. It even follows from arguments similar to the proof of Observation 1.3A that $\limsup r_n < 1$ in this case.

A similar argument shows that if $\tilde{r}_n > 1$ for one index n, then all $\tilde{r}_n > 1$, and even $\liminf \tilde{r}_n > 1$.

THEOREM 1.6 *If $\liminf |\zeta_n| > 1$, $(\zeta_n + k_n)$ is independent of n with $|\zeta_n + k_n| \neq 2$, and $\limsup r_n < 1$, then (S1.1) holds.*

If $\zeta_n = \infty$, then also $k_n = \infty$, and we set $(\zeta_n + k_n) := \infty$ in Theorem 1.6. Remark R1.2.2 gives a stronger result for this case, though.

2. THE EQUIVALENCE

In this section we shall prove that Lemma 1.1 is essentially equivalent to the following two lemmas by Jones and Thron. (Their notation is slightly changed.)

LEMMA A2.1, [4, Lemma 4.1].   *Let $\{T_n\}$ be a sequence of linear fractional transformations of the form*

$$T_n(z) = C_n + P_n \frac{z - Q_n}{1 - \overline{Q}_n z} \quad \text{for } n = 1, 2, 3, \ldots, \tag{2.1}$$

satisfying

$$|P_n| = R_n \searrow R > 0, \quad |C_n - C_{n-1}| \le R_{n-1} - R_n, \quad |Q_n| = g_n < 1. \tag{2.2}$$

(A) *Suppose that there exists a sequence $\{u_n\}$ in $\hat{\mathbf{C}}$ such that*

$$T_{2n+1}(u_{2n+1}) = T_{2n}(u_{2n}), \quad |u_{2n+1}| \ge 1, \quad |u_{2n}| \le 1, \quad n \ge 1. \tag{2.3}$$

If for some $\varepsilon > 0$, $|u_{2n+1}| \ge 1 + \varepsilon$, $n \ge 1$, then

$$\sum_{n=1}^{\infty} (1 - g_{2n+1}) < \infty. \tag{2.4}$$

If for some $\varepsilon > 0$, $|u_{2n}| \le 1 - \varepsilon$, $n \ge 1$, then

$$\sum_{n=1}^{\infty} (1 - g_{2n}) < \infty. \tag{2.5}$$

(B) *Suppose that there exists a sequence $\{v_n\}$ in $\hat{\mathbf{C}}$ such that*

$$T_{2n}(v_{2n}) = T_{2n-1}(v_{2n-1}), \quad |v_{2n}| \ge 1, \quad |v_{2n-1}| \le 1, \quad n \ge 1. \tag{2.6}$$

If for some $\varepsilon > 0$, $|v_{2n}| \ge 1 + \varepsilon$, $n \ge 1$, then (2.5) holds.
If for some $\varepsilon > 0$, $|v_{2n-1}| \le 1 - \varepsilon$, $n \ge 1$, then (2.4) holds.

LEMMA A2.2, [4, Lemma 4.2].   *Let $\{T_n\}$ be a sequence of linear fractional transformations of the form (2.1) – (2.2).*
   (A) *Suppose that there exist sequences $\{\xi_n\}$ and $\{\eta_n\}$ in $\hat{\mathbf{C}}$ such that for some $\varepsilon > 0$*

$$T_{2n+1}(\xi_n) = T_{2n-1}(\eta_n), \quad \big||\xi_n| - 1\big| \ge \varepsilon, \quad \big||\eta_n| - 1\big| \ge \varepsilon, \quad n \ge 1. \tag{2.7}$$

If $\sum_{n=1}^{\infty} (1 - g_{2n+1}) < \infty$, then $\{T_{2n+1}(z)\}$ converges for all z in the complex plane such that $|z| \ne 1$, and

$$\lim T_{2n+1}(z) = \lim \left(C_{2n+1} - P_{2n+1} / \overline{Q_{2n+1}} \right), \quad |z| \ne 1. \tag{2.8}$$

(B) Suppose that there exist sequences $\{\xi'_n\}$ and $\{\eta'_n\}$ in $\hat{C}$ such that for some $\varepsilon > 0$

$$T_{2n+2}(\xi'_n) = T_{2n}(\eta'_n), \quad \left||\xi'_n| - 1\right| \geq \varepsilon, \quad \left||\eta'_n| - 1\right| \geq \varepsilon, \quad n \geq 1. \qquad (2.9)$$

If $\sum_{n=1}^{\infty}(1 - g_{2n}) < \infty$, then $\{T_{2n}(z)\}$ converges for all z in the complex plane such that $|z| \neq 1$, and

$$\lim T_{2n}(z) = \lim \left(C_{2n} - P_{2n}/\overline{Q_{2n}}\right), \quad |z| \neq 1. \qquad (2.10)$$

Of course, (2.8) and (2.10) do not have analogues in Lemma 1.1, but the limits are mentioned in Remark R1.1.1. The uniformity of the convergence, and the convergence at $z = \infty$ are not mentioned in Lemma A2.1, Lemma A2.2, but they follow very easily. Apart from these minor points, we shall see that the results are equivalent.

Condition (2.1) and (2.2) just mean that $T_n(z)$ are linear fractional transformations which map the unit disk onto circular disks with centers C_n and radii R_n such that

$$T_n(U) \subseteq T_{n-1}(U) \quad \text{and} \quad T_n(\overline{U}) \to \Delta = \{w \in \mathbf{C} : |w - C| \leq R\}, \quad C := \lim C_n,$$

just as our $\{T_n\}$ in (1.2). Let $t_n := T_{n-1}^{-1} \circ T_n$ for $n = 2, 3, 4, \ldots$. Then t_n maps U into U since $t_n(U) = T_{n-1}^{-1} \circ T_n(U) \subseteq T_{n-1}^{-1}(T_{n-1}(U)) = U$. Without loss of generality we may assume that $T_1(U) \subseteq U$ in Lemma A2.1. (It is just a matter of a simple transformation $\tilde{T}_n := \varphi \circ T_n$, where $\varphi(T_1(U)) \subseteq U$.) Then also $t_1 := T_1$ maps U into U, and thus (2.1) – (2.2) is equivalent to (1.1) – (1.2).

<u>Lemma A2.1 $\Rightarrow$ Lemma 1.1A:</u>

In view of the arguments above, the sequences $\{u_n\}$ in Lemma A2.1(A) and $\{v_n\}$ in Lemma A2.1(B) satisfy

$$u_{2n} = t_{2n+1}(u_{2n+1}), \quad v_{2n-1} = t_{2n}(v_{2n}). \qquad (2.11)$$

Let first $\liminf \tilde{r}_n > 1$. Then there exist sequences $\{u_n\}$ and $\{v_n\}$ and an $\varepsilon > 0$ such that

$$|u_{2n+1}| \geq 1 + \varepsilon, \quad |u_{2n}| \leq 1 \quad \text{in Lemma A2.1(A)}$$

and

$$|v_{2n}| \geq 1 + \varepsilon, \quad |v_{2n-1}| \leq 1 \quad \text{in Lemma A2.1(B)}$$

from some n on.

Similarly, if $\limsup r_n < 1$, then there exist sequences $\{u_n\}$ and $\{v_n\}$ and an $\varepsilon > 0$ such that

$$|u_{2n+1}| \geq 1, \quad |u_{2n}| \leq 1 - \varepsilon \quad \text{in Lemma A2.1(A)}$$

$$|v_{2n}| \geq 1, \quad |v_{2n-1}| \leq 1 - \varepsilon \quad \text{in Lemma A2.1(B)} \qquad (2.12)$$

from some n on. Hence Lemma 1.1A follows.

Lemma 1.1A $\Rightarrow$ Lemma A2.1:

Let $\{T_n\}$ be given by (2.1) – (2.2), where we without loss of generality assume that $T_1(U) \subseteq U$. Let first $\{u_n\}$ satisfy (2.3) with all $|u_{2n+1}| \geq 1 + \varepsilon$, and let $\tau_n := T_{2n-1}^{-1} \circ T_{2n+1}$ for all n. Then $\tau_n = t_{2n} \circ t_{2n+1}$ with the notation above. Since $\tau_n^{-1}(U) = t_{2n+1}^{-1} \circ t_{2n}^{-1}(U) \supseteq t_{2n+1}^{-1}(U)$, we find that the "radius" of $\tau_n^{-1}(U)$ is greater than or equal to the "radius" of $t_{2n+1}^{-1}(U)$, which again is $\geq 1 + \varepsilon/2$ for all n. Since $t_1 \circ \tau_1 \circ \tau_2 \circ \cdots \circ \tau_n = T_{2n+1}$ and $R > 0$, it follows from Lemma 1.1A that

$$\sum (1 - |Q_{2n+1}|) = \sum (1 - g_{2n+1}) < \infty.$$

A similar argument shows that also Lemma A2.1(B) follows from Lemma 1.1A in the case where all $|v_{2n}| \geq 1 + \varepsilon$.

Next, let (2.3) hold with all $|u_{2n}| \leq 1 - \varepsilon$. That is, $t_{2n+1}(U)$ has radius $r_{2n+1} \leq 1 - \varepsilon/2$. Let $\tilde{\tau}_n := T_{2n}^{-1} \circ T_{2n+2}$ for all n. Then $\tilde{\tau}_n = t_{2n+1} \circ t_{2n+2}$, and thus the radius of $\tilde{\tau}_n(U) = t_{2n+1} \circ t_{2n+2}(U) \subseteq t_{2n+1}(U)$ is $\leq r_{2n+1} \leq 1 - \varepsilon/2$. Since $\tilde{\tau}_0 \circ \tilde{\tau}_1 \circ \cdots \circ \tilde{\tau}_{n-1} = T_{2n}$ and $R > 0$, it follows again from Lemma 1.1A that

$$\sum (1 - |Q_{2n}|) = \sum (1 - g_{2n}) < \infty.$$

The last case in Lemma A2.1(B) follows similarly.

Lemma A2.2 $\Rightarrow$ Lemma 1.1B:

As above we define $\tau_n := T_{2n-1}^{-1} \circ T_{2n+1}$ for $n = 1, 2, 3, \ldots$. Then $\tau_n = t_{2n} \circ t_{2n+1}$ and $\tau_n(U) \subseteq U$. Condition (2.7) can now be interpreted as follows: there exists a sequence $\{\xi_n\}$ in $\hat{\mathbf{C}}$ such that for some $\varepsilon > 0$

$$\big| |\xi_n| - 1 \big| \geq \varepsilon \quad \text{and} \quad \big| |\tau_n(\xi_n)| - 1 \big| \geq \varepsilon.$$

Hence, Lemma 1.1B is a consequence of Lemma A2.2(A). (Just choose all $t_{2n}(w) \equiv w$ in Lemma A2.2(A).)

Similarly, Lemma 1.1B is also a consequence of Lemma A2.2(B), this time with $\tilde{\tau}_n := T_{2n}^{-1} \circ T_{2n+2}$ ($= t_{2n+1} \circ t_{2n+2}$), where all $t_{2n+1}(w) \equiv w$ and $T_1(U) \subseteq U$ in Lemma A2.2(B).

Lemma 1.1B $\Rightarrow$ Lemma A2.2:

Let (2.7) hold. That is, $w_n := \xi_n$ is a sequence of numbers from $\hat{\mathbf{C}}$ such that

$$\liminf \big| |w_n| - 1 \big| > 0 \quad \text{and} \quad \liminf \big| |\tau_n(w_n)| - 1 \big| > 0.$$

Since $T_{2n+1} = t_1 \circ \tau_1 \circ \tau_2 \circ \cdots \circ \tau_n$, the convergence of $\{T_{2n+1}(z)\}$ follows from Lemma 1.1B. If (2.9) holds, then the convergence of $\{T_{2n}(z)\}$ follows similarly.

3. PROOFS

The central point is the mapping properties of linear fractional transformations. The following result will be useful. It is not original, of course, but we include its proof for completeness.

LEMMA 3.1 *Let J be a circle with center $\Gamma \in \mathbf{C}$ and radius $R > 0$, and let*

$$t(w) := \frac{a + bw}{c + dw}, \quad ad - bc \neq 0 \tag{3.1}$$

be a linear fractional transformation such that $t(J)$ is bounded. Then $t(J)$ is a circle with center γ and radius r given by

$$\gamma = \frac{a(\overline{c + d\Gamma}) + b\overline{c}\Gamma + \overline{d}b(|\Gamma|^2 - R^2)}{|c + d\Gamma|^2 - |d|^2 R^2}, \quad r = \left| \frac{ad - bc}{|c + d\Gamma|^2 - |d|^2 R^2} \right| R. \tag{3.2}$$

PROOF: Assume first that $d \neq 0$. Then $t(w)$ can be written

$$t(w) = \frac{b}{d} + \frac{ad - bc}{d(c + dw)}.$$

$c + dJ$ is a circle with center at $\Gamma_1 := c + d\Gamma$ and radius $R_1 := |d|R$. This circle does not pass through the origin since $t(J)$ is bounded. Hence $1/(c + dJ)$ is a circle with center at

$$\Gamma_2 := \frac{1}{2}\left(\frac{1}{|\Gamma_1| - R_1} \cdot \frac{\overline{\Gamma_1}}{|\Gamma_1|} + \frac{1}{|\Gamma_1| + R_1} \cdot \frac{\overline{\Gamma_1}}{|\Gamma_1|} \right) = \frac{\overline{\Gamma_1}}{|\Gamma_1|^2 - R_1^2}$$

(where $\overline{\Gamma_1}/|\Gamma_1| := 1$ if $\Gamma_1 = 0$), and radius

$$R_2 := \frac{1}{2}\left| \frac{1}{|\Gamma_1| - R_1} - \frac{1}{|\Gamma_1| + R_1} \right| = \frac{R_1}{\left| |\Gamma_1|^2 - R_1^2 \right|}.$$

Finally, $t(J)$ is therefore a circle with center and radius given by

$$\gamma := \frac{b}{d} + \frac{ad - bc}{d}\Gamma_2, \quad r := \left| \frac{ad - bc}{d} \right| R_2,$$

which gives the result when $d \neq 0$.

If $d = 0$, then $c \neq 0$ and $t(w) = (a/c) + (b/c)w$. Hence $t(J)$ is a circle with center at $\gamma := (a/c) + (b/c)\Gamma$ and radius $r := |b/c|R$, which again proves (3.2). ∎

Inspired by Hillam and Thron's proof of Theorem A1.1, we note that t_n can be written in the form

$$t_n(w) = c_n + p_n \frac{w - q_n}{1 - \overline{q}_n w}, \quad \text{where} \quad |p_n| = r_n, \ |c_n| + r_n \leq 1 \ \text{and} \ |q_n| < 1, \tag{3.3}$$

where $q_n = 1/\overline{\zeta_n}$. (Keep in mind that every linear fractional transformation which maps the unit disk onto itself can be written

$$\phi(w) = e^{i\alpha} \frac{w - q}{1 - \overline{q}w} \qquad \text{for some } \alpha \in \mathbf{R} \text{ and } |q| < 1. \quad)$$

If $\zeta_n \neq \infty$, which is equivalent to $k_n = t_n(\infty) \neq \infty$, then (3.3) has the equivalent formulation

$$t_n(w) = k_n + \pi_n \frac{|\zeta_n|^2 - 1}{\zeta_n - w} \qquad \text{where } \pi_n := p_n \zeta_n / \overline{\zeta_n}, \tag{3.4}$$

and if $\zeta_n = \infty$, then (3.3) takes the simple form (1.16). Evidently also T_n maps U into U. Hence, just as for t_n in (3.3), $T_n(w)$ can be written in the form (2.1); i.e.

$$T_n(w) = C_n + P_n \frac{w - Q_n}{1 - \overline{Q_n}w} = \frac{(C_n - P_n Q_n) + (P_n - \overline{Q_n}C_n)w}{1 - \overline{Q_n}w}, \tag{3.5}$$

$$\text{where } |P_n| = R_n, \ R_n + |C_n| \leq 1 \text{ and } |Q_n| < 1.$$

PROOF OF LEMMA 1.1:

A. Since $\Delta_{n+1} = T_{n+1}(\overline{U}) = T_n(t_{n+1}(\overline{U}))$, where $t_{n+1}(\overline{U})$ is a circular disk with center c_{n+1} and radius r_{n+1}, it follows from (3.5) and Lemma 3.1 that

$$\begin{aligned}
R_{n+1} &= \frac{|(C_n - P_n Q_n)(-\overline{Q_n}) - (P_n - \overline{Q_n}C_n) \cdot 1|}{|1 - \overline{Q_n}c_{n+1}|^2 - |Q_n|^2 r_{n+1}^2} r_{n+1} \\
&= \frac{R_n(1 - |Q_n|^2)r_{n+1}}{|1 - \overline{Q_n}c_{n+1}|^2 - |Q_n|^2 r_{n+1}^2}.
\end{aligned} \tag{3.6}$$

Let first $\limsup r_n < 1$, and let $r < 1$ and $n_0 \in \mathbf{N}$ be chosen such that $r_n \leq r$ for all $n \geq n_0$. Since $|c_n| + r_n \leq 1$, we find from (3.6) that

$$\begin{aligned}
\frac{R_{n+1}}{R_n} &\leq \frac{r_{n+1}(1 - |Q_n|^2)}{(1 - |Q_n|(1 - r_{n+1}))^2 - |Q_n|^2 r_{n+1}^2} \\
&= \frac{r_{n+1}(1 + |Q_n|)}{1 - |Q_n| + 2|Q_n|r_{n+1}} \\
&= 1 - \frac{(1 - r_{n+1})(1 - |Q_n|)}{1 - |Q_n| + 2|Q_n|r_{n+1}} \\
&\leq 1 - \frac{(1 - r)(1 - |Q_n|)}{1 - |Q_n| + 2r|Q_n|} =: 1 - \delta_n,
\end{aligned} \tag{3.7}$$

where $\delta_n > 0$. Since $\prod_{n=1}^{\infty}(R_{n+1}/R_n) = R/R_1 > 0$, it follows by (3.7) that $\sum \delta_n < \infty$, i.e. $\sum(1 - |Q_n|) < \infty$.

Next we let $\liminf \tilde{r}_n > 1$. Let $D_n := t_n^{-1}(U)$ for all n. Assume first that $\infty \notin \overline{D_{n+1}}$. Then D_{n+1} is a circular disk with center $\tilde{c}_{n+1}$ and radius $\tilde{r}_{n+1}$, such that $U \subseteq D_{n+1}$ and thus $\tilde{r}_{n+1} - |\tilde{c}_{n+1}| \geq 1$. Since $\Delta_n = T_n(\overline{U}) = T_{n+1}(\overline{D_{n+1}})$ is bounded, we know that

$$|1 - \overline{Q_{n+1}}\tilde{c}_{n+1}| > |Q_{n+1}|\tilde{r}_{n+1}. \tag{3.8}$$

According to Lemma 3.1 we therefore know that

$$
\begin{aligned}
R_n &= \frac{R_{n+1}(1 - |Q_{n+1}|^2)\tilde{r}_{n+1}}{|1 - \overline{Q_{n+1}}\tilde{c}_{n+1}|^2 - |Q_{n+1}|^2\tilde{r}_{n+1}^2} \\
&\geq \frac{R_{n+1}(1 - |Q_{n+1}|^2)\tilde{r}_{n+1}}{(1 + |Q_{n+1}|(\tilde{r}_{n+1} - 1))^2 - |Q_{n+1}|^2\tilde{r}_{n+1}^2} \\
&= \frac{R_{n+1}(1 + |Q_{n+1}|)\tilde{r}_{n+1}}{1 - |Q_{n+1}| + 2|Q_{n+1}|\tilde{r}_{n+1}} \\
&= R_{n+1}\left(1 + \frac{(\tilde{r}_{n+1} - 1)(1 - |Q_{n+1}|)}{1 - |Q_{n+1}| + 2|Q_{n+1}|\tilde{r}_{n+1}}\right) =: R_{n+1}(1 + \tilde{\delta}_n).
\end{aligned}
\tag{3.9}
$$

If $\infty \in \overline{D_{n+1}}$, then we can always find a circular disk $\tilde{D}_{n+1} \subseteq D_{n+1}$, $U \subseteq \tilde{D}_{n+1}$, such that $\tilde{D}_{n+1}$ has radius $\tilde{r}_{n+1} := 2$. Since then $\Delta_n = T_{n+1}(\overline{D_{n+1}}) \supseteq T_{n+1}(\tilde{D}_{n+1})$, we find that $R_n \geq$ the radius of $T_{n+1}(\tilde{D}_{n+1})$. Hence the inequality in (3.9) still holds with this choice for $\tilde{r}_n$. Since $\prod(R_{n+1}/R_n) = R/R_1 > 0$, and $R_{n+1}/R_n \leq 1/(1 + \tilde{\delta}_n)$ by (3.9), it follows that $\sum \tilde{\delta}_n < \infty$. Since $\liminf \tilde{r}_n > 1$, this means that $\sum(1 - |Q_{n+1}|) < \infty$.

B. Let $\varepsilon > 0$ and $n_0 \in \mathbf{N}$ be chosen such that $|1 - \overline{Q_n}t_{n+1}(w_{n+1})| \geq \varepsilon$, $|1 - \overline{Q_n}w_n| \geq \varepsilon$ and $|Q_n| \geq \varepsilon$ for all $n \geq n_0$. (This is always possible by (1.7), since $|Q_n| \to 1$ under our conditions.) From (3.5) it follows that

$$
|T_n(u) - T_n(v)| = \begin{cases} R_n \dfrac{(1 - |Q_n|^2)|u - v|}{|1 - \overline{Q_n}u| \cdot |1 - \overline{Q_n}v|}, & \text{if } u, v \neq \infty, \\[2ex] R_n \dfrac{1 - |Q_n|^2}{|Q_n| \cdot |1 - \overline{Q_n}v|}, & \text{if } u = \infty, \\[2ex] 0 & \text{if } u = v = \infty. \end{cases}
\tag{3.10}
$$

Hence there exists a constant $A > 0$ such that

$$
|T_{n+1}(w_{n+1}) - T_n(w_n)| = |T_n(t_{n+1}(w_{n+1})) - T_n(w_n)| < A(1 - |Q_n|)
\tag{3.11}
$$

for all $n \geq n_0$. This means that $\sum(T_{n+1}(w_{n+1}) - T_n(w_n))$ converges absolutely; i.e. the limit $c := \lim T_n(w_n)$ exists.

Let $w \in \hat{\mathbf{C}} \setminus \partial U$ be arbitrarily chosen. Then it follows from (3.10), just as in (3.11), that there is a positive constant $A(w) > 0$ such that

$$
|T_n(w) - T_n(w_n)| \leq A(w)(1 - |Q_n|) \to 0 \quad \text{as} \quad n \to \infty.
$$

This proves that also $T_n(w) \to c$. The local uniformity follows by the Stieltjes – Vitali Theorem. ∎

PROOF OF OBSERVATION 1.3:

A. Clearly $\zeta_n = \infty$ if and only if $k_n = \infty$. Hence (1.8) and (1.9) hold (with equality) in this case. Assume that $\zeta_n \neq \infty$.

Since $k_n = t_n(\infty) = c_n - p_n\zeta_n$, where $|c_n| + r_n = |k_n + p_n\zeta_n| + r_n \leq 1$, we find that

$$|k_n| - r_n|\zeta_n| \leq |k_n + p_n\zeta_n| \leq 1 - r_n,$$

which proves the upper bounds for $|k_n|$ in (1.8), where the equality $|k_n| = |\zeta_n|$ only can occur if $r_n = 1$. ($|\zeta_n| > 1$ since t_n is non–singular and $t_n(\overline{U}) \subseteq \overline{U}$.)

Similarly,

$$r_n|\zeta_n| - |k_n| \leq |k_n + p_n\zeta_n| \leq 1 - r_n$$

gives the lower bound for $|k_n|$ in (1.8).

To prove (1.9), we observe that for $|k_n| > 1$, $D_n := t_n^{-1}(U)$ is a circular disk. As in the introduction, we let $\tilde{c}_n$ be the center of this disk D_n, and $\tilde{r}_n$ be its radius. In accordance with (3.3), t_n^{-1} can then be written

$$t_n^{-1}(w) = \tilde{c}_n + \tilde{p}_n \frac{w - 1/\overline{k_n}}{1 - w/k_n}, \quad |\tilde{p}_n| = \tilde{r}_n. \tag{3.12}$$

Since $U \subseteq D_n$, we know that $|\tilde{c}_n| + 1 \leq \tilde{r}_n$. Moreover, $t_n^{-1}(\infty) = \zeta_n = \tilde{c}_n - \tilde{p}_n k_n$; i.e. $\tilde{c}_n = \zeta_n + \tilde{p}_n k_n$, so that

$$\left||\zeta_n| - \tilde{r}_n|k_n|\right| \leq |\zeta_n + \tilde{p}_n k_n| = |\tilde{c}_n| \leq \tilde{r}_n - 1,$$

which gives the desired result (1.9).

Let $r_n = 1$. Then $c_n = 0$ and $k_n = t_n(\infty) = c_n - p_n\zeta_n = -p_n\zeta_n$. That is, $|k_n| = |\zeta_n|$. That $r_n = 1$ if and only if $\tilde{r}_n = 1$ follows since $r_n = 1$ if and only if $t_n(U) = U$, and thus $t_n^{-1}(U) = U$.

B. Assume first that $\zeta_n \neq \infty$. Then it follows from (3.4) that

$$t_n^{-1}(w) = \zeta_n + \pi_n \frac{|\zeta_n|^2 - 1}{k_n - w}, \quad |\pi_n| = r_n. \tag{3.13}$$

Since $|k_n| > 1$, we know that $D_n := t_n^{-1}(U)$ is a bounded disk, and (1.10) is a simple consequence of Lemma 3.1 in this case.

If $\zeta_n = \infty$, then $t_n(w) = c_n + p_n w$, where $|p_n| = r_n$, and thus (1.10) still holds.

C. If $\zeta_n = \infty$, then $r_n = 1/\tilde{r}_n$ by Part B, and there is no problem. For the remaining indices n, $\zeta_n \neq \infty$. (Then also $k_n \neq \infty$.) We first observe that by Part A we then have

$$\frac{|k_n| - 1}{|\zeta_n| - 1} \leq r_n \leq \frac{|k_n| + 1}{|\zeta_n| + 1} \leq 1 \tag{3.14}$$

and

$$\frac{|\zeta_n| + 1}{|k_n| + 1} \leq \tilde{r}_n \leq \frac{|\zeta_n| - 1}{|k_n| - 1} \quad \text{if} \quad 1 < |k_n| < \infty. \tag{3.15}$$

For the indices n where $1 < |k_n| < \infty$, we therefore know that if $r_n \leq 1 - \varepsilon$ and $|\zeta_n| \geq 1 + \varepsilon$ for an $\varepsilon > 0$, then by (3.14)

$$\frac{|k_n| - 1}{|\zeta_n| - 1} \leq 1 - \varepsilon$$

$$|k_n| + 1 \leq |\zeta_n| + 1 - \varepsilon(|\zeta_n| - 1)$$

$$\frac{|k_n| + 1}{|\zeta_n| + 1} \leq 1 - \varepsilon \frac{|\zeta_n| - 1}{|\zeta_n| + 1} \leq 1 - \frac{\varepsilon^2}{2 + \varepsilon},$$

and thus, by (3.15)

$$\tilde{r}_n \geq \frac{|\zeta_n| + 1}{|k_n| + 1} \geq \frac{1}{1 - \varepsilon^2/(2 + \varepsilon)} > 1 + \frac{\varepsilon^2}{2 + \varepsilon}.$$

If $|k_n| \leq 1$, then $\tilde{r}_n = \infty$ by definition. This proves (1.11).

To prove (1.12) we observe that if $1 < |k_n| < \infty$ and $\tilde{r}_n \geq 1 + \varepsilon$ and $|\zeta_n| \geq 1 + \varepsilon$, then by (3.15)

$$\frac{|\zeta_n| - 1}{|k_n| - 1} \geq 1 + \varepsilon$$

$$|\zeta_n| + 1 \geq |k_n| + 1 + \varepsilon(|k_n| - 1)$$

$$\frac{|\zeta_n| + 1}{|k_n| + 1} \geq 1 + \varepsilon \frac{|k_n| - 1}{|k_n| + 1}.$$

Hence, by (3.14),

$$r_n \leq \frac{|k_n| + 1}{|\zeta_n| + 1} \leq \begin{cases} \dfrac{1}{1 + \varepsilon \dfrac{|k_n| - 1}{|k_n| + 1}} \leq \dfrac{1}{1 + \varepsilon \dfrac{\varepsilon}{4 + \varepsilon}} = 1 - \dfrac{\varepsilon^2}{4 + \varepsilon + \varepsilon^2} & \text{if } |k_n| \geq 1 + \frac{\varepsilon}{2}, \\[3ex] \dfrac{2 + \varepsilon/2}{2 + \varepsilon} = 1 - \dfrac{\varepsilon/2}{2 + \varepsilon} & \text{if } 1 < |k_n| < 1 + \varepsilon/2. \end{cases}$$

If $|k_n| \leq 1$ (and thus $\tilde{r}_n = \infty$) and $|\zeta_n| \geq 1 + \varepsilon$, then we have by (3.14) that

$$r_n \leq \frac{|k_n| + 1}{|\zeta_n| + 1} \leq \frac{2}{2 + \varepsilon} = 1 - \frac{\varepsilon}{2 + \varepsilon}.$$

Finally, if $|k_n| \leq 1 - \varepsilon$, then

$$r_n \leq \frac{|k_n| + 1}{|\zeta_n| + 1} \leq \frac{2 - \varepsilon}{2} = 1 - \frac{\varepsilon}{2}.$$

This proves (1.12).

D. Choose $w_n = \zeta_n$ if $|\zeta_n| \geq 1 + \varepsilon_1$, and $w_n = \infty$ if $|\zeta_n| < 1 + \varepsilon_1$.

E. Assume that (1.7) holds for a sequence $\{w_n\}$. Without loss of generality we assume that $\big||w_n| - 1\big| \geq \varepsilon$, $\big||t_n(w_n)| - 1\big| \geq \varepsilon$ for all n for an $\varepsilon > 0$, and that $\infty \neq \zeta_{n_m} \to \exp(i\theta)$. From (3.4) it follows that

$$t_n(w_n) - k_n = \pi_n \frac{|\zeta_n|^2 - 1}{\zeta_n - w_n}.$$

Since $\big||t_n(w_n)| - 1\big| \geq \varepsilon$, $r_n = |\pi_n| \leq 1$, and $|\zeta_{n_m}|^2 \to 1$, we thus have $w_{n_m} \to \exp(i\theta)$, which contradicts the fact that $\big||w_{n_m}| - 1\big| \geq \varepsilon$. Hence no such sequence $\{w_n\}$ exists. ∎

PROOF OF THEOREM 1.4:

A. Let $\mu_n := r_n + |c_n|$. Then $\mu_n \leq 1$, and it follows from (3.6) (as in (3.7)) that

$$
\begin{aligned}
\frac{R_{n+1}}{R_n} &\leq \frac{r_{n+1}(1 - |Q_n|^2)}{(1 - |Q_n c_{n+1}|)^2 - |Q_n|^2 r_{n+1}^2} \\
&= \frac{r_{n+1}(1 - |Q_n|^2)}{(1 - |Q_n|(\mu_{n+1} - r_{n+1}))^2 - |Q_n|^2 r_{n+1}^2} \\
&= \frac{1 - |Q_n|^2}{2|Q_n|(1 - |Q_n|\mu_{n+1}) + (1 - |Q_n|\mu_{n+1})^2/r_{n+1}} \\
&\leq \frac{1 - |Q_n|^2}{2|Q_n|(1 - |Q_n|\mu_{n+1}) + (1 - |Q_n|\mu_{n+1})^2/\mu_{n+1}} \\
&= \frac{\mu_{n+1}(1 - |Q_n|^2)}{1 - |Q_n|^2 \mu_{n+1}^2} \leq \mu_{n+1}.
\end{aligned}
$$

Since $0 < \mu_n \leq 1$ and $\sum(1 - \mu_n) = \infty$, it follows that $\prod_{n=1}^{\infty}(R_{n+1}/R_n) = 0$; i.e. $R_n \to 0$, and thus $\Delta_n \to \Delta = \{c\}$ for a $c \in \overline{U}$. Hence $T_n(w)$ converges uniformly to c in $\overline{U}$.

B. Let first n be an index such that $1 < |k_{n+1}| < \infty$. Then $T_n(U) = T_{n+1}(D_{n+1})$, where $D_{n+1} := t_{n+1}^{-1}(U)$ is a circular disk with center $\tilde{c}_{n+1}$ and radius $\tilde{r}_{n+1}$. Let $\tilde{\mu}_{n+1} := \tilde{r}_{n+1} - |\tilde{c}_{n+1}|$. Then $\tilde{\mu}_{n+1} = 1 + \text{dist}(\partial U, t_{n+1}^{-1}(\partial U))$, and it follows from the first line in (3.9) that

$$
\begin{aligned}
\frac{R_n}{R_{n+1}} &= \frac{\tilde{r}_{n+1}(1 - |Q_{n+1}|^2)}{|1 - \overline{Q_{n+1}}\tilde{c}_{n+1}|^2 - |Q_{n+1}|^2 \tilde{r}_{n+1}^2} \\
&\geq \frac{\tilde{r}_{n+1}(1 - |Q_{n+1}|^2)}{(1 + |Q_{n+1}\tilde{c}_{n+1}|)^2 - |Q_{n+1}|^2 \tilde{r}_{n+1}^2} \\
&= \frac{\tilde{r}_{n+1}(1 - |Q_{n+1}|^2)}{(1 + |Q_{n+1}|(\tilde{r}_{n+1} - \tilde{\mu}_{n+1}))^2 - |Q_{n+1}|^2 \tilde{r}_{n+1}^2} \\
&= \frac{1 - |Q_{n+1}|^2}{2|Q_{n+1}|(1 - |Q_{n+1}|\tilde{\mu}_{n+1}) + (1 - |Q_{n+1}|\tilde{\mu}_{n+1})^2/\tilde{r}_{n+1}} \\
&\geq \frac{1 - |Q_{n+1}|^2}{2|Q_{n+1}|(1 - |Q_{n+1}|\tilde{\mu}_{n+1}) + (1 - |Q_{n+1}|\tilde{\mu}_{n+1})^2/\tilde{\mu}_{n+1}} \\
&= \frac{\tilde{\mu}_{n+1}(1 - |Q_{n+1}|^2)}{1 - |Q_{n+1}|^2 \tilde{\mu}_{n+1}^2} \geq \tilde{\mu}_{n+1} =: 1 + e_{n+1}.
\end{aligned} \tag{3.16}
$$

Next, let n be an index such that $|k_{n+1}| \leq 1$. Then $D_{n+1} = t_{n+1}^{-1}(U)$ is a halfplane or the complement of a disk with $U \subseteq D_{n+1}$. One can always find a circular disk $\tilde{D}_{n+1} \subseteq D_{n+1}$, with center $\tilde{c}'_{n+1}$ and radius $\tilde{r}'_{n+1}$ such that $U \subseteq \tilde{D}_{n+1}$ and $\tilde{\mu}_{n+1} := \tilde{r}'_{n+1} - |\tilde{c}'_{n+1}|$ is equal to $1+$ the euclidean distance between ∂U and ∂D_{n+1}. Then $R_n \geq$ the radius of $T_{n+1}(\tilde{D}_{n+1})$, just as in the proof of Lemma 1.1A, and the inequality (3.16) still holds.

Since $\prod \tilde{\mu}_{n+1} = \infty$ under our conditions, we thus find that $\prod(R_n/R_{n+1}) = \infty$; i.e. $R_n \to 0$.

C. Evidently $R_n \to 0$ if $T_n'(w) \to 0$ uniformly in $\overline{U}$. Since $T_n = t_1 \circ t_2 \circ \cdots \circ t_n$, it follows by the chain rule that

$$T_n'(w) = \prod_{j=1}^{n} t_j'(w_{n,j}), \qquad \text{where} \quad w_{n,j} := t_{j+1} \circ t_{j+2} \circ \cdots \circ t_n(w), \quad w_{n,n} := w.$$

If $\zeta_j \neq \infty$, we find from (3.4) that

$$|t_j'(w)| = r_j \frac{|\zeta_j|^2 - 1}{|\zeta_j - w|^2} \leq \kappa_j \quad \text{for all } w \in t_{j+1}(\overline{U})$$

and

$$|t_j'(w)| \leq r_j \frac{|\zeta_j|^2 - 1}{(|\zeta_j| - 1)^2} = r_j \frac{|\zeta_j| + 1}{|\zeta_j| - 1} \quad \text{for all } w \in \overline{U}.$$

If $\zeta_j = \infty$, then $|t_j'(w)| = |p_j| = r_j = \kappa_j$ by (1.16) and (1.15). Hence $T_n'(w) \to 0$ uniformly in $\overline{U}$ if (1.15) holds. Evidently $\kappa_j \leq r_j(|\zeta_j| + 1)/(|\zeta_j| - 1)$ for all j. ∎

PROOF OF THEOREM 1.6: If $\zeta_n = \infty$ for an index n, then $\zeta_n + k_n = \infty$ for all n, which can only happen if $\zeta_n = \infty$ for all n. In this case the result is a simple consequence of Theorem 1.4C. (See Remark R1.2.2.)

Let all $\zeta_n \neq \infty$, and let $\zeta_n + k_n =: 2\eta$. Let $w^* \in U^* := 2\eta - U$. Then $w^* = \zeta_n + k_n - w$ for a $w \in U$, and thus, by (3.13),

$$t_n^{-1}(w^*) = \zeta_n + \pi_n \frac{|\zeta_n|^2 - 1}{-\zeta_n + w} = \zeta_n + k_n - t_n(w) \in 2\eta - U = U^*.$$

Hence $t_n^{-1}(U^*) \subseteq U^*$ for all n. Since $T_n^{-1} = t_n^{-1} \circ t_{n-1}^{-1} \circ \cdots \circ t_1^{-1}$, this means that $T_n^{-1}(w) \in U^*$ for all $w \in U^*$ and all n. Hence $\{T_n^{-1}(w)\}$ has all its limit points in $\overline{U^*}$ if $w \in U^*$.

Assume that $R_n \to R > 0$. Then $\sum(1 - |Q_n|) < \infty$ by Lemma 1.1A. In particular, the set L of limit points for $\{Z_n\}$ is a subset of ∂U, the boundary of U. Since $\liminf |\zeta_n| > 1$, it follows by Observation 1.3D and Lemma 1.1B that (S1.2) holds. In particular $K_n := T_n(\infty)$ converges to a point $c \in \Delta \subseteq \overline{U}$.

In accordance with (3.4) and (3.13) we also have

$$T_n^{-1}(w) = Z_n + P_n \frac{Z_n}{\overline{Z_n}} \cdot \frac{|Z_n|^2 - 1}{K_n - w} \tag{3.17}$$

for sufficiently large n, since then Z_n and K_n are finite. Since $|Z_n| \to 1$, this means that $(T_n^{-1}(w) - Z_n) \to 0$ locally uniformly in $\hat{\mathbf{C}} \setminus \Delta$. If $\eta = 0$, then $U = U^*$, and thus $t_n(U) = U$ which contradicts the fact that $\limsup r_n < 1$. Hence $\eta \neq 0$. This means that there exist points $w \in U^*$ such that $(T_n^{-1}(w) - Z_n) \to 0$. Hence $L \subseteq \overline{U^*}$.

Case 1: $|\eta| > 1$.

In this case $\overline{U} \cap \overline{U^*} = \emptyset$, which contradicts the fact that $L \subseteq \partial U \cap \overline{U^*}$. Hence, in this case we have $R_n \to 0$ and (S1.1) holds.

<u>Case 2:</u> $|\eta| < 1$.

Let $U_d := U \setminus \overline{U^*}$. Then U_d is a nonempty, open domain and

$$t_n(U_d) = t_n(U) \cap t_n(\hat{\mathbf{C}} \setminus \overline{U^*}) \subseteq U \cap \hat{\mathbf{C}} \setminus \overline{U^*} = U_d$$

for all n since $t_n^{-1}(\overline{U^*}) \subseteq \overline{U^*}$. But then it follows from [9, Thm. 5.1] that if $R > 0$, then $\{T_n^{-1}(w)\}$ has all its limit points in ∂U_d for every $w \notin \Delta$. In particular $\{Z_n\}$ has all its limit points in ∂U_d. Combined with the observations above, this means that L consists of at most the two points of intersection between the two circles ∂U and ∂U^*, which are $\alpha_j := \eta(1 \pm i\sqrt{1/|\eta|^2 - 1})$ for $j = 1, 2$; i.e. $L \subseteq \{\alpha_1, \alpha_2\}$.

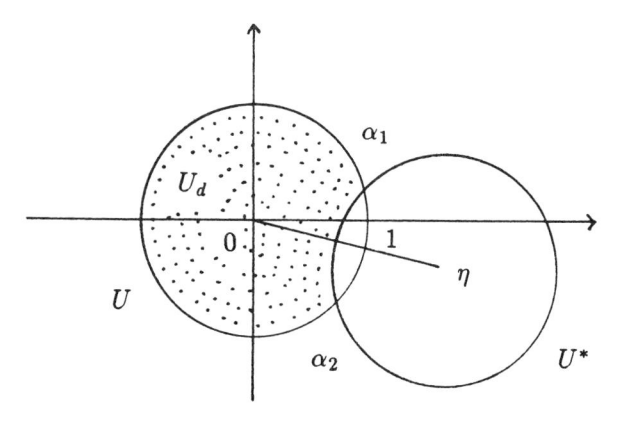

Figure 1.

If $\zeta_n \to \infty$, then it follows by Theorem 1.4 C that $R_n \to 0$. Hence $\{\zeta_n\}$ has a bounded subsequence. Let $\zeta_{n_m} \to \zeta \neq \infty$ as $m \to \infty$. Then also $k_{n_m} \to 2\eta - \zeta$.

Assume first that $Z_n \to \alpha_1$. Since

$$Z_n = T_n^{-1}(\infty) = t_n^{-1} \circ T_{n-1}^{-1}(\infty) = t_n^{-1}(Z_{n-1}) = \zeta_n + \pi_n \frac{|\zeta_n|^2 - 1}{k_n - Z_{n-1}} \tag{3.18}$$

when $\zeta_n \neq \infty$, and since $\alpha_1 + \alpha_2 = 2\eta$, we find that

$$\pi_{n_m}(|\zeta_{n_m}|^2 - 1) \to (\alpha_1 - \zeta)(2\eta - \zeta - \alpha_1) = (\zeta - \eta)^2 - (\alpha_1 - \eta)^2 = (\zeta - \alpha_1)(\zeta - \alpha_2).$$

Hence by (3.4)

$$t_{n_m}(w) \to t(w) := 2\eta - \zeta + \frac{(\zeta - \alpha_1)(\zeta - \alpha_2)}{\zeta - w}.$$

The convergence is locally uniform in $\mathbf{C} \setminus \{\zeta\}$, and thus uniform in $\overline{U}$, since $|\zeta| > 1$ under our conditions. Since $|\alpha_1| = |\alpha_2| = 1$ whereas $|\zeta| > 1$, we find that $t(w)$ is a non–singular linear fractional transformation with fixed points at α_1 and α_2. Hence $t(U) = U$. This contradicts the fact that $\limsup r_n < 1$. Hence $R_n \to 0$.

The same kind of contradiction occurs if there is a subsequence $\{n_m\}$ of $\mathbf{N}$ such that $\lim Z_{n_m} = \lim Z_{n_m+1} = \alpha_1$. Hence it only remains to consider the case where

$$Z_{2n-1} \to \alpha_1 \quad \text{and} \quad Z_{2n} \to \alpha_2 \tag{3.19}$$

or vice versa. Again we may assume that $\zeta_{n_m} \to \zeta \neq \infty$. Without loss of generality we assume that all n_m are even numbers. Then it follows from (3.18) that

$$\pi_{n_m}(|\zeta_{n_m}|^2 - 1) \to (\alpha_2 - \zeta)(2\eta - \zeta - \alpha_1) = (\zeta - \alpha_2)^2,$$

and thus

$$t_{n_m}(w) \to t(w) := 2\eta - \zeta + \frac{(\zeta - \alpha_2)^2}{\zeta - w}.$$

Again t is a non–singular linear fractional transformation, this time with $t(\alpha_2) = \alpha_1$.

Since $t(\alpha_2) = \alpha_1$, where α_1 and α_2 are the two corners of the moon–shaped region U_d (see Figure 1), and $t(U_d)$ is moon–shaped with one corner at α_1, and finally t preserves the orientation of ∂U_d (since $t(U_d) \subseteq U_d$), we find that $t(U_d) \not\subseteq U_d$. This is again a contradiction. ∎

4. REMARKS ON SHARPNESS

How sharp are the results in Section 1? It is evident that we need conditions on $\{t_n\}$ in (1.1) if we want (1.5) to hold. Otherwise one can easily find counterexamples. For instance, if t is an elliptic transformation, then iterations T_n of t (i.e. $T_n := t \circ t \circ \ldots \circ t,$) diverges as $n \to \infty$, except at its two fixed points. Similarly, if t is the identity transformation $t(w) \equiv w$, then iterations of t are also the identity transformation, and thus do not satisfy (1.5).

From [9] we know that if $t(U) \subseteq U$, then t is elliptic only if $t(U) = U$; that is, only if $t(w)$ has the form

$$t(w) = e^{i\alpha}\frac{w - q}{1 - \bar{q}w}, \quad |q| < 1, \ \alpha \in \mathbb{R}. \tag{4.1}$$

It is therefore easy to derive the following result:

THEOREM 4.1   *A linear fractional transformation which maps U into U is elliptic if and only if it can be written in the form (4.1) with $0 < \alpha < 2\pi$ and $|q| < \sin(\alpha/2)$.*

PROOF: It is well known (see for instance [**6**, Lemma 11]) that

$$t(w) = \frac{aw + b}{cw + d}, \quad ad - bc \neq 0, \quad c \neq 0$$

is elliptic if and only if $0 \leq E^2 < 1$, where

$$E^2 := \frac{(a + d)^2}{4(ad - bc)}.$$

We have $a = \exp(i\alpha)$, $b = -q\exp(i\alpha)$, $c = -\bar{q}$ and $d = 1$ in (4.1). Hence, if $q \neq 0$, then

$$E^2 = \frac{(e^{i\alpha} + 1)^2}{4(e^{i\alpha} - e^{i\alpha}|q|^2)} = \frac{(e^{i\alpha/2} + e^{-i\alpha/2})^2}{4(1 - |q|^2)} = \frac{\cos^2(\alpha/2)}{1 - |q|^2},$$

which is always ≥ 0. Moreover, $E^2 < 1$ iff $\cos^2(\alpha/2) < 1 - |q|^2$; i.e. iff $|q|^2 < 1 - \cos^2(\alpha/2) = \sin^2(\alpha/2)$, which proves the result in this case. If $q = 0$, then $t(w) = \exp(i\alpha)w$ which is elliptic except for $\exp(i\alpha) = 1$. $\blacksquare$

We also have to avoid that $\{t_n\}$ approaches an elliptic transformation or the identity transformation too fast. Let us look at some examples.

Example 4.1. Let

$$t_n(w) := r_n w, \qquad \text{where } 0 < r_n \leq 1 \text{ for } n = 1, 2, 3, \ldots.$$

Then

$$T_n(w) := t_1 \circ t_2 \circ \cdots \circ t_n(w) = w \prod_{k=1}^{n} r_k \to Rw,$$

where the limit $R = \prod_{k=1}^{\infty} r_k \geq 0$ exists since $r_k \leq 1$. Now, $R > 0$ if and only if $\sum(1 - r_n) < \infty$, in which case (1.5) does not hold. (Compare with Remark R1.2.2.) We may say that $\{t_n\}$ approaches the identity transformation too fast.

Example 4.2. Let

$$t_n(w) := c_n - r_n \frac{w - q}{1 - qw} \qquad \text{where } 0 < q < 1, \ c_n \geq 0, \ c_n + r_n = 1 \text{ for } n = 1, 2, 3, \ldots.$$

We shall prove that (1.5) does not hold if $\sum(1 - r_n) < \infty$. ($\{t_n\}$ approaches an elliptic transformation too fast.) Since $1/Q_n = T_n^{-1}(\infty)$, it follows that all Q_n are real. From (3.18) we know that

$$Z_n = \zeta - r_n \frac{\zeta^2 - 1}{k_n - Z_{n-1}} = \zeta - r_n \frac{\zeta^2 - 1}{c_n + r_n\zeta - Z_{n-1}} = \zeta + r_n \frac{\zeta^2 - 1}{Z_{n-1} - 1 - r_n(\zeta - 1)}, \quad (4.2)$$

where $\zeta = 1/q > 1$. Since $Z_1 = \zeta$, it follows from this, by induction, that all $Z_n \geq \zeta$. Hence $0 \leq Q_n = 1/Z_n \leq 1/\zeta < 1$ for all n. On the other hand, from (3.6) we find that

$$\frac{R_{n+1}}{R_n} = 1 - \frac{(1 - Q_n)(1 - r_{n+1})}{1 - Q_n + 2Q_n r_{n+1}}.$$

Assume that $\sum(1 - r_n) < \infty$. Then this implies that $R_n \to R > 0$. Let $x, y \in U$, $x > y > 0$. Then, by (3.10),

$$|T_n(x) - T_n(y)| = R_n \frac{(x - y)(1 - Q_n^2)}{(1 - Q_n x)(1 - Q_n y)},$$

which does not converge to 0 as $n \to \infty$; that is, (1.5) does not hold.

$\sum(1 - r_n) = \infty$ is not a necessary condition for convergence, though. This can be seen by the following example:

Example 4.3. Let

$$t_n(w) := t(w) = \frac{w - q}{1 - qw} \quad \text{where } 0 < q < 1 \quad \text{for all } n.$$

Then $r_n = 1$, $t_n(0) = -q$ and $t_n^{-1}(\infty) = -1/q$ for all n. Simple computation shows that t is a hyperbolic linear fractional transformation with fixed points at $w = \pm 1$. Hence iterations $T_n(w)$ of $t(w)$ converge locally uniformly to the attractive fixed point $w = -1$ in $\hat{\mathbf{C}} \setminus \{+1\}$. In particular this means that (1.5) holds.

5. AN APPLICATION TO CONTINUED FRACTIONS

A <u>continued fraction</u>

$$K(a_n/b_n) := \frac{a_1}{b_1} + \frac{a_2}{b_2} + \frac{a_3}{b_3} \cdots = \cfrac{a_1}{b_1 + \cfrac{a_2}{b_2 + \cfrac{a_3}{b_3 + \cdots}}}; \quad \text{all } a_n \neq 0, \quad (5.1)$$

with $a_n \in \mathbf{C}$, $b_n \in \mathbf{C}$, has <u>approximants</u>

$$S_n(w_n) := \frac{a_1}{b_1} + \frac{a_2}{b_2} + \cdots + \frac{a_n}{b_n + w_n} \quad \text{for } n = 1, 2, 3, \ldots. \quad (5.2)$$

Defining

$$s_n(w) := \frac{a_n}{b_n + w}, \quad a_n \neq 0 \text{ for } n = 1, 2, 3, \ldots, \quad (5.3)$$

we find that s_n is a (non–singular) linear fractional transformation, and $S_n = s_1 \circ s_2 \circ \cdots \circ s_n$ for all n. The continued fraction (5.1) <u>converges in the classical sense</u> if the limit $\lim S_n(0)$ exists in $\hat{\mathbf{C}}$. It <u>converges generally</u> if the two limits

$$\lim_{n \to \infty} S_n(w_n) = \lim_{n \to \infty} S_n(v_n) = f \in \hat{\mathbf{C}} \quad (5.4)$$

exist and are equal for two sequences $\{w_n\}$ and $\{v_n\}$ from $\hat{\mathbf{C}}$ which are "sufficiently distinct"; i.e.

$$\liminf_{n \to \infty} d(w_n, v_n) > 0. \quad (d(\cdot, \cdot) \text{ is the chordal metric in } \hat{\mathbf{C}}.) \quad (5.5)$$

The limit f of a generally convergent continued fraction is unique, and if $K(a_n/b_n)$ converges to f in the classical sense, then it also converges generally to f, [2], [10, p.43].

A set $V \neq \emptyset$ in $\hat{\mathbf{C}}$ is a <u>value set</u> for $K(a_n/b_n)$ if $s_n(V) \subseteq V$ for all n, [10, p.110]. (Note that this is often called a pre value set, in which case the notion "value set" is reserved for the case where also $a_n/b_n \in V$ for all n.) The connection to the results in Section 1 is clear. We can use these to prove convergence of continued fractions

which have a circular disk as a value set. This was how Hillam and Thron applied their Theorem A1.1, which gave the following beautiful result:

THEOREM A5.1, [1], [5, Thm. 4.37, p.94], [10, Thm. 12, p.119]. *Let $K(a_n/b_n)$ have a value set V which is an open, circular disk with $0 \in V$. Then $K(a_n/b_n)$ converges in the classical sense to a value $c \in \overline{V}$.*

The significant point in their proof is that $s_n(\infty) = 0$ is an interior point in V. We can now weaken this condition.

THEOREM 5.1   *Let $K(a_n/b_n)$ have a value set $V := \{w \in \mathbf{C} : |w - \Gamma| < \rho\}$ such that $\liminf |b_n + \Gamma| > |\Gamma|$. Then $K(a_n/b_n)$ converges generally to a value $f \in \overline{V}$, and $S_n(w) \to f$ for all $w \in V$.*

REMARKS R5.1

1. If $0 \in V$, then this means that $S_n(0) \to f$; i.e. $K(a_n/b_n)$ converges to f in the classical sense. If $0 \notin V$, we still know that $S_n(w) \to f$ for $w \in V$. In many applications the need for classical convergence is then minimal.

2. Of course, if $|\Gamma| < \rho$, then $\liminf |b_n+\Gamma| > |\Gamma|$ holds automatically since $a_n/(b_n+V) \subseteq V$ implies that $|b_n+\Gamma| > \rho$ for all n. In this case we also have convergence in the classical sense since $0 \in V$. Hence Theorem 5.1 generalizes Hillam and Thron's result, Theorem A5.1.

3. We always have $|b_n + \Gamma| \geq |\Gamma|$ when V is a value set for $K(a_n/b_n)$. If $|b_n + \Gamma| = |\Gamma|$, then

$$a_n := \Gamma(b_n + \Gamma)(1 - \rho^2/|b_n + \Gamma|^2)$$

is the only possible choice for a_n. This follows from the proof of Theorem 5.1. (More precisely, it follows from (5.12)–(5.14) to come.)

PROOF OF THEOREM 5.1: Let ρ_n and γ_n denote the radii and centers of the circular disks $s_n(V)$, and let

$$\varphi(w) := \Gamma + \rho w, \qquad t_n := \varphi^{-1} \circ s_n \circ \varphi \quad \text{for} \quad n = 1, 2, 3, \ldots. \tag{5.6}$$

Then $\varphi(U) = V$, $t_n(U) \subseteq U$ for all n, and

$$
\begin{aligned}
T_n &:= t_1 \circ t_2 \circ \cdots \circ t_n \\
&= \varphi^{-1} \circ s_1 \circ \varphi \circ \varphi^{-1} \circ s_2 \circ \varphi \circ \cdots \circ \varphi^{-1} \circ s_n \circ \varphi \\
&= \varphi^{-1} \circ S_n \circ \varphi.
\end{aligned}
\tag{5.7}
$$

This means that $\{S_n(w)\}$ converges to a constant value $f \in \overline{V}$ for all $w \in V$, if and only if (1.5) holds.

We first observe that

$$\zeta_n := t_n^{-1}(\infty) = \varphi^{-1} \circ s_n^{-1} \circ \varphi(\infty) = \varphi^{-1} \circ s_n^{-1}(\infty) = \varphi^{-1}(-b_n) = -\frac{b_n + \Gamma}{\rho} \qquad (5.8)$$

and

$$k_n := t_n(\infty) = \varphi^{-1} \circ t_n \circ \varphi(\infty) = -\Gamma/\rho. \qquad (5.9)$$

The transformation φ is just a translation of the unit circle and a change of scale. With the notation from Theorem 1.1 we therefore have

$$r_n = \rho_n/\rho \quad \text{and} \quad c_n = \varphi^{-1}(\gamma_n) = (\gamma_n - \Gamma)/\rho. \qquad (5.10)$$

From Lemma 3.1 it follows that $s_n(V)$ is a circular disk with center and radius given by

$$\gamma_n = \frac{a_n(\overline{b_n} + \overline{\Gamma})}{|b_n + \Gamma|^2 - \rho^2}, \qquad \rho_n = \frac{|a_n|\rho}{|b_n + \Gamma|^2 - \rho^2}. \qquad (5.11)$$

Since $s_n(V) \subseteq V$, we thus know that $|\gamma_n - \Gamma| + \rho_n \leq \rho$; i.e.

$$d_n \leq \rho(|b_n + \Gamma|^2 - \rho^2), \qquad (5.12)$$

where

$$d_n := \left| a_n(\overline{b_n} + \overline{\Gamma}) - \Gamma(|b_n + \Gamma|^2 - \rho^2) \right| + |a_n|\rho. \qquad (5.13)$$

This shows that

$$(|b_n + \Gamma| + \rho)(|\Gamma| - \rho) \leq |a_n| \leq (|b_n + \Gamma| - \rho)(|\Gamma| + \rho) \qquad (5.14)$$

for all n. (See for instance [3, formula (5.2), p.100].) Hence

$$\limsup \rho_n \leq \limsup \rho \cdot \frac{|\Gamma| + \rho}{|b_n + \Gamma| + \rho} < \rho$$

under our conditions. This means that $\limsup r_n < 1$. Moreover, if $|\Gamma| \geq \rho$, the choice $v_n := -b_n$ gives $s_n(v_n) = \infty \notin \partial V$. This means that $w_n := \varphi^{-1}(v_n)$ satisfies (1.7), since $\liminf |b_n + \Gamma| > |\Gamma| \geq \rho$. If $|\Gamma| < \rho$, the choice $v_n := \infty$ gives $s_n(v_n) = 0 \notin \partial V$, so, again, $w_n := \varphi^{-1}(v_n)$ satisfies (1.7). It follows therefore from Theorem 1.2 that $T_n(w) \to c \in \overline{U}$, locally uniformly in U. The result follows therefore from (5.7). ∎

In the particular cases where all $b_n = 1$ or all $a_n = 1$, we actually get:

COROLLARY 5.2

A. Let $K(a_n/1)$ have a value set $V := \{w \in \mathbf{C} : |w - \Gamma| < \rho\}$ with $\Re(\Gamma) \neq -\frac{1}{2}$. Then $K(a_n/1)$ converges in the classical sense to a value $f \in \overline{V}$.

B. Let $K(1/b_n)$ have a value set $V := \{w \in \mathbf{C} : |w - \Gamma| < \rho\}$ with $|\Gamma|^2 \neq 1 + \rho^2$. Then $K(1/b_n)$ converges in the classical sense to a value $f \in \overline{V}$.

This is a consequence of the following result:

THEOREM A5.2, [10, Lemma 18, p.126].   *Let $K(a_n/b_n)$ have a bounded value set V containing at least two points. Let all $a_n = 1$ or all $b_n = 1$. Then $K(a_n/b_n)$ converges in the classical sense to f if and only if $K(a_n/b_n)$ converges generally to f.*

REMARKS R5.2

1. A continued fraction $K(a_n/b_n)$ can always be brought to the form $K(1/d_n)$ by an equivalence transformation. If all $b_n \neq 0$, then it can also be brought to the form $K(c_n/1)$. (See for instance [5, p.35], [10, Cor.10, p.74].)

2. From Remark R5.1.3 it follows that V can not be a value set for a continued fraction of the form $K(a_n/1)$ unless $\Re(\Gamma) \geq -1/2$. If $\Re(\Gamma) = -1/2$, then there is only one $a \in \mathbf{C}$ such that $a/(1 + V) \subseteq V$, and thus $K(a_n/1)$ is 1–periodic. (See [9], [10, Thm. 2.5, p.141].)

3. V can not be a value set for continued fractions of the form $K(1/b_n)$, unless $|\Gamma|^2 - \rho^2 \leq 1$. This follows from (5.14), since

$$(|b_n + \Gamma| + \rho)(|\Gamma| - \rho) \leq 1 \quad \text{and} \quad 1 \leq (|b_n + \Gamma| - \rho)(|\Gamma| + \rho) \qquad (5.14')$$

can not both hold if $|\Gamma|^2 - \rho^2 > 1$. If $|\Gamma|^2 - \rho^2 = 1$, we need equalities in both relations in (5.14'), and thus, all b_n satisfy $|b_n + \Gamma| = \sqrt{1 + \rho^2} = |\Gamma|$. If $|\Gamma|^2 - \rho^2 < 1$, then $\liminf |b_n + \Gamma| > |\Gamma|$ always holds in this case.

Corollary 5.2A was proved in a stronger version in [9, Thm. 2.3]:

THEOREM A5.3, [9, Thm. 2.3].   *Let $K(a_n/1)$ have a bounded value set $V := \{w \in \mathbf{C} : |w - \Gamma| < \rho\}$ with $\Re(\Gamma) \neq -1/2$. Then $K(a_n/1)$ converges in the classical sense to a value $f \in \overline{V}$, and $\lim S_n(\overline{V}) = \{f\}$.*

In other words, the limit point case holds! We shall see that Theorem 1.5 is a simple consequence of this result:

PROOF OF THEOREM 1.5:   Clearly, $k = \infty$ if and only if $\zeta = \infty$. Hence k and ζ are finite under our conditions. Let $\alpha := \arg(k - \zeta)$, $\rho := 1/|k - \zeta|$, $\Gamma := -\rho k e^{-i\alpha}$ and $\varphi(w) := \Gamma + w \rho e^{-i\alpha}$. Then $\varphi(U) = V$, and thus

$$s_n := \varphi \circ t_n \circ \varphi^{-1} \quad \text{for} \quad n = 1, 2, 3, \ldots$$

are linear fractional transformations which map $V := \{w \in \mathbf{C} : |w - \Gamma| < \rho\}$ into V. Moreover $s_n(\infty) = \varphi \circ t_n(\infty) = \varphi(k) = \Gamma + k\rho e^{-i\alpha} = 0$ and $s_n^{-1}(\infty) = \varphi \circ t_n^{-1}(\infty) = \varphi(\zeta) = \Gamma + \zeta \rho e^{-i\alpha} = -k\rho e^{-i\alpha} + \zeta \rho e^{-i\alpha} = -1$. That is, all s_n have the form

$s_n(w) = a_n/(1+w)$. Therefore it follows from Theorem A5.3 that $\{S_n(w)\}$ converges uniformly in $\overline{V}$. Hence $T_n = \varphi^{-1} \circ S_n \circ \varphi$ (see (5.7)) converges uniformly in $\overline{U}$. ∎

If the condition $\liminf |b_n + \Gamma| > |\Gamma|$ does not hold in Theorem 5.1, there is still hope. For instance we have:

THEOREM 5.3 Let $K(a_n/b_n)$ have a value set $V := \{w \in \mathbf{C} : |w - \Gamma| < \rho\}$. If either

A. $\sum (\rho - d_n/(|b_n + \Gamma|^2 - \rho^2)) = \infty$, where d_n is given by (5.13),

or

B. $|\Gamma| > \rho$ and $\sum (\tilde{d}_n - \rho(|\Gamma|^2 - \rho^2)) = \infty$, where $\tilde{d}_n := |a_n|\rho - |a_n \overline{\Gamma} - (b_n + \Gamma)(|\Gamma|^2 - \rho^2)|$,

or

C.

$$\lim_{n \to \infty} \rho_n \frac{|b_n + \Gamma| + \rho}{|b_n + \Gamma| - \rho} \prod_{j=1}^{n-1} \kappa_j = 0, \quad \text{where } \kappa_j := \frac{\rho_j}{\rho} \cdot \frac{|b_j + \Gamma|^2 - \rho^2}{(|b_j + \gamma_{j+1}| - \rho_{j+1})^2} \quad (5.15)$$

and γ_n and ρ_n is the center and radius of $s_n(V)$, given by (5.11),

then $K(a_n/b_n)$ converges generally to a value $f \in \overline{V}$, and $S_n(w) \to f$, uniformly, for all $w \in \overline{V}$.

PROOF:

A. (5.10) shows that $\sum(1 - r_n - |c_n|) = \infty$ if and only if $\sum(\rho - \rho_n - |\gamma_n - \Gamma|) = \infty$, and the result follows directly from Theorem 1.4A.

B. This follows directly from Theorem 1.4B.

C. (5.15) is just a translation of (1.13), and the result follows from Theorem 1.4C. ∎

We can also have a sequence $\{V_n\}_{n=0}^{\infty}$ of value sets for $K(a_n/b_n)$; i.e. all V_n are non–empty sets from $\hat{\mathbf{C}}$ and $s_n(V_n) \subseteq V_{n-1}$ for all n. If these sets are all circular disks on the Riemann sphere, i.e. either circular disks, halfplanes or the exterior of circular disks in the extended complex plane, then we may still be able to use Theorem 1.1. We just let φ_n be a linear fractional transformation such that $\varphi_n(U) = V_n$ for each n. Then $t_n := \varphi_{n-1}^{-1} \circ s_n \circ \varphi_n$ is a linear fractional transformation mapping U into U, and

$$\begin{aligned} T_n &= t_1 \circ t_2 \circ \cdots \circ t_n \\ &= (\varphi_0^{-1} \circ s_1 \circ \varphi_1) \circ (\varphi_1^{-1} \circ t_2 \circ \varphi_2) \circ \cdots \circ (\varphi_{n-1}^{-1} \circ t_n \circ \varphi_n) \qquad (5.16) \\ &= \varphi_0^{-1} \circ S_n \circ \varphi_n. \end{aligned}$$

This idea is for instance used in [4], [8], [11].

REFERENCES

1. K. L. Hillam and W. J. Thron, A General Convergence Criterion for Continued Fractions $K(a_n/b_n)$, *Proc Amer. Math. Soc.* **16**: 1256 – 1262 (1965).

2. L. Jacobsen, General Convergence of Continued Fractions, *Trans. Amer. Math. Soc.*, **294**: 477 – 485 (1986).

3. L. Jacobsen and W. J. Thron, Oval Convergence Regions and Circular Limit Regions for Continued Fractions $K(a_n/1)$, *Lecture Notes in Math., Springer – Verlag*, **1199**: 90 – 126 (1986).

4. W. B. Jones and W. J. Thron, Twin – Convergence Regions for Continued Fractions $K(a_n/1)$, *Trans. Amer. Math. Soc.* **150**: 93 – 119 (1970).

5. W. B. Jones and W. J. Thron, *Continued Fractions. Analytic Theory and Applications*, Addison–Wesley Publishing Company, Encyclopedia of Mathematics and its Applications (1980).

6. L. Lorentzen, Bestness of the Parabola Theorem for Continued Fractions, *Journ. of Comp. and Appl. Math.*, **40**: 297 – 304 (1992).

7. L. Lorentzen, Properties of Limit Sets and Convergence of Continued Fractions, *Journ. of Math. Anal. and Appl.*. (To appear.)

8. L. Lorentzen, Circular Twin Value Sets for Continued Fractions and How They Imply Convergence. (This volume.)

9. L. Lorentzen and St. Ruscheweyh, Simple Convergence Sets for Continued Fractions $K(a_n/1)$, *Journ. of Math. Anal. and Appl.* (To appear.)

10. L. Lorentzen and H. Waadeland, *Continued Fractions with Applications*, North–Holland, Studies in Computational Mathematics, **3** (1992).

11. W. J. Thron, Convergence of Sequences of Linear Fractional Transformations and of Continued Fractions, *J. Indian Math. Soc.*, **27**: 103 – 127 (1963).

13

Circular Twin Value Sets for Continued Fractions and How They Imply Convergence

LISA LORENTZEN   Division of Mathematical Sciences, University of Trondheim, NTH, N–7034 Trondheim, Norway

Dedicated to the memory of ARNE MAGNUS

1. INTRODUCTION

1.1. DEFINITIONS

A continued fraction

$$\mathrm{K}\frac{a_n}{b_n} = K(a_n/b_n) := \frac{a_1}{b_1} + \frac{a_2}{b_2} + \frac{a_3}{b_3} \cdots = \cfrac{a_1}{b_1 + \cfrac{a_2}{b_2 + \cfrac{a_3}{b_3 + \cdots}}}; \quad \text{all } a_n \neq 0, \quad (1.1.1)$$

with elements $a_n \in \mathbf{C}$, $b_n \in \mathbf{C}$, has approximants

$$S_n(w_n) := \frac{a_1}{b_1} + \frac{a_2}{b_2} + \cdots + \frac{a_n}{b_n + w_n} \qquad \text{for } n = 1, 2, 3, \ldots; \quad w_n \in \hat{\mathbf{C}} := \mathbf{C} \cup \{\infty\}. \tag{1.1.2}$$

(The notion "approximants" is often reserved for the classical approximants $S_n(0)$, but we shall allow $S_n(w_n)$.) Defining

$$s_n(w) := \frac{a_n}{b_n + w}, \quad a_n \neq 0 \text{ for } n = 1, 2, 3, \ldots, \tag{1.1.3}$$

we find that s_n is a (non–singular) linear fractional transformation, and $S_n = s_1 \circ s_2 \circ \cdots \circ s_n$ for all n.

The author has changed her name from Lisa Jacobsen.

The continued fraction (1.1.1) <u>converges in the classical sense</u> to the <u>value</u> f, if the limit $f = \lim S_n(0)$ exists in $\hat{\mathbf{C}}$.

A less restrictive, but still very useful, definition of convergence, is the newer concept of <u>general convergence</u>, [2]. It is based on the wider notion (1.1.2) of approximants. We say that $K(a_n/b_n)$ <u>converges generally</u> to f, if the two limits

$$\lim_{n \to \infty} S_n(w_n) = \lim_{n \to \infty} S_n(v_n) = f \in \hat{\mathbf{C}} \tag{1.1.4}$$

exist and are equal, for two sequences $\{w_n\}$ and $\{v_n\}$ from $\hat{\mathbf{C}}$ which are "sufficiently distinct". By this we mean that the chordal distance $d(w_n, v_n)$ does not become "too small"; i.e. $\liminf_{n \to \infty} d(w_n, v_n) > 0$.

At first sight, this definition of convergence may seem strange. However, the linear fractional transformations (1.1.3), which generate the continued fraction, are normalized in the sense that $s_n(\infty) = 0$. Hence, when the classical definition requires that $S_n(0) \to f$, we automatically also have $S_n(\infty) \to f$, and the two sequences $w_n := 0$ and $v_n := \infty$ are definitely "sufficiently distinct" according to our definition. This also shows that if $K(a_n/b_n)$ converges to f in the classical sense, then it also converges generally to f.

But there is more to it than this. The limit f of a generally convergent continued fraction is unique. That is, if $\lim S_n(u_n) = \lim S_n(\tilde{u}_n) = g$, where $\liminf d(u_n, \tilde{u}_n) > 0$, and $\lim S_n(v_n) = \lim S_n(w_n) = f$, where $\liminf d(v_n, w_n) > 0$, then $f = g$. In fact, let $K(a_n/b_n)$ converge generally to f, and let $p \in \hat{\mathbf{C}} \setminus \{f\}$ be a fixed, arbitrarily chosen number. Then $S_n(w_n) \to f$ for every sequence $\{w_n\}$ from $\hat{\mathbf{C}}$, which stays bounded away from $S_n^{-1}(p)$ in the sence that $\liminf d(w_n, S_n^{-1}(p)) > 0$. In the setting of this paper, it is easy to find sequences $\{w_n\}$ which give convergence to the right value f. (See Lemma 1.3.1 and the subsequent remarks.) For more information on general convergence we refer to [2], [12, p.43].

A pair (V_0, V_1) of sets from $\hat{\mathbf{C}}$ is called a pair of <u>twin value sets</u> for $K(a_n/b_n)$, if $V_0 \neq \emptyset$ and

$$s_{2n-1}(V_1) \subseteq V_0 \quad \text{and} \quad s_{2n}(V_0) \subseteq V_1 \quad \text{for all } n. \tag{1.1.5}$$

(See for instance [12, p.111]. Note that this definition corresponds to what is often called pre value regions. The term value set or value region is then reserved for the special case, where also all $a_{2n+j}/b_{2n+j} \in V_{j-1}$ for $j = 1, 2$. Of course, the concepts of approximants, convergence and value sets must be seen in connection.)

1.2. THE PURPOSE OF THIS PAPER

Value sets are important tools to prove convergence for continued fractions. Fruitful <u>value set techniques</u> are for instance described in [12]. The more traditional forms date back at least to Scott and Wall in 1941, [14], and Paydon and Wall in 1942, [13], and have been used extensively by Thron, Jones and Thron, Lange, etc. in a number of papers. This was then based on the classical definitions of approximants, value regions and convergence. The more general setting that we use here, was put together by the author. (See for instance [1], [2].)

The purpose of this paper is to find sufficient conditions for general convergence of continued fractions. We shall require that $K(a_n/b_n)$ has twin value sets (V_0, V_1), and

we shall consider cases where V_0 and V_1 are open, <u>circular domains</u>; i.e. either circular disks, half–planes or complements of circular disks.

We shall consider the three cases

V_0 disk	V_1 disk	(Section 3)
V_0 disk	V_1 halfplane	(Section 4)
V_0 disk	V_1 complement of disk	(Section 5)

The halfplane – halfplane case has been thoroughly handled by Jones and Thron, [5], and shall be left out in this paper.

A result in this direction was proved by Jones and Thron in 1970. (The presentation here is slightly adjusted. See Remark R1.2.1.1.) As everywhere else in this paper, we use the notation $\overline{A}$ for the closure of a set A in $\hat{C}$ and ∂A for the boundary of A in $\hat{C}$.

THEOREM A1.2.1, [6, Thm. 4.3].   *Let $K(a_n/1)$ have a pair (V_0, V_1) of open twin value sets. Then the following statements hold:*

A. *If V_0 and V_1 are circular disks with*

$$0 \in \overline{V_0} \quad \text{and} \quad 0 \notin \partial V_1, \tag{1.2.1}$$

  then $K(a_n/1)$ converges in the classical sense to a value $f \in \overline{V_0}$.

B. *Let V_0 be a half–plane and V_1 be a circular disk. If*

$$-1 \notin \partial V_1 \quad \text{and} \quad 0 \notin \partial V_1, \tag{1.2.2}$$

  then $K(a_n/1)$ converges in the classical sense to f.

C. *Let V_0 be the complement of a disk and V_1 be a circular disk. If $-1 \notin \overline{V_1}$, then $\{S_{2n-1}(0)\}$ converges to a value $f \in \overline{V_0}$. If, on the other hand,*

$$0 \in \overline{V_0} \quad \text{and} \quad -1 \notin \partial V_1, \tag{1.2.3}$$

  then $K(a_n/1)$ converges in the classical sense to f.

This is a most beautiful result. Simple, clear, and still very general. Of course, when it comes to choosing the twin value sets (V_0, V_1), explicitly, there are many possibilities. That choice also depends on what properties one favours. But this is another question, which we shall not touch upon here.

REMARKS R1.2.1

1. It follows from [6, Table 1] that if (V_0, V_1) are twin value sets for a continued fraction $K(a_n/1)$, then

$-1 \notin \overline{V_0} \cup \overline{V_1}$	in Theorem A1.2.1A,
$-1 \notin \overline{V_0} \cup V_1$ and $0 \in \overline{V_1}$	in Theorem A1.2.1B,
$-1 \notin \overline{V_0}$ and $0 \in V_1$	in Theorem A1.2.1C.

In Theorem A1.2.1, we have removed such conditions from [6, Theorem 4.3] which are automatically satisfied.

2. There is probably a little misprint in [6, Theorem 4.3, Case 2(B)]. If I have interpreted it correctly, then it says that $\{f_{2n-1}\}$ always converges if $V_0 = V_2$ is a halfplane and V_1 is a circular disk, since then

$$-1 \notin \overline{V_0} \cup V_1$$

always holds. This can not be true. A counterexample is for instance

$$V_0 = V_2 := \{w \in \mathbf{C} : \Re(w) > p\}, \quad V_1 := \{w \in \mathbf{C} : |w + 1/2| < 1/2\},$$

where $p > 0$. Then $p/(1 + V_1) = V_0$ and $-(p+1)/(1 + V_0) = V_1$, but the 2–periodic continued fraction

$$\frac{p}{1} - \frac{p+1}{1} + \frac{p}{1} - \frac{p+1}{1} + \cdots$$

diverges, since the linear fractional transformation $S_2(w) := p/(1 - (p+1)/(1+w))$ is elliptic.

3. If the roles of V_0 and V_1 are interchanged in Theorem A1.2.1, then Theorem A1.2.1 holds for the first tail

$$\overset{\infty}{\underset{n=2}{\mathrm{K}}} \frac{a_n}{1} = \mathrm{K}_{n=2}^{\infty}(a_n/1) = \frac{a_2}{1} + \frac{a_3}{1} + \frac{a_4}{1} + \cdots \tag{1.2.4}$$

of $K(a_n/1)$. The approximants $S_{2n-1}^{(1)}(w_n)$ of this tail converge to $f_1 \in \overline{V_1}$, if and only if $S_{2n}(w_n) = a_1/(1 + S_{2n-1}^{(1)}(w_n)) \to f = a_1/(1 + f_1) \in \overline{V_0}$. A similar relation exists between $S_{2n}^{(1)}(w_n)$ and $S_{2n+1}(w_n)$.

4. It has been remarked (see for instance [15]), that a serious drawback with Theorem A1.2.1 is that it does not give truncation error estimates. That is true, of course. However, if convergence is already established, or if value sets are already known, the problem of deriving good truncation error estimates is often simplified.

    Since the value set technique is based on the mapping properties (1.1.5), it is natural that the convergence criteria we obtain, appear as conditions on the value sets, and how the elements a_n and b_n relate to these. For applications, however, it is often an advantage to have the conditions on the location of (a_n, b_n), alone. Jones and Thron achieved this by finding criteria on (a_n, b_n), which guarantee that s_n has the proper mapping properties (1.1.5). These criteria turn out to hold also under our milder conditions.

    Each of the sections 3, 4 and 5 are divided into four parts. The first parts contain the general convergence results. The second parts describe the location of the elements (a_n, b_n). The third parts treat the question of convergence in the classical sense, and, finally, the fourth parts contain the proofs.

1.3. CHOOSING APPROXIMANTS

Theorem A1.2.1 gives classical convergence for continued fractions $K(a_n/1)$. Hence, $\{S_n(0)\}$ converges to the value of $K(a_n/1)$. We shall prove general convergence of continued fractions $K(a_n/b_n)$. We shall do this by combining results from [6] with more recent results. Which approximants will then converge to the value of $K(a_n/b_n)$? The following observation provides an answer:

LEMMA 1.3.1   Let $K(a_n/b_n)$ have open twin value sets (V_0, V_1), let $\{n_m\}$ be a subsequence of the natural numbers, and let $j \in \{0,1\}$. If $\overline{V_0} \neq \hat{\mathbf{C}}$, and $\lim_{m\to\infty} S_{2n_m+j}(u_m)$ $= \lim_{m\to\infty} S_{2n_m+j}(v_m) = f$ for two sequences $\{u_m\}$ and $\{v_m\}$ from $\hat{\mathbf{C}}$ with $\liminf d(u_m, v_m) > 0$, then $\lim_{m\to\infty} S_{2n_m+j}(w) \to f$ for all $w \in V_j$. The convergence is locally uniform in V_j with respect to the chordal metric.

PROOF: Let $g \in \hat{\mathbf{C}} \setminus \overline{V_0}$ be a fixed number. Then

$$g_{2n} := S_{2n}^{-1}(g) = s_{2n}^{-1} \circ s_{2n-1}^{-1} \circ \cdots \circ s_1^{-1}(g) \in \hat{\mathbf{C}} \setminus \overline{V_0}, \quad g_{2n+1} := S_{2n+1}^{-1}(g) \in \hat{\mathbf{C}} \setminus \overline{V_1}$$

for all n, since by (1.1.5)

$$s_{2n-1}^{-1}(\hat{\mathbf{C}} \setminus \overline{V_0}) \subseteq \hat{\mathbf{C}} \setminus \overline{V_1}, \quad s_{2n}^{-1}(\hat{\mathbf{C}} \setminus \overline{V_1}) \subseteq \hat{\mathbf{C}} \setminus \overline{V_0}. \tag{1.3.1}$$

According to [4, Thm. 2.5], we know that $S_{2n_m+j}(w_m) \to f$ for every sequence $\{w_m\}$ such that $\liminf d(w_m, g_{2n_m+j}) > 0$. This means that $S_{2n_m+j}(w) \to f$ for all $w \in V_j$. The local uniformity follows by the Stieltjes – Vitali Theorem. ∎

If $K(a_n/b_n)$ converges generally to f, and (V_0, V_1) are open twin value sets for $K(a_n/b_n)$ with $\overline{V_0} \neq \hat{\mathbf{C}}$, then Lemma 1.3.1 means in particular that

$$S_{2n}(w) \to f \quad \text{for all } w \in V_0, \quad S_{2n+1}(w) \to f \quad \text{for all } w \in V_1, \tag{1.3.2}$$

where the convergence is locally uniform with respect to the chordal metric in V_0 and V_1, respectively. In most applications, the need for convergence in the classical sense is then minimal. However, in some special cases the classical convergence follows "almost for free". For instance, this is so if $0 \in V_0 \cap V_1$, or if the continued fraction has the form $K(a_n/1)$ or $K(1/b_n)$ and satisfies conditions described in Lemma 2.3.2 to come.

2. BASIC TOOLS

2.1. A CONVERGENCE RESULT FOR LINEAR FRACTIONAL TRANS- FORMATIONS

The following lemma, essentially due to Jones and Thron, [6, Lemmas 4.1, 4.2], is the basis for our arguments. The simpler form that we present here, can be found in [10, Thm. 1.2]:

LEMMA A2.1.1 *Let $\{t_n\}$ be a sequence of (non–singular) linear fractional transformations such that*

$$t_n(U) \subseteq U \quad for \quad n = 1, 2, 3, \ldots, \qquad where \quad U := \{w \in \mathbf{C} : |w| < 1\}, \tag{2.1.1}$$

and let

$$T_n := t_1 \circ t_2 \circ \cdots \circ t_n \qquad for \quad n = 1, 2, 3, \ldots. \tag{2.1.2}$$

Let further r_n denote the radius of $t_n(U)$, and let $\tilde{r}_n$ be the radius of $t_n^{-1}(U)$ if $\infty \notin t_n^{-1}(\overline{U})$, $\tilde{r}_n := \infty$ otherwise. If

$$either \quad \limsup r_n < 1 \quad or \quad \liminf \tilde{r}_n > 1, \tag{2.1.3}$$

and if there exists a sequence $\{w_n\}$ of numbers from $\hat{\mathbf{C}}$ such that

$$\liminf \big||w_n| - 1\big| > 0 \qquad and \qquad \liminf \big||t_n(w_n)| - 1\big| > 0, \tag{2.1.4}$$

then (at least) one of the following two statements hold:

A. *$\{T_n(w)\}$ converges uniformly in $\overline{U}$ to a constant $c \in \overline{U}$.*

B. *$\{T_n(w)\}$ converges locally uniformly in $\hat{\mathbf{C}} \setminus \partial U$ to a constant $c \in \overline{U}$.*

2.2. APPLICATION OF LEMMA A2.1.1

The application of Lemma A2.1.1 to our situation is very natural. Let (V_0, V_1) be twin value sets for the continued fraction $K(a_n/b_n)$. Let $s_n(w) := a_n/(b_n + w)$ for all n (as in (1.1.3)), let U be the open unit disk (as in (2.1.1)), and let φ_0 and φ_1 be linear fractional transformations such that $\varphi_j(U) = V_j$ for $j = 0, 1$. This can always be done, since V_j are circular domains. Then it follows from (1.1.5) that

$$t_{2n-1} := \varphi_0^{-1} \circ s_{2n-1} \circ \varphi_1 \quad and \quad t_{2n} := \varphi_1^{-1} \circ s_{2n} \circ \varphi_0,$$
$$\tau_{2n-1} := \varphi_0^{-1} \circ s_{2n-1} \circ s_{2n} \circ \varphi_0 \quad and \quad \tau_{2n} := \varphi_1^{-1} \circ s_{2n} \circ s_{2n+1} \circ \varphi_1 \tag{2.2.1}$$

are linear fractional transformations mapping U into U. We shall mainly apply Lemma A2.1.1 to the sequence $\{\tau_{2n-1}\}$ or $\{\tau_{2n}\}$. Let r_n and σ_n denote the radii of $t_n(U)$ and $\tau_n(U)$, respectively. Then $\sigma_{2n-1} \leq r_{2n-1}$, since $\tau_{2n-1}(U) = t_{2n-1} \circ t_{2n}(U) \subseteq t_{2n-1}(U)$. Hence, $\limsup \sigma_{2n-1} < 1$ if $\limsup r_{2n-1} < 1$. Similarly, $\limsup \sigma_{2n} < 1$ if $\limsup r_{2n} < 1$.

Let $\tilde{r}_n$ and $\tilde{\sigma}_n$ be the "radii" of $t_n^{-1}(U)$ and $\tau_n^{-1}(U)$ in the sense described in Lemma A2.1.1. That is, $\tilde{r}_n := \infty$ if $\infty \in t_n^{-1}(\overline{U})$, and $\tilde{\sigma}_n := \infty$ if $\infty \in \tau_n^{-1}(\overline{U})$. Then $\tilde{\sigma}_{2n-1} \geq \tilde{r}_{2n}$, since $\tau_{2n-1}^{-1}(U) = t_{2n}^{-1} \circ t_{2n-1}^{-1}(U) \supseteq t_{2n}^{-1}(U)$. This means that $\liminf \tilde{\sigma}_{2n-1} > 1$ if $\liminf \tilde{r}_{2n} > 1$. Finally, $\tilde{\sigma}_{2n} \geq \tilde{r}_{2n+1}$, and thus $\liminf \tilde{\sigma}_{2n} > 1$ if $\liminf \tilde{r}_{2n+1} > 1$.

We shall use the phrase "$\{c_n\}$ is bounded away from ∂V_j" to mean that the chordal distance between the point c_n and the curve ∂V_j in $\hat{\mathbf{C}}$ is $\geq \varepsilon$ for some $\varepsilon > 0$, for all n from some n on. Hence, if $\limsup r_{2n-1} < 1$ or $\liminf \tilde{r}_{2n} > 1$, and if there exists a

sequence $\{w_n\}$ bounded away from ∂V_0, such that also $\{s_{2n-1} \circ s_{2n}(w_n)\}$ is bounded away from ∂V_0, then either Lemma A2.1.1A or Lemma A2.1.1B holds for

$$\mathcal{T}_n = \tau_1 \circ \tau_3 \circ \cdots \circ \tau_{2n-1}$$

$$= (\varphi_0^{-1} \circ s_1 \circ s_2 \circ \varphi_0) \circ (\varphi_0^{-1} \circ s_3 \circ s_4 \circ \varphi_0) \circ \cdots \circ (\varphi_0^{-1} \circ s_{2n-1} \circ s_{2n} \circ \varphi_0) \quad (2.2.2)$$

$$= \varphi_0^{-1} \circ s_1 \circ s_2 \circ \cdots \circ s_{2n} \circ \varphi_0 = \varphi_0^{-1} \circ S_{2n} \circ \varphi_0.$$

This means that $S_{2n}(w) \to f \in \overline{V_0}$ for all $w \in V_0$. Similarly, we can use $\{\tau_{2n}\}$ to prove that

$$\tilde{\mathcal{T}}_n := \tau_2 \circ \tau_4 \circ \cdots \circ \tau_{2n} = \varphi_1^{-1} \circ s_2 \circ s_3 \circ \cdots \circ s_{2n+1} \circ \varphi_1 =: \varphi_1^{-1} \circ S_{2n}^{(1)} \circ \varphi_1 \quad (2.2.3)$$

converges; i.e. $S_{2n}^{(1)}(w)$, and thus also $S_{2n+1}(w) = s_1 \circ S_{2n}^{(1)}(w)$, converges for all $w \in V_1$.

The two cases A and B in Lemma A2.1.1 are essentially different. In case A we find that the radii of $T_n(\overline{U})$ converge to 0, such that the nested disks $\{T_n(\overline{U})\}$ converge to the one-point set $\{f\}$. This case is called the limit point case. If this limit point case occurs for $\{\tau_{2n-1}\}$; i.e. $S_{2n}(w) \to f$ uniformly in $\overline{V_0}$, then also $S_{2n+1}(w) = S_{2n}(s_{2n+1}(w)) \to f$ uniformly in $\overline{V_1}$, and thus $K(a_n/b_n)$ converges generally.

According to [10, Thm. 1.4], we know that this limit point case holds if

$$\sum (1 - r_n - |c_n|) = \infty \quad (2.2.4)$$

in Lemma A2.1.1, where c_n is the center of $t_n(U)$. Other sufficient conditions are for instance

$$\sum \text{dist}_e(\partial U, t_n^{-1}(\partial U)) = \infty, \quad [10, \text{Thm. } 1.4], \quad (2.2.5)$$

where $\text{dist}_e(J_1, J_2)$ denotes the euclidean distance between two curves J_1 and J_2 in $\mathbf{C}$, or

$$t_n(\infty) = k \quad \text{and} \quad t_n^{-1}(\infty) = \zeta \quad \text{for all} \quad n, \quad |k| \neq |\zeta|, \quad [10, \text{Thm. } 1.5]. \quad (2.2.6)$$

In these cases it is easy to derive truncation error estimates.

If case A in Lemma A2.1.1 does not hold, then $\{T_n(\overline{U})\}$ converges to a circular disk. This is called the limit circle case. Even if Lemma A2.1.1B still holds, such that $S_{2n}(w) \to f$ for all $w \in \hat{\mathbf{C}} \setminus \partial V_0$, we have to look more carefully on the behavior of $s_{2n+1}(w)$ to conclude general convergence. If $\{s_{2n+1}(w)\}$ has a limit point in ∂V_0, it is harder to determine whether $S_{2n+1}(w) = S_{2n}(s_{2n+1}(w))$ converges to f or not. Unfortunately, Lemma A2.1.1 does not distinguish between these two cases.

If Lemma A2.1.1B holds, the following result may be of use:

LEMMA 2.2.1   Let $K(a_n/b_n)$ have twin value sets (V_0, V_1), where V_0 and V_1 are circular domains (open disk, halfplane or complement of disk). Further, let $S_{2n}(w) \to f$ for all $w \in \hat{\mathbf{C}} \setminus \partial V_0$. Then the following statements hold.

A. If there exist two sequences $\{u_n\}$ and $\{v_n\}$ from $\hat{\mathbf{C}}$, such that $\liminf d(u_n, v_n) > 0$, and $\{s_{2n+1}(u_n)\}$ and $\{s_{2n+1}(v_n)\}$ are both bounded away from ∂V_0, then also $S_{2n+1}(w) \to f$ for all $w \in V_1$.

B. If $\{s_{2n+1}(w_0)\}$ is bounded away from ∂V_0 for a $w_0 \in V_1$, then $S_{2n+1}(w) \to f$ for all $w \in V_1$.

C. If $\{s_{2n+1}(w_0)\}$ is bounded away from ∂V_0 for a point $w_0 \notin \overline{V_1}$, and $\liminf(1-|c_{2n+1}|+r_{2n+1}) > 0$, where c_n and r_n are the center and radius of $t_n(U)$, respectively, then $S_{2n+1}(w) \to f$ for all $w \in V_1$.

PROOF:

A. Since $S_{2n+1}(u_n) = S_{2n}(s_{2n+1}(u_n)) \to f$ and $S_{2n+1}(v_n) = S_{2n}(s_{2n+1}(v_n)) \to f$ under our conditions, the result follows from Lemma 1.3.1 with $j = 1$ and $n_m = m$.

B. Obviously, $S_{2n+1}(w_0) \to f$. Let $\{t_n\}$ and φ_0, φ_1 be as in (2.2.1). Then $\{t_{2n+1}(u_0)\}$ is bounded away from ∂U for $u_0 := \varphi_1^{-1}(w_0)$. A linear fractional transformation $t_n(w)$, which maps U into U, can be written

$$t_n(w) = \begin{cases} k_n + \pi_n \dfrac{|\zeta_n|^2 - 1}{\zeta_n - w} & \text{if } \zeta_n \neq \infty, \\ c_n + p_n w & \text{if } \zeta_n = \infty, \end{cases} \tag{2.2.7}$$

where $\zeta_n := t_n^{-1}(\infty)$, $k_n := t_n(\infty)$, $|\pi_n| = |p_n| = r_n$, and r_n and c_n are the radius and center of $t_n(U)$. (See for instance [10, formulas (1.16) and (3.4)].) Clearly, $t_n^{-1}(\infty) \neq \infty$ if and only if $t_n(\infty) \neq \infty$. We also know from [10, Observation 1.3A] that $|k_n| \leq |\zeta_n|$.

If $\zeta_{2n+1} = \infty$, then (2.2.7) shows that

$$|t_{2n+1}(u_0) - t_{2n+1}(u_n)| = r_{2n+1}|u_0 - u_n|. \tag{2.2.8}$$

Otherwise, if $\zeta_{2n+1} \neq \infty$, and u_n is chosen such that $|u_n - u_0| < 1 - |u_0|$, then it follows from (2.2.7) that

$$\begin{aligned} |t_{2n+1}(u_0) - t_{2n+1}(u_n)| &= r_{2n+1} \frac{(|\zeta_{2n+1}|^2 - 1)|u_n - u_0|}{|\zeta_{2n+1} - u_0| \cdot |\zeta_{2n+1} - u_n|} \\ &= |t_{2n+1}(u_0) - k_{2n+1}| \frac{|u_n - u_0|}{|\zeta_{2n+1} - u_n|} \\ &\leq (1 + |\zeta_{2n+1}|) \frac{|u_n - u_0|}{|\zeta_{2n+1} - u_0| - |u_n - u_0|} \\ &< 2 \frac{|u_n - u_0|}{(1 - |u_0|) - |u_n - u_0|}. \end{aligned} \tag{2.2.9}$$

This shows that for every $\varepsilon > 0$, there exist sequences $\{u_n\}$, such that $\limsup |t_{2n+1}(u_n) - t_{2n+1}(u_0)| \leq \varepsilon$ and $\liminf |u_n - u_0| > 0$. In particular, $\{u_n\}$ can be chosen such that $\liminf |u_n - u_0| > 0$ and $\{t_{2n+1}(u_n)\}$ is bounded away from ∂U, and the result follows from Part A (with all $v_n := u_0$).

C. Let $v_0 \in U$ be symmetric to $u_0 := \varphi_1^{-1}(w_0)$ with respect to ∂U, and let $\varepsilon > 0$ be chosen such that $1 - |c_{2n+1}| + r_{2n+1} \geq \varepsilon$ for all n. Let further $d_n := \text{dist}_e(t_{2n+1}(u_0), \partial t_{2n+1}(U))$ and $e_n := \text{dist}_e(t_{2n+1}(v_0), \partial t_{2n+1}(U))$ (euclidean distance). Since $t_{2n+1}(u_0)$ and $t_{2n+1}(v_0)$ are symmetric points with respect to $\partial t_{2n+1}(U)$, we know that

$$(r_{2n+1} + d_n)(r_{2n+1} - e_n) = r_{2n+1}^2 \qquad \text{if } d_n \neq \infty,$$

$$e_n = r_{2n+1} \qquad \text{otherwise.}$$

Let $\{n_m\}$ be the natural numbers for which $r_{2n_m+1} \geq \varepsilon/4$. For these indices we have

$$\text{dist}_e(t_{2n_m+1}(v_0), \partial U) \geq e_{n_m} = r_{2n_m+1} - \frac{r_{2n_m+1}^2}{r_{2n_m+1} + d_{n_m}}$$

$$= r_{2n_m+1}\left(1 - \frac{1}{1 + d_{n_m}/r_{2n_m+1}}\right) \geq \frac{\varepsilon}{4}\left(1 - \frac{1}{1 + d_{n_m}}\right). \tag{2.2.10}$$

Assume that $\{n_m\}$ is an infinite sequence. Since $\{t_{2n_m+1}(v_0)\}$ is bounded away from ∂V_0, and thus $s_{2n_m+1}(\varphi_1(v_0))\}$ is bounded away from ∂V_0, we find that also $S_{2n_m+1}(\varphi_1(v_0)) \to f$. Finally, $\varphi_1(v_0) \neq w_0$. From Lemma 1.3.1, with $j = 1$, $u_n := w_0$ and $v_n := \varphi_1(v_0)$, it follows therefore that $S_{2n_m+1}(w) \to f$, locally uniformly in V_1.

Next, let $\{n_m\}$ be the indices for which $r_{2n_m+1} < \varepsilon/4$. If they are only finitely many, the convergence follows by the arguments above. Assume that they are infinitely many. Then $1 - |c_{2n_m+1}| - r_{2n_m+1} \geq \varepsilon - 2r_{2n_m+1} \geq \varepsilon - \varepsilon/2 = \varepsilon/2$. Hence, $\sum(1 - |c_n| - r_n) = \infty$, and the limit point case holds according to (2.2.4). So, again, the convergence follows trivially. ∎

2.3. CLASSICAL CONVERGENCE FOR $K(1/b_n)$ AND $K(a_n/1)$

If $K(a_n/b_n)$ converges generally, and if $0 \in V_0 \cap V_1$, where V_0 and V_1 are open sets and $\overline{V_0} \neq \hat{C}$, then $K(a_n/b_n)$ also converges in the classical sense by Lemma 1.3.1. For continued fractions of the form $K(a_n/1)$ and $K(1/b_n)$ we have:

LEMMA A2.3.1, [**7**, Proof of Thm. 4.6, p.70], [**12**, Proof of Lemma 17, p.125].   *Let* (V_0, V_1), *with* $\overline{V_0} \neq \hat{C}$, *be twin value sets for* $K(a_n/b_n)$. *Then the following statements hold.*
A. *If all* $a_n = 1$, *then* $(\hat{C} \setminus (-1/V_1), \hat{C} \setminus (-1/V_0))$ *are also twin value sets for* $K(1/b_n)$.
B. *If all* $b_n = 1$, *then* $(\hat{C} \setminus (-1 - V_1), \hat{C} \setminus (-1 - V_0))$ *are also twin value sets for* $K(a_n/1)$.

Of course, if (V_0, V_1) are twin value stes with non–empty interiors V_0° and V_1° for $K(a_n/b_n)$, then also (V_0°, V_1°) are twin value sets for $K(a_n/b_n)$. Moreover, the closures $(\overline{V_0}, \overline{V_1})$ are also twin value sets for $K(a_n/b_n)$. Combined with Lemma 1.3.1, this gives:

LEMMA 2.3.2   *Let* (V_0, V_1) *be a pair of open twin value sets for the generally convergent continued fraction* $K(a_n/b_n)$. *Then the following statements hold.*
A. *If all* $a_n = 1$, *and if either* $0 \in V_0$ *or* V_1 *is bounded, and either* $0 \in V_1$ *or* V_0 *is bounded, then* $K(1/b_n)$ *converges also in the classical sense.*
B. *If all* $b_n = 1$, *and if either* $0 \in V_0$ *or* $-1 \notin \overline{V_1}$, *and either* $0 \in V_1$ *or* $-1 \notin \overline{V_0}$, *then* $K(a_n/1)$ *converges also in the classical sense.*

In particular this means that if $V_0 \cup V_1$ is bounded, then $K(a_n/1)$ or $K(1/b_n)$ converges also in the classical sense.

2.4. NOTATION

In the rest of the paper, S_n and s_n shall always mean the linear fractional transformations (1.1.2), (1.1.3), and (V_0, V_1) denotes twin value sets for a continued fraction $K(a_n/b_n)$. Moreover, t_n, τ_n, φ_0 and φ_1 are always as in (2.2.1), and U is the unit disk (2.1.1). Finally, c_n, r_n; ξ_n, σ_n; $\tilde{c}_n, \tilde{r}_n$ and $\tilde{\xi}_n, \tilde{\sigma}_n$ are the centers and radii of $t_n(U)$, $\tau_n(U)$, $t_n^{-1}(U)$ and $\tau_n^{-1}(U)$, respectively, (if $\infty \notin t_n^{-1}(\overline{U})$, $\infty \notin \tau_n^{-1}(\overline{U})$, otherwise $\tilde{r}_n := \infty$, $\tilde{\sigma}_n := \infty$ and $\tilde{c}_n$, $\tilde{\xi}_n$ are undefined).

We also write $\bar{a}$, $\Re(a)$ and $\Im(a)$ to denote the complex conjugate, the real part and the imaginary part of a complex number a.

3. THE DISK – DISK CASE

3.1. CONVERGENCE RESULTS

Let V_0 and V_1 be open, circular disks. Let them have centers at $\Gamma_j \in \mathbf{C}$ and radii $\mathcal{R}_j > 0$ for $j = 0, 1$. That is, $V_j := D_j$, where

$$D_j := \{w \in \mathbf{C} : |w - \Gamma_j| < \mathcal{R}_j\} \tag{3.1.1}$$

for $j = 0, 1$. For simplicity we use the notation

$$V_{2n} = V_0, \quad V_{2n+1} = V_1, \quad \Gamma_{2n} = \Gamma_0, \quad \Gamma_{2n+1} = \Gamma_1, \quad \mathcal{R}_{2n} = \mathcal{R}_0 \text{ and } \mathcal{R}_{2n+1} = \mathcal{R}_1 \tag{3.1.2}$$

for all n. We get:

THEOREM 3.1.1   Let $K(a_n/b_n)$ have twin value sets $(V_0, V_1) = (D_0, D_1)$ given by (3.1.1), with $0 \in V_0$ and $0 \notin \partial V_1$. Then $K(a_n/b_n)$ converges generally to a value $f \in \overline{V_0}$.

REMARKS R3.1.1

1. Theorem 3.1.1 and Lemma 1.3.1 show that $K(a_n/b_n)$ converges in the classical sense, if $K(a_n/b_n)$ has twin value sets $(V_0, V_1) = (D_0, D_1)$ with $0 \in V_0 \cap V_1$. This particular case was also proved by Jones and Thron in [5, Thm. 3.2].

2. The condition $0 \notin \partial V_1$ has its background in the following observation. If $\{t_n\}$ has a subsequence $\{t_{n_m}\}$ converging to a singular transformation t, where t attracts all points in $\hat{\mathbf{C}}$, except one, (which then has to be $\in \partial U$,) to a point $w_0 \in \partial U$, and if the transformations t_{n_m+1} have repulsive fixed points converging to w_0, we may loose control. Condition (2.1.4) ensures that this does not happen.
   If $0 \notin \partial V_0 \cup \partial V_1$, then $s_n(\infty) = 0 \notin \partial V_{n-1}$ for all n, which also prevents this unfortunate situation to occur. It is a simple condition, but not a necessary one.

Theorem 3.1.1 takes care of the situation, where 0 is an interior point in at least one of the two disks V_0, V_1, as long as 0 is not a boundary point in the other. If $0 \notin \overline{V_0} \cup \overline{V_1}$, conditions are needed to ensure contraction in the sense of (2.1.3). We shall use

$$\limsup \frac{|a_{2n+1}|}{|b_{2n+1} + \Gamma_1|^2 - \mathcal{R}_1^2} < \frac{\mathcal{R}_0}{\mathcal{R}_1} \quad \text{or} \quad \liminf |a_{2n}| > \frac{\mathcal{R}_0}{\mathcal{R}_1}(|\Gamma_1|^2 - \mathcal{R}_1^2). \tag{3.1.3}$$

It was proved in [10, Section 4] that some conditions of this type are needed, although (3.1.3) is too restrictive. Still, it follows from (3.4.2) to come, and from Proposition 3.2.2A to come, that

$$\frac{|a_{2n+1}|}{|b_{2n+1} + \Gamma_1|^2 - \mathcal{R}_1^2} \leq \frac{\mathcal{R}_0}{\mathcal{R}_1} \quad \text{and} \quad |a_{2n}| \geq \frac{\mathcal{R}_0}{\mathcal{R}_1}(|\Gamma_1|^2 - \mathcal{R}_1^2)$$

always hold when (D_0, D_1) are twin value sets for $K(a_n/b_n)$. It follows from the proof section that the first inequality in (3.1.3) ensures that $\liminf \sigma_{2n-1} < 1$, whereas the second one ensures that $\limsup \tilde{\sigma}_{2n-1} > 1$. Hence, it suffices that one of them holds, to use Lemma A2.1.1.

The three conditions

$$\limsup \frac{|a_{2n+1}|}{|b_{2n+1} + \Gamma_1|^2 - \mathcal{R}_1^2} < \frac{\mathcal{R}_0}{\mathcal{R}_1}, \tag{3.1.4}$$

$$\liminf |b_{2n+1} + \Gamma_1| > \mathcal{R}_1 |\Gamma_0| / \mathcal{R}_0, \tag{3.1.5}$$

$$\liminf |a_{2n+1}| > \frac{\mathcal{R}_1}{\mathcal{R}_0}(|\Gamma_0|^2 - \mathcal{R}_0^2) \tag{3.1.6}$$

are closely related. We actually have:

LEMMA 3.1.2   *Let $K(a_n/b_n)$ have twin value sets $(V_0, V_1) = (D_0, D_1)$ given by (3.1.1). Then*

$$(3.1.6) \quad \Rightarrow \quad (3.1.5) \quad \Rightarrow \quad (3.1.4).$$

Moreover, if $|\Gamma_0| > \mathcal{R}_0$, then

$$(3.1.6) \quad \Leftrightarrow \quad (3.1.5) \quad \Leftrightarrow \quad (3.1.4).$$

In the case where $0 \notin \overline{V_0} \cup \overline{V_1}$, we can use:

THEOREM 3.1.3   *Let $K(a_n/b_n)$ have twin value sets $(V_0, V_1) = (D_0, D_1)$ given by (3.1.1). If $0 \notin \overline{V_1}$ and $0 \notin \partial V_0$, and (3.1.3) holds, then $K(a_n/b_n)$ converges generally to a value $f \in \overline{V_0}$.*

If exactly one of the two boundaries $\partial V_0, \in \partial V_1$ passes through the origin, we can use:

THEOREM 3.1.4   *Let $K(a_n/b_n)$ have twin value sets $(V_0, V_1) = (D_0, D_1)$ given by (3.1.1). If $0 \notin \partial V_0$ and $0 \in \partial V_1$, and*

$$\liminf \left(|b_{2n} + \Gamma_0| + \left| b_{2n} + \Gamma_0 + \frac{a_{2n}}{b_{2n-1}} \right| - |s_{2n-1} \circ s_{2n}(\Gamma_0) - \Gamma_0| \right) > \mathcal{R}_0, \tag{3.1.7}$$

then $K(a_n/b_n)$ converges generally to a value $f \in \overline{V_0}$.

Condition (3.1.7) prevents the unfortunate situation, described in Remark R3.1.1.2, to occur. It holds in particular, if either $\liminf |b_{2n} + \Gamma_0| > \mathcal{R}_0$, or $\liminf |b_{2n} + \Gamma_0 + a_{2n}/b_{2n-1}| > \mathcal{R}_0$, or $\limsup |s_{2n-1} \circ s_{2n}(\Gamma_0) - \Gamma_0| < \mathcal{R}_0$, since we always have $|b_{2n} + \Gamma_0| > \mathcal{R}_0$, $|(s_{2n-1} \circ s_{2n})^{-1}(\infty) - \Gamma_0| = |b_{2n} + \Gamma_0 + a_{2n}/b_{2n-1}| > \mathcal{R}_0$ and $|s_{2n-1} \circ s_{2n}(\Gamma_0) - \Gamma_0| < \mathcal{R}_0$.

By virtue of Lemma 3.1.2, we actually get:

THEOREM 3.1.5   Let $K(a_n/b_n)$ have twin value sets $(V_0, V_1) = (D_0, D_1)$ given by (3.1.1). If $0 \notin \partial V_1$, and (3.1.5) holds, then $K(a_n/b_n)$ converges generally to a value $f \in \overline{V_0}$.

It remains to look at the case where $0 \in \partial V_0 \cap \partial V_1$:

THEOREM 3.1.6   Let $K(a_n/b_n)$ have twin value sets $(V_0, V_1) = (D_0, D_1)$ given by (3.1.1). If $0 \in \partial V_0 \cap \partial V_1$, and

$$\liminf(|b_{2n} + \Gamma_0 + a_{2n}/b_{2n-1}| - |s_{2n-1} \circ s_{2n}(\Gamma_0) - \Gamma_0|) > 0, \qquad (3.1.8)$$

then $S_{2n}(w) \to f \in \overline{V_0}$ locally uniformly for $w \in V_0$.

If, in addition, $\liminf |b_{2n+1}| > 0$, then $K(a_n/b_n)$ converges generally to f.

We recognize (3.1.8) from (3.1.7). The next convergence result can also be applied to this situation:

THEOREM 3.1.7   Let $K(a_n/b_n)$ have twin value sets $(V_0, V_1) = (D_0, D_1)$ given by (3.1.1). If $0 \in \partial V_0 \cap \partial V_1$, and

$$\liminf \left(|b_n + \Gamma_n| - \mathcal{R}_n + \mathcal{R}_{n-1} - \left| \frac{a_n}{b_n + \Gamma_n} - \Gamma_{n-1} \right| \right) > 0, \qquad (3.1.9)$$

then $K(a_n/b_n)$ converges generally to a value $f \in \overline{V_0}$.

The next theorem is just a straight forward application of (2.2.4) and (2.2.5), which give sufficient conditions for the limit point case:

THEOREM 3.1.8   Let $K(a_n/b_n)$ have twin value sets $(V_0, V_1) = (D_0, D_1)$ given by (3.1.1). If either

$$\sum \left(\mathcal{R}_0 - \left| \frac{a_{2n+1}(\overline{b_{2n+1}} + \overline{\Gamma_1})}{|b_{2n+1} + \Gamma_1|^2 - \mathcal{R}_1^2} - \Gamma_0 \right| - \frac{|a_{2n+1}|\mathcal{R}_1}{|b_{2n+1} + \Gamma_1|^2 - \mathcal{R}_1^2} \right) = \infty \qquad (3.1.10)$$

or

$$R_1 < |\Gamma_1| \quad \text{and} \quad \sum \left(\frac{|a_{2n}|R_1}{|\Gamma_1|^2 - R_1^2} - \left| b_{2n} + \Gamma_0 + \frac{a_{2n}\overline{\Gamma_1}}{|\Gamma_1|^2 - R_1^2} \right| - R_0 \right) = \infty, \quad (3.1.11)$$

then $K(a_n/b_n)$ converges generally to a value $f \in \overline{V_0}$.

The terms in these two sums are all ≥ 0, since they are equal to the euclidean distances $\text{dist}_e(s_{2n+1}(\partial V_1), \partial V_0)$ and $\text{dist}_e(s_{2n}^{-1}(\partial V_1), \partial V_0)$, respectively. More theorems of this type can be constructed.

3.2. LOCATION OF THE ELEMENTS (a_n, b_n)

In 1945 Lane proved a result which also holds in our situation. It characterizes the elements (a_n, b_n) which give $a_n/(b_n + V_n) \subseteq V_{n-1}$:

PROPOSITION A3.2.1, [8], [5, Lemma 2.2.1], [7, Thm. 4.3, p.67], [12, Thm.26, p.145]. *The continued fraction* $K(a_n/b_n)$ *has twin value sets* $(V_0, V_1) = (D_0, D_1)$ *given by* (3.1.1), *if and only if* $(a_n, b_n) \in \Omega_n$ *for all* n, *where* $\Omega_{2n-1} = \Omega_1$ *and* $\Omega_{2n} = \Omega_2$ *are sets from* $\mathbf{C} \times \mathbf{C}$ *given by*

$$(a, b) \in \Omega_j \quad \Leftrightarrow \quad \begin{cases} a = \Gamma_{j-1}(b + \Gamma_j)(1 - R_j^2/|b + \Gamma_j|^2)\alpha \neq 0, \text{ where} \\[2mm] |\alpha - 1| + |\alpha|\dfrac{R_j}{|b + \Gamma_j|} \leq \dfrac{R_{j-1}}{|\Gamma_{j-1}|} \text{ if } \Gamma_{j-1} \neq 0, \quad (3.2.1) \\[2mm] 0 < |a| \leq R_{j-1}(|b + \Gamma_j| - R_j) \text{ if } \Gamma_{j-1} = 0 \end{cases}$$

for $j = 1, 2$.

For later reference we note that if $(a, b) \in \Omega_j$, then

$$(|\Gamma_{j-1}| + R_{j-1})(|b + \Gamma_j| - R_j) \geq |a| \geq (|\Gamma_{j-1}| - R_{j-1})(|b + \Gamma_j| + R_j). \quad (3.2.2)$$

This follows from (3.2.1) by straightforward computation. Of course, if $0 \in \overline{V_{j-1}}$; i.e. $|\Gamma_{j-1}| \leq R_{j-1}$, the lower bound in (3.2.2) is trivial.

It is evident that $(V_0, V_1) = (D_0, D_1)$ can be twin value sets for $K(a_n/b_n)$, only if $-b_{2n} \notin \overline{V_0}$ and $-b_{2n+1} \notin \overline{V_1}$. (Both Lane and Jones and Thron required that also $0 \in V_j$, but that was for other reasons.) We actually have:

PROPOSITION 3.2.2 *Let* $b \in \mathbf{C}$ *be a fixed number. Then the following statements hold:*

A. $s(V_1) = a/(b + V_1) \subseteq V_0$ for some $a \in \mathbf{C} \setminus \{0\}$, if and only if $|b + \Gamma_1| > \mathcal{R}_1$ and $|b + \Gamma_1|\mathcal{R}_0 \geq |\Gamma_0|\mathcal{R}_1$.

B. If $|b + \Gamma_1|\mathcal{R}_0 = |\Gamma_0|\mathcal{R}_1$, then there is only one $a \in \mathbf{C}$ such that $a/(b + V_1) \subseteq V_0$. It is given by

$$a = \Gamma_0(b + \Gamma_1)(1 - \mathcal{R}_1^2/|b + \Gamma_1|^2). \tag{3.2.3}$$

C. $a/(b + V_1) = V_0$, if and only if $|b + \Gamma_1|\mathcal{R}_0 = |\Gamma_0|\mathcal{R}_1 > \mathcal{R}_0\mathcal{R}_1$ and a is given by (3.2.3), which happens if and only if

$$|a|\mathcal{R}_0 = \mathcal{R}_1(|\Gamma_0|^2 - \mathcal{R}_0^2) > 0 \tag{3.2.4}$$

and

$$b = -\Gamma_1 + \frac{a\overline{\Gamma_0}}{|\Gamma_0|^2 - \mathcal{R}_0^2}. \tag{3.2.5}$$

REMARKS R3.2.1.

1. Since $|b + \Gamma_1| > \mathcal{R}_1 > 0$, we find that $|b + \Gamma_1|\mathcal{R}_0 = |\Gamma_0|\mathcal{R}_1$ only if $\Gamma_0 \neq 0$. Hence $a \neq 0$ in (3.2.3), as it should be.

2. Of course, changing the indices in Proposition 3.2.2, gives a characterization of when $s(V_0) \subseteq V_1$.

3.3. CLASSICAL CONVERGENCE FOR $K(1/b_n)$ AND $K(a_n/1)$

For continued fractions of the forms $K(1/b_n)$ and $K(a_n/1)$, the convergence results in Section 3.1 are actually results on classical convergence. This follows from Lemma 2.3.2, since V_0 and V_1 are bounded. These simpler forms also simplify the conditions involved. We get:

THEOREM 3.3.1   Let $K(1/b_n)$ have twin value sets $(V_0, V_1) = (D_0, D_1)$ given by (3.1.1). Then $K(1/b_n)$ converges in the classical sense to a value $f \in \overline{V_0}$, if one of the following sets of conditions holds.

A. $0 \in \overline{V_0}$.

B. $0 \notin \partial V_0$ and $\mathcal{R}_0 > \mathcal{R}_1(|\Gamma_0|^2 - \mathcal{R}_0^2)$.

C. $0 \notin \partial V_0$ and $\liminf |b_{2n+1} + \Gamma_1|^2 > \mathcal{R}_1^2 + \mathcal{R}_1/\mathcal{R}_0$.

REMARKS R3.3.1.

1. From (3.2.2) we find that

$$(|\Gamma_0| + \mathcal{R}_0)(|b_{2n+1} + \Gamma_1| - \mathcal{R}_1) \geq 1$$

$$|b_{2n+1} + \Gamma_1| \geq \mathcal{R}_1 + \frac{1}{|\Gamma_0| + \mathcal{R}_0} > \mathcal{R}_1, \tag{3.3.1}$$

which shows that $\{-b_{2n+1}\}$ is always bounded away from ∂V_1, when $K(1/b_n)$ has twin value sets $(V_0, V_1) = (D_0, D_1)$.

2. We also know by (3.2.2) that

$$1 \geq (|\Gamma_0| - \mathcal{R}_0)(|b_{2n+1} + \Gamma_1| + \mathcal{R}_1).$$

Combined with (3.3.1) this shows that

$$\mathcal{R}_1(|\Gamma_0|^2 - \mathcal{R}_0^2) \leq \mathcal{R}_0$$

when all $a_n = 1$. Hence, (D_0, D_1) can be twin value sets for a continued fraction of the form $K(1/b_n)$, if and only if

$$\mathcal{R}_0(|\Gamma_1|^2 - \mathcal{R}_1^2) \leq \mathcal{R}_1 \quad \text{and} \quad \mathcal{R}_1(|\Gamma_0|^2 - \mathcal{R}_0^2) \leq \mathcal{R}_0. \tag{3.3.2}$$

3. If

$$(|\Gamma_0|^2 - \mathcal{R}_0^2)\mathcal{R}_1 = \mathcal{R}_0 \quad \text{and} \quad (|\Gamma_1|^2 - \mathcal{R}_1^2)\mathcal{R}_0 = \mathcal{R}_1, \tag{3.3.3}$$

then it follows from Proposition 3.2.2C, that $K(1/b_n)$ is 2–periodic with

$$\begin{aligned} b_{2n-1} = b_1 &= -\Gamma_1 + \overline{\Gamma_0}/(|\Gamma_0|^2 - \mathcal{R}_0^2) = -\Gamma_1 + \overline{\Gamma_0}\mathcal{R}_1/\mathcal{R}_0, \\ b_{2n} = b_2 &= -\Gamma_0 + \overline{\Gamma_1}/(|\Gamma_1|^2 - \mathcal{R}_1^2) = -\Gamma_0 + \overline{\Gamma_1}\mathcal{R}_0/\mathcal{R}_1 \end{aligned} \tag{3.3.4}$$

for all n. This 2–periodic continued fraction $K(1/b_n)$ converges, unless $S_2(w) = 1/(b_1 + 1/(b_2 + w))$ is elliptic or the identity transformation. According to for instance [**8**, Lemma 11], we know that if $b_1 \neq 0$, then $S_2(w)$ is elliptic if and only if $0 \leq R^2 < 1$, where $R := 1 + b_1 b_2/2$, i.e.

$$\begin{aligned} R &= 1 + \frac{1}{2}(-\Gamma_1 + \overline{\Gamma_0}\mathcal{R}_1/\mathcal{R}_0)(-\Gamma_0 + \overline{\Gamma_1}\mathcal{R}_0/\mathcal{R}_1) \\ &= 1 + \Re(\Gamma_0\Gamma_1) - (1 + \mathcal{R}_0\mathcal{R}_1). \end{aligned}$$

On the other hand, if (3.3.3) holds, then

$$\frac{|\Gamma_0|^2\mathcal{R}_1}{\mathcal{R}_0} = \frac{|\Gamma_1|^2\mathcal{R}_0}{\mathcal{R}_1} = 1 + \mathcal{R}_0\mathcal{R}_1,$$

and thus $|\Gamma_0\Gamma_1| = 1 + \mathcal{R}_0\mathcal{R}_1$. Hence, both elliptic and non–elliptic cases may occur.

THEOREM 3.3.2 Let $K(a_n/1)$ have twin value sets $(V_0, V_1) = (D_0, D_1)$ given by (3.1.1). Then $K(a_n/1)$ converges in the classical sense to a value $f \in \overline{V_0}$, if one of the following sets of conditions holds.

A. $0 \in \overline{V_0}$.

B. $0 \notin \partial V_0$, and

$$\text{either} \quad |\Gamma_0|\mathcal{R}_1 < |1 + \Gamma_1|\mathcal{R}_0 \quad \text{or} \quad |\Gamma_1|\mathcal{R}_0 < |1 + \Gamma_0|\mathcal{R}_1. \tag{3.3.5}$$

REMARKS R3.3.2

1. In the special case where $V_0 = V_1$, Theorem 3.3.2 was proved in [11, Thm. 2.3].

2. Theorem 3.3.2A contains Theorem A1.2.1A. Actually, Theorem 3.3.2 extends Theorem A1.2.1A considerably.

3. In [1], the idea was to choose the centers Γ_0, Γ_1, such that Γ_0 is the value of a convergent 2–periodic continued fraction $K(\tilde{a}_n/1)$ and Γ_1 is the value of its first tail. It follows from Proposition 3.2.2B that then, either Theorem 3.3.2 applies, or $K(\tilde{a}_n/1)$ is the only continued fraction with these value sets.

4. For continued fractions $K(a_n/1)$, the element sets Ω_j in (3.2.1) have the form $\Omega_j = E_j \times \{1\}$. It was pointed out in [3] that ∂E_j is a cartesian oval if $\Gamma_{j-1} \neq 0$, unless E_j reduces to a point or the empty set. If $\Gamma_{j-1} = 0$, then E_j is a circular disk.

5. We can not choose the disks V_0 and V_1 freely, and expect them to be twin value sets for continued fractions of the form $K(a_n/1)$. By Proposition 3.2.2A it follows that we need

$$\mathcal{R}_0|1 + \Gamma_1| \geq \mathcal{R}_1|\Gamma_0| \quad \text{and} \quad \mathcal{R}_1|1 + \Gamma_0| \geq \mathcal{R}_0|\Gamma_1|. \tag{3.3.6}$$

Multiplying these to equations, further, shows that

$$|\Gamma_0\Gamma_1| \leq |1 + \Gamma_0| \cdot |1 + \Gamma_1|, \tag{3.3.7}$$

where equality occurs if and only if there is equality in both expressions in (3.3.6).

6. If $|1 + \Gamma_1|\mathcal{R}_0 = |\Gamma_0|\mathcal{R}_1$ and $|1 + \Gamma_0|\mathcal{R}_1 = |\Gamma_1|\mathcal{R}_0$, then it follows from Proposition 3.2.2B that $K(a_n/1)$ is 2–periodic with

$$\begin{aligned} a_{2n-1} &= a_1 = \Gamma_0(1 + \Gamma_1)(1 - \mathcal{R}_1^2/|1 + \Gamma_1|^2), \\ a_{2n} &= a_2 = \Gamma_1(1 + \Gamma_0)(1 - \mathcal{R}_0^2/|1 + \Gamma_0|^2) \end{aligned} \tag{3.3.8}$$

for all n. Hence, $K(a_n/1)$ converges, unless $S_2(w) = a_1/(1 + a_2/(1 + w))$ is an elliptic linear fractional transformation, [7, Thm. 3.1, p. 47], [12, Thm. 6, p.104].

3.4. PROOFS

Since we shall use Proposition 3.2.2 in some of the other proofs, we shall start by proving this result.

PROOF OF PROPOSITION 3.2.2:

A. Assume first that $a/(b + V_1) \subseteq V_0$ for a given $a \in \mathbf{C} \setminus \{0\}$. Since V_0 is bounded, it follows that 0 is not contained in the closure of $(b + V_1)$; i.e. $|b + \Gamma_1| > \mathcal{R}_1$. The inequality $|b + \Gamma_1|\mathcal{R}_0 > |\Gamma_0|\mathcal{R}_1$ holds trivially if $\Gamma_0 = 0$. Let $\Gamma_0 \neq 0$. Then it follows from (3.2.1), that the inequality

$$|\alpha - 1| + |\alpha|\frac{\mathcal{R}_1}{|b + \Gamma_1|} \leq \frac{\mathcal{R}_0}{|\Gamma_0|} \tag{3.4.1}$$

holds for some $\alpha \neq 0$. Since $|b + \Gamma_1| > \mathcal{R}_1$, we find that the left side of (3.4.1) attains its minimum for $\alpha = 1$. This means that $\alpha = 1$ has to satisfy (3.4.1); i.e. $\mathcal{R}_1/|b + \Gamma_1| \leq \mathcal{R}_0/|\Gamma_0|$, which is the desired inequality.

Conversely, let $|b + \Gamma_1| > \mathcal{R}_1$ and $|b + \Gamma_1|\mathcal{R}_0 \geq |\Gamma_0|\mathcal{R}_1$. If $\Gamma_0 \neq 0$, then $\alpha = 1$ is a solution of (3.4.1), and the existence of $a \neq 0$ follows. Let $\Gamma_0 = 0$. Then it follows from (3.2.1) that every a with $0 < |a| < \mathcal{R}_0(|b + \Gamma_1| - \mathcal{R}_1)$ can be used.

B. Let $|b + \Gamma_1|\mathcal{R}_0 = |\Gamma_0|\mathcal{R}_1$. Then $\Gamma_0 \neq 0$, and (3.4.1) only holds for $\alpha = 1$. That is, a, given by (3.2.3), is the only point in **C** such that $a/(b + V_1) \subseteq V_0$.

C. Since $|b + \Gamma_1| > \mathcal{R}_1$, we find by straightforward computation, that the set $s(V_1) = a/(b + V_1)$ is a circular disk with center γ and radius ρ given by

$$\gamma = \frac{a(\bar{b} + \overline{\Gamma_1})}{|b + \Gamma_1|^2 - \mathcal{R}_1^2}, \qquad \rho = \frac{|a|\mathcal{R}_1}{|b + \Gamma_1|^2 - \mathcal{R}_1^2}. \tag{3.4.2}$$

Clearly, $a/(b + V_1) = V_0$, if and only if $\gamma = \Gamma_0$ and $\rho = \mathcal{R}_0$. This can only occur if

$$|\Gamma_0| = |\gamma| = \frac{|a| \cdot |b + \Gamma_1|}{|b + \Gamma_1|^2 - \mathcal{R}_1^2} = \frac{\rho}{\mathcal{R}_1}|b + \Gamma_1| = \frac{\mathcal{R}_0}{\mathcal{R}_1}|b + \Gamma_1|,$$

which by Part B implies that a is given by (3.2.3). On the other hand, if $|\Gamma_0|\mathcal{R}_1 = |b + \Gamma_1|\mathcal{R}_0$, and a is given by (3.2.3), then it follows from (3.4.2) that

$$\gamma = \frac{a(\bar{b} + \overline{\Gamma_1})}{|b + \Gamma_1|^2 - \mathcal{R}_1^2} = \Gamma_0 \quad \text{and} \quad \rho = \frac{|a|\mathcal{R}_1}{|b + \Gamma_1|^2 - \mathcal{R}_1^2} = \mathcal{R}_1|\gamma|/|b + \Gamma_1| = \mathcal{R}_0.$$

This proves the first equivalence.

Next, we observe that if $|b + \Gamma_1|\mathcal{R}_0 = |\Gamma_0|\mathcal{R}_1 > \mathcal{R}_0\mathcal{R}_1$, and a is given by (3.2.3), then $\Gamma_0 \neq 0$, and

$$b + \Gamma_1 = \frac{a}{\Gamma_0(1 - \mathcal{R}_1^2/|b + \Gamma_1|^2)} = \frac{a\overline{\Gamma_0}}{|\Gamma_0|^2 - \mathcal{R}_0^2}, \tag{3.4.3}$$

and

$$|b + \Gamma_1| = \frac{|a| \cdot |\Gamma_0|}{|\Gamma_0|^2 - \mathcal{R}_0^2} = |\Gamma_0|\frac{\mathcal{R}_1}{\mathcal{R}_0}, \tag{3.4.4}$$

which proves (3.2.4) and (3.2.5). Conversely, if (3.2.4)–(3.2.5) hold, then (3.4.3)–(3.4.4) hold, which again proves that $|b + \Gamma_1|\mathcal{R}_0 = |\Gamma_0|\mathcal{R}_1$ and a is given by (3.2.3). ∎

PROOF OF THEOREM 3.1.1: Here, and in the rest of this section, we let ρ_n be the radius of $s_n(V_n)$, and $\tilde{\rho}_n$ be the radius of $s_n^{-1}(V_{n-1})$ in the sense of Lemma A2.1.1 (i.e. $\tilde{\rho}_n := \infty$ if $0 \in \overline{V_{n-1}}$) for all n. Since $s_{2n-1}(\infty) = 0$, where 0 is an interior point in V_0 and $\infty \notin \overline{V_1}$, it follows that $\limsup \rho_{2n-1} < \mathcal{R}_0$. Since $\varphi_0(U) = V_0$, this means that $\limsup r_{2n-1} < 1$, and thus also that $\limsup \sigma_{2n-1} < 1$.

Assume first that $0 \notin \overline{V_1}$. Then it follows from Proposition 3.2.2A, that $\liminf |b_{2n} + \Gamma_0| \geq |\Gamma_1|\mathcal{R}_0/\mathcal{R}_1 > \mathcal{R}_0$. Since $w_n := -b_{2n}$ therefore is bounded away from ∂V_0, and

since $s_{2n-1} \circ s_{2n}(w_n) = 0 \notin \partial V_0$, it follows that $S_{2n}(w) \to f \in \overline{V_0}$ for all $w \in V_0$, by Lemma A2.1.1.

If $S_{2n}(w) \to f$ uniformly in $\overline{V_0}$ (the limit point case), then also $S_{2n+1}(w) \to f$ uniformly in $\overline{V_1}$, and the result follows. Assume that we have the limit circle case. Then $S_{2n}(w) \to f$, locally uniformly in $\hat{\mathbf{C}} \setminus \partial V_0$, by Lemma A2.1.1B. In particular, $S_{2n}(\infty) = S_{2n-1}(0) \to f$ and $S_{2n}(0) = S_{2n+1}(\infty) \to f$. That is, $K(a_n/b_n)$ converges generally to f. (It even converges in the classical sense.)

Assume next that $0 \in V_1$. Then also $\limsup r_{2n} < 1$. Moreover, $s_{2n+1}(\infty) = 0 \notin \partial V_0$ and $s_{2n}(\infty) = 0 \notin \partial V_1$. Hence, $\{t_n\}$ and $w_n := \varphi_n^{-1}(\infty)$ satisfy the conditions of Lemma A2.1.1. Hence, $\{S_n(0)\}$ converges, and the result follows, since classical convergence implies general convergence. ∎

PROOF OF LEMMA 3.1.2: The implications $(3.1.6) \Rightarrow (3.1.5) \Rightarrow (3.1.4)$ follow by simple computation, using the upper bound for $|a_{2n+1}|$ which we get from $(3.2.2)$. The lower bound for $|a_{2n+1}|$, which we get from $(3.2.2)$, shows that if $|\Gamma_0| > \mathcal{R}_0$, then $(3.1.4)$ implies that

$$\limsup \frac{|\Gamma_0| - \mathcal{R}_0}{|b_{2n+1} + \Gamma_1| - \mathcal{R}_1} < \frac{\mathcal{R}_0}{\mathcal{R}_1},$$

which is equivalent to $(3.1.5)$. Moreover we find that $(3.1.5) \Rightarrow (3.1.6)$ when $|\Gamma_0| > \mathcal{R}_0$. This proves the equivalences. ∎

PROOF OF THEOREM 3.1.3: $s_{2n+1}(V_1)$ is a circular disk, with center γ_{2n+1} and radius ρ_{2n+1} given by $(3.4.2)$. Hence, $\liminf \rho_{2n+1} < \mathcal{R}_1$ if the first inequality in $(3.1.3)$ holds. Since $0 \notin \overline{V_1}$, it follows that $s_{2n}^{-1}(V_1) = -b_{2n} + a_{2n}/V_1$ is a circular disk, with center $\tilde{\gamma}_{2n}$ and radius $\tilde{\rho}_{2n}$ given by

$$\tilde{\gamma}_{2n} := -b_{2n} + \frac{a_{2n}\overline{\Gamma_1}}{|\Gamma_1|^2 - \mathcal{R}_1^2}, \qquad \tilde{\rho}_{2n} := \frac{|a_{2n}|\mathcal{R}_1}{|\Gamma_1|^2 - \mathcal{R}_1^2}. \tag{3.4.5}$$

Hence, $\liminf \tilde{\rho}_{2n} > \mathcal{R}_0$ if the second inequality in $(3.1.3)$ holds. In other words, $(3.1.3)$ ensures that either $\limsup \sigma_{2n-1} < 1$ or $\liminf \tilde{\sigma}_{2n-1} > 1$; i.e. $\{\tau_{2n-1}\}$ has the contraction property $(2.1.3)$.

Let $w_n := -b_{2n}$ for all n. From $(3.2.2)$ we find that

$$(|\Gamma_1| + \mathcal{R}_1)(|b_{2n} + \Gamma_0| - \mathcal{R}_0) \geq (|\Gamma_1| - \mathcal{R}_1)(|b_{2n} + \Gamma_0| + \mathcal{R}_0),$$

that is

$$|b_{2n} + \Gamma_0| \geq \mathcal{R}_0|\Gamma_1|/\mathcal{R}_1. \tag{3.4.6}$$

Since $|\Gamma_1| > \mathcal{R}_1$, this shows that $\{-b_{2n}\}$ is bounded away from ∂V_0. Moreover, $s_{2n-1} \circ s_{2n}(w_n) = s_{2n-1}(\infty) = 0 \notin \partial V_0$. Hence, it follows, by use of Lemma A2.1.1, that $S_{2n}(w) \to f$, locally uniformly in V_0.

If Lemma A2.1.1A holds, the convergence of $S_{2n+1}(w)$ to f, for all $w \in V_1$, follows immediately. If Lemma A2.1.1B holds, then $S_{2n}(0) \to f$ and $S_{2n}(\infty) \to f$. Since $S_{2n-1}(0) = S_{2n}(\infty)$, this proves that $K(a_n/b_n)$ converges in the classical sense, and thus also generally, to f. ∎

PROOF OF THEOREM 3.1.4: Since $0 \in \partial V_1$, it follows that $s_{2n}^{-1}(V_1)$ is a halfplane. Hence, $\infty \in s_{2n}^{-1}(\overline{V_1})$ for all n, whereas $\infty \notin \overline{V_0}$. Hence, $\liminf \tilde{r}_{2n} > 1$, and thus $\liminf \tilde{\sigma}_{2n-1} > 1$.

For each $n \in \mathbf{N}$, we compare the quantities $d_n := |b_{2n} + \Gamma_0| - \mathcal{R}_0$, $e_n := |b_{2n} + \Gamma_0 + a_{2n}/b_{2n-1}| - \mathcal{R}_0$ and $f_n := \mathcal{R}_0 - |s_{2n-1} \circ s_{2n}(\Gamma_0) - \Gamma_0|$. They are all positive, since $-b_{2n} \notin \overline{V_0}$, $-b_{2n} - a_{2n}/b_{2n-1} = s_{2n}^{-1} \circ s_{2n-1}^{-1}(\infty) = (s_{2n-1} \circ s_{2n})^{-1}(\infty) \notin \overline{V_0}$ and $s_{2n-1} \circ s_{2n}(\Gamma_0) \in V_0$. If d_n is largest, we choose $w_n := -b_{2n}$. If e_n is largest, we choose $w_n := -b_{2n} - a_{2n}/b_{2n-1}$, and finally, if f_n is largest, we choose $w_n := \Gamma_0$ (with the obvious modifications, if two or three are equal). Then (3.1.7) shows that $\{w_n\}$ and $\{s_{2n-1} \circ s_{2n}(w_n)\}$ are bounded away from ∂V_0. Therefore, $\{\varphi_0^{-1}(w_n)\}$ and $\{\tau_{2n-1}\}$ satisfy the conditions of Lemma A2.1.1. Hence, $S_{2n}(w) \to f$ for some $f \in \overline{V_0}$, for all $w \in V_0$. The general convergence follows as in the proof of Theorem 3.1.3. ∎

PROOF OF THEOREM 3.1.5: We first observe that by Lemma 3.1.2, also (3.1.4) holds. This means that also (3.1.3) holds. If $0 \notin \partial V_0$, the convergence follows therefore from Theorem 3.1.1, with the indices shifted, if $0 \in V_1$, and from Theorem 3.1.3 if $0 \notin \overline{V_1}$.

Let $0 \in \partial V_0$. Then the first tail

$$\frac{a_2}{b_2} + \frac{a_3}{b_3} + \frac{a_4}{b_4} + \cdots \tag{3.4.7}$$

of $K(a_n/b_n)$ is a continued fraction with twin value sets (V_1, V_0). Theorem 3.1.4 applies to this tail (3.4.7). Since $|\Gamma_0| = \mathcal{R}_0$, condition (3.1.5) implies (3.1.7) for this "shifted situation". Hence, (3.4.7) converges generally, and thus also $K(a_n/b_n)$. ∎

PROOF OF THEOREM 3.1.6: Since $0 \in \partial V_1$, it follows that $\liminf \tilde{\sigma}_{2n-1} > 1$, as in the proof of Theorem 3.1.4. Let

$$w_n := \begin{cases} \Gamma_0, & \text{if} \quad \mathcal{R}_0 - |s_{2n-1} \circ s_{2n}(\Gamma_0) - \Gamma_0| \geq |b_{2n} + \Gamma_0 + a_{2n}/b_{2n-1}| - \mathcal{R}_0, \\ -b_{2n} - a_{2n}/b_{2n-1}, & \text{otherwise.} \end{cases}$$

Then (3.1.8) shows that $\{w_n\}$ and $\{s_{2n-1} \circ s_{2n}(w_n)\}$ are bounded away from ∂V_0. Hence, $S_{2n}(w) \to f$, locally uniformly in V_0, by Lemma A2.1.1.

If this convergence is uniform in $\overline{V_0}$ (the limit point case), then $K(a_n/b_n)$ converges generally. Assume that this is not so. Then $S_{2n}(w) \to f$ for all $w \notin \partial V_0$. In particular, $S_{2n}(\infty) = S_{2n-1}(0) = S_{2n+1}(-b_{2n+1}) \to f$. Hence, $K(a_n/b_n)$ converges generally if $\liminf |b_{2n+1}| > 0$.

PROOF OF THEOREM 3.1.7: Since $0 \in \partial V_1$, it follows that $\liminf \tilde{r}_{2n} > 1$, as in the proof of Theorem 3.1.4. Similarly, since $0 \in \partial V_0$, we have $\liminf \tilde{r}_{2n+1} > 1$. That is, $\liminf \tilde{r}_n > 1$.

For each $n \in \mathbf{N}$, let

$$w_n := \begin{cases} -b_n, & \text{if} \quad |b_n + \Gamma_n| - \mathcal{R}_n > \mathcal{R}_{n-1} - |a_n/(b_n + \Gamma_n) - \Gamma_{n-1}|, \\ \Gamma_n, & \text{otherwise.} \end{cases}$$

Then $\{\varphi_n^{-1}(w_n)\}$ and $\{\varphi_{n-1}^{-1} \circ s_n(w_n)\}$ are bounded away from ∂U, and thus $K(a_n/b_n)$ converges generally. ∎

PROOF OF THEOREM 3.3.1: In view of Lemma 2.3.2A, it suffices to prove that $K(1/b_n)$ converges generally, since V_0 and V_1 are bounded. From (3.3.1) it follows that $\{-b_{2n+1}\}$ is bounded away from ∂V_1. Similarly, $\{-b_{2n}\}$ is always bounded away from ∂V_0. Hence, (3.1.7) and (3.1.9) always hold.

We shall prove the results in the "backwards order".

C. If $0 \in V_1$, the convergence follows from Theorem 3.1.1, with the indices shifted, and if $0 \in \partial V_1$, then it follows from Theorem 3.1.4. Let $0 \notin \overline{V_1}$. Since the first inequality in (3.1.3) holds under our conditions, the result follows then from Theorem 3.1.3.

B. This follows from Part C, since Lemma 3.1.2 shows that (3.1.6) $\Rightarrow$ (3.1.4); i.e.

$$\mathcal{R}_0 > \mathcal{R}_1(|\Gamma_0|^2 - \mathcal{R}_0^2) \quad \Rightarrow \quad \liminf |b_{2n+1} + \Gamma_1|^2 > \mathcal{R}_1^2 + \mathcal{R}_1/\mathcal{R}_0.$$

A. If $0 \notin \partial V_0$, the result follows from Part B. Let $0 \in \partial V_0$. If $0 \notin \partial V_1$, it follows from Theorem 3.1.4, with the indices shifted, since (3.1.7) still holds by Remark R3.3.1.1. Let also $0 \in \partial V_1$. Then the result follows from Theorem 3.1.7. ∎

PROOF OF THEOREM 3.3.2: In view of Lemma 2.3.2B, it suffices to prove that $K(a_n/1)$ converges generally, since $-1 \notin \overline{V_0} \cup \overline{V_1}$. We also note that $|1 + \Gamma_n| > \mathcal{R}_n$, such that (3.1.7) and (3.1.9) hold trivially. Finally,

$$|\Gamma_0|\mathcal{R}_1 < |1 + \Gamma_1|\mathcal{R}_0 \quad \Leftrightarrow \quad \limsup |a_{2n+1}| < \frac{\mathcal{R}_0}{\mathcal{R}_1}(|1 + \Gamma_1|^2 - \mathcal{R}_1^2).$$

The implication "$\Rightarrow$" corresponds to the implication (3.1.5) $\Rightarrow$ (3.1.4) in Lemma 3.1.2. The implication "$\Leftarrow$" follows by Proposition 3.2.2B.

A. If $0 \in V_0$, the result follows from Theorem 3.1.1 if $0 \notin \partial V_1$, and from Theorem 3.1.4 if $0 \in \partial V_1$. Let $0 \in \partial V_0$. Then it follows from Theorem 3.1.4 (with indices shifted), if $0 \notin \partial V_1$, and from Theorem 3.1.7 if $0 \in \partial V_1$.

B. If $0 \in \overline{V_1}$, the result follows from Part A. Let $0 \notin \overline{V_1}$. Then $0 \notin \partial V_0 \cup \partial V_1$, and the result follows from Theorem 3.1.5. ∎

4. THE DISK – HALFPLANE CASE

4.1. CONVERGENCE RESULTS

As in the previous section, we let $V_0 := D_0$, given by (3.1.1), but now we let $V_1 := H$ be the halfplane

$$H := \{w \in \mathbf{C} : \Re(we^{-i\beta}) > p\}; \quad \beta, p \in \mathbf{R}. \tag{4.1.1}$$

Again, we let $\varphi_{2n} := \varphi_0$, $\varphi_{2n+1} := \varphi_1$, $V_{2n} := V_0$ and $V_{2n+1} := V_1$ for all n. If (V_0, V_1) is a pair of twin value sets for a continued fraction $K(a_n/b_n)$, then it follows easily, by

(1.1.5), that $0 \in \overline{V_0}$ and $-b_{2n} \notin V_0$. That is, $|\Gamma_0| \leq \mathcal{R}_0$ and $|b_{2n} + \Gamma_0| \geq \mathcal{R}_0$. Similarly, $-b_{2n+1} \notin \overline{V_1}$; i.e.

$$q_{2n+1} := 2[p + \Re(b_{2n+1}e^{-i\beta})] > 0. \tag{4.1.2}$$

If $0 \in V_0$ and $0 \in V_1$, we get:

THEOREM 4.1.1   Let $K(a_n/b_n)$ have twin value sets $(V_0, V_1) = (D_0, H)$ given by (3.1.1) and (4.1.1), with $0 \in V_0 \cap V_1$. Then $K(a_n/b_n)$ converges in the classical sense to a value $f \in \overline{V_0}$.

If we have just $0 \in V_1$ and $0 \in \partial V_0$ (the only alternative to $0 \in V_0$), then we can use:

THEOREM 4.1.2   Let $K(a_n/b_n)$ have twin value sets $(V_0, V_1) = (D_0, H)$ given by (3.1.1) and (4.1.1), with $0 \in V_1$. If $\liminf q_{2n+1} > 0$, then $K(a_n/b_n)$ converges generally to a value $f \in \overline{V_0}$.

It remains to consider the cases where $0 \notin V_1$. If $0 \in V_0$ (we always have $0 \in \overline{V_0}$), we get:

THEOREM 4.1.3   Let $K(a_n/b_n)$ have twin value sets $(V_0, V_1) = (D_0, H)$ given by (3.1.1) and (4.1.1), with $0 \notin \partial V_0$. If

$$\liminf \left(|b_{2n} + \Gamma_0| + |b_{2n} + \Gamma_0 + a_{2n}/b_{2n-1}| - |s_{2n-1} \circ s_{2n}(\Gamma_0) - \Gamma_0| \right) > \mathcal{R}_0, \tag{4.1.3}$$

then $K(a_n/b_n)$ converges generally to a value $f \in \overline{V_0}$.

We can also use:

THEOREM 4.1.4   Let $K(a_n/b_n)$ have twin value sets $(V_0, V_1) = (D_0, H)$ given by (3.1.1) and (4.1.1), with $0 \notin \partial V_0 \cup \partial V_1$. If $\liminf q_{2n+1} > 0$, and $\limsup |a_{2n+1}| < \infty$, then $S_{2n+1}(w)$ converges, locally uniformly in V_1, to a value $f \in \overline{V_0}$.
   If, in addition, $\limsup |a_{2n+1}/b_{2n+1}| < \infty$, then $K(a_n/b_n)$ converges generally to f.

If $0 \notin \partial V_1$, we also have:

THEOREM 4.1.5   Let $K(a_n/b_n)$ have twin value sets $(V_0, V_1) = (D_0, H)$ given by (3.1.1) and (4.1.1), with $0 \notin \partial V_1$. If $\liminf q_{2n+1} > 0$, and either

$$\limsup \frac{|a_{2n}|}{|b_{2n} + \Gamma_0|^2 - \mathcal{R}_0^2} < \infty \tag{4.1.4}$$

or

$$0 \in \partial V_0 \quad \text{and} \quad \limsup |a_{2n+1}|/q_{2n+1} < \mathcal{R}_0, \tag{4.1.5}$$

then $S_{2n+1}(w)$ converges locally uniformly in V_1 to a value $f \in \overline{V_0}$.

If, in addition, $\limsup |a_{2n+1}/b_{2n+1}| < \infty$, then $K(a_n/b_n)$ converges generally to f.

The final result is convenient if $0 \in \partial V_0 \cap \partial V_1$:

THEOREM 4.1.6 *Let $K(a_n/b_n)$ have twin value sets $(V_0, V_1) = (D_0, H)$ given by (3.1.1) and (4.1.1). If*

$$\liminf \left(|b_{2n} + \Gamma_0 + a_{2n}/b_{2n-1}| - |s_{2n-1} \circ s_{2n}(\Gamma_0) - \Gamma_0| \right) > 0, \tag{4.1.6}$$

and either $p \leq 0$ or $\liminf |a_{2n}| > 2p\mathcal{R}_0$, then $S_{2n}(w)$ converges locally uniformly in V_0 to a value $f \in \overline{V_0}$.

If, in addition, $\liminf(|a_{2n+1}| + |b_{2n+1}|) > 0$, then $K(a_n/b_n)$ converges generally to f.

We could also have included some more results in this section. In particular, it is straight forward to use the limit point criteria (2.2.4) and (2.2.5).

4.2. LOCATION OF THE ELEMENTS (a_n, b_n)

As already mentioned, we always have

$$|\Gamma_0| \leq \mathcal{R}_0, \quad |b_{2n} + \Gamma_0| \geq \mathcal{R}_0 \quad \text{and} \quad q_{2n+1} := 2\big(p + \Re(b_{2n+1}e^{-i\beta})\big) > 0. \tag{4.2.1}$$

In harmony with Lane's result, Proposition A3.2.1, we can also characterize the elements (a_n, b_n) which satisfy the mapping property $a_n/(b_n + V_n) \subseteq V_{n-1}$. The following result is obtained by straight forward computation. We do not claim priority. For instance, Jones and Thron have a version for continued fractions $K(a_n/1)$, [**6**, Lemma 5.3]. Still, we include its proof for completeness.

PROPOSITION 4.2.1 *The continued fraction $K(a_n/b_n)$ has twin value sets $(V_0, V_1) = (D_0, H)$, if and only if $(a_n, b_n) \in \Omega_n$ for all n, where $\Omega_{2n-1} = \Omega_1$ and $\Omega_{2n} = \Omega_2$ are sets from $\mathbf{C} \times \mathbf{C}$ given by*

$$(a, b) \in \Omega_1 \quad \Leftrightarrow \quad |a - q\Gamma_0 e^{i\beta}| + |a| \leq \mathcal{R}_0 q,$$

$$(a, b) \in \Omega_2 \quad \Leftrightarrow \quad \begin{cases} \Re[a(\overline{b} + \overline{\Gamma_0})e^{-i\beta}] - |a|\mathcal{R}_0 \geq p(|b + \Gamma_0|^2 - \mathcal{R}_0^2), \\ \qquad \qquad \text{if } |b + \Gamma_0| > \mathcal{R}_0, \\ a = k(b + \Gamma_0)e^{i\beta}, \qquad \text{if } |b + \Gamma_0| = \mathcal{R}_0, \\ \qquad \text{where } k > 0 \text{ if } p \leq 0, \ k \geq 2p, \text{ otherwise,} \end{cases} \tag{4.2.2}$$

where

$$q := 2[p + \Re(be^{-i\beta})] > 0. \tag{4.2.3}$$

REMARKS R4.2.1

1. Ω_1 can be regarded as a domain for a/q. Then the boundary of this domain is an ellipse with foci at 0 and $\Gamma_0 e^{i\beta}$. This means in particular that

$$(a,b) \in \Omega_1 \quad \Rightarrow \quad |a|/q \leq (|\Gamma_0| + \mathcal{R}_0)/2. \tag{4.2.4}$$

2. If $|b + \Gamma_0| > \mathcal{R}_0$ and b is kept fixed, then Ω_2 can be regarded as a domain for a. The boundary of this domain is then a parabola. Evidently,

$$(a,b) \in \Omega_2 \quad \Rightarrow \quad |a| \geq p(|b + \Gamma_0| + \mathcal{R}_0). \tag{4.2.5}$$

The following result gives some additional information on the relations between the value sets and the elements (a_n, b_n):

PROPOSITION 4.2.2   *Let $b \in \mathbf{C}$ be a fixed number. Then the following statements hold.*

A. *There exist (infinitely many) $a \in \mathbf{C}$ such that $a/(b + V_0) \subseteq V_1$, if and only if $|b + \Gamma_0| \geq \mathcal{R}_0$. They satisfy the inequality $|a| \geq p(|b + \Gamma_0| + \mathcal{R}_0)$.*

B. *If $|\Gamma_0| = \mathcal{R}_0$, then $a/(b + V_1) \subseteq V_0$, if and only if (4.2.3) holds and $a = kq\Gamma_0 e^{i\beta}$, where $0 < k \leq 1$.*

C. *$a/(b + V_0) = V_1$ if and only if $p > 0$, $|b + \Gamma_0| = \mathcal{R}_0$ and a is given by*

$$a = 2p(b + \Gamma_0)e^{i\beta}. \tag{4.2.6}$$

D. *$a/(b + V_1) = V_0$, if and only if $|\Gamma_0| = \mathcal{R}_0$ and $a = q\Gamma_0 e^{i\beta}$.*

4.3. CLASSICAL CONVERGENCE FOR $K(1/b_n)$ AND $K(a_n/1)$

In the disk – halfplane case, we do not have both V_0 and V_1 bounded, so the classical convergence does not follow as easily as in Section 3.3. Still, we get:

THEOREM 4.3.1   *Let $K(1/b_n)$ have twin value sets $(V_0, V_1) = (D_0, H)$ given by (3.1.1) and (4.1.1). Then $K(1/b_n)$ converges in the classical sense to a value $f \in \overline{V_0}$, if one of the following sets of conditions holds.*

A. *$0 \notin \partial V_0$ and $0 \in V_1$.*

B. *$0 \notin \partial V_0 \cup \partial V_1$ and $\liminf |b_{2n+1}| > 0$.*

C. $0 \notin \partial V_0$ and (4.1.3) holds (with $a_{2n} := 1$).

D. $0 \notin \partial V_0$ and $p < 1/2\mathcal{R}_0$ and (4.1.6) holds (with all $a_n := 1$).

E. $K(1/b_n)$ converges generally and $\liminf |b_{2n} + \Gamma_0| > \mathcal{R}_0$.

F. $\liminf |b_{2n} + \Gamma_0| > \mathcal{R}_0$ and

$$\sum_{n=n_0}^{\infty} \left(\frac{\Re[(\overline{b_{2n}} + \overline{\Gamma_0})e^{-i\beta}] - \mathcal{R}_0}{|b_{2n} + \Gamma_0|^2 - \mathcal{R}_0^2} - p \right) = \infty, \tag{4.3.1}$$

where $|b_{2n} + \Gamma_0| \neq \mathcal{R}_0$ for all $n \geq n_0$.

REMARKS R4.3.1

1. From (4.2.4) it follows that $1/q_{2n+1} \leq (|\Gamma_0| + \mathcal{R}_0)/2$. Hence, $\liminf q_{2n+1} > 0$ always holds for continued fractions $K(1/b_n)$ in the disk – halfplane case.

2. From (4.2.5) and (4.2.2) it follows that if $0 \notin \overline{V_1}$, then $2\mathcal{R}_0 p \leq p(|b_{2n} + \Gamma_0| + \mathcal{R}_0) \leq 1$; i.e.

$$\mathcal{R}_0 \leq |b_{2n} + \Gamma_0| \leq \frac{1}{p} - \mathcal{R}_0.$$

This means that $\{b_{2n}\}$ is bounded, and that

$$2p\mathcal{R}_0 \leq 1 \quad \text{if } p > 0.$$

3. Part F just gives one example of how (2.2.4)–(2.2.5) can be applied in the disk–halfplane case.

THEOREM 4.3.2   Let $K(a_n/1)$ have twin value sets $(V_0, V_1) = (D_0, H)$ given by (3.1.1) and (4.1.1). Then $K(a_n/1)$ converges in the classical sense to a value $f \in \overline{V_0}$, if one of the following sets of conditions holds.

A. $0 \in V_1$.

B. $-1 \notin \partial V_0$ and $0 \notin \partial V_0$.

C. $-1 \notin \partial V_0$, $K(a_n/1)$ converges generally, and $\{a_{2n}\}$ is bounded.

REMARKS R4.3.2

1. For $K(a_n/1)$ we always have $q_{2n+1} =: q = 2[p + \cos\beta] > 0$. Moreover, it follows from (4.2.4) and (4.2.5) that

$$|a_{2n}| \geq p(|1 + \Gamma_0| + \mathcal{R}_0) \quad \text{and} \quad |a_{2n+1}| \leq q(|\Gamma_0| + \mathcal{R}_0)/2.$$

2. Theorem 4.3.2B is identical to Theorem A1.2.1B (except that the regions V_0 and V_1 are interchanged in Theorem A1.2.1B).

4.4. PROOFS

Just as in the disk – disk case, we shall prove the result from the second section first:

PROOF OF PROPOSITION 4.2.1: Since ∂V_1 is described by $w = e^{i\beta}(p+iy)$, $y \in \mathbf{R}$, it follows that $\partial(b+V_1)$ is described by $w = e^{i\beta}[p+\Re(be^{-i\beta})+iy'] = e^{i\beta}[q/2+iy']$, $y' \in \mathbf{R}$. Since $q > 0$, it follows that $0 \notin (b + \overline{V_1})$, and thus $1/(b + V_1)$ is a circular disk with boundary passing through the origin and the point $w_1 := e^{-i\beta}/(q/2)$. Since the ray $w = e^{i\beta}$ intersects $\partial(b + V_1)$ at an angle $\pi/2$ at the point $e^{i\beta}(q/2)$, we find that 0 and w_1 are diametrically opposite points in $\partial(1/(b + V_1))$. Hence $a/(b + V_1)$ is a circular disk, with center γ' and radius ρ' given by

$$\gamma' = \frac{ae^{-i\beta}}{q} \quad \text{and} \quad \rho' = \frac{|a|}{q}. \tag{4.4.1}$$

Hence, $a/(b + V_1) \subseteq V_0$, if and only if $|\gamma' - \Gamma_0| + \rho' \leq \mathcal{R}_0$. This proves the expression for Ω_1 in (4.2.2).

If $0 \notin (b + \overline{V_0})$, then $a/(b + V_0)$ is a circular disk, with center γ and radius ρ given as in (3.4.2), with indices 0 instead of 1. Hence, $a/(b + V_0) \subseteq V_1$, if and only if $\Re(\gamma e^{-i\beta}) - \rho \geq p$; i.e.

$$\frac{\Re(a(\overline{b} + \overline{\Gamma_0})e^{-i\beta})}{|b + \Gamma_0|^2 - \mathcal{R}_0^2} - \frac{|a|\mathcal{R}_0}{|b + \Gamma_0|^2 - \mathcal{R}_0^2} \geq p,$$

which proves the expression for Ω_2 in this case.

Finally, if $0 \in (b + \partial V_0)$, then $a/(b + V_0)$ is a halfplane. Since $w_2 := 2(b + \Gamma_0)$ is the point diametrically opposite to 0 on $\partial(b + V_0)$, this halfplane is given by

$$\Re(we^{-i\delta}) > \frac{|a|}{2|b + \Gamma_0|} = \frac{|a|}{2\mathcal{R}_0}, \quad \text{where } \delta := \arg(a) - \arg(b + \Gamma_0). \tag{4.4.2}$$

Hence, $a/(b + V_0) \subseteq V_1 = H$, if and only if $\delta = \beta_{(\mathrm{mod}2\pi)}$ and $|a| \geq 2p\mathcal{R}_0$, which concludes the proof of (4.2.2). ∎

PROOF OF PROPOSITION 4.2.2:

A. We know that $|b + \Gamma_0| \geq \mathcal{R}_0$ is a necessary condition. We shall see that it is also sufficient. If $|b + \Gamma_0| = \mathcal{R}_0$, there always exist admissible points a. (See (4.2.2).) Let $|b + \Gamma_0| > \mathcal{R}_0$. According to (4.2.2), a is admissible, if and only if it satisfies the inequality

$$\Re[a(\overline{b} + \overline{\Gamma_0})e^{-i\beta}] - |a|\mathcal{R}_0 \geq p(|b + \Gamma_0|^2 - \mathcal{R}_0^2). \tag{4.4.3}$$

Clearly, (4.4.3) has a solution, if and only if it has a solution $a := a_1$ with $\arg(a_1) = \beta + \arg(b + \Gamma_0)$; i.e. if and only if

$$|a_1|(|b + \Gamma_0| - \mathcal{R}_0) \geq p(|b + \Gamma_0|^2 - \mathcal{R}_0^2)$$
$$|a_1| \geq p(|b + \Gamma_0| + \mathcal{R}_0)$$

has a solution $|a_1|$. This is also always so.

B. Since V_0 is bounded, we obviously need that $q > 0$. From (4.2.2) we find that admissible a exist, if and only if the inequality

$$|a - q\Gamma_0 e^{i\beta}| + |a| \leq \mathcal{R}_0 q \qquad (4.4.4)$$

has a solution. Let k, $y \in \mathbf{R}$. Inserting $a = q\Gamma_0(k + iy)e^{i\beta}$ into (4.4.4), shows that a is a solution, if and only if $|k - 1 + iy| + |k + iy| \leq 1$, i.e. if and only if $y = 0$ and $0 \leq k \leq 1$. If $k = 0$, then $a = 0$, and $a/(b + w)$ is singular. (Actually, in this case $0/(b + w) = 0 \notin V_0$ for every $w \in V_1$.) Hence $0 < k \leq 1$.

C. If $|b + \Gamma_0| > \mathcal{R}_0$, then $a/(b + V_0)$ is a circular disk, and thus can not be equal to the halfplane V_1. Let $|b + \Gamma_0| = \mathcal{R}_0$. Then it follows from (4.4.2), that $a/(b + V_0) = V_1$ if and only if a is given by (4.2.6) with $p > 0$.

D. Assume that $a/(b + V_1) = V_0$. Then $\gamma' = \Gamma_0$ and $\rho' = \mathcal{R}_0$ in (4.4.1). Since $|\gamma'| = \rho'$, this means that $|\Gamma_0| = \mathcal{R}_0$. From Part B2 it thus follows that $a = kq\Gamma_0 e^{i\beta}$ for a $k \in (0, 1]$. Moreover, since $\rho' = |a|/q = \mathcal{R}_0$, we know that $|a| = q\mathcal{R}_0$. Hence, $a = q\Gamma_0 e^{i\beta}$.

    On the other hand, if $|\Gamma_0| = \mathcal{R}_0$ and $a = q\Gamma_0 e^{i\beta}$, then inserting these expressions into (4.4.1), shows that $\gamma' = \Gamma_0$ and $\rho' = |\Gamma_0| = \mathcal{R}_0$. ∎

PROOF OF THEOREM 4.1.1: Since $s_{2n}(V_0) \subseteq V_1$, it follows that $s_{2n}(V_0)$ is either a circular disk or a halfplane. Clearly, $0 \notin s_{2n}(V_0) = a_{2n}/(b_{2n} + V_0)$ for all n. Since 0 is an interior point in V_1, and thus has a positive distance $\text{dist}_e(0, \partial V_1)$ to ∂V_1, it follows that $V_1 \setminus s_{2n}(\overline{V_0})$ contains a circular disk with diameter $\geq \text{dist}_e(0, \partial V_1)$ for all n. This means that $\limsup r_{2n} < 1$.

    Similarly, $s_{2n+1}(V_1)$ is a circular disk with 0 on the boundary for all n, since $\infty \in \partial V_1$. So, again, since 0 is an interior point in V_0, we find that $\limsup r_{2n+1} < 1$.

    Let $\varepsilon := \frac{1}{2}\text{dist}_e(0, \partial V_0)$, and let $w_{2n} := \infty$ and

$$w_{2n+1} := \begin{cases} -b_{2n+1}, & \text{if } q_{2n+1} \geq \varepsilon|p|/\mathcal{R}_0, \\ 0, & \text{otherwise.} \end{cases}$$

Then $\{w_{2n}\}$ is bounded away from ∂V_0, and $\{w_{2n+1}\}$ is bounded away from ∂V_1. Moreover, $s_{2n}(w_{2n}) = 0 \notin \partial V_1$, and $s_{2n+1}(w_{2n+1})$ is either $\infty \notin \partial V_0$ or a_{2n+1}/b_{2n+1}, where, by (4.1.2), $|b_{2n+1}| \geq \Re(b_{2n+1}e^{-i\beta}) > -p = |p|$. Hence,

$$\left|\frac{a_{2n+1}}{b_{2n+1}}\right| < \frac{|a_{2n+1}|}{|p|} \leq \frac{\mathcal{R}_0 q_{2n+1}}{|p|} \leq \varepsilon,$$

since, by Proposition 4.2.1, $|a_{2n+1}| \leq \mathcal{R}_0 q_{2n+1}$. That is, also $\{s_{2n+1}(w_{2n+1})\}$ is bounded away from ∂V_0. Hence, $\{t_n\}$ and $\{\varphi_n^{-1}(w_n)\}$ satisfy the conditions of Lemma A2.1.1, and the result follows. ∎

PROOF OF THEOREM 4.1.2: Since $0 \in V_1$; i.e. $p < 0$, we know from the proof of Theorem 4.1.1 that $\limsup r_{2n} < 1$. Let $w_n := -b_{2n+1}$ for all n. Then $\{w_n\}$ is bounded away from ∂V_1 under our conditions, and $s_{2n} \circ s_{2n+1}(w_n) = 0 \notin \partial V_1$.

Hence $S_{2n+1}(w) \to f$ for $w \in V_1$ by Lemma A2.1.1. Assume that the limit point case does not hold. Then $S_{2n+1}(0) \to f$, and thus $S_{2n+2}(\infty) = S_{2n+1}(0) \to f$ and $S_{2n}(a_{2n+1}/b_{2n+1}) = S_{2n+1}(0) \to f$, where $a_{2n+1}/b_{2n+1} = s_{2n+1}(0) \in V_0$. That is, $\{a_{2n+1}/b_{2n+1}\}$ is bounded, and the general convergence follows. ∎

PROOF OF THEOREM 4.1.3: Since $0 \in V_0$, it follows that $\limsup r_{2n+1} < 1$. (See the proof of Theorem 4.1.1.) As in the proof of Theorem 3.1.4, we compare the quantities $d_n := |b_{2n} + \Gamma_0| - \mathcal{R}_0$, $e_n := |b_{2n} + \Gamma_0 + a_{2n}/b_{2n-1}| - \mathcal{R}_0$ and $f_n := \mathcal{R}_0 - |s_{2n-1} \circ s_{2n}(\Gamma_0) - \Gamma_0|$, and choose $w_n := -b_{2n}$ if d_n is largest, $w_n := -b_{2n} - a_{2n}/b_{2n-1}$ if e_n is largest, and $w_n := \Gamma_0$ if f_n is largest. Then it follows from (4.1.3) that $\{w_n\}$ is bounded away from ∂V_0. Moreover, $s_{2n-1} \circ s_{2n}(w_n)$ is either 0, ∞ or $s_{2n-1} \circ s_{2n}(\Gamma_0)$, and thus, also this sequence is bounded away from ∂V_0. Therefore, $\{\varphi_0^{-1}(w_n)\}$ satisfies the conditions for the sequence $\{w_n\}$ in Lemma A2.1.1, which means that $S_{2n}(w) \to f \in \overline{V_0}$, locally uniformly in V_0.

The general convergence of $K(a_n/b_n)$ is then trivial, if we have the limit point case. Otherwise, $S_{2n}(w) \to f$ for all $w \in \hat{\mathbf{C}} \setminus \partial V_0$. In particular, $S_{2n}(0) \to f$ and $S_{2n}(\infty) = S_{2n-1}(0) \to f$, which proves classical, and thus also general convergence of $K(a_n/b_n)$. ∎

PROOF OF THEOREM 4.1.4: Since $0 \in V_0$; i.e. $\mathcal{R}_0 > |\Gamma_0|$, it follows that $s_{2n+1}^{-1}(V_0)$ is the complement of a disk, with center γ_{2n+1}^* and radius ρ_{2n+1}^* given by

$$\gamma_{2n+1}^* := -b_{2n+1} - \frac{a_{2n+1}\overline{\Gamma_0}}{\mathcal{R}_0^2 - |\Gamma_0|^2}, \qquad \rho_{2n+1}^* := \frac{|a_{2n+1}|\mathcal{R}_0}{\mathcal{R}_0^2 - |\Gamma_0|^2}. \qquad (4.4.5)$$

(See (3.4.5).) Hence, $\liminf \tilde{r}_{2n+1} > 1$ when $\limsup |a_{2n+1}| < \infty$.

Let $w_n := -b_{2n+1}$ for all n. Then the convergence of $S_{2n+1}(w)$ and the general convergence of $K(a_n/b_n)$ follow as in the proof of Theorem 4.1.2. ∎

PROOF OF THEOREM 4.1.5: From (3.4.2) it follows that the radius of $s_{2n}(V_0)$ is given by

$$\rho_{2n} = \frac{|a_{2n}|\mathcal{R}_0}{|b_{2n} + \Gamma_0|^2 - \mathcal{R}_0^2} \quad \text{if } |b_{2n} + \Gamma_0| > \mathcal{R}_0.$$

Hence $\liminf r_{2n} < 1$ if (4.1.4) holds.

On the other hand, if $|\Gamma_0| = \mathcal{R}_0$, it follows that $s_{2n+1}^{-1}(V_0) = -b_{2n+1} + a_{2n+1}/V_0$ is the halfplane

$$\Re(we^{-i\beta}) > \left|\frac{a_{2n+1}}{2\Gamma_0}\right| - \Re(b_{2n+1}e^{-i\beta}),$$

where, by Proposition 4.2.2B, $a_{2n+1} = k_{2n+1}q_{2n+1}\Gamma_0 e^{i\beta}$ for a $k_{2n+1} \in (0, 1]$. Hence $\liminf \tilde{r}_{2n+1} > 1$ if

$$\limsup \left(\left|\frac{a_{2n+1}}{2\Gamma_0}\right| - \Re(b_{2n+1}e^{-i\beta})\right) < p$$

$$\limsup \left(\left|\frac{a_{2n+1}}{\Gamma_0}\right| - q_{2n+1}\right) < 0$$

$$\liminf \frac{q_{2n+1}}{\mathcal{R}_0}\left(\mathcal{R}_0 - \frac{|a_{2n+1}|}{q_{2n+1}}\right) > 0,$$

which holds if (4.1.5) holds. This means that either $\limsup \sigma_{2n} < 1$ or $\liminf \tilde{\sigma}_{2n} > 1$ under our conditions.

Let $w_n := -b_{2n+1}$ for all n. Again, the convergence results follow as in the proof of Theorem 4.1.2, since $0 \notin \partial V_1$. ∎

PROOF OF THEOREM 4.1.6: If $0 \notin \overline{V_1}$, i.e. $p > 0$, then $s_{2n}^{-1}(V_1) = -b_{2n} + a_{2n}/V_1$ is a circular disk, with center $\tilde{\gamma}_{2n}$ and radius $\tilde{\rho}_{2n}$ given by

$$\tilde{\gamma}_{2n} := -b_{2n} + \frac{a_{2n}}{2p}e^{-i\beta}, \qquad \tilde{\rho}_{2n} := \frac{|a_{2n}|}{2p}. \tag{4.4.6}$$

If $p \leq 0$, it is a halfplane or the complement of a disk. Hence, $\liminf \tilde{\rho}_{2n} > \mathcal{R}_0$ under our conditions. Let

$$w_n := \begin{cases} \Gamma_0 & \text{if } \mathcal{R}_0 - |s_{2n-1} \circ s_{2n}(\Gamma_0) - \Gamma_0| \geq |b_{2n} + \Gamma_0 + a_{2n}/b_{2n-1}| - \mathcal{R}_0, \\ -b_{2n} - a_{2n}/b_{2n-1} & \text{otherwise.} \end{cases}$$

Then (4.1.6) ensures that $\{w_n\}$ and $\{s_{2n-1} \circ s_{2n}(w_n)\}$ are bounded away from ∂V_0. The convergence of $S_{2n}(w)$ thus follows.

The general convergence of $K(a_n/b_n)$ is trivial if we have the limit point case. Assume that this is not so. Then $S_{2n}(w) \to f$ for all $w \in \hat{\mathbf{C}} \setminus \partial V_0$. In particular, $S_{2n}(\infty) \to f$. Since $S_{2n}(\infty) = S_{2n-1}(0) = S_{2n+1}(-b_{2n+1})$, the general convergence follows immediately if $\liminf |b_{2n+1}| > 0$.

Otherwise, let $v_0 \in \mathbf{C} \setminus \partial V_0$, $v_0 \neq 0$, and let

$$v_n := \begin{cases} -b_{2n+1}, & \text{if } |b_{2n+1}| \geq |-b_{2n+1} + a_{2n+1}/v_0|, \\ -b_{2n+1} + a_{2n+1}/v_0, & \text{otherwise} \end{cases}$$

for all n. Then $\liminf(|a_{2n+1}| + |b_{2n+1}|) > 0 \Rightarrow \liminf |v_n| > 0$, and $s_{2n+1}(v_n)$ is either $\infty \notin \partial V_0$ or $v_0 \notin \partial V_0$. Hence, $S_{2n+1}(v_n) \to f$ and $S_{2n+1}(0) \to f$, which imply general convergence. ∎

PROOF OF THEOREM 4.3.1:

A, B, C, D: Since $0 \in V_0$ and V_0 is bounded, it suffices to prove general convergence. (See Lemma 2.3.2A.) Part A follows from Theorem 4.1.1, Part B follows from Theorem 4.1.4, Part C follows from Theorem 4.1.3, and Part D follows from Theorem 4.1.6.

E. Let $\tilde{V}_1 := \cup_{n=1}^{\infty} 1/(b_{2n} + V_0)$. Then $(V_0, \tilde{V}_1)$ are also twin value sets for $K(1/b_n)$. Since V_0 and $\tilde{V}_1$ are bounded, the result follows from Lemma 2.3.2A.

F. If $|b_{2n} + \Gamma_0| > \mathcal{R}_0$, it follows that $s_{2n}(V_0) = 1/(b_{2n} + V_0)$ is a circular disk, with center γ_{2n} and radius ρ_{2n} given by

$$\gamma_{2n} := \frac{\overline{b_{2n} + \Gamma_0}}{|b_{2n} + \Gamma_0|^2 - \mathcal{R}_0^2}, \qquad \rho_{2n} := \frac{\mathcal{R}_0}{|b_{2n} + \Gamma_0|^2 - \mathcal{R}_0^2}. \tag{4.4.7}$$

Hence, the nth term in the series (4.3.1) is equal to the euclidean distance between ∂V_1 and $s_{2n}(V_0)$. It follows therefore from (2.2.4) that the limit point case holds. That

is, $S_{2n}(w) \to f$ uniformly in $\overline{V_0}$. In particular $S_{2n}(0) \to f$. Moreover, $S_{2n+1}(w) \to f$ uniformly in $\overline{V_1}$. If $0 \notin \overline{V_1}$, the convergence of $S_{2n+1}(0)$ follows by Part E. ∎

PROOF OF THEOREM 4.3.2: If $0 \in V_1$, then $K(a_n/1)$ converges in the classical sense if it converges generally, since $-1 \notin \overline{V_1}$. (See Lemma 2.3.2B.) So also if $-1 \notin \overline{V_0}$ and $0 \in V_0$. Since $\infty \in \partial V_1$, we know that $-1 \notin \overline{V_0}$.

A. The result follows from Theorem 4.1.2.

B. This is a direct consequence of Theorem 4.1.3, since (4.1.3) holds when $|1+\Gamma_0| > \mathcal{R}_0$.

C. Let $\tilde{V}_1 := \cup_{n=1}^{\infty} a_{2n}/(1+V_0)$. Then $(V_0, \tilde{V}_1)$ are bounded twin value sets for $K(a_n/1)$, and the result follows from Lemma 2.3.2B. ∎

5. THE DISK – COMPLEMENT OF DISK CASE

5.1. CONVERGENCE RESULTS

Let $V_0 := D_0$ and $V_1 := \hat{\mathbf{C}} \setminus \overline{D_1}$, where D_j is given by (3.1.1). For convenience, we define V_n as in (3.1.2). In this case it follows that, if (V_0, V_1) is a pair of twin value sets for a continued fraction $K(a_n/b_n)$, then $0 \in V_0$ and $-b_{2n+1} \notin \overline{V_1}$; i.e. $|\Gamma_0| < \mathcal{R}_0$ and $|b_{2n+1} + \Gamma_1| < \mathcal{R}_1$. In particular, this means that $\{b_{2n+1}\}$ is bounded.

The first result holds when also $0 \in V_1$:

THEOREM 5.1.1   Let $K(a_n/b_n)$ have twin value sets $(V_0, V_1) = (D_0, \hat{\mathbf{C}} \setminus \overline{D_1})$ given by (3.1.1) with $0 \in V_1$. If

$$either \quad \limsup |b_{2n+1} + \Gamma_1| < \mathcal{R}_1 \quad or \quad \limsup |a_{2n+1}| < \frac{\mathcal{R}_1}{\mathcal{R}_0}(\mathcal{R}_0^2 - |\Gamma_0|^2), \quad (5.1.1)$$

then $K(a_n/b_n)$ converges in the classical sense to a value $f \in \overline{V_0}$.

As already mentioned, we always have $|b_{2n+1} + \Gamma_1| < \mathcal{R}_1$. Moreover, it follows from the proof section that we always have $|a_{2n+1}| \leq \mathcal{R}_1(\mathcal{R}_0^2 - |\Gamma_0|^2)/\mathcal{R}_0$, with equality if and only if $s_{2n+1}(V_1) = V_0$. (See also Proposition 5.2.2C to come.) Hence, condition (5.1.1) is not so very restrictive.

We can also obtain classical convergence if $0 \in \partial V_1$:

THEOREM 5.1.2   Let $K(a_n/b_n)$ have twin value sets $(V_0, V_1) = (D_0, \hat{\mathbf{C}} \setminus \overline{D_1})$ given by (3.1.1) with $0 \in \overline{V_1}$. If

$$\limsup |a_{2n+1}| < \frac{\mathcal{R}_1}{\mathcal{R}_0}(\mathcal{R}_0^2 - |\Gamma_0|^2) \quad and \quad \limsup \left| b_{2n} + \Gamma_0 - \frac{a_{2n}}{\Gamma_1} \right| < \mathcal{R}_0, \quad (5.1.2)$$

then $K(a_n/b_n)$ converges in the classical sense to a value $f \in \overline{V_0}$.

THEOREM 5.1.3   Let $K(a_n/b_n)$ have twin value sets $(V_0, V_1) = (D_0, \hat{\mathbf{C}} \setminus \overline{D_1})$ given by (3.1.1) with $0 \in \overline{V_1}$. If

$$\liminf \left(\left| |b_{2n} + \Gamma_0| - \mathcal{R}_0 \right| + |b_{2n} + \Gamma_0 + a_{2n}/b_{2n-1}| \right) > \mathcal{R}_0, \tag{5.1.3}$$

then $K(a_n/b_n)$ converges in the classical sense to a value $f \in \overline{V_0}$.

If $0 \notin \overline{V_1}$, we can still conclude general convergence:

THEOREM 5.1.4   Let $K(a_n/b_n)$ have twin value sets $(V_0, V_1) = (D_0, \hat{\mathbf{C}} \setminus \overline{D_1})$ given by (3.1.1). If (5.1.3) holds, and either

$$\limsup \frac{|a_{2n+1}| \mathcal{R}_1}{\mathcal{R}_1^2 - |b_{2n+1} + \Gamma_1|^2} < \mathcal{R}_0 \qquad \text{or} \qquad \liminf |a_{2n}| > \frac{\mathcal{R}_0}{\mathcal{R}_1}(\mathcal{R}_1^2 - |\Gamma_1|^2), \tag{5.1.4}$$

then $K(a_n/b_n)$ converges generally to a value $f \in \overline{V_0}$.

The first condition in (5.1.4) is closely related to the first condition in (5.1.2):

LEMMA 5.1.5   Let $K(a_n/b_n)$ have twin value sets $(V_0, V_1) = (D_0, \hat{\mathbf{C}} \setminus \overline{D_1})$ given by (3.1.1). Then

$$\limsup |a_{2n+1}| < \frac{\mathcal{R}_1}{\mathcal{R}_0}(\mathcal{R}_0^2 - |\Gamma_0|^2) \quad \Rightarrow \quad \limsup \frac{|a_{2n+1}| \mathcal{R}_1}{\mathcal{R}_1^2 - |b_{2n+1} + \Gamma_1|^2} < \mathcal{R}_0. \tag{5.1.5}$$

The last three theorems can also be used if $0 \notin \overline{V_1}$:

THEOREM 5.1.6   Let $K(a_n/b_n)$ have twin value sets $(V_0, V_1) = (D_0, \hat{\mathbf{C}} \setminus \overline{D_1})$ given by (3.1.1) with $0 \notin \partial V_1$. If $\limsup |b_{2n+1} + \Gamma_1| < \mathcal{R}_1$, and either

$$\limsup \frac{\mathcal{R}_0^2 - |b_{2n} + \Gamma_0|^2}{|a_{2n}|} < \frac{\mathcal{R}_0}{\mathcal{R}_1} \qquad \text{or} \qquad \limsup |a_{2n+1}| < \frac{\mathcal{R}_1}{\mathcal{R}_0}(\mathcal{R}_0^2 - |\Gamma_0|^2), \tag{5.1.6}$$

then $K(a_n/b_n)$ converges generally to a value $f \in \overline{V_0}$.

THEOREM 5.1.7   Let $K(a_n/b_n)$ have twin value sets $(V_0, V_1) = (D_0, \hat{\mathbf{C}} \setminus \overline{D_1})$ given by (3.1.1) with $0 \notin \partial V_1$. If

$$\limsup \left(|b_{2n+1} + \Gamma_1| + \left| b_{2n+1} + \Gamma_1 + \frac{a_{2n+1}}{b_{2n} - a_{2n}/\Gamma_1} \right| \right) < 2\mathcal{R}_1, \tag{5.1.7}$$

and (5.1.6) hold, then $S_{2n+1}(w)$ converges locally uniformly in V_1 to a value $f \in \overline{V_0}$. If, in addition, $\liminf |b_{2n}| > 0$, then $K(a_n/b_n)$ converges generally to f.

THEOREM 5.1.8   Let $K(a_n/b_n)$ have twin value sets $(V_0, V_1) = (D_0, \hat{\mathbf{C}} \setminus \overline{D_1})$ given by (3.1.1). If

$$\limsup \left| b_{2n+1} + \Gamma_1 + \frac{a_{2n+1}}{b_{2n} - a_{2n}/\Gamma_1} \right| < \mathcal{R}_1, \tag{5.1.8}$$

and (5.1.6) hold, then $S_{2n+1}(w)$ converges locally uniformly in V_1 to a value $f \in \overline{V_0}$. If, in addition, $\liminf |b_{2n}| > 0$, then $K(a_n/b_n)$ converges generally to f.

5.2. LOCATION OF THE ELEMENTS (a_n, b_n)

Again, we characterize the elements (a, b), which have the proper mapping properties, and we do not claim priority of the result. For instance, Jones and Thron had the result for continued fractions $K(a_n/1)$, [6, Lemma 5.5]. Still, we include its proof for completeness.

PROPOSITION 5.2.1 The continued fraction $K(a_n/b_n)$ has twin value sets $(V_0, V_1) = (D_0, \hat{\mathbf{C}} \setminus \overline{D_1})$ if and only if $(a_n, b_n) \in \Omega_n$ for all n, where $\Omega_{2n-1} = \Omega_1$ and $\Omega_{2n} = \Omega_2$ are sets from $\mathbf{C} \times \mathbf{C}$ given by

$$(a, b) \in \Omega_1 \quad \Leftrightarrow \quad \begin{cases} a = \Gamma_0(b + \Gamma_1)(1 - \mathcal{R}_1^2/|b + \Gamma_1|^2)\alpha \neq 0, \quad \text{where} \\ |\alpha - 1| + |\alpha|\mathcal{R}_1/|b + \Gamma_1| \leq \mathcal{R}_0/|\Gamma_0| \quad \text{if } \Gamma_0(b + \Gamma_1) \neq 0, \\ 0 < |a| \leq (\mathcal{R}_0 - |\Gamma_0|)(\mathcal{R}_1 - |b + \Gamma_1|) \quad \text{if } \Gamma_0(b + \Gamma_1) = 0, \end{cases} \tag{5.2.1a}$$

and

$$(a, b) \in \Omega_2 \quad \Leftrightarrow \quad \begin{cases} a = \Gamma_1(b + \Gamma_0)(1 - \mathcal{R}_0^2/|b + \Gamma_0|^2)\alpha \neq 0 \quad \text{where} \\ |\alpha - 1| - |\alpha|\mathcal{R}_0/|b + \Gamma_0| \geq \mathcal{R}_1/|\Gamma_1| \quad \text{if } \Gamma_1 \neq 0, \\ |a| \geq \mathcal{R}_1(|b + \Gamma_0| + \mathcal{R}_0) \quad \text{if } \Gamma_1 = 0 \end{cases} \tag{5.2.1b}$$

if $|b + \Gamma_0| > \mathcal{R}_0$,

$$(a, b) \in \Omega_2 \quad \Leftrightarrow \quad |a| \geq 2\mathcal{R}_0(\Re(\Gamma_1 e^{-i\psi}) + \mathcal{R}_1) \quad \text{where } \psi = \arg(a) - \arg(b + \Gamma_0) \tag{5.2.1c}$$

if $|b + \Gamma_0| = \mathcal{R}_0$, and finally

$$(a, b) \in \Omega_2 \quad \Leftrightarrow \quad \begin{cases} a = \Gamma_1(b + \Gamma_0)(1 - \mathcal{R}_0^2/|b + \Gamma_0|^2)\alpha \neq 0 \quad \text{where} \\ |\alpha|\mathcal{R}_0/|b + \Gamma_0| - |\alpha - 1| \geq \mathcal{R}_1/|\Gamma_1| \quad \text{if } \Gamma_1(b + \Gamma_0) \neq 0, \\ |a| \geq (\mathcal{R}_0 + |b + \Gamma_0|)(|\Gamma_1| + \mathcal{R}_1) \quad \text{if } \Gamma_1(b + \Gamma_0) = 0 \end{cases} \tag{5.2.1d}$$

if $|b + \Gamma_0| < \mathcal{R}_0$.

PROPOSITION 5.2.2 *Let $b \in \mathbf{C}$ be a fixed number. Then the following statements hold.*

A. $a/(b + V_1) \subseteq V_0$ *for some* $a \in \mathbf{C} \setminus \{0\}$ *if and only if* $|\Gamma_0| < \mathcal{R}_0$ *and* $|b + \Gamma_1| < \mathcal{R}_1$, *and then*

$$|a| \leq \mathcal{R}_0 \mathcal{R}_1 - |\Gamma_0(b + \Gamma_1)| - \big||\Gamma_0|\mathcal{R}_1 - |b + \Gamma_1|\mathcal{R}_0\big|. \tag{5.2.2}$$

B. *If* $a/(b + V_1) \subseteq V_0$ *for an* $a \in \mathbf{C} \setminus \{0\}$, *then* $a/(b + V_1) \subseteq V_0$ *for infinitely many* $a \in \mathbf{C}$.

C. $a/(b + V_1) = V_0$, *if and only if* $|\Gamma_0|\mathcal{R}_1 = |b + \Gamma_1|\mathcal{R}_0$ *and* a *is given by*

$$a = \begin{cases} \Gamma_0(b + \Gamma_1)(1 - \mathcal{R}_1^2/|b + \Gamma_1|^2) & \text{if } b + \Gamma_1 \neq 0, \\ \mathcal{R}_0\mathcal{R}_1 e^{i\theta}; \ \theta \in \mathbf{R} & \text{if } b + \Gamma_1 = 0. \end{cases} \tag{5.2.3}$$

D. *There always exist (infinitely many)* $a \in \mathbf{C}$ *such that* $a/(b + V_0) \subseteq V_1$. *If* $0 \notin V_1$, *then*

$$|a| \geq \mathcal{R}_0 \mathcal{R}_1 - |\Gamma_1(b + \Gamma_0)| + \big||\Gamma_1|\mathcal{R}_0 - |b + \Gamma_0|\mathcal{R}_1\big|. \tag{5.2.4}$$

E. $a/(b + V_0) = V_1$ *if and only if* $|b + \Gamma_0| < \mathcal{R}_0$, $\mathcal{R}_1|b + \Gamma_0| = \mathcal{R}_0|\Gamma_1|$ *and* a *is given by*

$$a = \begin{cases} \Gamma_1(b + \Gamma_0)(1 - \mathcal{R}_0^2/|b + \Gamma_0|^2) & \text{if } b + \Gamma_0 \neq 0, \\ \mathcal{R}_0\mathcal{R}_1 e^{i\theta}; \ \theta \in \mathbf{R} & \text{if } b + \Gamma_0 = 0. \end{cases} \tag{5.2.5}$$

5.3. CLASSICAL CONVERGENCE FOR $K(1/b_n)$ AND $K(a_n/1)$

THEOREM 5.3.1 *Let* $K(1/b_n)$ *have twin value sets* $(V_0, V_1) = (D_0, \hat{\mathbf{C}} \setminus \overline{D_1})$ *given by* (3.1.1). *Then* $K(1/b_n)$ *converges in the classical sense to a value* $f \in \overline{V_0}$, *if one of the following sets of conditions holds.*

A. $K(1/b_n)$ *converges generally.*

B. $0 \in V_1$.

C. $\mathcal{R}_1 > \mathcal{R}_0(\mathcal{R}_1^2 - |\Gamma_1|^2)$ *and*

$$\limsup \big(\big||b_{2n} + \Gamma_0| - \mathcal{R}_0\big| + |b_{2n} + \Gamma_0 + 1/b_{2n-1}|\big) > \mathcal{R}_0. \tag{5.3.1}$$

D. $0 \notin \partial V_1$ *and* $\mathcal{R}_0 < \mathcal{R}_1(\mathcal{R}_0^2 - |\Gamma_0|^2)$.

E. $0 \notin \partial V_1$ *and* $\liminf |b_{2n} + \Gamma_0|^2 > \mathcal{R}_0^2 - \mathcal{R}_0/\mathcal{R}_1$.

F. $\limsup |b_{2n+1} + \Gamma_1|^2 < \mathcal{R}_1^2 - \mathcal{R}_1/\mathcal{R}_0$ *and* (5.3.1) *hold.*

REMARKS R5.3.1

1. Note that by Theorem 5.3.1A, general convergence implies classical convergence for these continued fractions. The other parts of this theorem just illustrate that the conditions for general convergence are simplified for continued fractions of the form $K(1/b_n)$. Observe in particular, that the condition $\mathcal{R}_1 > \mathcal{R}_0(\mathcal{R}_1^2 - |\Gamma_1|^2)$ always holds if $0 \in \overline{V_1}$.

2. There are restrictions on D_0 and D_1 for $(D_0, \hat{C} \setminus \overline{D_1})$ to be twin value sets for a continued fraction of the form $K(1/b_n)$. It follows from the proof of Theorem 5.1.1 that

$$\frac{|a_{2n+1}|\mathcal{R}_0}{\mathcal{R}_0^2 - |\Gamma_0|^2} \leq \mathcal{R}_1, \tag{5.3.2}$$

where equality holds if and only if $s_{2n+1}^{-1}(V_0) = V_1$; i.e. $s_{2n+1}(V_1) = V_0$. (See (5.4.5).) Similarly, it follows from the proof of Theorem 5.1.4 that

$$\frac{|a_{2n}|\mathcal{R}_1}{\mathcal{R}_1^2 - |\Gamma_1|^2} \geq \mathcal{R}_0 \qquad \text{if } |\Gamma_1| \leq \mathcal{R}_1, \tag{5.3.3}$$

where equality holds if and only if $s_{2n}(V_0) = V_1$. (See (5.4.8).) In the case where all $a_n = 1$, we therefore have

$$\frac{\mathcal{R}_0}{\mathcal{R}_1}(\mathcal{R}_1^2 - |\Gamma_1|^2) \leq 1 \leq \frac{\mathcal{R}_1}{\mathcal{R}_0}(\mathcal{R}_0^2 - |\Gamma_0|^2)$$

which also holds if $\mathcal{R}_1 \leq |\Gamma_1|$. That is,

$$1 - \frac{|\Gamma_1|^2}{\mathcal{R}_1^2} \leq \frac{1}{\mathcal{R}_0\mathcal{R}_1} \leq 1 - \frac{|\Gamma_0|^2}{\mathcal{R}_0^2}, \tag{5.3.4}$$

which shows that $\mathcal{R}_0\mathcal{R}_1 \geq 1$ and that $|\Gamma_1|/\mathcal{R}_1 \geq |\Gamma_0|/\mathcal{R}_0$, with equality if and only if $s_{2n}(V_0) = V_1$ and $s_{2n+1}(V_1) = V_0$ for all n.

THEOREM 5.3.2 Let $K(a_n/1)$ have twin value sets $(V_0, V_1) = (D_0, \hat{C} \setminus \overline{D_1})$ given by (3.1.1). Then $K(a_n/1)$ converges in the classical sense to a value $f \in \overline{V_0}$, if one of the following sets of conditions holds.

A. $0 \in V_1$.

B. $0 \in \overline{V_1}$ and $-1 \notin \partial V_0$.

C. $0 \in \overline{V_1}$ and $\liminf |a_{2n} + 1 + \Gamma_0| > \mathcal{R}_0$.

D. $-1 \notin \overline{V_0}$ and $0 \notin \partial V_1$.

E. $-1 \notin \overline{V_0}$ and either

$$\limsup |a_{2n+1}| < \frac{\mathcal{R}_0}{\mathcal{R}_1}(\mathcal{R}_1^2 - |1 + \Gamma_1|^2) \quad \text{or} \quad \liminf |a_{2n}| > \frac{\mathcal{R}_0}{\mathcal{R}_1}(\mathcal{R}_1^2 - |\Gamma_1|^2). \tag{5.3.5}$$

F. $-1 \notin \overline{V_0}$ and $0 \notin \partial V_1$ and either

$$\limsup |a_{2n+1}| < \frac{\mathcal{R}_1}{\mathcal{R}_0}(\mathcal{R}_0^2 - |\Gamma_0|^2) \quad \text{or} \quad \liminf |a_{2n}| > 0. \tag{5.3.6}$$

G. $-1 \notin \overline{V_0}$ and

$$\limsup \left| 1 + \Gamma_1 + \frac{a_{2n+1}}{1 - a_{2n}/\Gamma_1} \right| < \mathcal{R}_1. \tag{5.3.7}$$

H. $-1 \notin \overline{V_0}$ and (5.3.6) and (5.3.7) hold.

REMARKS R5.3.2.

1. At least one of the two conditions on $\limsup |a_{2n+1}|$ holds if $\mathcal{R}_0|1 + \Gamma_1| \neq \mathcal{R}_1|\Gamma_0|$. Similarly, at least one of the two conditions on $\liminf |a_{2n}|$ holds if $\mathcal{R}_1|1 + \Gamma_0| \neq \mathcal{R}_0|\Gamma_1|$.

2. The second part of Theorem A1.2.1C is equivalent to Theorem 5.3.2B. The first part follows from Theorem 5.1.7, since

$$-1 \notin \overline{V_0} \quad \Rightarrow \quad \mathcal{R}_0^2 - |1 + \Gamma_0|^2 < 0 \quad \text{in (5.1.6)},$$

and (5.1.7) holds always, since $|1 + \Gamma_1| < \mathcal{R}_1$ always holds. Actually, we find that $K(a_n/1)$ converges generally to f in this case.

5.4. PROOFS

As earlier, we start by proving the results from Section 5.2.

PROOF OF PROPOSITION 5.2.1: $a/(b + V_1)$ is a subset of the circular disk V_0 only if $-b \in D_1 = \hat{\mathbf{C}} \setminus \overline{V_1}$, such that $a/(b + V_1)$ is a circular disk, with center γ^* and radius ρ^* given by

$$\gamma^* := -\frac{a(\bar{b} + \overline{\Gamma_1})}{\mathcal{R}_1^2 - |b + \Gamma_1|^2}, \qquad \rho^* := \frac{|a|\mathcal{R}_1}{\mathcal{R}_1^2 - |b + \Gamma_1|^2}. \tag{5.4.1}$$

Clearly, $a/(b + V_1) \subseteq V_0$ if and only if $|\gamma^* - \Gamma_0| + \rho^* \leq \mathcal{R}_0$, which proves (5.2.1a).

If $-b \notin \overline{V_0}$; i.e. $|b + \Gamma_0| > \mathcal{R}_0$, then $a/(b + V_0)$ is a circular disk, with center γ and radius ρ given by

$$\gamma := \frac{a(\bar{b} + \overline{\Gamma_0})}{|b + \Gamma_0|^2 - \mathcal{R}_0^2}, \qquad \rho := \frac{|a|\mathcal{R}_0}{|b + \Gamma_0|^2 - \mathcal{R}_0^2},$$

such that $a/(b + V_0) \subseteq V_1$ if and only if $|\gamma - \Gamma_1| - \rho \geq \mathcal{R}_1$. This proves (5.2.1b).

Let $|b + \Gamma_0| = \mathcal{R}_0$. Then $1/(b + V_0)$ is the halfplane not containing the origin, whose boundary is perpendicular to the ray $\arg w = -\arg(b + \Gamma_0)$, and passes through the point $1/(2(b + \Gamma_0))$. That is, $a/(b + V_0)$ is the halfplane given by $\Re(we^{-i\psi}) > |a|/(2|b + \Gamma_0|) = |a|/(2\mathcal{R}_0)$. Hence, $a/(b + V_0) \subseteq V_1$ if and only if $\Re(\Gamma_1 e^{-i\psi}) + \mathcal{R}_1 \leq |a|/(2\mathcal{R}_0)$, which proves (5.2.1c).

Finally, let $|b + \Gamma_0| < \mathcal{R}_0$. Then $a/(b + V_0)$ is the complement of the circular disk $\overline{D}$, where D has center $\gamma^\dagger$ and radius $\rho^\dagger$ given by

$$\gamma^\dagger := -\frac{a(\bar{b} + \overline{\Gamma_0})}{\mathcal{R}_0^2 - |b + \Gamma_0|^2}, \qquad \rho^\dagger := \frac{|a|\mathcal{R}_0}{\mathcal{R}_0^2 - |b + \Gamma_0|^2}. \tag{5.4.2}$$

Hence, $a/(b + V_0) \subseteq V_1$ if and only if $D_1 \subseteq D$; i.e. $|\gamma^\dagger - \Gamma_1| + \mathcal{R}_1 \leq \rho^\dagger$, which proves (5.2.1d). ∎

PROOF OF PROPOSITION 5.2.2:

A. The existence of a follows immediately from (5.2.1a) if $\Gamma_0(b + \Gamma_1) = 0$. Let $\Gamma_0(b + \Gamma_1) \neq 0$, and assume that $a/(b + V_1) \subseteq V_0$. Since V_0 is bounded, we have $|b + \Gamma_1| < \mathcal{R}_1$, and thus, it follows that the left side of the inequality

$$|\alpha - 1| + |\alpha|\mathcal{R}_1/|b + \Gamma_1| \leq \mathcal{R}_0/|\Gamma_0| \tag{5.4.3}$$

has its minimum for $\alpha = 0$. Hence, (5.4.3) has a solution $\alpha \neq 0$ if and only if $1 < \mathcal{R}_0/|\Gamma_0|$.

Conversely, if $|b + \Gamma_1| < \mathcal{R}_1$ and $1 < \mathcal{R}_0/|\Gamma_0|$, the existence of a solution $\alpha \neq 0$ of (5.4.3) is obvious.

The upper bound (5.2.2) follows trivially from (5.2.1a) if $\Gamma_0(b + \Gamma_1) = 0$. Let $\Gamma_0(b+\Gamma_1) \neq 0$. The domain for α, given by (5.4.3), is then bounded by a cartesian oval. (See for instance Remark R3.3.2.4.) This domain always contains the origin. Straight forward computation shows that it also contains $\alpha = 1$, if and only if $\mathcal{R}_1/|b + \Gamma_1| \leq \mathcal{R}_0/|\Gamma_0|$. Hence, we find from (5.4.3) that

$$|\alpha| \leq \begin{cases} \dfrac{|b + \Gamma_1|}{|\Gamma_0|} \cdot \dfrac{\mathcal{R}_0 - |\Gamma_0|}{\mathcal{R}_1 - |b + \Gamma_1|} & \text{if } \dfrac{|\Gamma_0|}{\mathcal{R}_0} \geq \dfrac{|b + \Gamma_1|}{\mathcal{R}_1}, \\[2ex] \dfrac{|b + \Gamma_1|}{|\Gamma_0|} \cdot \dfrac{\mathcal{R}_0 + |\Gamma_0|}{\mathcal{R}_1 + |b + \Gamma_1|} & \text{if } \dfrac{|\Gamma_0|}{\mathcal{R}_0} \leq \dfrac{|b + \Gamma_1|}{\mathcal{R}_1}. \end{cases}$$

It follows therefore that

$$|a| \leq \begin{cases} (\mathcal{R}_0 - |\Gamma_0|)(\mathcal{R}_1 + |b + \Gamma_1|) & \text{if } \dfrac{|\Gamma_0|}{\mathcal{R}_0} \geq \dfrac{|b + \Gamma_1|}{\mathcal{R}_1}, \\[2ex] (\mathcal{R}_0 + |\Gamma_0|)(\mathcal{R}_1 - |b + \Gamma_1|) & \text{otherwise,} \end{cases}$$

which proves (5.2.2).

B. This follows immediately from the arguments in the proof of Part A.

C. Let $a/(b + V_1) = V_0$. Then $-b \in D_1$, and the center γ^* and radius ρ^* of $a/(b + V_1)$, as given in (5.4.1), satisfy $\gamma^* = \Gamma_0$ and $\rho^* = \mathcal{R}_0$. Let first $b + \Gamma_1 \neq 0$. Then this means that

$$|\Gamma_0| = |\gamma^*| = \frac{|b + \Gamma_1|}{\mathcal{R}_1}\rho^* = \frac{|b + \Gamma_1|}{\mathcal{R}_1}\mathcal{R}_0 \neq 0,$$

and $|a| = \mathcal{R}_0(\mathcal{R}_1^2 - |b + \Gamma_1|^2)/\mathcal{R}_1$, so that by (5.2.1a), α is a solution of (5.4.3) with $|\alpha| = 1$. Since $\mathcal{R}_1/|b + \Gamma_1| = \mathcal{R}_0/|\Gamma_0|$, this can only happen if $\alpha = 1$, that is, a is given by (5.2.3).

Next, we let $b + \Gamma_1 = 0$. Then $\gamma^* = 0$. Hence, this can only happen if $\Gamma_0 = 0$. Moreover, since $\rho^* = |a|/\mathcal{R}_1$, this means that $|a| = \mathcal{R}_0\mathcal{R}_1$.

On the other hand, if $\mathcal{R}_0|b + \Gamma_1| = \mathcal{R}_1|\Gamma_0|$, and if a is given by (5.2.3), then it follows from (5.4.1) that $\gamma^* = \Gamma_0$ and $\rho^* = \mathcal{R}_0$; i.e. $a/(b + V_1) = V_0$.

D. The existence of a is a simple consequence of (5.2.1b), (5.2.1c) and (5.2.1d).

If $\Gamma_1 = 0$, then (5.2.4) follows trivially from (5.2.1b,c,d). If $b + \Gamma_0 = 0$, it follows from (5.2.1d). Let $\Gamma_1(b + \Gamma_0) \neq 0$. If $|b + \Gamma_0| > \mathcal{R}_0$, then it follows from (5.2.1b) that $1 + |\alpha| - |\alpha|\mathcal{R}_0/|b + \Gamma_0| \geq \mathcal{R}_1/|\Gamma_1|$; i.e.

$$|\alpha| \geq \frac{-1 + \mathcal{R}_1/|\Gamma_1|}{1 - \mathcal{R}_0/|b + \Gamma_0|} \quad \text{if } |b + \Gamma_0| > \mathcal{R}_0.$$

This shows that $|a| \geq (\mathcal{R}_1 - |\Gamma_1|)(\mathcal{R}_0 + |b + \Gamma_0|)$ if $|b + \Gamma_0| > \mathcal{R}_0$. Hence, (5.2.4) follows if $|b + \Gamma_0|/\mathcal{R}_0 \geq |\Gamma_1|/\mathcal{R}_1$. This holds in particular if $\mathcal{R}_1 \geq |\Gamma_1|$; i.e. $0 \notin V_1$.

If $|b + \Gamma_0| = \mathcal{R}_0$, the lower bound $(\mathcal{R}_0 + |b + \Gamma_0|)(\mathcal{R}_1 - |\Gamma_1|)$ follows trivially from (5.2.1c), and thus (5.2.4) follows if $|\Gamma_1| \leq \mathcal{R}_1$. If $|b + \Gamma_0| < \mathcal{R}_0$, the inequality for α in (5.2.1d) implies that

$$|\alpha| \geq \begin{cases} \dfrac{-1 + \mathcal{R}_1/|\Gamma_1|}{-1 + \mathcal{R}_0/|b + \Gamma_0|} & \text{if } \dfrac{|\Gamma_1|}{\mathcal{R}_1} \leq \dfrac{|b + \Gamma_0|}{\mathcal{R}_0} < 1, \\[3mm] \dfrac{1 + \mathcal{R}_1/|\Gamma_1|}{1 + \mathcal{R}_0/|b + \Gamma_0|} & \text{if } \dfrac{|\Gamma_1|}{\mathcal{R}_1} > \dfrac{|b + \Gamma_0|}{\mathcal{R}_0} \text{ and } \dfrac{|b + \Gamma_0|}{\mathcal{R}_0} < 1. \end{cases}$$

Hence, this proves (5.2.4).

E. If $a/(b + V_0) = V_1$, then $a/(b + V_0)$ must be the complement of the disk $\overline{D_1}$; that is, $|b + \Gamma_0| < \mathcal{R}_0$ and $\gamma^\dagger = \Gamma_1$, $\rho^\dagger = \mathcal{R}_1$, where $\gamma^\dagger$ and $\rho^\dagger$ are given by (5.4.2). Let first $|b + \Gamma_0| \neq 0$. Then this means that

$$|\Gamma_1| = |\gamma^\dagger| = \rho^\dagger |b + \Gamma_0|/\mathcal{R}_0 = \mathcal{R}_1 |b + \Gamma_0|/\mathcal{R}_0,$$

so that $\Gamma_1 \neq 0$ and $|a| = \mathcal{R}_1(\mathcal{R}_0^2 - |b + \Gamma_0|^2)/\mathcal{R}_0$. Hence, by (5.2.1d), α is a solution of

$$|\alpha|\mathcal{R}_0/|b + \Gamma_0| - |\alpha - 1| \geq \mathcal{R}_1/|\Gamma_1| \tag{5.4.4}$$

with $|\alpha| = 1$. Since $\mathcal{R}_0/|b + \Gamma_0| = \mathcal{R}_1/|\Gamma_1|$, we find from (5.4.4) that the only such solution is $\alpha = 1$; i.e. a is given by (5.2.5).

If $|b + \Gamma_0| = 0$, then $\Gamma_1 = \gamma^\dagger = 0$ and $\mathcal{R}_1 = \rho^\dagger = |a|/\mathcal{R}_0$, i.e. $|a| = \mathcal{R}_0 \mathcal{R}_1$.

Conversely, if $\mathcal{R}_1|b + \Gamma_0| = \mathcal{R}_0|\Gamma_1|$ and a is given by (5.2.5), then it follows from (5.4.2) that $\gamma^\dagger = \Gamma_1$ and $\rho^\dagger = \mathcal{R}_1$; i.e. $a/(b + V_0) = V_1$. ∎

PROOF OF THEOREM 5.1.1: Assume first that $\limsup |b_{2n+1} + \Gamma_1| < \mathcal{R}_1$. Since $0 \in V_1$ (interior point), and $0 \notin s_{2n}(V_0)$ for all n, it follows that $\limsup r_{2n} < 1$, and thus that $\limsup \sigma_{2n} < 1$. Let $w_n := -b_{2n+1}$ for all n. Then $\{w_n\}$ is bounded away from ∂V_1 under our conditions, and $s_{2n} \circ s_{2n+1}(w_n) = 0 \notin \partial V_1$. Hence, $S_{2n+1}(w) \to f$ for all $w \in V_1$ by Lemma A2.1.1. In particular, $S_{2n+1}(0) \to f$ and $S_{2n+1}(\infty) = S_{2n}(0) \to f$, which proves the result.

Next, let the second inequality in (5.1.1) hold. Since $0 \in V_1$, it follows that $s_{2n}^{-1}(V_1)$ is the complement of a circular disk. Hence, $\liminf \tilde{r}_{2n} > 1$.

Since $s_{2n+1}^{-1}(V_0)$ is the complement of a circular disk, with center $\tilde{\gamma}_{2n+1}$ and radius $\tilde{\rho}_{2n+1}$ given by

$$\tilde{\gamma}_{2n+1} := -b_{2n+1} - \frac{a_{2n+1}\overline{\Gamma_0}}{\mathcal{R}_0^2 - |\Gamma_0|^2}, \qquad \tilde{\rho}_{2n+1} := \frac{|a_{2n+1}|\mathcal{R}_0}{\mathcal{R}_0^2 - |\Gamma_0|^2}, \tag{5.4.5}$$

it follows by (5.1.1) that $\limsup \tilde{\rho}_{2n+1} < \mathcal{R}_1$, and thus $\liminf \tilde{r}_{2n+1} > 1$. Hence $\liminf \tilde{r}_n > 1$. Let $w_n := \infty$ for all n. Then $w_n \notin \partial V_n$ and $s_n(w_n) = 0 \notin \partial V_{n-1}$ for all n. The convergence follows then by Lemma A2.1.1, since $0 \in V_0 \cap V_1$. ∎

PROOF OF THEOREM 5.1.2: Since $0 \in \overline{V_1}$, it follows that $s_{2n}^{-1}(V_1)$ is a halfplane or the complement of a circular disk. Hence $\liminf \tilde{r}_{2n} > 1$. Further, we know that $\liminf \tilde{r}_{2n+1} > 1$ by the second part of the proof of Theorem 5.1.1. Hence, $\liminf \tilde{r}_n > 1$.

Let $w_{2n-1} := \infty$ and $w_{2n} := -b_{2n} + a_{2n}/\Gamma_1$ for all n. Since also now $\{w_{2n}\}$ is bounded away from ∂V_0 and $s_{2n}(w_{2n}) = \Gamma_1 \notin \partial V_1$, it follows that $K(a_n/b_n)$ converges generally by Lemma A2.1.1.

If the limit point case holds, then the classical convergence follows directly. Assume that the limit circle case holds. Then $S_{2n}(0) \to f$ and $S_{2n}(\infty) = S_{2n-1}(0) \to f$, which again proves the classical convergence. ∎

PROOF OF THEOREM 5.1.3: As in the previous proof we have $\liminf \tilde{r}_{2n} > 1$. We shall apply Lemma A2.1.1 to the sequence $\tau_{2n-1} = \varphi_0^{-1} \circ s_{2n-1} \circ s_{2n} \circ \varphi_0$. Since $\liminf \tilde{r}_{2n} > 1$, we know that "the radii" $\tilde{\sigma}_{2n-1}$ of $\tau_{2n-1}^{-1}(U)$ satisfy $\liminf \tilde{\sigma}_{2n-1} > 1$. Let

$$w_n := \begin{cases} -b_{2n} & \text{if } \left| |b_{2n} + \Gamma_0| - \mathcal{R}_0 \right| > |b_{2n} + \Gamma_0 + a_{2n}/b_{2n-1}| - \mathcal{R}_0, \\ -b_{2n} - a_{2n}/b_{2n-1} & \text{otherwise.} \end{cases} \tag{5.4.6}$$

Then $\{w_n\}$ is bounded away from ∂V_0, and $s_{2n-1} \circ s_{2n}(w_n)$ is either $0 \notin \partial V_0$ or $\infty \notin \partial V_0$. It follows therefore from Lemma A2.1.1 that $S_{2n}(w) \to f$ for all $w \in V_0$. In the limit point case, the classical convergence follows trivially. In the limit circle case, we have $S_{2n}(0) \to f$ and $S_{2n}(\infty) \to f$. Hence, the result follows, since $S_{2n-1}(0) = S_{2n}(\infty)$. ∎

PROOF OF THEOREM 5.1.4: Assume first that the first inequality in (5.1.4) holds. The radius ρ_{2n+1} of the circular disk $s_{2n+1}(V_1)$ is given by (5.4.1), i.e.

$$\rho_{2n+1} = \frac{|a_{2n+1}|\mathcal{R}_1}{\mathcal{R}_1^2 - |b_{2n+1} + \Gamma_1|^2}. \tag{5.4.7}$$

Hence, it follows that $\limsup \rho_{2n+1} < \mathcal{R}_0$.

Let $\{w_n\}$ be given by (5.4.6). Then (5.1.3) shows that $\{w_n\}$ is bounded away from ∂V_0. Moreover, $s_{2n-1} \circ s_{2n}(w_n)$ is either 0 or ∞ which are $\notin \partial V_0$. It follows therefore from Lemma A2.1.1 that $S_{2n}(w) \to f$ for all $w \in V_0$. If the limit point case holds, the general convergence follows trivially. Otherwise, it is a consequence of Lemma 2.2.1B, since $\infty \in V_1$ and $s_{2n+1}(\infty) = 0$ which is an interior point in V_0.

Next, we assume that the second inequality in (5.1.4) holds. If $0 \in \overline{V_1}$, the result follows from Theorem 5.1.3. Let $0 \notin \overline{V_1}$. Then $s_{2n}^{-1}(V_1) = -b_{2n} + a_{2n}/V_1$ is a circular disk with center $\tilde{\gamma}_{2n}$ and radius $\tilde{\rho}_{2n}$ given by

$$\tilde{\gamma}_{2n} := -b_{2n} - \frac{a_{2n}\overline{\Gamma_1}}{\mathcal{R}_1^2 - |\Gamma_1|^2}, \qquad \tilde{\rho}_{2n} := \frac{|a_{2n}|\mathcal{R}_1}{\mathcal{R}_1^2 - |\Gamma_1|^2}. \tag{5.4.8}$$

Hence $\liminf \tilde{r}_{2n} > 1$, and the result follows as above. ∎

PROOF OF LEMMA 5.1.5: Let $\mu := \mathcal{R}_1(\mathcal{R}_0^2 - |\Gamma_0|^2)/\mathcal{R}_0 - \limsup |a_{2n+1}|$ and $\varepsilon := \mu/3\mathcal{R}_0\mathcal{R}_1$, and assume that $\mu > 0$. Clearly, $\mu \le \mathcal{R}_0\mathcal{R}_1$, and thus $0 < \varepsilon \le 1/3 < 1$.

We consider first the indices n for which $|b_{2n+1} + \Gamma_1|/\mathcal{R}_1 \leq |\Gamma_0|/\mathcal{R}_0 + \varepsilon$. If they are only finitely many, they contribute nothing to the asymptotic behavior (5.1.5). Assume that they are infinitely many. For n running through these indices, we then have

$$\limsup \frac{|a_{2n+1}|}{\mathcal{R}_1^2 - |b_{2n+1} + \Gamma_1|^2} \leq \limsup \frac{\dfrac{\mathcal{R}_1}{\mathcal{R}_0}(\mathcal{R}_0^2 - |\Gamma_0|^2) - \mu}{\mathcal{R}_1^2 \left(1 - \dfrac{|b_{2n+1} + \Gamma_1|^2}{\mathcal{R}_1^2}\right)}$$

$$\leq \frac{\dfrac{\mathcal{R}_1}{\mathcal{R}_0}\left(\mathcal{R}_0^2 - |\Gamma_0|^2 - \dfrac{\mathcal{R}_0}{\mathcal{R}_1}\mu\right)}{\mathcal{R}_1^2 \left(1 - \left(\dfrac{|\Gamma_0|}{\mathcal{R}_0} + \varepsilon\right)^2\right)} = \frac{\mathcal{R}_0}{\mathcal{R}_1} \cdot \frac{\mathcal{R}_0^2 - |\Gamma_0|^2 - 3\mathcal{R}_0^2\varepsilon}{\mathcal{R}_0^2 - |\Gamma_0|^2 - 2\mathcal{R}_0|\Gamma_0|\varepsilon - \mathcal{R}_0^2\varepsilon^2} < \frac{\mathcal{R}_0}{\mathcal{R}_1}.$$

(Keep in mind that $0 \in V_0$; i.e. $|\Gamma_0| < \mathcal{R}_0$.) For the remaining indices we have $|b_{2n+1} + \Gamma_1|/\mathcal{R}_1 > |\Gamma_0|/\mathcal{R}_0 + \varepsilon$. From (5.2.2) we find that for these indices, $|a_{2n+1}| \leq (|\Gamma_0| + \mathcal{R}_0)(\mathcal{R}_1 - |b_{2n+1} + \Gamma_1|)$, and thus

$$\frac{|a_{2n+1}|}{\mathcal{R}_1^2 - |b_{2n+1} + \Gamma_1|^2} \leq \frac{|\Gamma_0| + \mathcal{R}_0}{\mathcal{R}_1 + |b_{2n+1} + \Gamma_1|} < \frac{|\Gamma_0| + \mathcal{R}_0}{\mathcal{R}_1 \left(1 + \dfrac{|\Gamma_0|}{\mathcal{R}_0} + \varepsilon\right)} < \frac{\mathcal{R}_0}{\mathcal{R}_1}.$$

∎

PROOF OF THEOREM 5.1.6: By the same kind of arguments as used in the proof of Theorem 5.1.4, we find that either $\limsup r_{2n} < 1$ or $\liminf \tilde{r}_{2n+1} > 1$ when (5.1.6) holds. Hence, either $\limsup \sigma_{2n} < 1$ or $\liminf \tilde{\sigma}_{2n} > 1$. Let $w_n := -b_{2n+1}$ for all n. Then $\{w_n\}$ is bounded away from ∂V_1 and $s_{2n} \circ s_{2n+1}(w_n) = 0 \notin \partial V_1$. Hence, $S_{2n+1}(w) \to f \in \overline{V_0}$ for all $w \in V_1$ by Lemma A2.1.1.

If the limit point case holds, the general convergence follows trivially. Assume that the limit circle case holds. Then $S_{2n+1}(\infty) \to f$ and $S_{2n+1}(w_n) \to f$. Since $S_{2n+1}(\infty) = S_{2n}(0)$ and $S_{2n+1}(-b_{2n+1}) = S_{2n}(\infty) = S_{2n-1}(0)$, $K(a_n/b_n)$ converges even in the classical sense. ∎

PROOF OF THEOREM 5.1.7: As above, we find that either $\limsup r_{2n} < 1$ or $\liminf \tilde{r}_{2n+1} > 1$ by (5.1.6), and thus, either $\limsup \sigma_{2n} < 1$ or $\liminf \tilde{\sigma}_{2n} > 1$. Let

$$w_n := \begin{cases} -b_{2n+1} & \text{if } |b_{2n+1} + \Gamma_1| < |b_{2n+1} + \Gamma_1 + a_{2n+1}/(b_{2n} - a_{2n}/\Gamma_1)|, \\ -b_{2n+1} + a_{2n+1}/(-b_{2n} + a_{2n}/\Gamma_1) & \text{otherwise} \end{cases}$$

for all n. Then (5.1.7) shows that $\{w_n\}$ is bounded away from ∂V_1. Since $s_{2n} \circ s_{2n+1}(w_n)$ is either $0 \notin \partial V_1$ or $\Gamma_1 \notin \partial V_1$, the convergence of $S_{2n+1}(w)$ follows from Lemma A2.1.1.

If the limit point case holds, then the general convergence follows trivially. In the limit circle case, we have $S_{2n+1}(\infty) \to f$, and thus, $S_{2n}(0) \to f$ and $S_{2n+2}(-b_{2n+2}) \to f$. ∎

PROOF OF THEOREM 5.1.8: As before, we have either $\limsup \sigma_{2n} < 1$ or $\liminf \tilde{\sigma}_{2n} > 1$. The choice $w_n := -b_{2n+1} + a_{2n+1}/(-b_{2n} + a_{2n}/\Gamma_1)$ gives the convergence of $S_{2n+1}(w)$. The general convergence follows as in the previous proof. ∎

PROOF OF THEOREM 5.3.1:

A. This follows from Lemma 2.3.2A.

B. Since the radius ρ_{2n+1} of $s_{2n+1}(V_1)$, as given by (5.4.7) with $a_{2n+1} := 1$, is $\leq \mathcal{R}_0$, we have $|b_{2n+1} + \Gamma_1|^2 \leq \mathcal{R}_1^2 - \mathcal{R}_1/\mathcal{R}_0$, and thus $\limsup |b_{2n+1} + \Gamma_1| < \mathcal{R}_1$. Hence, the result follows from Theorem 5.1.1.

C. This is a simple consequence of Part A and Theorem 5.1.4.

D, E. This follows from Theorem 5.1.6, since $\limsup |b_{2n+1} + \Gamma_1| < \mathcal{R}_1$ always holds.

F. This follows from Theorem 5.1.4. ∎

PROOF OF THEOREM 5.3.2: In view of Lemma 2.3.2B, it suffices to prove general convergence if $0 \in V_0$ and $0 \in V_1 \cup [\hat{\mathbf{C}} \setminus (-1 - \overline{V_0})]$. Of course, $0 \in V_0$ always holds. So, if $0 \in V_1$ or $-1 \notin \overline{V_0}$, then general convergence implies classical convergence. If $0 \in \partial V_1$ or $-1 \in \partial V_0$, classical convergence is still a consequence of general convergence in the limit point case. Otherwise, if Lemma A2.1.1B holds, then $S_{2n}(0) \to f$ and $S_{2n}(\infty) = S_{2n-1}(0) \to f$; i.e. classical convergence.

A. This follows immediately from Theorem 5.1.1, since $|1 + \Gamma_1| < \mathcal{R}_1$ always holds.

B, C. This is a consequence of Theorem 5.1.3.

D. This follows from Theorem 5.1.6, since $|1 + \Gamma_1| < \mathcal{R}_1$, and, moreover, $|1 + \Gamma_0| > \mathcal{R}_0$ so that the first inequality in (5.1.6) holds trivially.

E. This is a consequence of Theorem 5.1.4.

F. This follows from Theorem 5.1.6.

G, H. This follows from Theorem 5.1.8. ∎

REFERENCES

1. L. Jacobsen, Some Periodic Sequences of Circular Convergence Regions, *Lecture Notes in Math., Springer – Verlag*, **932**: 87 – 98 (1982).
2. L. Jacobsen, General Convergence of Continued Fractions, *Trans. Amer. Math. Soc.*, **294**: 477 – 485 (1986).
3. L. Jacobsen and W. J. Thron, Oval Convergence Regions and Circular Limit Regions for Continued Fractions $K(a_n/1)$, *Lecture Notes in Math., Springer – Verlag*, **1199**: 90 – 126 (1986).
4. L. Jacobsen and W. J. Thron, Limiting Structures for Sequences of Linear Fractional Transformations, *Proc. Amer. Math. Soc.*, **99** No 1: 141 – 146 (1987).

5. W. B. Jones and W. J. Thron, Convergence of Continued Fractions, *Canad. J. Math.* **20**: 1037 – 1055 (1968).

6. W. B. Jones and W. J. Thron, Twin – Convergence Regions for Continued Fractions $K(a_n/1)$, *Trans. Amer. Math. Soc.* **150**: 93 – 119 (1970).

7. W. B. Jones and W. J. Thron, *Continued Fractions. Analytic Theory and Applications*, Addison–Wesley Publishing Company, Encyclopedia of Mathematics and its Applications (1980).

8. R. E. Lane, The Convergence and Values of Periodic Continued Fractions, *Bull. Amer. Math. Soc.* **51**: 246 – 250 (1945).

9. L. Lorentzen, Bestness of the Parabola Theorem for Continued Fractions, *Journ. of Comp. and Appl. Math.*, **40**: 297 – 304 (1992).

10. L. Lorentzen, A Convergence Property for Sequences of Linear Fractional Transformations. (This volume.)

11. L. Lorentzen and St. Ruscheweyh, Simple Convergence Sets for Continued Fractions $K(a_n/1)$, *Journ. of Math. Anal. and Appl.* (To appear.)

12. L. Lorentzen and H. Waadeland, *Continued Fractions with Applications*, North–Holland, Studies in Computational Mathematics, **3** (1992).

13. J. F. Paydon and H. S. Wall, The Continued Fraction as a Sequence of Linear Fractional Transformations, *Duke Math. J.*, **9**: 360 – 372 (1942).

14. W. T. Scott and H. S. Wall, Value Regions for Continued Fractions, *Bull. Amer. Math. Soc.*, **47**: 580 – 585 (1941).

15. W. J. Thron, Convergence of Sequences of Linear Fractional Transformations and of Continued Fractions, *J. Indian Math. Soc.*, **27**: 103 – 127 (1963).

14

A Szegö Quadrature Formula Arising from q-Starlike Functions

FRODE RØNNING, Trondheim College of Education, Rotvoll allé, N–7050 Charlottenlund, Norway

Dedicated to the memory of Arne Magnus

1 SOME BACKGROUND

In a similar way as quadrature formulas on a real interval are constructed using zeros of orthogonal polynomials as nodes, Jones et al. [3] showed how to construct a quadrature formula based on polynomials orthogonal on the unit circle (Szegö polynomials). The Szegö polynomials have their zeros inside the unit circle, but by combining the Szegö polynomials and the reciprocal Szegö polynomials in a certain way, they obtained polynomials with zeros on the unit circle. By using these zeros as nodes, and by choosing weights in a certain manner, which will be described later, they showed that the resulting quadrature formula would have a maximal domain of validity in the set of Laurent polynomials. This type of quadrature formula is referred to as a Szegö quadrature formula. Further, there is a close connection between Szegö polynomials and Carathéodory functions via the elements of a certain continued fraction (positive PC–fraction) as shown by Jones et al. in [2]. Now, given the connection between q–starlike functions and Carathéodory functions which will be established below, we are justified in talking about a quadrature formula arising from q–starlike functions.

DEFINITION. *PS_q is the set of functions $f(z) = z + a_2 z^2 + \ldots$, analytic in $|z| < 1$, with the property*

$$\left| \frac{z(D_q f)(z)}{f(z)} - \frac{1}{1-q} \right| \leq \frac{1}{1-q}, \quad |z| < 1, \tag{1.1}$$

where for $0 < q < 1$

$$(D_q f)(z) = \frac{f(z) - f(qz)}{z - qz}, \quad z \neq 0$$

$$(D_q f)(0) = f'(0).$$

PS_q is known as the class of q–starlike functions and was introduced by Ismail et al. in [1]. $D_q f$ is the q–difference operator, known from the theory of hypergeometric series. We note that $D_q f \to df/dz$ as $q \to 1$.

The function

$$G_q(z) = \frac{1-z}{1+qz} = 1 + \frac{1+q}{q} \sum_{n=1}^{\infty} (-q)^n z^n \tag{1.2}$$

is a Carathéodory function mapping the open unit disk onto the disk

$$\left| w - \frac{1}{1-q} \right| < \frac{1}{1-q}, \tag{1.3}$$

so from (1.1) we see that G_q is in a natural way connected to the class PS_q. The function $k_q(z)$, which is the solution of the difference equation

$$\frac{z(D_q k_q)(z)}{k_q(z)} = G_q(z), \tag{1.4}$$

is extremal in the class PS_q in the sense that the mapping G_q is onto the disk (1.3). Note that for $q \to 1$ the equation (1.4) will read

$$\frac{zk'(z)}{k(z)} = \frac{1-z}{1+z}$$

which means that the function $k_q(z)$ tends to the classical Koebe function $k(z) = z(1-z)^{-2}$ when $q \to 1$.

We briefly state some of the other results that we shall use. A positive PC–fraction is a continued fraction of the form

$$\delta_0 - \frac{2\delta_0}{1} + \frac{1}{\bar{\delta}_1 z} + \frac{(1-|\delta_1|^2)z}{\delta_1} + \frac{1}{\bar{\delta}_2 z} + \frac{(1-|\delta_2|^2)z}{\delta_2} + \cdots \tag{1.5}$$

where $\delta_0 > 0$ and $|\delta_n| < 1$, $n = 1, 2, \ldots$. There is a one-to-one correspondence between (1.5) and a Carathéodory function

$$F(z) = \mu_0 + 2 \sum_{n=1}^{\infty} \mu_n z^n, \quad \mu_0 > 0, \ |\mu_n| \le 1, \ n \ge 1$$

in the sense that the even approximants of (1.5) correspond to $F(z)$ in the open unit disk. Let $\{Q_{2n+1}(z)\}_{n=1}^{\infty}$ and $\{Q_{2n}(z)\}_{n=1}^{\infty}$ be the sequences of odd and even denominators of

(1.5). These sequences of polynomials are orthogonal on the unit circle, and they are known as the Szegö polynomials and reciprocal Szegö polynomials, respectively. We use the notation $\rho_n(z) = Q_{2n+1}(z)$ and $\rho_n^*(z) = Q_{2n}(z)$. All this is established in [**2**]. The coefficients μ_n of the Carathéodory function are often referred to as the moments and the δ_n's in the PC–fraction as the reflection coefficients.

2 THE PC–FRACTION AND SZEGÖ POLYNOMIALS FOR $G_q(z)$

It is not difficult to compute the reflection coefficients in the PC–fraction expansion for $G_q(z)$ directly. We shall, however, rely on a result from earlier work, because this will give us the opportunity to present and use a very simple, but general observation. We state this as a lemma.

LEMMA. Let $F(z) = \mu_0 + 2\sum_{n=1}^{\infty} \mu_n z^n$ ($|\mu_n| \leq 1$, $n \geq 1$) be a Carathéodory function with the positive PC–fraction expansion $PPC\{\gamma_n\}$. Then, $1/F(z)$ has the expansion $PPC\{\delta_n\}$ with $\delta_0 = 1/\gamma_0$ and $\delta_n = -\gamma_n$, $n = 1, 2, \ldots$.

Proof: There is an algorithm for computing the reflection coefficients when the Carathéodory function is given. (See [**4**, p.137].) With $F_0 := F$ and $\gamma_0 := F_0(0) = \mu_0$, we define

$$F_1 := \frac{\gamma_0 - F_0}{\gamma_0 + F_0}, \quad \overline{\gamma}_1 := F_1'(0),$$

and recursively

$$F_{n+1} := \frac{\overline{\gamma}_n z - F_n}{\gamma_n F_n - z}, \quad \overline{\gamma}_n := F_n'(0).$$

With $G_0 := 1/F_0$, $\delta_0 := G_0(0) = 1/\gamma_0$, we get

$$G_1 := \frac{\delta_0 - G_0}{\delta_0 + G_0} = \frac{F_0 - \gamma_0}{F_0 + \gamma_0} = -F_1$$

and $\overline{\delta}_1 := G_1'(0) = -F_1'(0) = -\overline{\gamma}_1$. Assuming $G_n = -F_n$ (i.e. $\delta_n = -\gamma_n$) we get

$$G_{n+1} := \frac{\overline{\delta}_n z - G_n}{\delta_n G_n - z} = -\frac{\overline{\gamma}_n z - F_n}{\gamma_n F_n - z} = -F_{n+1},$$

and hence $\delta_{n+1} = -\gamma_{n+1}$. ∎

For

$$P_q(z) = \frac{1}{G_q(z)} = \frac{1 + qz}{1 - z} = 1 + (1 + q)\sum_{n=1}^{\infty} z^n$$

we proved in [**5**] that in the corresponding PC–fraction $PPC\{\gamma_n\}$ we have

$$\gamma_0 = 1, \quad \gamma_n = -\frac{1}{n + \frac{1-q}{1+q}}, \quad n = 1, 2, \ldots$$

Using the lemma, it is clear that

$$G_q(z) = \frac{1-z}{1+qz} = 1 + \frac{1+q}{q} \sum_{n=1}^{\infty} (-1)^n q^n z^n$$

has the PPC–fraction expansion

$$\delta_0 = 1, \quad \delta_n = \frac{1}{n + \frac{1-q}{1+q}}, \quad n = 1, 2, \dots. \tag{2.1}$$

In what follows we shall denote by $\{\delta_n\}_{n=0}^{\infty}$ the reflection coefficients of $G_q(z)$ as given by (2.1). In [5] we computed the Szegö polynomials corresponding to $P_q(z)$ and found them to be

$$\rho_n(z) = z^n - \delta_n(z^{n-1} + \cdots + z + 1), \quad n = 1, 2, \dots.$$

Although there is a very simple relationship between the reflection coefficients of P_q and $G_q = 1/P_q$, the structure of the PC–fraction implies that there will not be an equally simple relationship between the Szegö polynomials arising from the two Carathéodory functions. Following the procedure in [2] we find the Szegö polynomials corresponding to $G_q(z)$ to be

$$\rho_n(z) = z^n + \delta_n[(n + (n-1)q)z^{n-1} + (n-1 + (n-2)q)z^{n-2} + \cdots + (2+q)z + 1], \tag{2.2}$$

$n = 1, 2, \dots.$

3   THE QUADRATURE FORMULA

It is known from the general theory that all the zeros of $\rho_n(z)$ are inside the open unit disk. To construct a quadrature formula on the unit circle, we want polynomials with zeros on the boundary, $|z| = 1$. This is achieved by introducing the *paraorthogonal polynomials* (see [3])

$$\mathcal{B}_n(z, w_n) = \rho_n(z) + w_n \rho_n^*(z), \quad |w_n| = 1. \tag{3.1}$$

These polynomials satisfy the orthogonality relations

$$\langle \mathcal{B}_n, 1 \rangle \neq 0, \quad \langle \mathcal{B}_n, z^m \rangle = 0, \quad 1 \leq m \leq n-1, \quad \langle \mathcal{B}_n, z^n \rangle \neq 0$$

where

$$\langle X, Y \rangle := \mu(X(z)\overline{Y(1/z)}),$$

and

$$\mu\left(\sum_{m=p}^{q} c_m z^m\right) = \sum_{m=p}^{q} c_m \mu_{-m}, \tag{3.2}$$

$c_m \in \mathbf{C}$, $-\infty < p \le q < \infty$, is the linear functional corresponding to the moment sequence $\{\mu_n\}_{-\infty}^{\infty}$. Finally, we introduce the fundamental polynomials

$$L_{n,m}(z, w_n) = \frac{\mathcal{B}_n(z, w_n)}{(z - \zeta_m^{(n)})\mathcal{B}_n'(\zeta_m^{(n)}, w_n)}$$

where $\zeta_m^{(n)}$ are the zeros of $\mathcal{B}_n(z, w_n)$. With

$$\lambda_m^{(n)} = \mu(L_{n,m}(z, w_n)), \tag{3.3}$$

μ as in (3.2) and $\zeta_m^{(n)}$ as above it was shown in [3] that

$$I_\Psi(f) = \int_{-\pi}^{\pi} f(e^{i\theta}) \, d\Psi(\theta) = \sum_{m=1}^{n} \lambda_m^{(n)} f(\zeta_m^{(n)}) \tag{3.4}$$

will be a quadrature formula valid for all $f \in \Lambda_{-(n-1),n-1}$, and that this is a maximal domain of validity of the form $\Lambda_{p,q}$ for a formula of this type. $\Psi(\theta)$ is the distribution function corresponding to the moment sequence $\{\mu_n\}_{-\infty}^{\infty}$. Choosing $\rho_n(z)$ as in (2.2) and $w_n = 1$ in (3.1) we get paraorthogonal polynomials with a particularly simple form,

$$\mathcal{B}_n(z) = (1 + \delta_n)(z^n + (1 + q)(z^{n-1} + \cdots + z) + 1), \tag{3.5}$$

where the δ_n's are as in (2.1).

We see that the zeros of $\mathcal{B}_n(z)$ are found by solving the equation

$$z^n = \frac{1 + qz}{z + q}, \tag{3.6}$$

omitting the false solution $z = 1$. What is known in general about polynomials like (3.1) is that they have n simple zeros lying on the unit circle [3]. For the polynomials in (3.5) we can prove an interlacing property. A similar situation was studied in [6] where also an interlacing property was proved for the zeros of the polynomials in question, but this is not known to be true in general.

THEOREM. Let $\mathcal{B}_n(z)$ be as in (3.5), and denote the zeros of $\mathcal{B}_n$ by $\zeta_m^{(n)} = e^{i\theta_m^{(n)}}$. Then we have

$$\frac{2\pi m}{n + 1} < \theta_m^{(n)} < \frac{2\pi m}{n} \tag{3.7}$$

which in particular gives

$$\theta_m^{(n+1)} < \theta_m^{(n)} < \theta_{m+1}^{(n+1)}.$$

Proof: We have mentioned before that the polynomials (3.1) have n simple zeros on the unit circle. We see that $z = -1$ is a zero of $\mathcal{B}_n$ for n odd, but not for n even. This means that we have $[\frac{n}{2}]$ nonreal zeros in the upper half plane. ([ ] denotes the integer part.) Denote by $\Gamma_m^{(n)}$ the part of the unit circle from $\theta = (2m - 1)\pi/n$ to $\theta = 2m\pi/n$,

$m = 1, 2, \ldots, [\frac{n}{2}]$, and denote by ∂U^+ and ∂U^- the halves of the unit circle in the upper and lower half plane, respectively. Then the mapping $z \mapsto z^n$ will map $\Gamma_m^{(n)}$ onto ∂U^- and the mapping $z \mapsto \frac{1+qz}{z+q}$ will map ∂U^+ onto ∂U^-. Hence, because of (3.6) and the fact that we know the number of zeros, we conclude that each arc $\Gamma_m^{(n)}$ must contain exactly one zero of $\mathcal{B}_n(z)$, $\zeta_m^{(n)}$. To prove (3.7) we proceed to show that the zero $\zeta_m^{(n)} \in \Gamma_m^{(n)}$ cannot be on $\Gamma_m^{(n+1)}$. Since $z \mapsto z^{n+1}$ maps $\Gamma_m^{(n+1)}$ onto ∂U^-, this will be established if we can show that $\mathrm{Im}\,(\zeta_m^{(n)})^{n+1} > 0$.

Let $\zeta_m^{(n)} = e^{i\theta}$. Taking real and imaginary parts on each side of (3.6) we get the equations

$$\cos n\theta = \frac{(1+q^2)\cos\theta + 2q}{1 + q^2 + 2q\cos\theta}$$

$$\sin n\theta = -\frac{(1-q^2)\sin\theta}{1 + q^2 + 2q\cos\theta}.$$

From this we get

$$\mathrm{Im}\,(\zeta_m^{(n)})^{n+1} = \sin(n+1)\theta$$

$$= -\frac{(1-q^2)\sin\theta\cos\theta}{1 + q^2 + 2q\cos\theta} + \frac{((1+q^2)\cos\theta + 2q)\sin\theta}{1 + q^2 + 2q\cos\theta}$$

$$= \frac{2q\sin\theta(1 + q\cos\theta)}{1 + q^2 + 2q\cos\theta} > 0.$$

For $m = 1, 2, \ldots, [\frac{n}{2}]$ we easily see that

$$\frac{2\pi m}{n} < \frac{2\pi(m+1)}{n+2}$$

and therefore

$$\theta_m^{(n+1)} < \frac{2\pi m}{n+1} < \theta_m^{(n)} < \frac{2\pi m}{n} < \theta_{m+1}^{(n+1)}.$$

This ends the proof of the theorem. ∎

Computing the weights from (3.3) we see that we get

$$\lambda_m^{(n)} = \mu(L_{n,m}(z)) = \sum_{k=0}^{n-1} c_{m,k}\mu_{-k}$$

$$= c_{m,0} + \frac{1+q}{2q}\sum_{k=1}^{n-1} c_{m,k}(-q)^k$$

with $\mu_{-k} = \mu_k$ given by (1.2). This can be written as

$$\lambda_m^{(n)} = L_{n,m}(0) + \frac{1+q}{2q}(L_{n,m}(-q) - L_{n,m}(0))$$

$$= \frac{1+q}{2q}L_{n,m}(-q) - \frac{1-q}{2q}L_{n,m}(0). \tag{3.8}$$

To compute the values of $L_{n,m}(z)$ in $z = -q$ and $z = 0$ we write (3.5) as

$$\frac{1}{1 + \delta_n}(1 - z)\mathcal{B}_n(z) = 1 + qz - qz^n - z^{n+1}. \tag{3.9}$$

Differentiating (3.9) with respect to z and substituting $z = \zeta_m^{(n)}$, we get

$$\frac{1}{1 + \delta_n}(1 - \zeta_m^{(n)})\mathcal{B}'_n(\zeta_m^{(n)}) = q - nq(\zeta_m^{(n)})^{n+1} - (n + 1)(\zeta_m^{(n)})^n. \tag{3.10}$$

By repeated application of the equation

$$q - q(\zeta_m^{(n)})^{n-1} - (\zeta_m^{(n)})^n = -\frac{1}{\zeta_m^{(n)}} = -\overline{\zeta_m^{(n)}}, \tag{3.11}$$

and the equation (3.6), (3.10) is seen to be equal to

$$-n\overline{\zeta_m^{(n)}} - (n - 1)q - \frac{1 + q\,\zeta_m^{(n)}}{\zeta_m^{(n)} + q}.$$

The equation (3.11) follows by substituting $z = \zeta_m^{(n)}$ in (3.9).
 Hence,

$$
\begin{aligned}
L_{n,m}(-q) &= \frac{-\mathcal{B}_n(-q)}{(q + \zeta_m^{(n)})\mathcal{B}'_n(\zeta_m^{(n)})} \\
&= \frac{(1 - q^2)(1 - \zeta_m^{(n)})}{(1 + q)[n\overline{\zeta_m^{(n)}}(\zeta_m^{(n)} + q) - (n - 1)q(\zeta_m^{(n)} + q) - (1 + q\,\zeta_m^{(n)})]} \\
&= \frac{(1 - q^2)(1 - \zeta_m^{(n)})}{(1 + q)[1 - q^2 + n(1 + q^2) + 2nq\,\mathrm{Re}\;\zeta_m^{(n)}]}
\end{aligned}
$$

and

$$L_{n,m}(0) = \frac{-\mathcal{B}_n(0)}{\zeta_m^{(n)}\mathcal{B}'_n(\zeta_m^{(n)})} = \frac{(1 - \zeta_m^{(n)})(\zeta_m^{(n)} + q)}{\zeta_m^{(n)}[1 - q^2 + n(1 + q^2) + 2nq\,\mathrm{Re}\;\zeta_m^{(n)}]}.$$

Putting all this into (3.8) we get

$$\lambda_m^{(n)} = \frac{(1 - q)(1 - \cos\theta_m^{(n)})}{1 - q^2 + n(1 + q^2) + 2nq\cos\theta_m^{(n)}}. \tag{3.12}$$

What remains in order to write down the quadrature formula (3.4) explicitly is an expression for $d\Psi(\theta)$. Since the Carathéodory function (1.2) is holomorphic in $|z| \leq 1$, it is known that the corresponding distribution function Ψ is absolutely continuous, and therefore

$$d\Psi(\theta) = \Psi'(\theta)d\theta = \frac{1}{2\pi}\,\mathrm{Re}\;G_q(e^{i\theta})d\theta$$

which is seen to be equal to

$$d\Psi(\theta) = \frac{1}{2\pi} \frac{(1-q)(1-\cos\theta)}{1+q^2+2q\cos\theta} d\theta. \tag{3.13}$$

Summing up, we now have a quadrature formula

$$\int_{-\pi}^{\pi} f(e^{i\theta}) \, d\Psi(\theta) = \sum_{m=1}^{n} \lambda_m^{(n)} f(\zeta_m^{(n)})$$

valid for all $f \in \Lambda_{-(n-1),n-1}$, and where $d\Psi(\theta)$ is given by (3.13), the $\zeta_m^{(n)}$'s are the solutions of (3.6) and the $\lambda_m^{(n)}$'s are given by (3.12). For a similar construction of a quadrature formula, we refer also to [6].

REFERENCES

1. M.E.H. Ismail, E. Merkes and D. Styer, *A Generalization of Starlike Functions*, Complex Variables **14** (1990), 77–84.
2. W.B. Jones, O. Njåstad and W.J. Thron, *Continued Fractions Associated with Trigometric and Other Strong Moment Problems*, Constr. Appr. **2** (1986), 197–211.
3. _____, *Moment Theory, Orthogonal Polynomials, Quadrature and Continued Fractions Associated with the Unit Circle*, Bull. London Math. Soc. **21** (1989), 113–152.
4. _____, *Schur Fractions, Perron–Carathéodory Fractions and Szegö Polynomials, a Survey*, Analytic Theory of Continued Fractions II (ed. W.J. Thron), Lecture Notes in Math. 1199, Springer (1986), 127–158.
5. F. Rønning, *PC-fractions and Szegö polynomials associated with starlike univalent functions*, Numerical Algorithms **3** (1992), 383–392.
6. H. Waadeland, *A Szegö quadrature formula for the Poisson formula*, C. Brezinski and U. Kulish (Eds.), Elsevier Sci. Publ. B.V., Comp. and Appl. Math. I (1992), 479–486.

15

Truncation Error for L.F.T. Algorithms $\{T_n(w)\}$

W. J. THRON   Department of Mathematics, University of Colorado, Boulder, CO
80309–0395, U.S.A.

Dedicated to the memory of Arne Magnus

1   INTRODUCTION

We are here concerned with convergence and truncation errors of what we shall
call, limit periodic *L.F.T. algorithms*. Let a sequence of linear fractional transformations (ℓ.f.t.)

$$t_n(w) := \frac{a_n w + b_n}{c_n w + d_n}, \tag{1.1}$$

where a_n, b_n, c_n, d_n, $n \geq 0$ are complex numbers satisfying

$$a_n d_n - b_n c_n \neq 0, \quad b_n \neq 0, \quad n \geq 0. \tag{1.2}$$

be given.

For $t_n^{-1}(n)$ we then have

$$t_n^{-1}(n) = \frac{d_n u - b_n}{-c_n u + a_n}, \qquad n \geq 0. \tag{1.3}$$

Condition (1.2) insures that the t_n are nonsingular and that

$$t_n(0) \neq 0, \quad t_n^{-1}(0) \neq 0, \quad n \geq 0.$$

We next define, inductively, for $n \geq 0$, $m \geq 0$

$$T_n^{(n)}(w) := t_n(w), \quad T_{n+m}^{(n)}(w) := T_{n+m-1}^{(n)}(t_{n+m}(w)), \quad m \geq 1. \tag{1.4}$$

Further, we set

$$T_n(w) := T_n^{(0}(w), \qquad n \geq 0. \tag{1.5}$$

The mapping

$$\langle \{a_n\}, \{b_n\}, \{c_n\}, \{d_n\} \rangle \rightarrow \{T_n(w)\}$$

will be called an *L.F.T.-algorithm*. The algorithm will be called *limit periodic* if the sequences $\{a_n\}, \{b_n\}, \{c_n\}, \{d_n\}$ all have finite limits. We set

$$\lim_{n \to \infty} a_n =: a, \quad \lim_{n \to \infty} b_n =: b, \quad \lim_{n \to \infty} c_n =: c, \quad \lim_{n \to \infty} d_n = d, \tag{1.6}$$

and write

$$\alpha_n := a_n - a, \quad \beta_n := b_n - b, \quad \gamma_n := c_n - c, \quad \delta_n := d_n - d. \tag{1.7}$$

We assume that $\{a_n\}, \{b_n\}, \{c_n\}, \{d_n\}$ have been so chosen that $t(w)$ defined by

$$t(w) := \frac{aw + b}{cw + d} \tag{1.8}$$

is not the identity mapping (so that it has at most two distinct fixed points) and that $cd \neq 0$ (so that $t(w)$ is well defined and not identically equal to ∞). The mapping may however be singular.

The fixed points of $t(w)$ or equivalently the zeros of

$$t(w) = w$$

shall be denoted by w_1 and w_2, provided $t(w)$ is nonsingular. Hence

$$w_i(cw_i + d) = aw_i + b, \qquad i = 1, 2. \tag{1.9}$$

and, since $w_i = t^{-1}(w_i)$ also holds,

$$w_i(a - cw_i) = dw_i - b. \tag{1.10}$$

Further, since w_1, w_2 are the solutions of

$$cw_i^2 + (d - a)w_i - b = 0, \tag{1.11}$$

we also have

$$\begin{align} \text{(a)} \quad & w_1 w_2 = -b/c \\ \text{(b)} \quad & w_1 + w_2 = (a - d)/c. \end{align} \tag{1.12}$$

If $t(w)$ is singular, that is if $ad = bc$, one has

$$t(w) = \frac{b(cw + d)}{d(cw + d)} = \frac{b}{d}, \qquad w \neq d/c.$$

The fixed point b/d of this transformation will be called w_2 and the point $-d/c$ for which $t(w)$ is not defined will be denoted by w_1. With these definitions the formulas (1.10), (1.11) and (1.12) are valid in this case also.

We now restrict ourselves to the case where $w_1 \neq w_2$ that is we exclude the parabolic mappings. The condition for this is

$$(d-a)^2 + 4bc \neq 0.$$

Also in the singular case this condition can be shown to insure $w_1 \neq w_2$. With this restriction we now analyze $(t(w) - w_2)/(t(w) - w_1)$. We have, using (1.12),

$$
\begin{aligned}
\frac{t(w) - w_2}{t(w) - w_1} &= \frac{(1 - dw_2)w + b - dw_2}{(a - cw_1)w + b - dw_1} \\
&= \left(\frac{(d + c(w_1 + w_2) - cw_1)w - (cw_1 + d)w_2}{d(w_1 + w_2) - cw_2)w - (cw_2 + d)w_1} \right) \\
&= \left(\frac{cw_1 + d}{cw_2 + d} \right) \left(\frac{w - w_2}{w - w_1} \right).
\end{aligned}
$$

Set

$$r := \frac{cw_1 + d}{cw_2 + d}. \tag{1.13}$$

If we exclude the elliptic case for $t(w)$, that is if $(bd - ad)/(a + d)^2$ is not on the real axis from $-\infty$ to $-\frac{1}{4}$, so that $|r| = 1$ is excluded, then the subscripts on w_1 and w_2 can be so chosen that

$$|r| = \left| \frac{cw_1 + d}{cw_2 + d} \right| < 1. \tag{1.14}$$

Set

$$\lambda(w) := \frac{w - w_2}{w - w_1}, \qquad \tau(w) := rw.$$

We have shown above that

$$t(w) = \lambda^{-1} \circ \tau \circ \lambda(w)$$

and hence

$$(t(w))^n = \lambda^{-1} \circ \tau^n \circ \lambda(w), \tag{1.15}$$

from which it follows that $(t(w))^n \to w_2$ for all $w \neq w_1$. Thus *pure* periodic L.F.T. algorithms converge to w_2, except for $w = w_1$. This motivates calling w_2 the *attractive fixed point* of $t(w)$. For $(t(0))^n - w_2$ one easily computes

$$T_n(0) - \lim_{n \to \infty} T_n(0) = (t(0))^n - w_2 = \frac{w_2(w_2 - w_1)r^n}{w_1 - w_2 r^n} \sim \frac{w_2}{w_1}(w_2 - w_1)r^n.$$

While it is not true in general that limit periodic L.F.T. algorithms converge to w_2 we shall show that in that case also, subject to the conditions imposed above,

$$\left| T_n(0) - \lim_{n \to \infty} T_n(0) \right| < K(r')|r'|^n, \qquad 0 < |r| < |r'| < 1.$$

If, as sometimes may be the case, we want to study $\{T_n(w_0)\}$, $w_0 \neq 0$, then the condition $b_n \neq 0$, which insures $t_n(0) \neq 0$ and $t_n^{-1}(0) \neq 0$, may have to be modified.

We note that continued fraction algorithms, that is algorithms where

$$t_n(w) = \frac{b_n}{w + d_n}, \qquad b_n \neq 0,$$

satisfy our requirements since

$$a_n d_n - b_n c_n \neq 0, \quad b_n \neq 0 \quad \text{and} \quad a_n = 0 \Rightarrow c_n \neq 0.$$

For $t(w)$ the requirements are that the mapping not be the identity nor be parabolic (one fixed point) nor elliptic ($|r| = 1$) and that it have no infinite fixed point ($c \neq 0$). The additional assumption $d \neq 0$ insures that in the nonsingular case $t(w) \not\equiv \infty$ and thus in particular also $w_2 \neq \infty$.

The pattern of the proofs used in this article is present in very rudimentary form in [9], though at that time we did not see how the results obtained there could be extended even to more general continued fractions than $K(a_n/1)$, $\lim a_n = a \neq 0$. The pattern becomes clearer in [6] and in [7] so that the author felt encouraged to attempt the general case even though it involved three complex number sequences, while in the cases successfully treated up to then only one such sequence $\{a_n\}$ or $\{\gamma_n\}$, respectively, occurred.

In addition, even though convergence of general $\{T_n\}$ was studied by Magnus and Mandell [3] and Gill [2], we present here another approach to establishing convergence of limit periodic L.F.T. algorithms with few restrictions.

Our result extends not only to continued fractions and Schur algorithms but also to subsequences of continued fraction approximants and thus in particular, also to k-limit periodic continued fractions, $k > 1$.

2   CONSEQUENCES OF THE INVARIANCE OF THE CROSSRATIO UNDER L.F.T.

Let $S(w)$ be an L.F.T. Then it is well known that

$$\left(\frac{S(u) - S(v)}{S(u) - S(w)}\right)\left(\frac{S(w) - S(z)}{S(v) - S(z)}\right) = \left(\frac{u - v}{u - w}\right)\left(\frac{w - z}{v - z}\right), \tag{2.1}$$

that is the *crossratio* of the four points u, v, w, z is *invariant* under any L.F.T. In [5] the author showed that many identities for continued fractions can be derived using this basic relationship. Clearly, the same must be true for L.F.T. algorithms.

We shall now establish a very general truncation error formula which is particularly useful for limit periodic algorithms. In addition we shall present a formula, which is an extension of a formula first given in [8], comparing the behavior of $\{T_n(0)\}$ to that of $\{T_n(w_2)\}$ for limit periodic L.F.T. algorithms.

In (2.1) set $S = T_n$, $u = u_n$, $v = v_n$, $w = w_n$, $z = g_n$, where

$$g_n := T_n^{-1}(\infty). \tag{2.2}$$

This leads to

$$\frac{T_n(u_n) - T_n(v_n)}{T_n(u_n) - T_n(w_n)} = \left(\frac{u_n - v_n}{u_n - w_n}\right)\left(\frac{w_n - g_n}{v_n - g_n}\right). \tag{2.3}$$

Now choose $u_n = 0$, $v_n = T_{n+m}^{(n+1)}(0)$, $w_n = t_n^{-1}(0) = -b_n/a_n$. This substitution is valid if $a_n = 0$ $(t_n^{-1}(0) = \infty)$ but requires $b_n \neq 0$ (which we are assuming) so that $u_n \neq w_n$. Then

$$\frac{T_n(0) - T_{n+m}(0)}{T_n(0) - T_{n-1}(0)} = \frac{T_{n+m}^{(n+1)}(0)}{-b_n}\left(\frac{-b_n - g_n a_n}{T_{n+m}^{(n+1)}(0) - g_n}\right). \tag{2.4}$$

From (2.4) with $m = 1$, $n = k$ one obtains

$$\frac{T_{k+1}(0) - T_k(0)}{T_k(0) - T_{k-1}(0)} = \frac{b_{k+1}}{b_k}\left(\frac{b_k + a_k g_k}{b_{k+1} - d_{k+1} g_k}\right). \tag{2.5}$$

since $T_{k+1}^{(k+1)}(0) = b_{k+1}/d_{k+1}$. Note that $b_k \neq 0 \Rightarrow t_k(0) \neq 0 \Rightarrow T_k(0) - T_{k-1}(0) \neq 0$, $k \geq 1$. Combining the results above we arrive at

$$T_{n+m}(0) - T_n(0)$$
$$= \frac{-T_{n+m}^{(n+1)}(0)(b_n + a_n g_n)}{b_n(T_{n+m}^{(n+1)}(0) - g_n)}\left(\prod_{k=1}^{n-1}\frac{T_{k+1}(0) - T_k(0)}{T_k(0) - T_{k-1}(0)}\right)(T_1(0) - T_0(0))$$
$$= \frac{-T_{n+m}^{(n+1)}(0)(b_{n+1} - d_{n+1} g_{n+1})(d_0 b_1 - b_0 d_1)}{(T_{n+m}^{(n+1)}(0) - g_n) d_0 d_1}\prod_{k=1}^{n}\left(\frac{b_k + a_k g_k}{b_{k+1} - d_{k+1} g_k}\right).$$

This formula is valid for all L.F.T. algorithm but becomes of special interest for limit periodic algorithm, for in this case, as we shall show in the next section,

$$g_k \to w_1.$$

With this assumption

$$\frac{b_k + a_k k g_k}{b_{k+1} - d_{k+1} g_k} \to \frac{b + a w_1}{b - d w_1} = \frac{-c w_1 w_2 + w_1(c(w_1 + w_2) + d)}{-c w_1 w_2 - d w_1}$$
$$= -\left(\frac{c w_1 + d}{c w_2 + d}\right) = -r.$$

Set $|r| < |r'| < 1$, then

$$|T_{n+m}(0) - T_n(0)| < \left|\frac{T_{n+m}^{(n+1)}(0)}{T_{n+m}^{(n+1)}(0) - g_n}\right| K(r')|r'|^n. \tag{2.7}$$

Restricting ourselves to the limit periodic case so that w_2 is well defined and assuming that

$$T^{(n)} := \lim_{m \to \infty} T_{n+m}^{(n)}(0), \quad n \geq 0, \quad T := T^{(0)}, \tag{2.8}$$

exist, we return to the formula (2.3) and choose $u_n = T^{(n)}$, $w_n = 0$, $v_n = w_2 \neq w_1$. (If not all of u_n, v_n, w_n, z_n are distinct, (2.3) still holds but becomes trivial.) Then

$$\frac{T - T_n(w_2)}{T - T_n(0)} = \frac{T^{(n)} - w_2}{T^{(n)}} \frac{g_n}{g_n - w_2}. \tag{2.9}$$

3 CONVERGENCE OF $\{T_n^{-1}(\infty)\}$ TO w_1

Since $T_{n+m}^{(n)}(w)$, defined in (1.4), is always an L.F.T. it can be written as

$$T_{n+m}^{(n)}(w) = t_n \circ \cdots \circ t_{n+m}(w) =: \frac{A_N^{(n)} w + B_N^{(n)}}{C_N^{(n)} w + D_N^{(n)}}$$

$$= \frac{A_{N-1}^{(n)} \frac{a_N w + b_N}{c_N w + d_N} + B_{N-1}}{C_{N-1}^{(n)} \frac{a_N w + b_N}{c_N w + d_N} + D_{N-1}^{(n)}}. \tag{3.1}$$

Here we have set

$$N := n + m. \tag{3.2}$$

Hence, with suitable normalization given by

$$A_n^{(n)} := a_n \quad B_n^{(n)} := b_n, \quad C_n^{(n)} := c_n, \quad D_n^{(n)} := d_n, \tag{3.3}$$

we have

$$\begin{pmatrix} A_N^{(n)} \\ C_N^{(n)} \end{pmatrix} = a_n \begin{pmatrix} A_{N-1}^{(n)} \\ C_{N-1}^{(n)} \end{pmatrix} + c_N \begin{pmatrix} B_{N-1}^{(n)} \\ D_{N-1}^{(n)} \end{pmatrix},$$

$$\begin{pmatrix} B_N^{(n)} \\ D_N^{(n)} \end{pmatrix} = b_N \begin{pmatrix} A_{N-1}^{(n)} \\ C_{N-1}^{(n)} \end{pmatrix} + d_N \begin{pmatrix} B_{N-1}^{(n)} \\ D_{N-1}^{(n)} \end{pmatrix}, \tag{3.4}$$

for $N > n \geq 0$.

It follows that

$$w_i A_N^{(n)} + B_N^{(n)} = (a_N w_i + b_N) A_{N-1}^{(n)} + (c_N w_i + d_N) B_{N-1}^{(n)}$$

$$= (a w_i + b) A_{N-1}^{(n)} + (c w_i + d) B_{N-1}^{(n)}$$

$$+ (\alpha_N w_i + \beta_N) + (\gamma_N w_i + \delta_N) B_{N-1}^{(n)} \tag{3.5}$$

$$= (c w_i + d)(w_i A_{N-1}^{(n)} + B_{N-1}^{(n)})$$

$$+ (\alpha_N w_i + \beta_N) A_{N-1}^{(n)} + (\gamma_N w_i + \delta_N) B_{N-1}^{(n)}.$$

Set

$$u_i := c w_i + d, \quad i = 1, 2, \tag{3.6}$$

then

$$
\begin{aligned}
w_i A_N^{(n)} + B_N^{(n)} &= (u_i(w_i A_{N-2}^{(n)} + B_{N-2}^{(n)}) + (\alpha_{N-1}w_i + \beta_{N-1})A_{N-2}^{(n)} \\
&\quad + (\gamma_{N-1}w_i + \delta_{N-1})B_{N-2}^{(n)} + (\alpha_N w_i + \beta_N)A_{N-1}^{(n)} \\
&\quad + (\gamma_N w_i + \delta_N)B_{N-1}^{(n)} \\
&= u_i^m(w_i A_n^{(n)} + B_n^{(n)}) \\
&\quad + \sum_{k=1}^{m-1} u_i^k((\alpha_{N-k}w_i + \beta_{N-k})A_{N-k-1}^{(n)} + (\gamma_{N-k}w_i + \delta_{N-k})B_{N-k-1}^{(n)}) \\
&= u_i^m(aw_i + b) + u_i^m(\alpha_n w_i + \beta_n) + \sum_{k=1}^{m-1} u_i^k((\alpha_N w_i + \beta_N) \\
&\quad + (\gamma_N w_i + \delta_N)B_{N-1}^{(n)}) \\
&= u_i^{m+1}w_i + \sum_{k=1}^{m} u_i^k((\alpha_{N-k}w_i + \beta_{N-k})A_{N-k-1}^{(n)} \\
&\quad + (\gamma_{N-k}w_i + \delta_{N-k})B_{N-k-1}^{(n)}).
\end{aligned}
\tag{3.7}
$$

Here we have used (3.3) and have defined

$$
A_{n-1}^{(n)} := 1, \quad B_{n-1}^{(n)} := 0, \quad C_{n-1}^{(n)} := 0, \quad D_{n-1}^{(n)} := 1.
\tag{3.8}
$$

Similarly one obtains

$$
w_i C_N^{(n)} + D_N^{(n)} = u_i^{m+1}w_i + \sum_{k=1}^{m} u_i^k((\alpha_{N-k}w_i + \beta_{N-k})C_{N-k-1}^{(n)} + (\gamma_{N-k}w_i + \delta_{N-k})D_{N-k-1}^{(n)})
\tag{3.9}
$$

Note that

$$
\frac{u_1}{u_2} = \frac{cw_1 + d}{cw_2 + d} = r, \qquad |r| < 1,
$$

with the r given in (1.13). Define

$$
\mathcal{A}_N^{(n)} := \frac{A_N^{(n)}}{u_2^{m+1}}, \quad \mathcal{B}_N^{(n)} := \frac{B_N^{(n)}}{u_2^{m+1}}, \quad \mathcal{C}_N^{(n)} := \frac{C_N^{(n)}}{u_2^{m+1}}, \quad \mathcal{D}_N^{(n)} := \frac{D_N^{(n)}}{u_2^{m+1}}.
\tag{3.10}
$$

Then

$$
\begin{aligned}
w_1 \mathcal{A}_N^{(n)} + \mathcal{B}_N^{(n)} &= w_1 r^{m+1} + \frac{1}{u_2} \sum_{k=1}^{m} r^k((\alpha_{N-k}w_1 + \beta_{N-k})\mathcal{A}_{N-k-1}^{(n)} \\
&\quad + (\gamma_{N-k}w_1 + \delta_{N-k})\mathcal{B}_{N-k-1}^{(n)}) \\
w_2 \mathcal{A}_N^{(n)} + \mathcal{B}_N^{(n)} &= w_2 + \frac{1}{u_2} \sum_{k=1}^{m} ((\alpha_{N-k}w_2 + \beta_{N-k})\mathcal{A}_{N-k-1}^{(n)} \\
&\quad + (\gamma_{N-k}w_2 + \delta_{N-k})\mathcal{B}_{N-k-1}^{(n)})
\end{aligned}
\tag{3.11}
$$

From these expressions we obtain

$$(w_1 - w_2)\mathcal{A}_N^{(n)} = w_1 r^{m+1} - w_2 + \frac{1}{u_2} \sum_{k=1}^{m} ((\alpha_{N-k}(r^k w_1 - w_2)$$
$$+ \beta_{N-k}(r^k - 1))\mathcal{A}_{N-k-1}^{(n)})$$
$$+ (\gamma_{N-k}(r^k w_1 - w_2) + \delta_{N-k}(r^k - 1))\mathcal{B}_{N-k-1}^{(n)})$$
$$(w_2 - w_1)\mathcal{B}_N^{(n)} = w_1 w_2 (r^{m+1} - 1) + \frac{1}{u_2} \sum_{k=1}^{m} ((\alpha_{N-k} w_1 w_2 (r^k - 1)$$
$$+ \beta_{N-k}(w_2 r^k - w_1))\mathcal{A}_{N-k-1}^{(n)}$$
$$+ (\gamma_{N-k} w_1 w_2 (r^k - 1) + \delta_{N-k}(w_2 r^k - w_1))\mathcal{B}_{N-k-1}^{(n)}).$$

(3.12)

Now, in terms of

$$s_k := \sum_{v=0}^{k} r^v = \frac{r^{k+1} - 1}{r - 1},$$

(3.13)

we have

$$r^k w_1 - w_2 = (w_1 - w_2)\left(\frac{u_2 s_k - d s_{k-1}}{u_2}\right),$$
$$r^k - 1 = c\frac{(w_1 - w_2)}{u_2} s_{k-1},$$
$$w_1 w_2 (r^k - 1) = -b(w_1 - w_2)s_{k-1} \mid u_2,$$
$$w_2 r^k - w_1 = (w_1 - w_2)\left(\frac{u_1 s_{k-2} - d s_{k-1}}{u_2}\right).$$

Hence

$$u_2^2 \mathcal{A}_N^{(n)} = u_2(c w_2 s_m + r^{m+1} u_2) + \sum_{k=1}^{m} ((\alpha_{N-k}(u_2 s_k - d s_{k-1})$$
$$+ \beta_{N-k} c s_{k-1})\mathcal{A}_{N-k-1}^{(n)} + (\gamma_{N-k}(u_2 s_k - d s_{k-1}) + \delta_{N-k} c s_{k-1})\mathcal{B}_{N-k-1}^{(n)})$$
$$u_2^2 \mathcal{B}_N^{(n)} = u_2 b s_m + \sum_{k=1}^{m} ((\alpha_{N-k} b s_{k-1} + \beta_{N-k}(d s_{k-1} - u_1 s_{k-2}))\mathcal{A}_{N-k-1}^{(n)}$$
$$+ (\gamma_{N-k} b s_{k-1} + \delta_{N-k}(d s_{k-1} - u_1 s_{k-2}))\mathcal{B}_{N-k-1}^{(n)}).$$

Introduce

$$\sum_{m}^{(1)}(X, Y) := \sum_{k=1}^{m} ((\alpha_{N-k}(u_2 s_k - d s_{k-1}) + \beta_{N-k} c s_{k-1})X_{N-k-1}$$
$$+ (\gamma_{N-k}(u_2 s_k - d s_{k-1}) + \delta_{N-k} c s_{k-1})Y_{N-k-1})$$
$$\sum_{m}^{(2)}(X, Y) := \sum_{k=1}^{m} ((\alpha_{N-k} b s_{k-1} + \beta_{N-k}(d s_{k-1} - u_1 s_k))X_{N-k-1}$$
$$+ (\gamma_{N-k} b s_{k-1} + \delta_{N-k}(d s_{k-1} - u_1 s_{k-2}))Y_{N-k-1}).$$

(3.14)

Then

$$u_2^2 A_N^{(n)} = u_2(cw_2 s_m + r^{m+1} u_2) + \overset{(1)}{\underset{m}{\sum}}(\mathcal{A}, \mathcal{B}),$$

$$u_2^2 B_N^{(n)} = u_2 bs_m + \overset{(2)}{\underset{m}{\sum}}(\mathcal{A}, \mathcal{B}). \tag{3.15}$$

In a similar way one shows that

$$u_2^2 C_N^{(n)} = u_2 cs_m + \overset{(1)}{\underset{m}{\sum}}(\mathcal{C}, \mathcal{D}),$$

$$u_2^2 D_N^{(n)} = u_2(-cw_1 s_m + u_1 r^m) + \overset{(2)}{\underset{m}{\sum}}(\mathcal{C}, \mathcal{D}). \tag{3.16}$$

We next apply these formulas to the computation of $g_N = T_N^{-1}(\infty)$. We have

$$g_N = g_{n+m} = T_{n+m}^{-1}(\infty) = t_{n+m}^{-1} \circ \cdots \circ t_n^{-1} \circ t_{n-1}^{-1} \circ \cdots \circ t_0^{-1}(\infty)$$

$$= (T_{n+m}^{(n)})^{-1}(g_{n-1})$$

$$= \frac{D_N^{(n)} g_{n-1} - B_N^{(n)}}{-C_N^{(n)} g_{n-1} + A_N^{(n)}} = \frac{D_N^{(n)} g_{n-1} - B_N^{(n)}}{-C_N^{(n)} g_{n-1} + A_N^{(n)}}.$$

Hence

$$g_N = \frac{-u_2 cw_1 s_m g_{n-1} + (r^m u_1 u_2 + \sum_m^{(2)}(\mathcal{C}, \mathcal{D})) g_{n-1} - u_2 bs_m - \sum_m^{(2)}(\mathcal{A}, \mathcal{B})}{-u_2 cs_m g_{n-1} - g_{n-1}\sum_m^{(1)}(\mathcal{C}, \mathcal{D}) + u_2 cw_2 s_m + r^{m+1} u_2^2 + \sum_m^{(2)}(\mathcal{A}, \mathcal{B})}.$$

It follows that, with (A) and (B) as defined in (3.19), one obtains

$$g_N = w_1 \frac{g_{n-1} - w_2 - \frac{1}{w_1 u_2 cs_m}(A)}{g_{n-1} - w_2 - \frac{1}{u_2 cs_m}(B)}.$$

From this one deduces

$$g_n - w_1 = \frac{-\frac{1}{u_2 cs_m}(A) + \frac{w_1}{u_2 cs_m}(B)}{g_{n-1} - w_2 - \frac{1}{u_2 cs_m}(B)}. \tag{3.17}$$

For this argument to be valid we need $u_2 c \neq 0$, which follows from (1.8) and $|r| < 1$, as well as

$$|f_{n-1} - w_2| > G > 0 \quad \text{for} \quad n > n_0$$

Before proceeding with the evaluation of $g_N - w_1$ we obtain bounds for $|A_N^{(n)}|, |B_N^{(n)}|, |C_N^{(n)}|, |D_N^{(n)}|$. From (3.15) one has

$$|u_2^2|(|A_N^{(n)}| + |B_N^{(n)}|)$$

$$\leq |u_2|(|cw_2 + b||s_m| + |r|^{m+1}|u_2|)$$

$$+ \overset{m}{\underset{k=1}{\sum}}(|\alpha_{N-k}|(|u_2| + |bs_k| + |ds_{k-1}|) + |\beta_{N-k}|(|cs_{k-1}| + |ds_{k+1}|)$$

$$+ |\gamma_{N-k}|(|u_2 s_k| + |dd + b||s_{k-1}|)$$

$$+ |\delta_{N-k}|(|c + d||s_{k-1}| + |u_1 s_{k-2}|))(|A_{N-k-1}^{(n)}| + |B_{N-k-1}^{(n)}|).$$

Introduce

$$p_n := \max_{k \geq n}[|\alpha_k|, |\beta_k|, |\gamma_k|, |\delta_k|].$$

There then exist constants $f, g > 0$, independent of n, m and k such that

$$|\mathcal{A}_N^{(n)}| + |\mathcal{B}_N^{(n)}| \leq f + g \sum_{k=1}^{m} p_{N-k}(|\mathcal{A}_{N-k-1}^{(n)}| + |\mathcal{B}_{N-k-1}^{(n)}|).$$

From the discrete version of Gronwall's inequality (see, for example [1])

$$|\mathcal{A}_N^{(n)}| + |\mathcal{B}_N^{(n)}| \leq f \prod_{k=1}^{m}(1 + gp_k) \leq f(1 + gp_n)^m.$$

A fortiori

$$\left.\begin{array}{c} |\mathcal{A}_N^{(n)}| \\ |\mathcal{B}_N^{(n)}| \end{array}\right\} \leq f(1 + gp_n)^m, \quad N = n + m, \tag{3.18}$$

In the formula (3.17) we have

$$(A) := g_{n-1}r^m u_1 u_2 + g_{n-1} \sum_{m}^{(2)}(\mathcal{C}, \mathcal{D}) - \sum_{m}^{(2)}(\mathcal{A}, \mathcal{B}).$$
$$(B) := g_{n-1} \sum_{m}^{(1)}(\mathcal{C}, \mathcal{D}) + r^{m-1}u_2^2 + \sum_{m}^{(1)}(\mathcal{A}, \mathcal{B}). \tag{3.19}$$

In a limit periodic algorithm one has no control over t_n for small n. The same is then also true for g_{n-1}. In particular $g_{n-1} = w_2$, for some n, is certainly possible. However, if we exclude the possibility $g_n = w_2$ for all $n > n_0$, then one can show that for n sufficiently large

$$|g_{n-1} - w_1| < M,$$

from which both the boundedness of g_{n-1} as well as the fact that $|g_{n-1} - w_2| > G$ follows.

Note that $g_n = w_2$, $n \geq n_0$ is equivalent to

$$g_{n_0} = w_2 \quad \text{and} \quad t_n(w_2) = w_2 \quad \text{for all} \quad n > n_0.$$

Assuming that (3.20) does not hold one then has

$$|g_N - w_1| < \frac{K_1(|r|^m + \hat{p}_n m(1 + \hat{p}_n)^m)}{G - K_2(|r|^m + \hat{p}_n m(1 + \hat{p}_n)^m)}$$

Here $\hat{p}_n = gp_n \downarrow 0$. We can agree that whatever m we pick we can choose n so large that

$$\hat{p}_n < \frac{1}{m^2}.$$

Then

$$|r|^m + \hat{p}_n m(1 + \hat{p}_n)^m < |r|^m + e^{1/m}/m.$$

Let $\varepsilon > 0$ be given. Since $|r| < 1$ we can find an m_ε so that

$$|r|^{m_\varepsilon} + e^{1/m_\varepsilon}/m_\varepsilon < \frac{G}{2} \min\left(\frac{\varepsilon}{K_1}, \frac{1}{K_2}\right).$$

Next we determine n_ε so that

$$\hat{p}_{n_\varepsilon} < \frac{1}{m_\varepsilon^2}.$$

Then for

$$N \geq n_\varepsilon + m_\varepsilon$$

$$|g_N - w_1| < \varepsilon$$

and it follows that g_N converges to w_1.

4 BOUNDEDNESS OF $|T_{n+m}^{(n+1)}(0)/(T_{n+m}^{(n+1)}(0) - g_N)|$

Let $t(w)$ be an L.F.T. with an attractive fixed point w_2 and a repulsive fixed point w_1. The family of circles perpendicular to all circles passing through both w_1 and w_2 is a one parameter family $[K_\rho(t) : 0 < \rho < \infty]$. We can set

$$K_\rho = \lambda^{-1}([u : |u| = \rho]). \tag{4.1}$$

Further define

$$D_\rho(t) := \lambda^{-1}([u : |u| < \rho]),$$

that is $D_\rho(t)$ is that region bounded by $K_\rho(t)$ which contains w_2.

Since $t(w) = \lambda^{-1}(r\lambda(w))$, where $|r| = |r(t)| < 1$, one has

$$t(D_\rho(t)) = D_{|r|\rho}(t). \tag{4.2}$$

Since $\rho_1 < \rho_2$ implies $D_{\rho_1}(t) \subset D_{\rho_2}(t)$, it follows that

$$t(D_\rho(t)) \subset D_\rho(t). \tag{4.3}$$

Let r' be such that $|r| < r' < 1$ and choose ρ_0 so that

$$0 \in D_{\rho_0}(t).$$

Now define

$$\mathcal{F}(\varepsilon_1, \varepsilon_2) := [\hat{t} : |\hat{w}_i - w_i| < \varepsilon_1, \; |\hat{r} - r| < \varepsilon_2].$$

One can then pick ε_1 and ε_2 small enough so that

$$\hat{t}(D_{\rho_0}(t)) \subset D_{r'\rho_0}(t) \subset D_{\rho_0}(t).$$

It follows that for arbitrary $m > 0$

$$T_{n+m}^{(n+1)}(0) \in D_{\rho_0}(t),$$

provided n is chosen large enough so that

$$t_{n+v} \in \mathcal{F}(\varepsilon_1, \varepsilon_2) \quad \text{for all} \quad v \geq 1.$$

Since $w_1 \notin \mathrm{cl}\, D_{\rho_0}(t)$ and since $g_n - w_1 \to 0$, one then can conclude that, though $D_{\rho_0}(t)$ may be unbounded, the set

$$\frac{D_{\rho_0}(t)}{D_{\rho_0}(t) - g_n}$$

must be bounded for sufficiently large n.

It follows that there is an $M(n_0) > 0$ so that

$$\left| \frac{T_{n+m}^{(n+1)}(0)}{T_{n+m}^{(n+1)}(0) - g_n} \right| < M(n_0) \quad \text{for all} \quad n > n_0,\ m > 0. \tag{4.4}$$

5   CONCLUSIONS

Since we have shown in Section 3 that $g_n \to w_1$ and in Section 4 that (4.4) is bounded, we can conclude that, subject to the conditions listed below, the truncation error formula (2.7) can be written as

$$|T_{n+m}(0) - T_n(0)| < K(r', n_0)|r'|^n \quad \text{for} \quad n > n_0. \tag{5.1}$$

Theorem 5.1. *Let $\{T_n(w)\}$ be an L.F.T. algorithm satisfying:*
(a) $t_n(w)$ is nonsingular and $b_n \neq 0$, $n \geq 0$,
(b) w_1 is not a fixed point of $t_n(w)$ for all $n \geq n_1$,
(c) $t(w)$ has exactly two fixed points, or fixed point and distinct exceptional point in case it is singular, in $\mathbb{C}$,
(d) with suitable numbering of the subscripts

$$\left| \frac{cw_1 + d}{cw_2 + d} \right| < 1,$$

(that is $t(w)$ is not elliptic).
Then for

$$\left| \frac{cw_1 + d}{cw_2 + d} \right| < r' < 1,$$

and an arbitrary positive integer m, there exists a positive integer n_0 such that (5.1) is valid. It follows from (5.1) by use of the Cauchy criterion that $\{T_n(0)\}$ converges.

Remark. If

$$\sum_{n=1}^{\infty} \max |\alpha_n, \beta_n, \gamma_n, \delta_n| < \infty$$

one can show that for $n > n^*$

$$|T_{n+m}(0) - T_n(0)| < K^*(n^*) \left| \frac{cw_1 + d}{cw_2 + d} \right|^n.$$

The truncation error formula can be applied to any $\{T_{n+m}^{(n)}(0)\}$, for fixed n, to conclude that

$$\lim_{m \to \infty} T_{n+m}^{(n)}(0) = T^{(n)},$$

exists for all $n \geq 0$. Moreover the convergence can be shown to be uniform with respect to n. The uniformity is the key ingredient in the proof that

$$T^{(n)} \to w_2,$$

which therefore can be completed following the pattern suggested by Perron [4, pp. 93–94] and also used in [6, Section 6].

With this taken care of (2.9) is seen to be a formula from which one can deduce that $\{T_n(w_2)\}$ converges substantially faster to T than $\{T_n(0)\}$ does.

The speed of convergence of $\{T_n(0)\}$ depends largely on $t(w)$ and only incidentally on the speed of approach of $t_n(w)$ to $t(w)$. In $\{T_n(w_2)\}$ the latter plays an important role.

REFERENCES

1. R. Beesack, *Gronwall Inequalities*, Carleton Mathematical Lecture Notes, No. 11, 1975.
2. John Gill, Infinite Composition of Möbius Transformations, *Trans. Amer. Math. Soc.* **176** (1973), 479–487.
3. M. Mandell and Arne Magnus, On convergence of sequences of linear fractional transformations, *Math. Zeitschrift* **115** (1970), 11–17.
4. O. Perron, *Die Lehre von den Kettenbrüchen*, 3rd Ed., vol. 2, Teubner, Stuttgart, 1957.
5. W. J. Thron, Continued fraction identities derived from the invariance of the crossratio under ℓ.f.t., *Analytic Theory of Continued Fractions III*, (L. Jacobsen, ed.), Lecture Notes in Math. **1406**, Springer, Berlin, 1989, pp. 124–134.
6. W. J. Thron, Truncation error for limit periodic Schur algorithms, *SIAM J. Math Analysis*, to appear.
7. W. J. Thron, Limit periodic Schur algorithms, the case $|\gamma| = 1$, $\sum d_n < \infty$, *Numerical Algorithms*.
8. W. J. Thron and Haakon Waadeland, Accelerating convergence of limit periodic continued fractions $K(a_n/1)$, *Numer. Math.*, **34** (1980), 155–170.
9. W. J. Thron and Haakon Waadeland, Truncation error bounds for limit periodic continued fractions, *Math. of Comp.*, **40** (1983), 589–597.

16

A Limit Theorem in Frequency Analysis

HAAKON WAADELAND, Department of Mathematics and Statistics, University of Trondheim (AVH), N–7055 Dragvoll, Norway

Dedicated to the memory of Arne Magnus

1. INTRODUCTION.

In the paper [7] signals of the form

$$
X_N(m) = \begin{cases} \displaystyle\sum_{j=-I}^{I} \alpha_j e^{i\omega_j m}, & m = 0, 1, \ldots, N-1, \\ 0 & \text{for } m < 0 \text{ and } m \geq N, \end{cases} \tag{1.1}
$$

where $X_N(0) \neq 0$, $\alpha_0 \geq 0$, $\alpha_j = \bar{\alpha}_{-j} \neq 0$, $\omega_{-j} = -\omega_j \in \mathbb{R}$, $j = 1, 2, \ldots, I$, $0 = \omega_0 < \omega_1 < \cdots < \omega_I < \pi$, were studied. The main part of the paper was the solution of the *frequency analysis problem*, i.e. to determine the frequencies $\omega_1, \omega_2, \ldots, \omega_I$ from signal values with a sample of size N. The frequency analysis problem has recently been dealt with in several papers [2], [3], [7], [8],[9],[10],[12] by the Wiener-Levinson method [11],[13]. For a detailed description of the method we refer to [7], but we shall need some steps from the method. The autocorrelation formulas

$$
\mu_m^{(N)} = \begin{cases} \displaystyle\sum_{k=0}^{N-m-1} X_N(k)X_N(k+m), & m = 0, 1, 2, \ldots, \\ \mu_{-m}^{(N)} & \text{for } m = -1, -2, -3, \ldots \end{cases} \tag{1.2}
$$

produce a *positive definite hermitian sequence*. This is (in the present case) a sequence of moments with respect to an absolutely continuos distribution function $\psi_N(\theta)$, easily expressable in terms of $X_N(m)$. The monic polynomials, orthogonal on the unit circle w.r. t. ψ_N, i.e. the *Szegö polynomials* $\rho_n(\psi_N; z)$, and the *reciprocal polynomials* $\rho_n^*(\psi_N; z) = z^n \overline{\rho_n(1/\overline{z})}$ can be expressed by determinant formulas in terms of the moments $\mu_m^{(N)}$ see [1] and [6]. An alternative, and for practical purposes better way, is to use the *Levinson algorithm*. These polynomials are the denominators $Q_k(z)$ of the approximants of the *positive Perron-Carathéodory continued fraction*

$$\delta_0^{(N)} - \frac{2\delta_0^{(N)}}{1} + \frac{1}{\overline{\delta_1^{(N)}}z} + \frac{(1-|\delta_1^{(N)}|^2)z}{\delta_1^{(N)}} + \frac{1}{\overline{\delta_2^{(N)}}z} + \frac{(1-|\delta_2^{(N)}|^2)z}{\delta_2^{(N)}} + \cdots, \tag{1.3}$$

$\delta_0^{(N)} > 0$, $|\delta_n^{(N)}| < 1$, introduced by Jones, Njåstad, Thron [4],[5], whose sequence of even approximants corresponds at $z = 0$ to the series

$$L_0^{(N)} = \mu_0^{(N)} + 2\sum_{m=1}^{\infty} \mu_m^{(N)} z^m, \tag{1.4}$$

and whose sequence of odd approximants corresponds at $z = \infty$ to the series

$$L_\infty^{(N)}(z) = -\overline{L_0^{(N)}}\left(\frac{1}{z}\right) = -\mu_0^{(N)} - 2\sum_{m=1}^{\infty} \overline{\mu_m^{(N)}} z^{-m}. \tag{1.5}$$

The function defined by (1.4) is holomorphic in the open unit disk $|z| < 1$ and maps it into the open right half plane, such that 0 is mapped onto $\mu_0^{(N)} > 0$, i.e. it is a *Carathéodory function*. Similarly (1.5) is holomorphic in $|z| > 1$ and maps it into the open left half plane, such that ∞ is mapped onto $-\mu_0^{(N)} < 0$.

We also know that the even approximants of (1.3) are Carathéodory-functions, and that the odd approximants of (1.3) are holomorphic in $|z| > 1$ and map this region into the open left half plane.

Remark.

The signal is real, and hence all $\mu_m^{(N)}$ and $\delta_m^{(N)}$ are real. We may therefore skip absolute value sign and conjugate sign in (1.3) and (1.5), and also in the definition of reciprocal polynomials.

Following the notation in [7] we shall write $P_n(\psi_N; z)$ and $Q_n(\psi_N; z)$ for the numerator and denominator of the nth approximants of (1.3), normalized in the standard way. The earlier mentioned connection to Szegö polynomials and reciprocal polynomials is given by

$$\begin{aligned} Q_{2n}(\psi_N; z) &= \rho_n^*(\psi_N; z) \\ Q_{2n+1}(\psi_N; z) &= \rho_n(\psi_N; z) \end{aligned}, n = 0, 1, 2, \dots \tag{1.6}$$

in particular $Q_0(\psi_N; z) = Q_1(\psi_N; z) = \rho_0^*(\psi_N, z) = \rho_0(\psi_n; z) = 1$.

The next step is to use the moments $\mu_m^{(N)}/N$, and let N tend to ∞. It is proved, that

$$\frac{1}{N}\mu_m^{(N)} = \mu_m + O\left(\frac{1}{N}\right), \qquad \text{as } N \to \infty, \tag{1.7}$$

where

$$\mu_m = \sum_{j=-I}^{I} |\alpha_j|^2 e^{i\omega_j m}. \tag{1.8}$$

The sequence $\{\mu_m\}$ gives rise to the Carathéodory function

$$F(z) = \mu_0 + 2 \sum_{m=1}^{\infty} \mu_m z^m = \sum_{j=-I}^{I} |\alpha_j|^2 \frac{e^{i\omega_j} + z}{e^{i\omega_j} - z}, \qquad |z| < 1, \tag{1.9}$$

and to the terminating PC-function

$$\delta_0 - \frac{2\delta_0}{1} + \frac{1}{\delta_1 z} + \frac{(1 - \delta_1^2)z}{\delta_1} + \cdots + \frac{1}{\delta_{n_0-1} z} + \frac{(1 - \delta_{n_0-1}^2)z}{\delta_{n_0-1}} + \frac{1}{\delta_{n_0} z}, \tag{1.10}$$

where $n_0 = 2I + L$, and $L = 0$ for $\alpha_0 = 0$ and $L = 1$ for $\alpha_0 \neq 0$. Here

$$\delta_0 = \lim_{N \to \infty} \frac{1}{N} \delta_0^{(N)} \quad \text{and} \quad \delta_n = \lim_{N \to \infty} \delta_n^{(N)}$$

for $1 \leq n \leq n_0$, and $|\delta_{n_0}| = 1$. If $P_n(z)$ and $Q_n(z)$ are numerator and denominator of (1.10) we have, for $k = 0, 1, 2, \ldots, 2n_0 + 1$

$$\lim_{N \to \infty} \frac{1}{N} P_k(\psi_N; z) = P_k(z),$$
$$\lim_{N \to \infty} \frac{1}{N} Q_k(\psi_N; z) = Q_k(z). \tag{1.11}$$

Furthermore (1.10) is equal to

$$F(z) = -F(\frac{1}{z}),$$

and on the other hand to $P_{2n_0}(z)/Q_{2n_0}(z)$, but also to $P_{2n_0+1}(z)/Q_{2n_0+1}$, since

$$\frac{(1 - \delta_{n_0}^2)z}{\delta_{n_0}} = 0.$$

Hence

$$\lim_{N \to \infty} \frac{1}{N} \frac{P_{2n_0}(\psi_N; z)}{Q_{2n_0}(\psi_N; z)} = \frac{P_{2n_0}(z)}{Q_{2n_0}(z)} = F(z) \qquad \text{for } |z| < 1,$$
$$\lim_{N \to \infty} \frac{1}{N} \frac{P_{2n_0+1}(\psi_N; z)}{Q_{2n_0+1}(\psi_N; z)} = \frac{P_{2n_0+1}(z)}{Q_{2n_0+1}(z)} = -F(1/z) \qquad \text{for } |z| > 1, \tag{1.12}$$

where in each case the convergence is uniform on compact subsets of the given region.

The limits above are rather easily established by using (1.7), (1.8) and determinant formulas. Of particular interest is the limit of $Q_{2n_0}(\psi_N; z) = \rho_{n_0}(\psi_N; z)$. Since

$$\rho_{n_0}(z) = Q_{2n_0}(z) = (z - 1)^L \prod_{j=1}^{I} (z - e^{i\omega_j})(z - e^{-i\omega_j}), \tag{1.13}$$

where L is as defined above, the limit of $Q_{2n_0}(\psi_N; z)$ is a monic polynomial with exactly the zeros $e^{i\omega_j}$, where the numbers ω_j are the frequencies of the signal. Since one normally does not know how many frequencies there are in the signal (if at all finite), we do not know n_0. Therefore, it seems appropriate to ask about possible limits of $\rho_n(\psi_N; z)$ (and other things) for n-values $> n_0$. The attempts to use the determinant formulas fail, since we get ratios where numerator and denominator both tend to 0 when $N \to \infty$. Nevertheless, even if $\{\rho_n(\psi_N; z)\}_{N=1}^\infty$ does not converge it turns out that n_0 of the zeros, properly chosen, converge to exactly the numbers $e^{i\omega_j}$. This is the main result in the paper [7], and it answers affirmatively the essential part of a conjecture by Jones, Njåstad and Saff [3]. The proof is established by going to subsequences $\{N_k\}$ and to use the normality of the sequence $\{\rho_n(\psi_N; z)\}_{N=1}^\infty$ $(|\rho_n(\psi_N; z)| \le (R+1)^n$ for $|z| \le R)$. An additional result from [7] is that (1.12) also holds when on the lefthand side n_0 is replaced by any fixed $n \ge n_0$, without going to subsequences. The main purpose of the present note is to give a rather direct proof of this, so we state it properly, in terms of notation already introduced.

Theorem 1.

$$\lim_{N \to \infty} \frac{1}{N} \frac{P_{2n}(\psi_N; z)}{Q_{2n}(\psi_N; z)} = F(z) \qquad \text{for } |z| < 1,$$

$$\lim_{N \to \infty} \frac{1}{N} \frac{P_{2n+1}(\psi_N; z)}{Q_{2n+1}(\psi_N; z)} = -F(\frac{1}{z}) = F(z) \qquad \text{for } |z| > 1. \tag{1.14}$$

The convergence is uniform on compact subsets of respectively $|z| < 1$ and $|z| > 1$ (or actually of the upper Riemann sphere).

The proof is almost immediate by using a mapping property of tails of a positive Perron-Carathéodory fraction (or even approximants of such tails). The significance of these properties justifies a separate section.

2. A PROPERTY OF THE TAILS OF PPC-FRACTIONS.

What will here be discussed is not limited to the use for the present problem. Nevertheless we shall use (1.3) as a reference PPC-fraction for practical purposes. With

$$\omega = \delta_0^{(N)} - \frac{2\delta_0^{(N)}}{1} \frac{1}{+w}, \qquad \delta_0^{(N)} > 0, \tag{2.1}$$

we have $|w| < 1$ if and only if $\Re\omega > 0$. This, combined with the mentioned properties of the positive Perron-Carathéodory fraction (1.3) shows that the even approximants of the continued fraction

$$\overline{\delta_1^{(N)}}z + \frac{(1 - |\delta_1^{(N)}|^2)z}{\delta_1^{(N)}} + \frac{1}{\overline{\delta_2^{(N)}}z} + \frac{(1 - |\delta_2^{(N)}|^2)z}{\delta_2^{(N)}} + \cdots \tag{2.2}$$

(approximant number 0 being $\overline{\delta_1^{(N)}}z$) as well as their uniform limit on compact subsets of $|z| < 1$, all are functions g, holomorphic in $|z| < 1$ and such that $|g(z)| < 1$ there. The

same also holds of the continued fractions

$$\overline{\delta_{n+1}^{(N)}}z + \cfrac{(1 - |\delta_{n+1}^{(N)}|^2)z}{\delta_{n+1}^{(N)}} + \cdots, \tag{2.3}$$

since the continued fraction obtained from (1.3) by replacing (2.2) by (2.3) is also a PPC-fraction.

Similarly we can find that the even approximants of the continued fraction

$$\cfrac{1}{\overline{\delta_{n+1}^{(N)}}z} + \cfrac{(1 - |\delta_{n+1}^{(N)}|^2)z}{\delta_{n+1}^{(N)}} + \cfrac{1}{\overline{\delta_{n+2}^{(N)}}z} + \cfrac{(1 - |\delta_{n+2}^{(N)}|^2)z}{\delta_{n+2}} + \cdots, \tag{2.4}$$

as well as their uniform limit on compact subsets of $|z| > 1$, are holomorphic in $|z| > 1$ and map this region into the open unit disk. (Approximant number 2 is

$$\cfrac{1}{\overline{\delta_{n+1}^{(N)}}z} + \cfrac{(1 - |\delta_{n+1}^{(N)}|^2)z}{\delta_{n+1}^{(N)}} = \frac{\delta_{n+1}^{(N)}}{z}.)$$

3. PROOF OF THE THEOREM.

Keeping in mind that the signal and hence all $\delta_n^{(N)}$ are real, we skip conjugate sign and absolute value sign. We have

$$\frac{P_{2(n_0+m)}^{(N)}(z)}{N \cdot Q_{2(n_0+m)}^{(N)}(z)} = \frac{1}{N}\left(\delta_0^{(N)} - \cfrac{2\delta_0^{(N)}}{1} + \cfrac{1}{\delta_1^{(N)}z} + \cfrac{(1 - \delta_1^{(N)2})z}{\delta_1^{(N)}} + \cdots \right. \tag{3.1}$$

$$\left. + \cfrac{1}{\delta_{n_0}^{(N)}z} + \cfrac{(1 - \delta_{n_0}^{(N)2})z}{\delta_{n_0}^{(N)}} + \cfrac{1}{\delta_{n_0+1}^{(N)}z} + \cdots + \cfrac{1}{\delta_{n_0+m}^{(N)}z}\right)$$

$$= \frac{\frac{1}{N}\left[P_{2n_0}^{(N)}(z) + wP_{2n_0-1}^{(N)}(z)\right]}{Q_{2n_0}^{(N)}(z) + wQ_{2n_0-1}^{(N)}(z)}$$

$$= \frac{\frac{1}{N}\left(P_{2n_0}^{(N)}(z) + \frac{(1-\delta_{n_0}^{(N)2})zP_{2n_0-1}^{(N)}(z)}{g^{(N)}(z)}\right)}{Q_{2n_0}^{(N)}(z) + \frac{(1-\delta_{n_0}^{(N)2})zQ_{2n_0-1}^{(N)}(z)}{g^{(N)}(z)}},$$

where

$$g^{(N)}(z) = \delta_{n_0}^{(N)} + \cfrac{1}{\delta_{n_0+1}^{(N)}z} + \cdots + \cfrac{1}{\delta_{n_0+m}^{(N)}}.$$

From the previous section we have

$$\left|\delta_{n_0+1}^{(N)}z + \cfrac{(1 - \delta_{n_0+1}^{(N)2})z}{\delta_{n_0+1}} + \cdots + \cfrac{1}{\delta_{n_0+m}^{(N)}}\right| < 1$$

in $|z| < 1$ and, by Schwarz's lemma, $\leq R$ in $|z| \leq R < 1$. Hence, since $|\delta_{n_0}^{(N)}| < 1$, we get

$$\left| g^{(N)}(z) \right| \geq \frac{1}{R} - 1 > 0 \qquad \text{for all } N.$$

Since $|\delta_{n_0}^{(N)}| \to 1$ when $N \to \infty$ it then follows from (1.11) for $k = 2n_0$ that the lefthand side of (3.1) tends to $P_{2n_0}(z)/Q_{2n_0}(z)$ when $N \to \infty$, and the first part of Theorem 1 is thus proved. The proof of the second part can be·established in the same way only by using properties of the odd order approximants of (2.4). But it can also be seen from the relations

$$\frac{P_{2n}(\psi_N; z)}{Q_{2n}(\psi_N; z)} = -\overline{\left(\frac{P_{2n+1}(\psi_N; 1/\overline{z})}{Q_{2n+1}(\psi_N; 1/\overline{z})} \right)}, \quad F(z) = -\overline{F(1/\overline{z})}.$$

REFERENCES

1. U. Grenander and G. Szegö, "Toeplitz Forms and their Applications," University of California Press, Berkeley, 1958.
2. W.B. Jones and O. Njåstad, *Applications of Szegö Polynomials to Digital Signal Processing*, Rocky Mountain Journal of Mathematics 21, 387-436 (1992).
3. W.B. Jones, O. Njåstad and E.B. Saff, *Szegö Polynomials Associated with Wiener-Levinson Filters*, Journal of Computational and Applied Mathematics 32, 387-406 (1990).
4. W.B. Jones, O. Njåstad and W.J. Thron, *Continued Fractions Associated with Trigonometric and Other Strong Moment Problems*, Constructive Approximation 2, 197-211 (1986).
5. W.B. Jones, O. Njåstad and W.J. Thron, *Schur Fractions, Perron-Carathéodory Fractions, a Survey*, Lecture Notes in Mathematics, Springer-Verlag 1199, 127-158 (1986).
6. W.B. Jones, O. Njåstad and W.J. Thron, *Moment Theory, Orthogonal Polynomials, Quadrature, and Continued Fractions associated with the Unit Circle*, Bulletin of the London Mathematical Society 21, 113-152 (1989).
7. W.B. Jones, O. Njåstad, W.J. Thron and H. Waadeland, *Szegö Polynomials Applied to Frequency Analysis*, Journal of Computational and Applied Mathematics (to appear).
8. W.B. Jones, O. Njåstad and H. Waadeland, *Asymptotics for Szegö Polynomial Zeros*, Numerical Algorithms (1992) (to appear).
9. W.B. Jones, O. Njåstad and H. Waadeland, *Application of Szegö Polynomials to Frequency Analysis*, (to appear).
10. W.B. Jones, O. Njåstad and Haakon Waadeland, *An Alternative Way of Using Szegö Polynomials in Frequency Analysis*, These Proceedings.
11. N. Levinson, *The Wiener RMS (root mean square) Error Criterion in Filter Design and Prediction*, Journal of Mathematics and Physics 25, 261-278 (1947).
12. K. Pan and E.B. Saff, *Asymptotics for Zeros of Szegö Polynomials Associated with Trigonometric Polynomial Signals*, ICM- Report Number 91-014, Department of Mathematics, University of South Florida..
13. N. Wiener, "Extrapolation, Interpolation and Smoothing of Stationary Time Series," The Technology Press of Massachusetts Institute of Technology/ John Wiley and Sons, Inc., New York, 1949.

Index

about the book . . .

This **outstanding** reference—the proceedings of a research conference held in Loen, Norway—contains **up-to-date** information on the analytic theory of continued fractions and their application to moment problems and orthogonal sequences of functions.

Uniting the research efforts of over **15** international experts, *Continued Fractions and Orthogonal Functions* treats strong moment problems, orthogonal polynomials, and Laurent polynomials . . . analyzes sequences of linear fractional transformations . . . presents convergence results, including truncation error bounds . . . considers discrete distributions and limit functions arising from indeterminate moment problems . . . discusses Szëgo polynomials and their application to frequency analysis . . . describes a quadrature formula arising from q-starlike functions . . . covers continued fractional representations for functions related to the gamma function . . . and much more.

Providing more than **1700** equations, references, and drawings, *Continued Fractions and Orthogonal Functions* is a vital resource for mathematical and numerical analysts, applied mathematicians, physicists, chemists, engineers, and upper-level undergraduate and graduate students in these disciplines.

about the editors . . .

S. CLEMENT COOPER is an Assistant Professor in the Department of Pure and Applied Mathematics at Washington State University, Pullman. The author or coauthor of numerous professional papers, Dr. Cooper is a member of the Society for Industrial and Applied Mathematics, the Mathematical Association of America, and the Association for Women in Mathematics, among others. She received the B.A. degree (1978) from the University of Florida, Gainesville, and the M.S. (1984) and Ph.D. (1988) degrees from Colorado State University, Fort Collins.

W. J. THRON is Professor Emeritus at the University of Colorado, Boulder. The author, coauthor, editor, or coeditor of over 100 research publications and several books, Dr. Thron was elected to the Royal Norwegian Society of Science and Letters in 1980. He is a member of the American Mathematical Society and the Mathematical Association of America. Dr. Thron received the A.B. degree (1939) in mathematics from Princeton University, New Jersey, and the M.A. degree (1942) and the Ph.D. degree (1943) in mathematics from the Rice Institute (now Rice University), Houston, Texas.

Printed in the United States of America   ISBN: 0−8247−9071−5

marcel dekker, inc./new york · basel · hong kong